Focus On Intermediate Algebra

Robert Hackworth
Robert Alwin

H&H Publishing Company, Inc.
Clearwater, Florida
(813) 442-7760

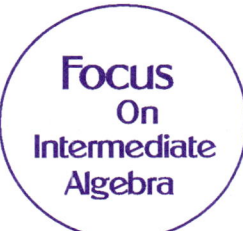

by
Robert Hackworth, Robert Alwin
of St. Petersburg Junior College, Clearwater Campus

Copyright 1993
H&H Publishing Company, Inc.
1231 Kapp Drive
Clearwater, Florida
telephone (813) 442-7760

All rights reserved. No part of this book may be reproduced in any form or by any means without written permission of the publisher.

Editorial/Production Supervision
  Robert Hackworth

Editor
  Karen H. Davis

Production Assistants
  Priscilla Trimmier, Mike Ealy

Business Management
  Sally Marston, Mike Ealy

ISBN 0-943202-44-2

Library of Congress Card Catalog Number 93-077816

Printing is the lowest number: 10 9 8 7 6 5 4 3 2 1

Preface

Focus on Intermediate Algebra was written for students who have recently completed a first course in Algebra or those who once had good skills in Algebra but now need some serious review before continuing other pursuits.

The text contains all the information its student readers are expected to learn. There are no topics left unexplained, no problems which will unduly surprise a well-prepared student, and no prior knowledge of Algebra that is not completely reviewed in Chapter 0. The text presents mathematics in a simple, logical way. At the same time, it is true to the nature of mathematics itself as it moves from one topic to another with a consistent adherence to logical principles.

The content of the book goes beyond the mathematics and presents ways to study the subject effectively. For students who have learned to do mathematics in ways that are so ineffective that failure is almost a natural consequence, assistance is given for learning new ways of approaching the subject. This text, beginning with its introduction on study skills, provides its student readers with methods for studying that will maximize the efforts of the student.

Professors who use this text will also find that it offers them numerous benefits in their teaching of the course. The professor can skim through the FOCUS statements of a unit and quickly determine the content of each section along with its expected level of rigor. The numerous exercises provide ample opportunity for class assignments and those particularly suited to individuals with other needs. Answers for most of the student exercises relieve the professor from the burden of being the source for that type of feedback and, where answers are not included, the individual problems are often keyed to the particular chapter and unit where the student can look for help. An extensive testing program with multiple forms of chapter tests, cumulative tests, and course finals gives the professor an opportunity for testing in a variety of ways, easily giving make-up exams, and implementing a mastery testing program.

The presentation of this text makes it possible to successfully implement cooperative instruction in the classroom. Students working in small groups can be reasonable expected to learn most of the concepts and skills without "chalk and talk" from their professor. This frees the professor from dispensing information and allows for interactions that are primarily directed by questions. Students confronting the material will learn their mathematics with the development of plenty of "good" questions that begin with the words "why," "where," "what," and "who." Professors will find a satisfying learning environment when the majority of their time is spent asking/answering the same kinds of questions.

Among the unique features of this text is the fact that its introduction is on study skills for mathematics. This introduction can be covered in class or assigned as homework. In either case, the student will have a source to refer to when problems with learning mathematics occur. There are many myths about mathematics and how it is best studied. Some students can greatly improve their progress in mathematics by properly addressing the subject. Learning to learn mathematics is a crucial factor for learning mathematics.

Another unique feature of this text is the fact that it is written to a specific learning theory. Its title, in fact, comes from that learning theory. FOCUS is the idea that a student always needs to know exactly what needs to be attended to at any moment. Concurrently, the student needs to know that portion of the problem that can be ignored (at least for the moment). Both the writing and the formatting are designed to give the student a focus for each new concept or skill that needs to be learned. The design of the text does this by stating in a short sentence the particular focus of any presentation and placing this sentence in an oval on the left side of the page. The text beside the oval FOCUS statement gives a clear example or description of the particular idea that needs to be learned at that time.

Mathematicians will be pleased to find that this text carefully defines the words for the course. Each definition is displayed prominently with a grey background. Each definition is written as an "if and only if" statement. Students are strongly encouraged to clearly establish the meaning of all words and symbols the text defines. Students are also discouraged from memorizing other material in the course. The constant message of the text is encouragement for "figuring out" the processes and the rules that might govern those processes. In other words, the text constantly reinforces the fact that learning mathematics is a process of making connections between topics. The student who learns the connections will find it unnecessary to memorize separate bits of information.

Each unit ends with an exercise that is separated into three sections: Part A, Part B, and Part C. Part A problems are intended to continue the instructional process by reviewing the foci of the unit and enhancing the material already presented. All answers for Part A problems are given at the back of the text. Instruction is incomplete without feedback.

Part B sections contain the type of drill and practice problems common to all mathematics texts. Answers for odd-numbered Part B problems are also in the back of the text.

Part C sections review all the material previously learned in the course. For example, a Part C exercise in Chapter 4 will contain problems from Chapters 1-3. These exercises are aimed at counteracting the tendency of many students to learn mathematics just long enough to get through a test. This text demands at the end of each unit that the student show a knowledge of all previous material. This constant review improves the likelihood that the student will begin to integrate new concepts and skills with those previously learned. As this integration occurs, the student's memory greatly improves. Part C problems do not have answers in the back of the text, but each problem is keyed to show the Chapter and Unit in which it was presented. In a way, each Part C section is a final examination for the course to that point in the book.

Word problems are considered of great importance in this text. They are found in every chapter and are included whenever the skill being learned can be enhanced by that type of problem.

Many people have aided us in the development of this text. Our students and colleagues from St. Petersburg Junior College, Clearwater Campus, have played the most crucial role and we wish to publicly thank them for that support. At H&H Publishing, we have had the wonderful assistance of Karen Davis as our editor.

To the students and faculty who use this text we express the hope that you will let us know your praises and your criticisms. We have tried to do this text perfectly, but know that with your help we can make it even better.

Robert Hackworth
Robert Alwin

Clearwater, Florida March 1993

Contents

Introduction		
Focus on Study Skills for Mathematics		xi

Chapter 0
Review of Elementary Algebra — 1

Unit 1	Sets, Elements, and Symbols	2
Unit 2	Simplifying Algebraic Expressions	8
Unit 3	Solving Linear Equations	14
Unit 4	Solving Linear Inequalities	21
Unit 5	Factoring Polynomials	27
Test		34

Chapter 1
System of Rational Numbers — 35

Unit 1	The System of Counting Numbers	35
Unit 2	The System of Integers	42
Unit 3	The System of Rational Numbers	48
Unit 4	Factoring Polynomials	55
Unit 5	Solving Quadratic Equations	62
Unit 6	Solving Quadratic Inequalities	67
Unit 7	Applications: Value, Number, and Cost	71
Test		79

Chapter 2
The Real Number System — 81

Unit 1	The System of Real Numbers	81
Unit 2	Simplifying Square Root Expressions	90
Unit 3	Simplifying 3rd and 4th Root Expressions	97
Unit 4	Decimal Representations of Real Numbers	102
Unit 5	Solving Linear Equations and Inequalities	110
Unit 6	Solving Quadratic Equations	116
Unit 7	Fractional Equations and Inequalities	126
Unit 8	Applications: Mixtures and Ingredients	134
Test		141

Chapter 3
System of Complex Numbers — 143

Unit 1	Imaginary Numbers	143
Unit 2	Solving Quadratic Equations Without Real Number Solutions	149
Unit 3	The Complex Numbers	154
Unit 4	Solving Equations With Complex Numbers	159
Unit 5	Applications: Solving Problems Using Two Variables	164
Test		170

Chapter 4
Absolute Value — 171

Unit 1	The Meaning of Absolute Value	171
Unit 2	Solving Absolute Value Equations	177
Unit 3	Solving Absolute Value Inequalities	184
Unit 4	Applications: Absolute Value, Distance, and Neighborhoods	192
Test		198

Chapter 5
Relations and Functions — 199

Unit 1	Sets of Ordered Pairs	199
Unit 2	The Set Selector Method of Showing Sets of Ordered Pairs	205
Unit 3	Relations	211
Unit 4	Functions	217
Unit 5	Applications: Venn Diagrams	223
Test		228

Chapter 6
Graphing Linear Functions — 229

Unit 1	Equations With Two Variables	229
Unit 2	Linear Functions	237
Unit 3	Slopes of Linear Functions	242
Unit 4	Finding Slopes and Y-Intercepts	249
Unit 5	Graphing Linear Inequalities in Two Variables	257
Test		263

Chapter 7
Graphing Polynomial Functions and Conic Sections — 265

Unit 1	Simple Transformations on the x,y Plane	265
Unit 2	Point Plotting Graphs	277
Unit 3	Distance Between Two Points	286
Unit 4	Conic Sections – Circles	294
Unit 5	Conic Sections – Ellipses	302
Unit 6	Conic Sections – Parabolas	311
Unit 7	Conic Sections – Hyperbolas	320
Test		329

Chapter 8
Exponents — 331

Unit 1	Counting Numbers as Exponents	331
Unit 2	Simplifying Power Expressions	336
Unit 3	Integer Exponents	343
Unit 4	Simplifying With Integer Exponents	348
Unit 5	Rational Number Exponents	354
Unit 6	Simplifying With Rational Number Exponents	360
Unit 7	Applications: Rate, Time, and Production	366
Test		374

Chapter 9
Exponential and Logarithm Functions — 375

Unit 1	Exponential Functions	375
Unit 2	Logarithm Functions	382
Unit 3	Place Value and Scientific Notation	387
Unit 4	Common Logarithms	392
Unit 5	Computing Using Logarithms	398
Unit 6	Applications: Approximating Solutions and Finding Equivalent Units	406
Test		412

Chapter 10
Systems of Equations   413

Unit 1	Systems of Two Linear Equations	413
Unit 2	Systems With Quadratic Equations	420
Unit 3	Systems With Three Equations	426
Unit 4	Matrices and Determinants	435
Unit 5	Cramer's Rule	442
Test		449

Logarithm Table   **450**
Square Root Table   **452**
Answers   **453**
Appendix   **489**
Index   **491**

Focus on Study Skills for Mathematics

There are two major topics for you to learn in this course. One is Intermediate Algebra. The other is learning how to learn mathematics successfully and that will be crucial if you are going to learn Intermediate Algebra.

Some students find learning mathematics to be easy and fun; others find it difficult and boring. The differnece between such students is not their intelligence or interests. The difference is the ways they approach studying the subject. Those that study mathematics effectively will find it makes sense and is relatively easy. Those that find mathematics difficult usually have developed study habits that make it seem that way.

If you are a student who has had a problem with mathematics you may think something is inherently wrong with you. The tendency is to give up before you start and make statements like:

> "I've never been good at math."
> "I don't have a math mind."
> "I don't like math."
> "My mother can't do math either."

All of these statements, and others like them, really mean that the student doesn't believe he/she can be successful in a math course. Not surprisingly, once a student accepts that belief the likelihood of doing well in mathematics is slight indeed.

Attitudes of Students Who Do Well in Mathematics

Students that do well in mathematics believe they will succeed and then do the common sense things needed to support that belief.

1. They attend class without fail.

2. They listen carefully and take good notes.

3. They always do all of their assignments.

4. They show all their work and they arrange it very neatly so they can refer to it later and read it easily.

5. They ask questions — lots of questions.

6. They work hard because nothing worthwhile is ever gained without a struggle.

Attitudes of Students Who Don't Do Well in Mathematics

Students who expect to fail in mathematics don't believe they can succeed and frequently do such silly things that their belief comes true.

1. They frequently miss class and make little effort to find out what was missed.

2. They don't listen carefully and rarely take notes that would help them remember.

3. They give up on any assignment that is difficult and skip some problems because they seem easy.

4. They write far less than their teacher would and they write so small that hardly anyone could read it anyway.

5. They hardly ever ask questions and are usually afraid that they will ask a "stupid" question. There really is no such thing as a "stupid" question.

6. They believe they are destined to do poorly anyway so they don't work hard. What's the use?

Today Is a New Day for You and Mathematics

The first objective you need to accomplish in this course is to establish the belief that you can, and will, succeed. This is not an easy task. You need to carefully consider your past experiences in mathematics and make plans to change those behaviors which did not work for you in other courses. Today is a new day in your mathematics life, a chance to start over and do it right, and an opportunity to find the easy ways to study and learn mathematics.

The authors of this book do not believe you are destined to fail. In fact, our many years experience with students like you have convinced us that there is nothing biologically or intellectually deficient about you. More commonly, our students have come to us with beliefs about mathematics and study habits in the subject which have made the subject very hard to learn. The aim of this section of the book is to present efficient and effective ways of dealing with your learning of mathematics. Many study skills have been preached constantly at you over the years. You have probably been told:

> "Read mathematics carefully."
>
> "Keep paper and pencil handy so you can constantly try the problems."
>
> "Arrange your work neatly so you or someone else can find any errors."
>
> "Do all your assignments and never get behind."

All of these statements are excellent advice. If you have ignored them in the past, make certain that you follow them in this course.

In this section, however, we want to go beyond some of the past advice you have been given and expand on four major themes for organizing your study of mathematics. Our aim here is to encourage you to change the ways you study mathematics. It is the way you have studied in the past that is the problem. The mathematics will be far easier if you find new ways to study it.

The four themes for organizing your study of mathematics are from the work of Claire E. Weinstein, Professor of Educational Psychology, University of Texas at Austin. These four themes are:

1. Work and study in quality environments.

2. Learn what is to be memorized and what must not be memorized.

3. Keep your mind active.

4. Learn to evaluate your own work and study habits.

1. Always work and study in a quality environment

A student does better work, stays at it longer, and feels better about the learning when that work is done in a setting in which the student is comfortable and physically relaxed. Your body and your mind should not be the source of any distractions. You want to be able to devote full energies, physical and mental, to your study.

Focus on building a quality study environment

A good restaurant earns its reputation on much more than the quality of its food. A restaurant is judged as much on its atmosphere (ambiance) as on its menu. Customers expect fine food to be served in a comfortable environment.

Make a list of the physical attributes of a fine restaurant. Some of these attributes may be different than you want in a good study. The important new idea to incorporate in your behavior is the selection, by you, of an environment that will allow you to study more effectively. Don't settle for less in your study environment than you demand in your other activities. Otherwise, you will not be your best when studying.

Physical Attributes of Quality Environments

Physical attributes such as lighting, temperature, and noise level are obvious factors to be carefully regulated by the student. Yet these obvious factors are sometimes ignored by students who are having trouble in a subject. Serious study cannot occur while working the drive-in window at MacDonald's or watching the Super Bowl on television. Although these examples may seem extreme, you need to look at your own study sessions and make certain they eliminate all possible physical distractions.

Focus on sounds in your study environment

The sounds in your study environment would be a good place for you to start altering the way you work. Complete silence is not necessarily the best condition, but loud rock and roll is definitely not helpful. You should try to find sound that is soft enough not to be intrusive and with a slow beat. Try playing classical music at such a low volume that you cannot pay attention to it. That music will over-ride other sounds in your environment that might be distracting.

Psychological Attributes of a Quality Environment

Equally important to the obvious physical attributes of your learning environment are those qualities which are psychological in nature. Fear of mathematics or anxiety over mathematics interferes greatly with the study habits of some students. Those fears or anxieties must be dealt with prior to a study session if they are to be overcome. Your mind, as well as your body, must be free from distractions. A mind that is full of fear or worry has no capacity to also learn mathematics. If your mind is distracted by other thoughts, find ways to clear it before studying. Otherwise, it will make your study time inefficient and distasteful. The result will be a lessening of your efforts and eventual failure.

Focus on relaxation techniques

Relaxation techniques can be learned and they will lower the level of fear or anxiety. You can learn methods like Benson's Relaxation Technique from books in the school library. You can also ask for some personal assistance from a school counselor.

All students should learn, as a minimum, to pay attention to their breathing. Do not attempt to alter your breathing. Just pay attention to it and its rhythm. That simple focusing will, by itself, have beneficial results in any anxiety producing situation. Try it. The more you try it, the better will be your results.

Play the Role of a Scholar

A final aspect of your environment needs close attention. Education is more than simply learning a list of facts. It is a different lifestyle with a value system that respects books, music, art, and beauty from an altered perspective. Put good literature, classical music, fine art, and flowers into your environment. Live with these elements in your study environment.

Imagine the private study of a learned professor at Yale and try to create the same study conditions for yourself. Playing the role of a scholar is helpful in becoming a good student.

2. Learn what part of mathematics needs to be memorized and those portions of the subject that must be figured out.

A major problem for many students is the belief that all mathematics should be memorized. This is false and trying to memorize everything always leads to failure. If you have been memorizing much in mathematics, that is not only inefficient – it is destructive. Read this section carefully and develop new ways of dealing with mathematics information.

Memorize Definitions and Symbols

In general, the **only** material that should be memorized in a mathematics course is the meaning of words and symbols. A good mathematics student must know the meaning of a word like "percent" and its symbol "%." **Memorize words and symbols** because their meanings are crucial to working out problems that contain them.

Focus on the meaning of words and symbols

Words and symbols play a crucial role in mathematics which is often overlooked by students. At the same time, words and symbols do not have meanings that can be figured out or discovered. Your teacher or your textbook needs to give you clear definitions for all the words that are relevant and clear descriptions of any new symbols that are introduced.

Words like "percent" and symbols like the radical sign, $\sqrt{}$, must be clearly understood before problems that contain them can be learned. Such words and symbols must be memorized. Don't skip past them and attempt to do problems without first understanding them.

Do Not Memorize Rules, Laws, or Processes

In general, do not memorize the steps involved in working a problem. **Do not memorize** rules or laws. Such memorization clutters your mind unnecessarily and tends to misdirect all your study. Blindly memorizing a "bunch of rules" is not the same as learning mathematics. Find reasons why rules, laws, processes, or steps work. If you understand why, there will be no need to memorize more than words or symbols.

Focus on not memorizing rules or processes

Treat rules or laws in mathematics just as you treat the legal laws in your community. Nobody memorizes all the state laws. Instead, the good citizen tries to make sense of them and learns how they apply to his/her needs.

Rules and laws of mathematics also need to make sense. Your study of them needs to uncover the logic that supports them. If you learn the reasons for different steps in a problem then you will find it is both unnecessary and unwise to learn the steps themselves!

Focus

Finally, and very importantly, you need to direct your attention to the particular idea or skill that is to be learned. **FOCUS** on each new idea or skill. That is the theme of this text and every effort is made, by the book's format and presentation, to constantly alert you to the new material that must be the object of your attention.

You need to **FOCUS** specifically on the purpose of each new idea or skill as it is presented. Do not be distracted by other parts of a problem or its accompanying material.

Focus on the objective of each lesson

Each new idea or skill in mathematics will be presented to you in a context which contains many other ideas with other skills. It is essential that you recognize both the new ideas and the older ones. You must learn to concentrate on the new ideas. This is what is meant by **focusing**.

Poor mathematics students generally try to learn with only a vague idea of the particular new skill being taught at any moment. Good mathematics students practice the habit of giving their undivided attention to the one new aspect of a problem that needs to be learned while treating the other parts of the problem as background information.

3. Keep your mind active

Learning is an activity of the mind and no one is learning when they aren't thinking.

> **There is no way to divorce learning from thinking.**

You can improve your ability to learn by constantly making certain that you are thinking. Good learning of mathematics occurs when the student is thinking about the mathematics. Poor learning of mathematics occurs when the student's mind is elsewhere.

Focus on thinking while learning

It may come as quite a surprise, but even the best learners encounter difficulties at trying to maintain their thinking. All students, good and bad, have their minds wander. Good students have learned methods for constantly bringing their minds back to the activity. Poor learners have frequently seemed pleased when they could do a lot of problems without thinking.

If you find yourself doing mathematics without thinking, stop and bring yourself back to learning.

Inner Dialogues Keep Your Mind Active

An excellent way to keep your mind active is to carry on a conversation with yourself as you study. This "inner dialogue" might be structured by always asking the following two questions:

> **"What is the meaning of this problem?"**
>
> **"Why is the problem worked this way?"**

Neither of these questions is simple. Both of them force the learner to think and that is their purpose. The first question will keep the student on track in constantly knowing the meaning of words and symbols. The second question will emphasize the need to understand the reasoning behind a problem rather than memorizing the steps used to solve it.

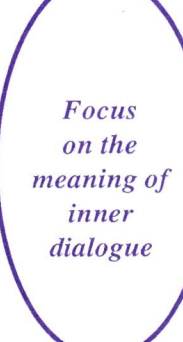

Focus on the meaning of inner dialogue

An "*inner dialogue*" is a form of talking to oneself and such conversations are valuable. It is also valuable to talk with someone else when studying. Find another student who will participate in a common learning experience. Again, the two major questions you can build into your study sessions are:

"What is the meaning of this problem?"

"Why is the problem worked this way?"

You want to constantly look at both meaning and reasoning. The questions, whether in an "inner dialogue" or with a learning "buddy," will structure your thinking and keep you on track for learning mathematics effectively.

4. Learn to evaluate your own learning

Many students believe that it is the task of the teacher to give them a grade and, consequently, accept little responsibility for making accurate judgments about their own progress in a mathematics course. When such students are asked, "How did you do on the test?" they are likely to give very vague or inaccurate responses. This is a characteristic that is drastically different from good students. When good students are asked, "How did you do on the test?" they are likely to be very specific in their comments and will accurately predict the grade they will receive.

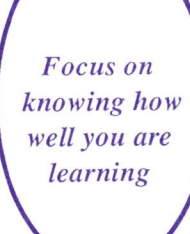

Focus on knowing how well you are learning

Most math programs and texts provide students with opportunities for evaluating the extent to which learning has occurred. For example, most texts contain answers for many of the problem sets. Students who check the answers to problems carefully and restudy those problems that are missed gain valuable insight into both the mathematics and their learning of the mathematics.

To make any evaluation of your work pay its best dividends, be certain that enough is written to provide a clear trail of your thinking on each problem. Finding wrong answers is not nearly as helpful as finding the point in a problem where you went wrong.

Two Kinds of Knowledge

Knowing how well you are learning in a mathematics course is very different from knowing the mathematics. In other words, there are always two types of knowledge for the mathematics student:

1. Knowledge of the mathematics, and
2. Knowledge about how well the mathematics has been learned.

Knowledge about the mathematics is not enough. The most effective student needs to have a knowledge about his/her learning of the mathematics. In studying mathematics, always pay attention to both the subject and the learning of that subject.

A simple example involves the checking of a problem. The student who does a problem, but skips the check of the problem is accepting responsibility for the mathematics, but ignoring the equally important responsibility for evaluating their learning.

Focus on knowing about yourself

The best of students make mistakes, misunderstand directions, and encounter real difficulties with some topics. But these students have developed mechanisms for telling them when those errors in learning occur.

Knowing when the required material has not been learned gives direction and meaning to further study. Excellence is achieved for the mathematics student when he/she knows the mathematics and also knows that he/she knows the mathematics.

Failures in Learning Are Correctable

A major difference between good and poor mathematics students is the fact that poor students do not know whether they know the mathematics. After studying for a mathematics test, poor students are unable to determine correctly whether they know or don't know the material. This is truly tragic because it is almost always correctable. The tragedy is that poor students generally do not take advantage of the opportunities they have for accurately assessing how much they know.

Use your textbook, your instructor, and other sources for help. If you look for them there are ample opportunities for you to determine how much and how well you know the mathematics. Problem sets, review exercises, practice tests, supplementary teaching aids (television, computers, etc.) and learning centers are usually available. Good students take advantage of them. Poor students may not realize their value in the learning process and either ignore them or give them only slight attention.

As a final word of advice consider the following. Whenever you think the mathematics is hard, stop and try to figure out which of the following aspects of the topic are giving you difficulty.

1. Do you know WHAT you are trying to learn?
   a. Sometimes a word or symbol deserves closer attention. When the word or symbol is more clearly understood, then it will be easier to proceed.
   b. Sometimes the problem looks far more complex than intended because you are not focusing on just that part of the problem where the new learning is to occur. Read any directions or clues you were given again. See if you can ignore parts of the problems and begin work on a smaller portion.
2. Try to do something with the problem.
   a. Ask WHERE you have seen something similar.
   b. Ask WHY this problem is related to the last problem or topic you worked on in the course.
3. Consciously learn from whatever you did in step 2.
   a. Ask WHY it worked or didn't work.
   b. Ask WHAT you could change so the result might be better.

If these steps fail, you haven't. Now you can take your work to a friend, a parent, or a teacher and talk intelligently about your difficulty. You may be surprised that your explanation suddenly leads directly to a solution. If not, you can expect a good conversation in which you can fully participate while receiving the help you need.

Chapter 0 Pre-test

The problems of this test will indicate your knowledge of a first course in Algebra. If you cannot answer at least 25 of these questions correctly you need to study this chapter before beginning Intermediate Algebra.

1. True or false?
   $9 \in \{3,7,9,17\}$

2. List the real numbers from the following:
   $12, \sqrt{3}, \frac{7}{0}, -15, \sqrt{-6}, \frac{4}{9}, \sqrt{\frac{3}{10}}$

3. $\{7, \pi, -5, 14\} \cap \{-5, 11, 13, 14\} = $ _____

4. List the factors in the expression
   $2(x-5)(3x+2)$.

5. Simplify: $3xy^2 - 4 - 3x^2y + 5$

6. Multiply: $(2x-1)(2x+1)$

7. Multiply: $(3x-7)(2x+3)$

8. Multiply: $(5x+1)(3x^2 - 2x - 1)$

9. Simplify: $\frac{3}{4}(10x - \frac{2}{5}) - 3x$

10. What is the reciprocal of -2?

11. Solve: $9x = 28$

12. Solve: $13 + x = 7$

13. Solve: $-4(2x-1) = 16 - 5x$

14. Solve: $3(x-4) = 4(2x+7)$

15. Solve: $\frac{-3}{4}x = \frac{2}{15}$

16. Solve: $\frac{2}{x} + 3 = \frac{3}{x} - 5$

17. Which of the numbers 15, 7, 18, -5, 9 make $x \geq 9$ true?

18. Is -4 in the truth set of $x > 2$ or $x < -4$?

19. Simplify: $3x - 5 > x + 4$

20. Simplify: $-2x + 3 < x - 3$

Factor each of the following or label it as prime.

21. $-15x - 3$

22. $4x(3x-2) - 3(3x-2)$

23. $rc + rd + sc + sd$

24. $x^3 - 3x^2 + x$

25. $ab + ac - xb - xc$

26. $x^2 - 14x + 40$

27. $x^2 + x - 42$

28. $x^2 - 49$

29. $4x^2 - 16x - 9$

30. $3x^2 + 10x + 8$

0
Review of Elementary Algebra

This introductory chapter reviews the types of skills and knowledge needed to be successful in this textbook. In general, each reader is expected to already know the material of a beginning algebra course. For some students, the material was learned some time ago, is incomplete, or otherwise lacking. This chapter provides the opportunity to begin the course with the information necessary to make intermediate algebra a successful academic adventure.

The presentation in this chapter is informal. Definitions are not rigorously stated so some words and symbols may appear without explanation. Justifications for processes in this chapter may not be given with the examples. In all later chapters this informality with the mathematics will be dropped and you will be expected to know, without hesitation, meanings of words and symbols. You will also be expected to know the reasons why processes work (or don't work). In short, this introductory chapter is an opportunity to get your feet on the ground before the course begins in chapter 1.

Unit 1: SETS, ELEMENTS, AND SYMBOLS

The terminology of sets, members of sets, and set symbols will be used throughout the text.

A set is a collection of objects, names, or numbers. The collection of numbers 2, 5, 9, 15, and 23 is a set. Similarly, the collection of a chair, a desk, and a pencil is a set of objects.

Focus on writing sets using braces

The set of numbers 5, 12, and 17 can be written with braces as {5,12,17}.

The set {2,7,11,44} is made up of elements or members. The elements of the set {2,7,11,44} are 2, 7, 11, and 44.

Elements of Sets

6 is in the set {1,4,6,9} and we say that 6 "is an element" of the set or 6 "belongs" to the set or the set "contains" 6.

The symbol "$\in$" means "is an element of." "$5 \in \{1,5,9\}$" is read as "5 is an element of the set {1,5,9}."

"$\notin$" is the symbol for "is not an element." $3 \notin \{1,5,8\}$ is a true statement.

Focus on the use of three dots in describing a set

The elements of some sets may be shown by a partial listing which shows a pattern by which all other elements can be found. For example, the set of even numbers less than 20 can be shown as {2,4,6,. . . ,18} where the three dots mean "and so forth" and indicate the next number is 8, then 10, etc., until the last element, 18, is reached.

Some sets have no last number (they keep going forever). For example, the set of odd numbers is {1,3,5, . . .}. The three dots indicate that the pattern established by 1, 3, and 5 is to continue and the lack of a last or closing element in the set indicates that the pattern is to continue forever.

Set Selector Notation

{6,7,8} may be shown as {x | x is a whole number between 5 and 9}. The open sentence "x is a whole number between 5 and 9" has a blank space now occupied by the letter x. Anything may replace the x in the open sentence. Some replacements will result in true statements. Other replacements will result in false statements.

The set {x | x is a whole number between 5 and 9} contains those replacements for x that result in true statements.

{6,7,8} = {x | x is a whole number between 5 and 9}.

Focus on use of the set selector method

The set {x | x is a whole number greater than 7} is described by the **set selector method**. The open sentence "x is a whole number greater than 7" allows us to select those replacements for x that make true statements and reject those replacements that make false statements.

The set {x | x is a whole number greater than 7} = {8,9,10, ... }.

Set Operations

Two sets can be combined into one by the operations of union and intersection.

The union of two sets (shown by the symbol ∪) is the set containing all the elements from either set.

For example, the union of {1,2,3,4} with {4,5,6} is {1,2,3,4,5,6}. Notice that 4 is an element of both sets, but appears only once in the union.

{1,2,3,4} ∪ {4,5,6} = {1,2,3,4,5,6}

Focus on the intersection of two sets

The intersection of two sets is the set containing only those elements that are in both sets.

For example, the intersection of {1,2,3,4} with {4,5,6} is {4}. Notice that 4 is the only element shared by the two sets.

{1,2,3,4} ∩ {4,5,6} = {4}

Chapter 0

The Numbers of Elementary Algebra

Four particular sets of numbers are studied in an elementary algebra course.

1. The set of **counting** numbers $\{1,2,3,\ldots\}$ contains the whole numbers of arithmetic which are learned in the early grades of elementary school.
2. The set of **integers** $\{\ldots,-3,-2,-1,0,1,2,3,\ldots\}$ contains the counting numbers, zero, and the negatives (opposites) of the counting numbers. These numbers are often introduced to a student when the study of algebra begins.
3. The set of **rational** numbers. The positive rational numbers are the fractions that are studied in the primary grades.
4. The set of **real** numbers. Each point on the number line is associated with a unique real number. Some of these real numbers are also rational numbers, but others like $\sqrt{2}$ and π are not rational. $\sqrt{2}$ and π are irrational numbers.

Focus on sets of numbers studied in beginning algebra

The numbers 7, 38, 417, and 89,254 are counting numbers. They are also integers, rational numbers, and real numbers.

The numbers 7, -8, 65, and -73 are integers. Two of the numbers are counting numbers, but all of them are integers, rational numbers, and real numbers.

The numbers 7, -8, $\frac{1}{2}$, and $\frac{-3}{4}$ are rational numbers. The first two are also integers, but all of them are both rational and real numbers.

The numbers 7, -8, $\frac{2}{3}$, $\frac{-5}{3}$, $\sqrt{13}$, and π are real numbers. The first four are also rational numbers, but all of them are real numbers.

Adding/Subtracting Real Numbers

Below are listed some examples of the addition/subtraction of real numbers.

Addition Examples
$8 + 13 = 21$
$-7 + 5 = -2$
$\frac{1}{2} + \frac{5}{7} = \frac{17}{14}$
$6\sqrt{3} + 5\sqrt{3} = 11\sqrt{3}$

Subtraction Examples
$9 - 4 = 5$
$8 - 17 = -9$
$\frac{1}{4} - \frac{-5}{8} = \frac{7}{8}$
$-3\pi - 5\pi = -8\pi$

Focus on multiplying/dividing real numbers

Below are listed some examples of the multiplication/division of real numbers.

Multiplication Examples	Division Examples
$7 \cdot 3 = 21$	$8 \div 12 = \frac{2}{3}$
$-7 \cdot 5 = -35$	$-20 \div -8 = \frac{5}{2}$
$\frac{9}{10} \cdot \frac{5}{7} = \frac{9}{14}$	$\frac{1}{4} \div \frac{-1}{6} = \frac{-3}{2}$
$6\sqrt{3} \cdot -2\sqrt{3} = -36$	$8\pi \div -3\pi = \frac{-8}{3}$

Terminology for the Operations of Arithmetic

In the addition problem $9 + 7 = 16$, the numbers 9 and 7 are called **addends** and the answer is the **sum** or total. The addends of open expressions such as $9x + 8$ are commonly called **terms**. For example, the open expression $8x + 5 + 3y$ has three terms which are $8x$, 5, and $3y$.

In the subtraction problem $13 - 5 = 8$, the **minuend** is 13, the **subtrahend** is 5, and the answer is the **difference**.

In the multiplication problem $5 \cdot 8 = 40$, the numbers 5 and 8 are called **factors** and the answer, 40, is the **product**.

In the division problem $54 \div 9 = 6$, the **dividend** is 54, the **divisor** is 9, and the answer is the **quotient**.

Focus on multiplication expressions

The open expression $6x$ is a product. Its factors are 6 and x and the multiplication dot is normally not written in such expressions. Similarly,

    The factors of $7(3x + 2)$ are 7 and $3x + 2$.
    The factors of $(2x + 9)(x + 4)$ are $2x + 9$ and $x + 4$.

Order of Operations

A numerical expression may contain more than one operation. The following examples all involve both addition and multiplication.

$7 + 5 \cdot 6$
$8 \cdot 4 + 3 \cdot 7$
$3 \cdot 5 + 7$

When a numerical expression contains two or more operations without parentheses (grouping symbols) to indicate the order in which the operations are to be completed, the following procedure applies.
1. From left to right do any multiplication and/or division.
2. From left to right do any addition and/or subtraction.

Using this procedure on the three sample expressions shown above gives:
$7 + 5 \cdot 6 = 7 + (5 \cdot 6) = 7 + 30 = 37$
$8 \cdot 4 + 3 \cdot 7 = (8 \cdot 4) + (3 \cdot 7) = 32 + 21 = 53$
$3 \cdot 5 + 7 = (3 \cdot 5) + 7 = 15 + 7 = 22$

Focus on the order of operations

Each of the numerical expressions shown below has been properly evaluated by the order of operations.

$12 \div 8 \cdot 16 - 5 \cdot 4 + 3$
$\frac{3}{2} \cdot 16 - 5 \cdot 4 + 3$
$24 - 5 \cdot 4 + 3$
$24 - 20 + 3$
$4 + 3$
7

$13 - 8 \cdot 3 \div 6 - 4 \cdot 2$
$13 - 24 \div 6 - 4 \cdot 2$
$13 - 4 - 4 \cdot 2$
$13 - 4 - 8$
$9 - 8$
1

Unit 1 Exercise

Part A: Answers for all Part A problems are at the back of the book.

1. Write the collection of numbers 2, 15, and 37 as a set using braces.

2. Which of the following statements is true?
   $5 \in \{1,3,7,8\}$
   $7 \notin \{1,3,5,6\}$

3. Is -12 an element of the set of counting numbers $\{1,2,3,\ldots\}$?

4. Is 80,371 an element of $\{1,2,3,\ldots\}$?

5. If $f \in \{1,2,3,\ldots\}$ and $z \in \{1,2,3,\ldots\}$ and $f + z = a$, then $a \in$ _____.

6. Is the difference of two counting numbers always a counting number?

7. If w and y are counting numbers and w • y = z, then z is a _____.

8. Is it possible to add two counting numbers correctly and find that the sum is not a counting number?

9. Is $\{1,4,9,10\} \cup \{9,10,12\} = \{1,4,9,10,12\}$ a true statement?

10. Is $\{1,5,8,11\} \cap \{1,8,12,14\} = \{1,8\}$ a true statement?

For problems 11-14 use the numbers
$17, -3, 31, \sqrt{3}, \frac{4}{5}, \pi, -9$

11. List the counting numbers.

12. List the integers.

13. List the rational numbers.

14. List the real numbers.

15. Write a set selector notation for $\{3,4,5,6\}$.

16. Write a set selector notation for $\{6,8,10,12,14\}$.

Part B: Answers for odd-numbered problems of Part B are at the back of the book.

For problems 1-6, true or false?

1. $3 \notin \{1,4,7,9\}$
2. $5 \notin \{1,4,5,17\}$
3. $3 \in \{1,3\ 6,8\}$
4. $7 \notin \{1,3,5,7,10\}$
5. $19 \in \{1,2,3,\ldots\}$
6. $-3 \notin \{1,2,3,\ldots\}$
7. Is $\frac{13}{4}$ an element of $\{1,2,3,\ldots\}$?
8. List the numbers in the set $\{x \mid x \text{ is a whole number between 3 and 8}\}$.
9. List the numbers in the set $\{x \mid x \text{ is a whole number greater than 10}\}$.
10. $\{3,5,7\} \cup \{2,3,6,7,9\} = $ _____

11. $\{2,6,9,12,17\} \cap \{3,9,14,17\} = $ _____

12. $\{1,5,10,19,25\} \cap \{2,4,10,17,27\} = $ _____

13. $\{1,2,3,4,5\} \cup \{1,5,6,7\} = $ _____

14. Which of the following numbers are counting numbers? $\pi, -2, 7, \frac{1}{2}, 5, \frac{-4}{3}, \sqrt{5}, \frac{3}{4}$

15. Which of the following numbers are integers? $\frac{-7}{10}, 0, -4, \pi, \sqrt{7}, 19, -13, \frac{2}{3}$

16. Which of the following numbers are rational numbers? $-9, 5, \sqrt{7}, \frac{-7}{8}, \frac{2}{3}, \pi, \frac{0}{2}$

17. Which of the following numbers are real numbers? $\pi, 4, \frac{-2}{3}, \sqrt{11}, -17, \frac{0}{5}$

18. List the addends in: $14 + 9 = 23$

19. List the factors in: $27 \cdot 7 = 189$

20. List the factors in: $9(2x - 1)(x + 1)$

Unit 2: SIMPLIFYING ALGEBRAIC EXPRESSIONS

Simplifying

The expressions $6 + x + 5$ and $x + 11$ are **equivalent** expressions which means that for any replacement of x, the two expressions will have the same numerical evaluation. $6 + x + 5$ is simplified to $x + 11$ because $6 + x + 5$ requires two additions and $x + 11$ only requires one addition.

In general, an open expression is simplified when it involves less addition and/or multiplication.

Focus on simplifying open expressions

To simplify an addition expression, group and combine its **like terms**.
$8x + 9 - 3y - 10 + x - 7y = (8x + x) + (9 - 10) + (-3y - 7y) = 9x - 1 - 10y$

To simplify a multiplication expression, multiply the numerical factors (**coefficients**) and list the variable factors. If like variable factors are multiplied, add their exponents.
$5xy(-7xz) = (5 \cdot -7)(x \cdot x)(y \cdot z) = -35x^2yz$

Removing Parentheses

Parentheses are removed as shown in the following example.

$-7x(5x - 3) = -7x \cdot 5x - (-7x \cdot 3) = -35x^2 - (-21x) = -35x^2 + 21x$

In general, parentheses are removed by multiplication.

$7(6x - 5) - (3x - 4) = 42x - 35 - 3x + 4 = 39x - 31$

Focus on removing parentheses

To simplify $(8x + 3) - 3(2x - 7)$ first remove the parentheses and then group and combine like terms.

$(8x + 3) - 3(2x - 7) = 1(8x + 3) - 3(2x - 7) = 8x + 3 - 6x + 21 = 2x + 24$

Simplifying Addition Expressions

To simplify $5x - \frac{2}{3} + \frac{7}{8}x$, the order and grouping of the terms is arranged so that like terms are added.

$$5x - \frac{2}{3} + \frac{7}{8}x = \left(5x + \frac{7}{8}x\right) - \frac{2}{3} = \left(\frac{40}{8}x + \frac{7}{8}x\right) - \frac{2}{3} = \frac{47}{8}x - \frac{2}{3}$$

Focus on simplifying multiplication expressions

To simplify $\left(\frac{-2}{3}y\right) \cdot \left(\frac{3}{4}w\right)$, alter the order and grouping of the factors so that numerical factors are together.

$$\left(\frac{-2}{3}y\right) \cdot \left(\frac{3}{4}w\right) = \left(\frac{-2}{3} \cdot \frac{3}{4}\right)(yw) = \frac{-1}{2}yw$$

Simplifying Expressions with Parentheses

Parentheses are removed as in the two examples shown below. In each case the number preceding the parentheses is multiplied by each term within the parentheses.

$$\frac{2}{3}\left(x + \frac{4}{5}\right) = \frac{2}{3} \cdot x + \frac{2}{3} \cdot \frac{4}{5} = \frac{2}{3}x + \frac{8}{15}$$

$$\frac{6}{7}\left(4x - \frac{3}{5}\right) = \frac{6}{7} \cdot 4x - \frac{6}{7} \cdot \frac{3}{5} = \frac{24}{7}x - \frac{18}{35}$$

In general, the simplification of open expressions with parentheses begins with the removal of the parentheses.

To simplify $\frac{-3}{14} - \frac{2}{5}\left(3x - \frac{3}{7}\right)$, first remove the parentheses. Then rearrange and add like terms.

Focus on simplifying an expression with parentheses

$$\frac{-3}{14} - \frac{2}{5}\left(3x - \frac{3}{7}\right) = \frac{-3}{14} - \frac{2}{5} \cdot 3x + \frac{2}{5} \cdot \frac{3}{7}$$

$$= \frac{-3}{14} - \frac{6}{5}x + \frac{6}{35}$$

$$= \left(\frac{-3}{14} + \frac{6}{35}\right) - \frac{6}{5}x$$

$$= \left(\frac{-15}{70} + \frac{12}{70}\right) - \frac{6}{5}x$$

$$= \frac{-3}{70} - \frac{6}{5}x$$

> **Definition** — **Polynomial Over the Integers**
>
> An open expression is a polynomial over the integers **if and only if** it is of the form $a_1x^n + a_2x^{n-1} + \ldots + a_{n-1}x^2 + a_nx + a$ where each a_i is an integer and at least one a_i is not zero.

Examples of polynomials over the integers are:

$$5x^4 - 3x^3 + 7x^2 + 9x - 10$$

$$-7x^5 + 4x^4 - 9x^2 + 6$$

$$2x^2 + 9x - 5$$

Adding Polynomials

The addition of polynomials is accomplished by grouping and adding like terms. The concept of like terms requires that the variables and the exponents on the variables must be identical.

An example of the addition of two polynomials is shown below.

$$(7x^8 - 4x^5 + 3x^2 - 7x + 4) + (x^6 - 3x^5 + 3x^3 - 5x - 7)$$
$$7x^8 + x^6 + (-4x^5 - 3x^5) + 3x^3 + 3x^2 + (-7x - 5x) + (4 - 7)$$
$$7x^8 + x^6 - 7x^5 + 3x^3 + 3x^2 - 12x - 3$$

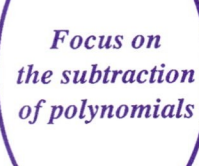
Focus on the subtraction of polynomials

The subtraction of polynomials is equivalent to adding the opposite of the second polynomial or multiplying the second polynomial by -1. An example is shown below.

$$(5x^4 - 2x^3 + 3x^2 - 8) - (4x^3 - 3x^2 - 2x + 3)$$
$$5x^4 - 2x^3 + 3x^2 - 8 - 4x^3 + 3x^2 + 2x - 3$$
$$5x^4 + (-2x^3 - 4x^3) + (3x^2 + 3x^2) + 2x + (-8 - 3)$$
$$5x^4 - 6x^3 + 6x^2 + 2x - 11$$

Multiplying Polynomials

To multiply 5 and (3x − 2), multiply 5 by 3x and then multiply 5 by -2.

$$5(3x - 2) = 5 \cdot 3x - 5 \cdot 2 = 15x - 10$$

To multiply (5x − 3) by (2x + 7), multiply 5x by (2x + 7) and then multiply -3 by (2x + 7).

$$\begin{aligned}(5x - 3)(2x + 7) &= 5x(2x + 7) - 3(2x + 7)\\ &= 10x^2 + 35x - 6x - 21\\ &= 10x^2 + 29x - 21\end{aligned}$$

Focus on multiplying binomials

The multiplication of two binomials with like terms is of sufficient importance that it should be practiced until it is unnecessary to write out the steps involved in the process. The first term of the **product** (answer) is the product of the first terms of the binomials. The last term of the product is the product of the second terms of the binomials. The middle term of the product is the sum of two multiplications.

$$(8x + 3)(3x - 1) = 24x^2 + (-8x + 9x) - 3 = 24x^2 + x - 3$$

Multiplying and Simplifying

The multiplication of $(x - 2)(x^2 + 2x + 4)$ is completed as shown below, but the product can be greatly simplified. Pay special attention to this multiplication because the result will be valuable in the next unit.

$$(x - 2)(x^2 + 2x + 4)$$
$$x(x^2 + 2x + 4) - 2(x^2 + 2x + 4)$$
$$x^3 + 2x^2 + 4x - 2x^2 - 4x - 8$$
$$x^3 + (2x^2 - 2x^2) + (4x - 4x) - 8$$
$$x^3 - 8$$

Focus on multiplying and simplifying

Another example of the multiplication of a binomial and trinomial is shown below. The product can be simplified. Notice the relationship between the terms.

$$(x + 5)(x^2 - 3x + 7)$$
$$x(x^2 - 3x + 7) + 5(x^2 - 3x + 7)$$
$$x^3 - 3x^2 + 7x + 5x^2 - 15x + 35$$
$$x^3 + (-3x^2 + 5x^2) + (7x - 15x) + 35$$
$$x^3 + 2x^2 - 8x + 35$$

Unit 2 Exercise

Part A: Answers for all Part A problems are at the back of the book.

1. Simplify $3x - 7y + 5x - 9y$ by combining like terms.
2. Simplify $6a - 3b + 8 + 4b - 8 - 5a$ by combining like terms.
3. Multiply $-3xy \cdot 5yz$ by reordering its factors.
4. Multiply $7a^2bc \cdot 4a^2b^2c$ by reordering its factors.
5. Remove the parentheses of $7(y + 5)$ by multiplying each term in the parentheses by 7.
6. Simplify $5(3x - 7) + 9(x + 4)$ by first removing parentheses and then combining like terms.
7. Simplify $-2(5x - 1) + 3(2x + 3)$ by first removing parentheses and then combining like terms.
8. Simplify $4(5x - 3) - (6x - 5)$ by first removing parentheses and then combining like terms.
9. Add $(3x^3 - 4x^2 + x - 6) + (5x^4 + 4x^2 + 5x - 2)$ by adding like terms.
10. $(8x^3 - 5x^2 - 7x + 3) - (6x^4 - x^3 - x + 6)$ by first changing the signs of the terms in the subtrahend.
11. Multiply $(x - 3)(3x + 5)$
12. Multiply $(4x + 5)(x - 2)$
13. Multiply $(7x - 3)(9x - 7)$
14. Simplify $3x - \frac{3}{5} + \frac{3}{4}x$
15. Simplify $7x - \frac{1}{6} + \frac{4}{7}$
16. Simplify $\frac{2}{5}(x + \frac{2}{3}) - \frac{3}{10}$
17. Simplify $5x + \frac{2}{5} - \frac{1}{3}x$
18. Simplify $\frac{2}{3}(6x - \frac{1}{2}) - 4x$
19. Simplify $\frac{5}{6} - \frac{3}{4}(8x - \frac{1}{2})$
20. Simplify $\frac{7}{8}x - \frac{3}{5} - \frac{1}{4}x - \frac{7}{10}$
21. Multiply $(x - 3)(x^2 - 7x + 3)$
22. Multiply $(x + 6)(x^2 - x + 3)$
23. Multiply $(x - 5)(x^2 + 5x + 25)$
24. Multiply $(x + 1)(x^2 - x + 1)$

Part B: Answers for odd-numbered problems of Part B are at the back of the book.

Perform the indicated operation and simplify.

1. $3x + 2y - 7x + x$
2. $7x - 3y - 5 - y + 5$
3. $4x - 5xy^2 + 8xy^2 - 9x$
4. $-6xy \cdot -8x^2y$
5. $y \cdot -5x$
6. $-4x \cdot 9x$
7. $5x^3y^2z \cdot -4xy^3z$
8. $5(2x + 3)$
9. $10(3 + 2x)$
10. $3(5x + 7)$
11. $(x + 2) \cdot 5$

12. $(3 + 2x) \cdot 7$
13. $(x - 8)(x + 3)$
14. $(2x - 3)(3x - 5)$
15. $(5x - 1)(5x - 1)$
16. $(2 - 7x)(1 + x)$
17. $(2x - 5)(2x - 5)$
18. $(x - 9)(2x + 9)$
19. $(6x - 5)(5x + 6)$
20. $(3x^2 - 5x - 7) + (4x^3 + 5x + 3)$
21. $(x^3 - 7x^2 + 4x - 7) - (2x^3 + 4x - 5)$
22. $5x - \frac{4}{7} + \frac{2}{3}x$
23. $3x - \frac{5}{8} + \frac{1}{6}$
24. $\frac{3}{8}(x + \frac{1}{3}) - \frac{2}{5}$
25. $9x + \frac{5}{6} - \frac{2}{3}x$
26. $\frac{3}{4}(8x - \frac{4}{9}) - 3x$
27. $\frac{1}{5} - \frac{1}{3}(5x - \frac{3}{4})$
28. $\frac{2}{5}x - \frac{1}{2} + \frac{2}{3}x - \frac{7}{8}$
29. $(x - 5)(x^2 - 2x + 3)$
30. $(x + 4)(x^2 + 3x - 2)$
31. $(2x - 3)(3x^2 - x + 1)$
32. $(4x - 5)(16x^2 + 20x + 25)$
33. $(x - 6)(2x^2 - 3x - 4)$
34. $(2x - 3)(3x^2 - x + 1)$
35. $(2x + 3)(4x^2 - 6x + 9)$

Unit 3: SOLVING LINEAR EQUATIONS

Solving Equations with Counting Numbers

The **truth set** of $x + 3 = 10$ is $\{7\}$ because 7 is the only counting number that can be added to 3 to give 10.

When elements of the set of counting numbers are the only allowable replacements for x, the truth set of $x + 7 = 2$ is the **empty set**, $\{\ \}$. There is no counting number that can be added to 7 to give 2. $-5 + 7 = 2$, but -5 is not a counting number.

Focus on solving equations with the set of counting numbers

Because there is no counting number that can be multiplied by 7 to give a product of 20, the truth set of $7x = 20$ with respect to the counting numbers is the empty set, $\{\ \}$. $7 \cdot \frac{20}{7} = 20$, but $\frac{20}{7}$ is not a counting number.

The truth set of $6x + 30 = 6(x + 5)$ is $\{1,2,3,\ldots\}$, because every counting number will make the equality true. $6x + 30$ and $6(x + 5)$ are equivalent and this means the equation is a true statement for any counting number replacement of x.

Finding Truth Sets Using Integers

To **solve** (find its truth set) $x + 5 = 1$ using the integers, note that -4 makes a true numerical statement $-4 + 5 = 1$. The **solution (root)** -4 is written in set braces, $\{-4\}$, to show the truth set of the equation.

As another example, the equation $-8x = 40$ is solved by finding the integer -5 and showing that the numerical statement $-8 \cdot -5 = 40$ is true. The solution -5 is written in set braces, $\{-5\}$, to show the truth set.

Focus on solving equations with the set of integers

When only integers are allowed as replacements for the variable:
  The truth set of $x + 4 = 1$ is $\{-3\}$.
  The truth set of $-6x = -42$ is $\{7\}$.
  The truth set of $4x = -21$ is $\{\ \}$.

There is no integer replacement for x that will make the open sentence $4x = -21$ a true statement. $\frac{-21}{4}$ makes $4x = -21$ a true statement, but $\frac{-21}{4}$ is not an integer.

The Additive Property Of Equality

$6x - 1 = 11$ is **equivalent** to $6x = 12$, because both equations have the same truth set, $\{2\}$. Similarly, $5x = 15$ is equivalent to $7x = 15 + 2x$, because $\{3\}$ is the truth set for both equations.

One way of constructing equivalent equations is to add the same number or open expression to both members of an equation. $3x + 19 = 12$ is equivalent to $(3x + 19) - 19 = 12 - 19$, because -19 was added to both sides of $3x + 19 = 12$.

The Additive Property of Equality: Any number or open expression may be added to both members of an equation to produce an equivalent equation.

Focus on the use of the Additive Property of Equality

If 9 is added to both members of $2x - 9 = 5$, the equivalent equation $2x = 14$ is obtained. Both $2x - 9 = 5$ and $2x = 14$ will have the same truth set, $\{7\}$.

One way of finding the truth set of $2x - 3 = -13$ is to use the Additive Property of Equality and add 3 to both sides of the equation.

$2x - 3 = -13$ becomes $(2x - 3) + 3 = -13 + 3$ and simplifies to $2x = -10$

$2x = -10$ is equivalent to $2x - 3 = -13$. Since the truth set of $2x = -10$ is obviously $\{-5\}$, then the truth set of $2x - 3 = -13$ is also $\{-5\}$.

Opposites

Every integer has an opposite. This fact is used to construct simpler equations that have the same truth set. For the equation $3x - 2 = 16$, the opposite of -2 is added to both sides of the equality as shown below.

$$3x - 2 = 16$$
$$(3x - 2) + 2 = 16 + 2$$
$$3x + (-2 + 2) = 18$$
$$3x + 0 = 18$$
$$3x = 18$$

The truth set of $3x - 2 = 16$ is $\{6\}$.

Focus on solving an equation by adding opposites

To solve 7x – 4 = 2x + 16, the following steps are used.

Add -2x to both sides of the equation.

$$\begin{array}{r} 7x - 4 = 2x + 16 \\ \underline{-2x \quad\quad -2x} \\ 5x - 4 = 0x + 16 \end{array}$$

Add 4 to both sides of the equation.

$$\begin{array}{r} 5x - 4 = 16 \\ \underline{+4 \quad +4} \\ 5x \quad = 20 \end{array}$$

The truth set of 7x – 4 = 2x + 16 is {4}.

Simplifying Equations Before Solving

The first step in solving any equation should be the simplification of each of its sides. When each side is completely simplified, the addition of opposites is used to construct a simpler equation with the same truth set.

To solve the equation -3(x + 5) + 7 = -2 first simplify the expression on the left side of the equation. -3(x + 5) + 7 = -2 is solved as follows.

$$\begin{array}{r} -3(x + 5) + 7 = -2 \\ -3x - 15 + 7 = -2 \\ -3x - 8 = -2 \\ -3x = 6 \end{array}$$

The truth set of -3(x + 5) + 7 = -2 is {-2}.

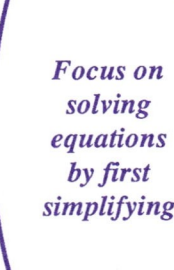
Focus on solving equations by first simplifying

To solve 5(x + 4) – 17 = 2x – (x + 13), first remove the parentheses and simplify each side of the equation separately. Then add opposites until the truth set is obvious.

$$\begin{array}{r} 5(x + 4) - 17 = 2x - (x + 13) \\ 5x + 20 - 17 = 2x - x - 13 \\ 5x + 3 = x - 13 \end{array}$$

$$\begin{array}{r} 5x + 3 = x - 13 \\ \underline{-x \quad\quad -x} \\ 4x + 3 = 0x - 13 \end{array}$$

$$\begin{array}{r} 4x + 3 = -13 \\ \underline{-3 \quad -3} \\ 4x \quad = -16 \end{array}$$

The truth set of 5(x + 4) – 17 = 2x – (x + 13) is {-4}.

Solving Equations Using Rational Numbers

Every equation of the form ax + b = c where a ≠ 0, such as 5x − 7 = 21 and $\frac{2}{3}x + \frac{5}{6} = \frac{1}{5}$, has a unique number in its truth set.

To solve $x + \frac{3}{7} = \frac{1}{3}$, first note that $\frac{3}{7}$ is being added to x. The x can be isolated in a new equivalent equation by adding the opposite of $\frac{3}{7}$ to both sides of the equation. The steps in solving $x + \frac{3}{7} = \frac{1}{3}$ are shown below.

$x + \frac{3}{7} = \frac{1}{3}$ becomes $\left(x + \frac{3}{7}\right) - \frac{3}{7} = \frac{1}{3} - \frac{3}{7}$ or $x + 0 = \frac{-2}{21}$ or $x = \frac{-2}{21}$

$\left\{\frac{-2}{21}\right\}$ is the truth set of $x + \frac{3}{7} = \frac{1}{3}$.

Focus on solving an equation using reciprocals

The equation $\frac{-3}{5}x = \frac{1}{4}$ is solved by noting that the expression $\frac{-3}{5}x$ indicates that $\frac{-3}{5}$ is to be multiplied by the replacement for x. The x can be isolated in a new equivalent equation by multiplying both sides of the equation by $\frac{-5}{3}$ which is the **reciprocal** of $\frac{-3}{5}$.

$\frac{-3}{5}x = \frac{1}{4}$ becomes $\frac{-5}{3} \cdot \frac{-3}{5} x = \frac{-5}{3} \cdot \frac{1}{4}$ or $1x = \frac{-5}{12}$ or $x = \frac{-5}{12}$

$\left\{\frac{-5}{12}\right\}$ is the truth set of $\frac{-3}{5}x = \frac{1}{4}$.

Multiplying All Terms by the Denominators' LCM

The **least common multiple (LCM)** of the denominators of $\frac{2}{5}x - \frac{2}{3} = \frac{1}{2}$ is 30. Before adding opposites, it is often easier to multiply each term by the denominators' LCM. This process assures that each term will have an integer as its coefficient as shown in the third line below.

$$\frac{2}{5}x - \frac{2}{3} = \frac{1}{2}$$

$$30 \cdot \frac{2}{5}x - 30 \cdot \frac{2}{3} = 30 \cdot \frac{1}{2}$$

$$12x - 20 = 15$$

$$12x = 35$$

$$x = \frac{35}{12}$$

Focus on using the LCM of the denominators

Equations which have the variable in one or more denominators are best solved by using the LCM of those denominators. The LCM of the denominators of $\frac{9}{x} + 3 = \frac{7}{x} - \frac{3}{4}$ is 4x. Notice how the multiplication of each term by 4x quickly simplifies the solution shown below.

$$\frac{9}{x} + 3 = \frac{7}{x} - \frac{3}{4}$$

$$4x \cdot \frac{9}{x} + 4x \cdot 3 = 4x \cdot \frac{7}{x} - 4x \cdot \frac{3}{4}$$

$$36 + 12x = 28 - 3x$$

$$36 + 15x = 28$$

$$15x = -8$$

$$x = \frac{-8}{15}$$

Solving Pairs of 2-Variable Equations

The equations $x + 4y = -7$ and $3x - 2y = 21$ have two variables. Any solution to these equations is an ordered pair (x,y). Many ordered pairs are solutions for each equation, but solving the pair of equations means finding any ordered pair that is a solution of both (**common solution**).

To find the common solution for $x + 4y = -7$ and $3x - 2y = 21$, add in vertical columns to eliminate one of the variables.

$\quad\quad x + 4y = -7$ remains $\quad x + 4y = -7$
$\quad\quad 3x - 2y = 21$ becomes $\underline{6x - 4y = 42}$
$\quad\quad\quad\quad\quad\quad\quad\quad\quad\quad\quad\quad 7x \quad\quad = 35$

Since x = 5, either equation can be used to find y.
$\quad\quad x + 4y = -7$ becomes $5 + 4y = -7$ or $4y = -12$ or $y = -3$.
The common solution is (5,-3).

Focus on finding a common solution

To solve $y = 3x + 1$ and $x - y = 3$ by substitution,

1. Since y equals (3x + 1), make a substitution in the other equation.
   $\quad x - y = 3$ becomes $x - (3x + 1) = 3$
2. Solve $x - (3x + 1) = 3$.
   $\quad x - (3x + 1) = 3$ becomes $x - 3x - 1 = 3$ or $-2x = 4$ or $x = -2$
3. Since x = -2, $y = 3x + 1$ becomes $y = -6 + 1$ or $y = -5$.

(-2,-5) is the common solution.

Review of Elementary Algebra

Unit 3 Exercise

Part A: Answers for all Part A problems are at the back of the book.

1. Using the set of counting numbers, what is the truth set of $x + 12 = 21$?
2. Using the set of counting numbers, what is the truth set of $5x = 12$?
3. Using the set of integers, what is the truth set of $-3x = 15$?
4. Using the set of integers, what is the truth set of $6x = -56$?
5. Using the set of rational numbers, what is the truth set of $3x - 7 = 1$?
6. Using the set of rational numbers, what is the truth set of $\frac{1}{5}x + 2 = \frac{7}{8}$?
7. In solving $3x - 5 = 10$ what term of the left side of the equation needs to be eliminated?
8. In solving $7x - 3 = 10 + 2x$, if $-2x$ is added to both sides as a first step, what term then should be added to both sides?
9. What is the opposite of $-9x$?
10. What is the reciprocal of -12?
11. What is the LCM for the denominators of the terms of $\frac{2}{3} + \frac{3}{4}x = \frac{1}{5}$?
12. What is the LCM for the denominators of the terms of $\frac{3}{8} + \frac{-3}{x} = \frac{1}{7} - \frac{2}{x}$?
13. Find the common solution of $4x + y = 15$ and $3x - 2y = -8$ by multiplying each term of the first equation by 2 and then eliminating the y by addition.
14. Find the common solution of $x = 3y$ and $4x - y = 1$ by substituting $3y$ for x in the second equation.

Part B: Answers for odd-numbered problems of Part B are at the back of the book.

For problems 1-10 use the set of counting numbers to find truth sets.

1. $x + 7 = 15$
2. $6x = 13$
3. $4y = 36$
4. $x + 17 = 36$
5. $4x = 28$
6. $x + 3 = 12$
7. $5x = 1$
8. $23 + x = 19$
9. $x + 17 = 9$
10. $x - 3 = 8$

For problems 11-20 use the set of integers to find truth sets.

11. $5x - 7 = 28$
12. $-6x + 3 = 21$
13. $3x - 2 = 4 - 3x$
14. $-6(x + 3) + 15 = 3$
15. $3x - 5 = x + 6$
16. $x + 3 = 5x - 1$
17. $-3x = 12 + 2x$
18. $2(3x - 4) + 6 = 10$
19. $-2x + 4 = -4x - 10$
20. $3(x - 7) - x = -9$

For problems 21-30 use the set of rational numbers to find truth sets.

21. $x + \frac{5}{4} = \frac{1}{2}$
22. $\frac{2}{9}x = \frac{1}{5}$
23. $\frac{3}{2}x - \frac{1}{7} = \frac{3}{5}$
24. $\frac{-2}{3}x - \frac{1}{4} = \frac{5}{6}$
25. $\frac{7}{x} = -2$
26. $\frac{-5}{x} = \frac{2}{3}$
27. $\frac{3}{x} - \frac{1}{5} = \frac{2}{7}$
28. $\frac{4}{x} + 3 = \frac{6}{x} - 5$
29. $\frac{3}{8} + \frac{1}{4}x = x - \frac{5}{6}$
30. $\frac{2}{3}x - \frac{7}{12} = \frac{1}{6}x - \frac{1}{2}$

For problems 31-35 find the common solution.

31. $3x + y = 14$ and $x - 2y = 7$
32. $x + 5y = 37$ and $3x + y = 13$
33. $y = x - 7$ and $3x - y = 15$
34. $x = 2y - 3$ and $x - 2y = 5$
35. $2x + 3y = -11$ and $5x - 2y = 20$

Unit 4: SOLVING LINEAR INEQUALITIES

Inequality Statements

The four symbols for **inequality** are:

Symbol	Translation
>	greater than
<	less than
≥	greater than or equal to
≤	less than or equal to

On the number ray shown below, the counting numbers are to the right of zero. With this orientation, the greater of two counting numbers will be the one farther to the right.

Focus on comparing the sizes of two counting numbers

$12 > 5$ is a true statement. 12 is to the right of 5 on the number ray.

$6 < 11$ is a true statement. Of the two numbers, 11 is farther to the right on the number line.

$8 \geq 3$ is true because $8 > 3$ is true and the symbol $\geq$ expresses an "or" condition that is true when one part of it is true. For a similar reason, $9 \geq 9$ is true because $9 = 9$ is true.

Finding Truth Sets Using Counting Numbers

$x > 5$ is an open sentence that is neither true nor false. For some counting number replacements of x, $x > 5$ will become true, but for other counting number replacements $x > 5$ will become false.

For $x > 5$, if $x = 9$ then $9 > 5$ is true.

For $x > 5$, if $x = 2$ then $2 > 5$ is false.

With respect to the set of counting numbers {1,2,3, . . .} the truth set of
x > 5 is {6,7,8, . . .} because each element in that set makes the inequality
true and each counting number not in the set makes the inequality false.

Focus on truth sets of inequalities

The inequality x ≤ 4 becomes true for any counting number less than or equal to 4. It becomes false for any counting number greater than 4.

For x ≤ 4, if x = 9 then 9 ≤ 4 is false.

For x ≤ 4, if x = 2 then 2 ≤ 4 is true.

For x ≤ 4, if x = 4 then 4 ≤ 4 is true.

With respect to the set of counting numbers, the truth set of x ≤ 4 is {1,2,3,4}.

Inequalities Involving Integers

On the number line shown below, the positive integers are to the right of zero and the negatives are to the left of zero. With this orientation, the greater of two integers will be the one farther to the right and the less of two integers will be the one farther to the left.

Focus on comparing the sizes of two integers

12 > 5 is a true statement. 12 is to the right of 5 on the number line.

-3 > -7 is a true statement. Of the two numbers, -3 is farther to the right on the number line.

4 is to the right of -2 on the number line, so 4 ≥ -2 is a true statement.

Since 1 is to the right of -5 on the number line, 1 ≥ -5 is also a true statement.

Truth Sets Using Integers

x + 9 > 14 is an open sentence that is neither true nor false. To determine the integers that will make the inequality true add -9 to both sides of the inequality.

1) Add -9 to both sides of the inequality.

$$\begin{aligned} x + 9 &> 14 \\ -9 & -9 \\ \hline x + 0 &> 5 \end{aligned}$$

2) Simplify.

$$x > 5$$

With respect to the set of integers { . . . ,-3,-2,-1,0,1,2,3,. . . } the truth set of x + 9 > 14 is {6,7,8, . . .} because each integer in that set makes the inequality true and each integer not in the set makes the inequality false.

Focus on truth sets of inequalities

The inequality x − 7 ≤ 11 is solved by adding 7 to both sides of the inequality.

$$\begin{aligned} x - 7 &\leq 11 \\ +7 & +7 \\ \hline x + 0 &\leq 18 \end{aligned}$$

$$x \leq 18$$

With respect to the set of integers, the truth set of x − 7 ≤ 11 is {. . . ,-3,-2,-1,0,1,2,3, . . . ,18}

Inequalities Joined by "Or" and "And"

"x > 2 or x < -5" is an open sentence combining two inequalities with the connective "or." The truth set of "x > 2 or x < -5" is the union of the separate truth sets. Therefore,

Since {3,4,5, . . .} is the truth set of x > 2,
and {-6,-7,-8, . . .} is the truth set of x < -5,
the truth set of x > 2 or x < -5 is:

{3,4,5, . . .} ∪ {-6,-7,-8, . . .}

This means that the elements of {-5,-4,-3,-2,-1,0,1,2} are the only integers that make x > 2 or x < -5 **false**.

Focus on the connective "and"

"x < 6 and x > -4" is an open sentence using the word "and". To make the statement true, an integer must satisfy **both** conditions (x < 6 and x > -4). This means that the truth set of x < 6 and x > -4 is the intersection of the truth sets of the separate inequalities. Therefore,

Since {5,4,3,2,1,0,-1 . . .} is the truth set of x < 6,
and {-3,-2,-1,0,1, . . .} is the truth set of x > -4,
the truth set of x < 6 and x > -4 is:
{5,4,3,2,1,0,-1 . . .} ∩ {-3,-2,-1,0,1, . . .} = {-3,-2,-1, . . .,5}

Inequalities Involving Rational Numbers

To compare the sizes of two rational numbers, write them with a common denominator and compare the sizes of their integer numerators. For example, to determine whether $\frac{5}{12}$ is greater than $\frac{4}{9}$:

$$\frac{5}{12} = \frac{15}{36} \text{ and } \frac{4}{9} = \frac{16}{36}$$

Since 16 > 15, $\frac{4}{9} > \frac{5}{12}$.

Focus on comparing the sizes of two rational numbers

The relative sizes of $\frac{-8}{15}$ and $\frac{-23}{40}$ are determined by finding a common denominator, 120.

$$\frac{-8}{15} = \frac{-64}{120} \text{ and } \frac{-23}{40} = \frac{-69}{120}$$

Since -64 > -69, $\frac{-8}{15} > \frac{-23}{40}$.

Finding Truth Sets of Linear Inequalities

There are three methods for constructing equivalent inequalities.
1. Any real number may be added to both sides of an inequality.
2. Both sides of an inequality may be multiplied by any positive number.
3. Both sides of an inequality may be multiplied by a negative number if the symbols are reversed (< to >; ≥ to ≤).

$5x - 7 > x + 1$ is simplified to $x > 2$ in the following steps.

$$5x - 7 > x + 1$$
$$(5x - 7) + 7 > (x + 1) + 7$$
$$5x > x + 8$$
$$5x - x > (x + 8) - x$$
$$4x > 8$$
$$\tfrac{1}{4} \cdot 4x > \tfrac{1}{4} \cdot 8$$
$$x > 2$$

Focus on constructing an equivalent inequality

To simplify $6 - 5x \leq x + 7$, the following steps are used. Pay special attention to the step in which $\tfrac{-1}{6}$ is multiplied by each term of the inequality and the symbol of inequality is reversed.

$$6 - 5x \leq x + 7$$
$$(6 - 5x) - 6 \leq (x + 7) - 6$$
$$-5x \leq x + 1$$
$$-5x - x \leq (x + 1) - x$$
$$-6x \leq 1$$
$$\tfrac{-1}{6} \cdot -6x \geq \tfrac{-1}{6} \cdot 1$$
$$x \geq \tfrac{-1}{6}$$

Unit 4 Exercise

Part A: Answers for all Part A problems are at the back of the book.

1. Is $15 > 11$ a true statement?
2. Is $-3 > -7$ a true statement?
3. Is $10 \leq 10$ a true statement?
4. Which of the numbers 11, 6, 15, 9, 23 make $x \leq 9$ true?
5. Which of the numbers 10, 6, 7, 8, 3 make $x > 7$ true?
6. Is 2 in the truth set of "$x < 6$ and $x > -4$"?
7. Is -9 in the truth set of "$x < 6$ and $x > -4$"?
8. Is 3 in the truth set of "$x \geq -7$ and $x < 10$"?
9. Is -5 in the truth set of "$x < 9$ and $x \geq -2$"?
10. What integers are in the truth set of $x + 12 \geq 7$?
11. What integers are in the truth set of $x - 8 > 2$?
12. What integers are in the truth set of $x + 3 < -6$?
13. What integers are in the truth set of $x - 9 \leq 4$?
14. What inequality is obtained by adding 7 to both sides of $5x - 7 > x + 1$?
15. What inequality is obtained by adding -x to both sides of $5x > x + 8$?

16. What inequality is obtained by multiplying both sides of $4x > 8$ by $\frac{1}{4}$?

17. Simplify $3x + 5 < x + 13$

18. Simplify $4x - 7 \leq 13$

19. Simplify $2 - 3x \geq x + 7$

20. Simplify $-5x + 3 > x - 15$

Part B: Answers for odd-numbered problems of Part B are at the back of the book.

Simplify each inequality.

1. $2x \leq 6x - 8$
2. $3x + 5 > 17$
3. $-4x > 12$
4. $-3x \leq -15$
5. $4x + 9 > x - 7$
6. $7x + 3 \leq 2x - 6$
7. $2x - 1 \leq 5 - x$
8. $5x - 1 \geq 3 - x$
9. $13 - 4x < x - 2$
10. $12 - 3x < 4x + 5$
11. $2x + 3 \leq 6x - 5$
12. $5 - 3x < 2x + 3$
13. $4x - 1 < x - 4$
14. $x - 3 > 5 - 6x$
15. $3x + 7 < 5x - 3$
16. $5 - 2x > x + 5$
17. $x + 9 < 5$
18. $-6x < 12$
19. $3 - 2x > x + 7$
20. $5x + 4 < 3x - 2$

Unit 5: FACTORING POLYNOMIALS

Addition Expressions Versus Multiplication Expressions

The polynomial $6x^2 - 7x + 2$ is an **addition expression** because if x is replaced by an integer then the **last** operation to be performed would be addition.

The polynomial $(3x - 2)(2x - 1)$ is a **multiplication expression** because if x is replaced by an integer then the **last** operation to be performed would be multiplication. A polynomial written as a multiplication expression is in **factored form**.

Focus on multiplication expressions

The polynomial $6x^2 - 7x + 2$ equals $(3x - 2)(2x - 1)$. One form of the polynomial is an addition expression. The other form is a multiplication expression.

$(3x - 2)(2x - 1)$ is the factored form of $6x^2 - 7x + 2$.

The polynomial $8x - 12$ equals $4(2x - 3)$. One form of the polynomial is an addition expression. The other form is a multiplication expression. $4(2x - 3)$ is the factored form of the addition expression $8x - 12$.

Factoring by the Common Factor Method

The simplest type of factoring is called the common factor method. 5 is a common factor for 5x and 10. Consequently, $5x + 10$ is factored as $5(x + 2)$. $4xy^2 - 2xy$ is factored as $2xy(2y - 1)$ using 2xy as the common factor of $4xy^2$ and $2xy$.

Focus on factoring polynomials by the common factor method

Each of the examples below shows a polynomial as both an addition expression and in its factored form.

$12x + 6 = 6(2x + 1)$
$8x^2 + 12x - 24 = 4(2x^2 + 3x - 6)$
$x^3 - 6x^2 + 7x = x(x^2 - 6x + 7)$
$7x^3 - 21x^2 + 35x = 7x(x^2 - 3x + 5)$

Focus on the use of 1 and -1 as common factors

1 and -1 are common factors for any polynomial.

$$3x^2 - 5y = 1(3x^2 - 5y)$$
$$3x^2 - 5y = -1(-3x^2 + 5y)$$
$$7xy + 3z^2 = 1(7xy + 3z^2)$$
$$7xy + 3z^2 = -1(-7xy - 3z^2)$$

Prime Polynomials

A polynomial is **prime** when its only factors are itself, its opposite, 1, and -1. For example, $3x^2 - 5y^3$ is a prime polynomial. The only common factors for $3x^3 - 5y^3$ are 1 and -1. The use of these common factors, results in the polynomial itself with 1 or its opposite with -1.

$$3x^2 - 5y^3 = 1(3x^2 - 5y^3)$$
$$3x^2 - 5y^3 = -1(-3x^2 + 5y^3)$$

Binomial Common Factors

The common factor of a polynomial can be a binomial. $(x - 3)$ is the common factor of $5x(x - 3) + 7(x - 3)$.

$2x(x - 3) + 5(x - 3)$ is factorable using the common factor method, because the binomial $(x - 3)$ is a factor of $2x(x - 3)$ and also of $5(x - 3)$.

$$2x(x - 3) + 5(x - 3) = (x - 3)(2x + 5)$$

Focus on a binomial used as a common factor

Four examples of using a binomial as a common factor are shown below.

$$x^2(3x - 5) - 2(3x - 5) = (3x - 5)(x^2 - 2)$$

$$(x - 5)(x + 4) + 4y(x + 4) = (x + 4)[(x - 5) + 4y] = (x + 4)(x - 5 + 4y)$$

$$3x^2(x + 3) - (x + 3)(x - 1) = (x + 3)[3x^2 - (x - 1)] = (x + 3)(3x^2 - x + 1)$$

$$x^2(x - 3) + (x - 3)^2 = (x - 3)[x^2 + (x - 3)] = (x - 3)(x^2 + x - 3)$$

Factoring by Grouping

To factor $6x^2 - 3x - 8x + 4$ the terms may be grouped.

$$6x^2 - 3x - 8x + 4$$
$$(6x^2 - 3x) + (-8x + 4)$$
$$3x(2x - 1) - 4(2x - 1)$$
$$(2x - 1)(3x - 4)$$

Focus on factoring by grouping

The following steps are used to factor $8x^3 + 12x^2 - 14x - 21$.

$$8x^3 + 12x^2 - 14x - 21$$
$$(8x^3 + 12x^2) + (-14x - 21)$$
$$4x^2(2x + 3) - 7(2x + 3)$$
$$(2x + 3)(4x^2 - 7)$$

Factoring Trinomials of the Form $x^2 + bx + c$

The trinomial $x^2 + 9x + 20$ has no common factors other than 1 and -1. Consequently, if the trinomial is to be factored its factors must be two binomials.

$$x^2 + 9x + 20 = (a + b)(c + d)$$

with x^2 as the product of the first terms and 20 as the product of the second terms.

To factor $x^2 + 9x + 20$, list possibilities that will produce x^2 as the product of the first terms of the binomials and 20 as the product of the second terms.

(1) $x^2 + 9x + 20$? $(x + 1)(x + 20)$
(2) $x^2 + 9x + 20$? $(x + 2)(x + 10)$
(3) $x^2 + 9x + 20$? $(x + 4)(x + 5)$

From this short list of possibilities, multiplication can be used to show that the correct factors are $(x + 4)$ and $(x + 5)$.
$$(x + 4)(x + 5) = x^2 + 5x + 4x + 20 = x^2 + 9x + 20$$

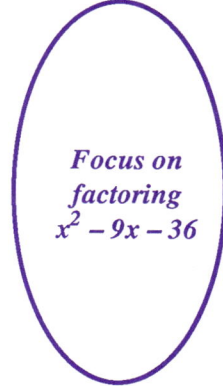

Focus on factoring $x^2 - 9x - 36$

To factor $x^2 - 9x - 36$, there are nine possibilities when the signs (+ or −) of the factors of -36 are considered.

(1-2) $x^2 - 9x - 36$? $(x + 1)(x - 36)$ or $(x - 1)(x + 36)$

(3-4) $x^2 - 9x - 36$? $(x + 2)(x - 18)$ or $(x - 2)(x + 18)$

(5-6) $x^2 - 9x - 36$? $(x + 3)(x - 12)$ or $(x - 3)(x + 12)$

(7-8) $x^2 - 9x - 36$? $(x + 4)(x - 9)$ or $(x - 4)(x + 9)$

(9) $x^2 - 9x - 36$? $(x + 6)(x - 6)$

$(x + 3)(x - 12)$ is correct because:

$$(x + 3)(x - 12) = x^2 - 12x + 3x - 36 = x^2 - 9x - 36$$

Factoring Trinomials of the Form $ax^2 + bx + c$

To factor $6x^2 + 13x - 15$ notice that the product of the first terms of the binomials must be $6x^2$ and the product of the second terms must be -15.

$$6x^2 + 13x - 15 = \underbrace{(a}_{} + b)\underbrace{(c}_{-15} + d)$$
(with $6x^2$ bracketing the outer terms a and c)

There are sixteen different possibilities for replacements of a, b, c, and d. However, eight of those possibilities are derived by switching the signs of c and d. Since the middle term of the trinomial is positive, use only those possibilities that will result in a positive middle term.

(1) $6x^2 + 13x - 15$? $(x - 1)(6x + 15)$

(2) $6x^2 + 13x - 15$? $(2x - 1)(3x + 15)$

(3) $6x^2 + 13x - 15$? $(3x - 1)(2x + 15)$

(4) $6x^2 + 13x - 15$? $(6x - 1)(x + 15)$

(5) $6x^2 + 13x - 15$? $(x + 3)(6x - 5)$

(6) $6x^2 + 13x - 15$? $(2x - 3)(3x + 5)$

(7) $6x^2 + 13x - 15$? $(3x - 3)(2x + 5)$

(8) $6x^2 + 13x - 15$? $(6x - 3)(x + 5)$

If you are now adept at multiplying binomials, you will soon find that the correct factors are $(x + 3)$ and $(6x - 5)$.

$$(x + 3)(6x - 5) = 6x^2 - 5x + 18x - 15 = 6x^2 + 13x - 15$$

Eliminating Possibilities

Factoring by listing all possibilities for producing the first and last terms can be greatly shortened by applying some of the knowledge you already have.

For example, we have already seen that the list of possibilities for factoring $6x^2 + 13x - 15$ should not include $(3x + 1)(2x - 15)$ because this product would have a negative middle term: $-45x + 2x = -43x$.

As another example, the possibility of $(x - 1)(6x + 15)$ as a factor of $6x^2 + 13x - 15$ should quickly be discarded because $6x + 15$ has a common factor of 3. Polynomials which have no common factors other than 1 and -1 cannot have factors with other common factors. For practice, look again at the possibilities listed on the previous page and you will find three more that can be quickly discarded.

Focus on factoring trinomials by listing all possibilities

To factor $8x^2 - 17x + 6$, list all the possibilities. Eight possibilities are listed below. Since 6 is positive and -17x has a negative coefficient it is only necessary to consider negative factors of 6.

(1) $8x^2 - 17x + 6$? $(x - 1)(8x - 6)$
(2) $8x^2 - 17x + 6$? $(2x - 1)(4x - 6)$
(3) $8x^2 - 17x + 6$? $(4x - 1)(2x - 6)$
(4) $8x^2 - 17x + 6$? $(8x - 1)(x - 6)$
(5) $8x^2 - 17x + 6$? $(x - 2)(8x - 3)$
(6) $8x^2 - 17x + 6$? $(2x - 2)(4x - 3)$
(7) $8x^2 - 17x + 6$? $(4x - 2)(2x - 3)$
(8) $8x^2 - 17x + 6$? $(8x - 2)(x - 3)$

Although 8 different possibilities are listed, most of them should be quickly discarded because they contain common factors. This leaves only (4) and (5) to check by multiplication.

But neither (4) nor (5) has a product of $8x^2 - 17x + 6$. This indicates **the trinomial is prime.**

Chapter 0

Unit 5 Exercise

Part A: Answers for all Part A problems are at the back of the book.

1. Factor $7x - 21$ by finding a common factor for $7x$ and -21.
2. Factor $3x^2y - 12xy^2$ by finding a common factor for $3x^3y$ and $12xy^2$.
3. Factor $x^3 - 2x^2 - x$ by finding a common factor for x^3, $2x^2$, and x.
4. Factor $-10x - 6$ by finding a negative number that is a common factor for $-10x$ and -6.
5. What is the common factor of $2x(x + 5) - 9(x + 5)$?
6. Factor $3x(x + 2) + 7(x + 2)$ by finding a common factor.
7. Factor $5x^2(x - 6) - (x - 4)(x - 6)$ by finding a common factor.
8. Factor $rt + rs + qt + qs$ by separately grouping the first two terms and the last two terms.
9. Factor $x^2 + 3x - 18$ by first listing all the possible factors of x^2 and -18.
10. Factor $6x^2 - x - 2$ by first listing all the possible factors of $6x^2$ and -2.

Part B: Answers for odd-numbered problems of Part B are at the back of the book.

Factor each of the following or label it prime.

1. $8x^3y^2z - 18xy^2z^2$
2. $x^2 - 3x$
3. $x^2y + xy$
4. $6x^2 - 12xy$
5. $8x^3y - 10z$
6. $-15x - 3$
7. $-9xy^3 + 15xy^2$
8. $4x^3 - 10x^2 - 16x$
9. $2x^2 - 10x + 14$
10. $6x^2y - 3xy + 9xy^2$
11. $x^2(4x - 1) - 6(4x - 1)$
12. $7x(3x - 2) + 5(3x - 2)$
13. $4x(2x - 1) + (2x - 1)$
14. $7x(x - 3) - (x - 3)$
15. $x^2(x + 4) + (x + 4)^2$
16. $3x(x - 5)^2 - 2(x - 5)$
17. $2x(x + 3)^2 - 5(x + 3)$
18. $ac + ad + bc + bd$
19. $mu + mr + tu + tr$
20. $xa + xb - ya - yb$

21. $x^2a + x^2b + y^2a + y^2b$
22. $x^3 - 2x^2 - x$
23. $-28a^2bc^4 + 42ab^2c^3$
24. $8x^3y - 20x^2y^2 + 22xy^3$
25. $15x^2z^3 - 25xz^4$
26. $7a^2 - 10b^2$
27. $8x(x - 3) + 5(x - 3)$
28. $8y(y + 1) + (y + 1)^2$
29. $xy + xz - ay - az$
30. $3x^2(5x - 2) - (5x - 2)$
31. $x^2 + 12x + 35$
32. $x^2 - 9x + 20$
33. $x^2 + 5x - 14$
34. $x^2 - 8x - 33$
35. $x^2 + 8x + 14$
36. $x^2 - 9$
37. $x^2 - 15x + 54$
38. $x^2 + 21x + 20$
39. $x^2 + 4$
40. $x^2 - 10x - 39$
41. $6x^2 + 23x + 7$
42. $4x^2 - 11x + 6$
43. $3x^2 + x - 10$
44. $2x^2 - 9x - 45$
45. $4x^2 + 17x + 10$
46. $4x^2 - 25$
47. $3x^2 - 13x + 12$
48. $6x^2 + 23x + 20$
49. $36x^2 + 1$
50. $6x^2 - 11x - 10$

Chapter 0 Test

«0,U» shows the unit in which this problem was studied in this chapter.

1. True or false?
   $4 \notin \{1,4,7,12\}$ «0,1»

2. Is the difference of two counting numbers always a counting number? «0,1»

3. $\{1,5,7\} \cup \{2,4,5,7,11\} = $ _____ «0,1»

4. List the terms in the expression $5x - 3$. «0,1»

5. Simplify: $7x - 2y - x + y$ «0,2»

6. Multiply: $5(7x - 8)$ «0,2»

7. Multiply: $(x - 9)(x + 5)$ «0,2»

8. Multiply: $(3x - 2)(4x^2 - x + 2)$ «0,2»

9. Simplify: $6x - \frac{1}{5} + \frac{2}{3}x$ «0,2»

10. What is the opposite of $\frac{-3}{5}$? «0,3»

11. Solve: $x + 13 = 31$ «0,3»

12. Solve: $5x = 35$ «0,3»

13. Solve: $5x - 3 = 2x - 9$ «0,3»

14. Solve: $-4x = 12 + x$ «0,3»

15. Solve: $5(2x - 3) = -(x + 3)$ «0,3»

16. Solve: $\frac{2}{3}x - \frac{1}{6} = x + \frac{3}{4}$ «0,3»

17. Is $-7 > -15$ a true statement? «0,4»

18. Is 7 in the truth set of $x \geq 7$ and $x < 11$? «0,4»

19. Simplify: $6x \leq 2x - 8$ «0,4»

20. Simplify: $-5x \geq 20$ «0,4»

Factor each of the following or label it as prime.

21. $x^2y + yx$ «0,5»

22. $2x^2 - 8x - 12$ «0,5»

23. $x(x + 5)^2 + (x + 5)$ «0,5»

24. $9x^2 - 5y^2$ «0,5»

25. $-28a^2bc^5 + 42abc$ «0,5»

26. $x^2 - 4x - 32$ «0,5»

27. $x^2 + 12x + 11$ «0,5»

28. $6x^2 - x - 12$ «0,5»

29. $4x^2 - 11x - 6$ «0,5»

30. $3x^2 + 17x + 10$ «0,5»

1
System of Rational Numbers

Unit 1: THE SYSTEM OF COUNTING NUMBERS

Definition		Variable
A letter such as x or y is a **variable for the counting numbers**	**if and only if**	it represents a position where a counting number can replace the letter.

Expressions written with one or more variables are **open expressions**. This means that each variable represents a position that is open for replacement by a counting number.

The Laws of the Counting Numbers

The laws for the counting numbers are listed below. The letters a, b, and c are used in this listing as variables for the counting numbers.

1. For addition:
   a. Closure Law. $a + b$ is a unique counting number.
   b. Commutative Law. $a + b = b + a$.
   c. Associative Law. $(a + b) + c = a + (b + c)$.

2. For multiplication:
   a. Closure Law. $a \cdot b$ is a unique counting number.
   b. Commutative Law. $a \cdot b = b \cdot a$.
   c. Associative Law. $(a \cdot b) \cdot c = a \cdot (b \cdot c)$.
   d. Multiplication Law of One. $a \cdot 1 = 1 \cdot a = a$.

3. Distributive Law of Multiplication over Addition:
   $a \cdot (b + c) = (a \cdot b) + (a \cdot c)$.

Definition		Equivalent Open Expressions
Two open expressions are **equivalent**	**if and only if**	when their variable(s) are replaced by the same counting number(s) the evaluations are the same.

Addition Laws for the Counting Numbers

Addition is a **closed** operation on the set of counting numbers, because the sum of two counting numbers is a counting number. This is not the case with subtraction because the difference of two counting numbers is not always a counting number.

$$6 - 2 = 4 \quad \text{and} \quad 5 - 8 = -3$$

Subtraction is **not** a closed operation for the set of counting numbers. Some subtraction problems do not have counting number answers.

Focus on equivalent addition expressions

According to the Commutative Law of Addition,
 x + 9 is equivalent to 9 + x
 7x + 3 is equivalent to 3 + 7x
 4x + 5y is equivalent to 5y + 4x

According to the Associative Law of Addition,
 (x + 5) + 3 is equivalent to x + (5 + 3)
 3x + (x + 9) is equivalent to (3x + x) + 9
 (5x + 8x) + 3x is equivalent to 5x + (8x + 3x)

Subtraction is neither commutative nor associative.
 ≠ means "is not equal to."
 5 − 3 ≠ 3 − 5 and (9 − 4) − 3 ≠ 9 − (4 − 3)

Multiplication Laws for the Counting Numbers

Multiplication is a **closed** operation for the set of counting numbers. When two counting numbers are multiplied, the product is a unique counting number. Division is not a closed operation on the counting numbers. For example, the quotient of 7 divided by 3 (7 ÷ 3) is not a counting number.

The counting number 1 is the identity element for multiplication. The product of 1 with any other counting number gives that same counting number. This means that:
 1x is equivalent to x.
 1(8x + 7) is equivalent to 8x + 7.

Focus on equivalent multiplication expressions

According to the Commutative Law of Multiplication
 a • 7 is equivalent to 7a
 (9x) • 4 is equivalent to 4(9x)
 (x + 3)(x + 7) is equivalent to (x + 7)(x + 3)

According to the Associative Law of Multiplication
 (5x) • 7 is equivalent to 5(x • 7)
 7(3x) is equivalent to (7 • 3)x

Division is neither a commutative nor an associative operation.
 8 ÷ 2 ≠ 2 ÷ 8   and   (20 ÷ 10) ÷ 2 ≠ 20 ÷ (10 ÷ 2)

The Distributive Law of Multiplication over Addition

The Distributive Law of Multiplication over Addition is the only property involving both addition and multiplication in a single equality.

$$x(y + z) = xy + xz$$

The Distributive Law allows **some** multiplication expressions to be written as additions. For example, $9(x + 7)$ is a multiplication expression that is equivalent to $9x + 63$ which is an addition expression. Similarly:

$6(x + 4)$ is equivalent to $6x + 24$.
$3(2x + 5)$ is equivalent to $6x + 15$.

Definition	**Like Terms**	
Two addends are **like terms**	**if and only if**	the addends contain exactly the same variable factors.

For example, $3x$ and $11x$ are like terms because both contain the variable factor x. $9xy$ and $3y$ are not like terms because $9xy$ has x and y as its variable factors, but $3y$ only has y as a variable factor.

Focus on the use of the Distributive Law to add like terms

The Distributive Law allows some addition expressions to be written as multiplications. For example, $7x + 2x$ is an addition expression that is equivalent to $9x$ and $9x$ is a multiplication expression. Similarly,

$4x + 8x$ is equivalent to $12x$. $(4 + 8)x = 12x$
$4x + x$ is equivalent to $5x$. $(4 + 1)x = 5x$

The Distributive Law is the basis for simplifying addition expressions involving like terms.

Solving Equations with the Counting Numbers

To solve an equation is the same as finding its **truth set**.

The truth set of $x + 5 = 11$ is found by inspection because 6 is the only counting number that can be added to 5 to give 11.

Similarly, the truth set of $4x = 20$ is {5} because 5 is the only counting number that can be multiplied by 4 to give 20.

Focus on sets with no unique counting number solution

When elements of the set of counting numbers are the only allowable replacements for x, the truth sets of $x + 9 = 3$ and $5x = 13$ are the empty set, { }. There are no counting numbers that make these equations true.

On the other hand, the truth sets of $8 + 5x = 5x + 8$ and $8x + 12 = 4(2x + 3)$ are the set of counting numbers. Every counting number makes each equation a true statement.

Truth Sets of Inequalities

Using the set of counting numbers as the replacement set, the truth set of $x > 9$ is $\{10, 11, 12, \ldots\}$.

Similarly, the truth set of $5x \geq 35$ can be determined by inspection to be $\{7, 8, 9, \ldots\}$.

Focus on truth sets of inequalities

With respect to the set of counting numbers:
    the truth set of $x \leq 31$ is $\{1, 2, 3, \ldots, 31\}$
    the truth set of $4x \leq 45$ is $\{1, 2, 3, \ldots, 11\}$

Unit 1 Exercise

Part A: Answers for all Part A problems are at the back of the book.

1. If $f \in \{1,2,3,\ldots\}$ and $z \in \{1,2,3,\ldots\}$ and $f + z = a$, then $a \in$ _____.

2. Is the difference of two counting numbers always a counting number?

3. If w and y are counting numbers and $w \cdot y = z$, then z is a _____ number.

4. $a + b = b + a$ is a statement of the _____ Law of Addition.

5. Are 15w and $w \cdot 15$ equivalent expressions?

6. $9x \cdot 17 =$ _____ by the Commutative Law of Multiplication.

7. Are $(5 + 6x) + 9x$ and $5 + (6x + 9x)$ equivalent expressions?

8. $(6 \cdot 5) \cdot 2 = 6 \cdot (5 \cdot 2)$ is a numerical statement of the _____ Law of Multiplication.

9. Subtraction (is, is not) an associative operation on the set of counting numbers.

10. Simplify $9x + 4x$ using the Distributive Law of Multiplication over Addition.

11. Remove the parentheses of $7(y + 5)$ by multiplying each term of $(y + 5)$ by 7.

12. Remove the parentheses of $(3 + 2x) \cdot 8$ by multiplying each term of $(3 + 2x)$ by 8.

13. Using the set of counting numbers, what is the truth set of $x + 7 = 15$?

14. Using the set of counting numbers, find the truth set of $x + 9 = 6$.

15. Using the set of counting numbers, solve $4y = 36$.

16. Using the set of counting numbers, find the truth set of $3x + 7 = 7 + 3x$.

17. What counting numbers make $x > 19$ true?

18. What counting numbers make $x \leq 35$ true?

Part B: Answers for odd-numbered problems of Part B are at the back of the book.

1. True or False? $19 \in \{1,2,3,\ldots\}$

2. True or False? $3 \notin \{1,2,3,\ldots\}$

3. True or False? $-7 \notin \{1,2,3,\ldots\}$

4. Is $\frac{13}{4}$ an element of $\{1,2,3,\ldots\}$?

5. If $t + s = d$, and t and s are counting numbers, then d is a _____ number.

6. Does every division problem involving counting numbers have a counting number answer?

7. Division (is, is not) a closed operation for the set of counting numbers.

8. Is division a commutative operation on the set of counting numbers?

9. Division (is, is not) an associative operation on the set of counting numbers.

10. ___ $\cdot x = x$

11. $1 \cdot a =$ _____

12. Is $3(2 + 4) = (3 \cdot 2) + (3 \cdot 4)$ a true statement?

13. $5(z + 6) =$ _____

14. $10(3 + 2x) =$ _____

15. Simplify $5xy^2 + 8xy^2$

16. The Distributive Law of Multiplication over Addition states that $3x + 3y =$ _____

In problems 17-27, solve the equations using the set of counting numbers as the replacement set.

17. $x + 17 = 36$
18. $4x = 28$
19. $7x = 21$
20. $x + 3 = 12$
21. $5x = 1$
22. $23 + x = 25$
23. $7x = 33$
24. $3 + a = 13$
25. $5(2x + 3) = 10x + 15$
26. $b + 4 = 1$
27. $c + 5 = 5$

For problems 28-30 what counting numbers make the inequalities true?

28. $x \geq 18$
29. $x < 21$
30. $x + 1 > x$

Part C: No answers are given for these problems. However, each is accompanied by an ordered pair «C,U» showing the chapter and unit in which it was taught.

1. If $x \in \{1,2,3,...\}$ and $z \in \{1,2,3,...\}$ and $x + z = t$, then $t \in$ _____.   «0,1»

2. List the rational numbers in the following list.
   $9, \frac{-2}{3}, \pi, -7, 0, \sqrt{5}, \frac{13}{4}$ «0,1»

3. Simplify $\frac{3}{4}(8x - \frac{1}{3}) - 5x$ «0,2»

4. Multiply $(x - 3)(x^2 + 3x + 9)$ «0,2»

5. Using the set of integers, what is the truth set of $-6x = 32$?   «0,3»

6. What is the LCM for the denominators of the terms of $\frac{2}{5} - \frac{3}{x} = \frac{1}{4}$?   «0,3»

7. Simplify $9x - 5 > -3x + 31$ «0,4»

8. Simplify $x + 17 \leq 4x + 5$ «0,4»

9. Factor $3x(5x - 1) - (5x - 1)$ «0,5»

10. Factor $rt + rs - gt - gs$ «0,5»

Unit 2: THE SYSTEM OF INTEGERS

The Set of Integers

The set of integers is $\{\ldots,-3,-2,-1,0,1,2,3,\ldots\}$.

Definition	Opposites	
Two numbers x and y are **opposites**	**if and only if**	their sum is zero, $x + y = 0$.

Definition	Variable	
A letter x or y is a **variable for the integers**	**if and only if**	it represents a position where an integer can replace the letter.

Expressions written with one or more variables are **open expressions** which means that each variable represents a position that is open for replacement by an integer.

The Laws of the Integers

The properties for the integers include all of the properties of the counting numbers and some new properties which are gained by a system containing zero and opposites of the counting numbers. The properties for the integers that were not properties of the counting numbers are shown in boldface type. When a, b, and c are variables for integers, the following laws apply.

1. For addition:
    a. Closure Law. $a + b$ is a unique integer.
    b. Commutative Law. $a + b = b + a$
    c. Associative Law. $(a + b) + c = a + (b + c)$
    d. **Identity Law.** $a + 0 = 0 + a = a$
    e. **Inverse Law.** $a + -a = -a + a = 0$

2. For multiplication:
   a. Closure Law. a • b is a unique integer.
   b. Commutative Law. a • b = b • a
   c. Associative Law. (a • b) • c = a • (b • c)
   d. Multiplication Law of One. a • 1 = 1 • a = a
   e. **Multiplication Law of Negative One.** a • -1 = -1 • a = -a
   f. **Multiplication Law of Zero.** a • 0 = 0 • a = 0

3. Distributive Law of Multiplication over Addition.
   a(b + c) = ab + ac

4. a. If a = b, then a + c = b + c
   b. If a = b, then ac = bc
   c. If ab = 0, then a = 0 or b = 0

5. a. If a > b, then a + c > b + c
   b. If a > b, then ac > bc when c > 0 and ac < bc when c < 0

Definition	Subtraction	
The subtraction of y from x (x − y) is z	if and only if	x + (-y) = z

Subtraction means that the opposite of the subtrahend is added to the minuend.

15 − 7 means 15 + (-7) and the difference is 8
9 − 11 means 9 + (-11) and the difference is -2

The New Properties for Adding Integers

Although the properties of the system of counting numbers are similar to the properties of the system of integers, the integers gain a great deal of power by the inclusion of zero and the opposites for each counting number.

For example, recall that there is no closure property for subtraction of counting numbers because problems such as 5 – 7 do not have counting number answers. The set of integers does have closure for subtraction. For any pair of integers x and y, the problem x – y will have a unique integer answer.

Focus on addition properties of the integers related to zero

Zero is called the **identity element for addition** because zero added to any integer has a sum of that same integer. $15 + 0 = 15$ and $-123 + 0 = -123$.

For integers, the Inverse Law of Addition states that the sum of any integer and its opposite is zero. Frequently, the opposite of an integer is called its **additive inverse**.

$$8 + (-8) = -8 + 8 = 0 \text{ and } (5x - 7) + (-5x + 7) = 0$$

The New Properties for Multiplying Integers

The number zero is too often overlooked as if it were nothing. Besides being the identity element for addition, zero has two valuable multiplication properties.

1. $a \cdot 0 = 0 \cdot a = 0$
2. If $ab = 0$, then $a = 0$ or $b = 0$.

The first multiplication property of zero states that zero times any integer gives a product of zero.

$$-7 \cdot 0 = 0, \ 0x = 0, \text{ and } 0 \cdot (5x - 7y) = 0$$

The second multiplication property of zero states that if the product of two integers is zero, then at least one of the integers is zero. If $xy = 0$, then x may be zero or y may be zero or both x and y may be zero.

If $x(y - 3) = 0$, then $x = 0$ or $(y - 3) = 0$.

Focus on the Multiplication Law of Negative One

The Multiplication Law of Negative One states that any integer multiplied by negative one will be the opposite of that integer.

$$-1 \cdot -37 = 37, \ 12 \cdot -1 = -12, \ x \cdot -1 = -x, \text{ and } -1 \cdot -p = p.$$

$-1(3x - 2y)$ is equivalent to $-3x + 2y$ because of the Multiplication Law of Negative One.

Solving Equations with Integers

The integers are a much more powerful set than the counting numbers for solving equations. The increase in power is due to the fact that every integer has an opposite.

It is not obvious that the truth set of $5x - 3 = 9x + 25$ is $\{-7\}$, but by successively adding -9x and 3 to both sides of the equation the solution is greatly simplified.

$$\begin{array}{r} 5x - 3 = 9x + 25 \\ \underline{-9x \quad\quad -9x} \\ -4x - 3 = 0x + 25 \end{array}$$

$$\begin{array}{r} -4x - 3 = 25 \\ \underline{+3 \quad +3} \\ -4x + 0 = 28 \end{array}$$

$$-4x = 28$$

$$x = -7$$

Focus on sets with no unique integer solution

When elements of the set of integers, $\{\ldots,-3,-2,-1,0,1,2,3,\ldots\}$, are the only allowable replacements for x, the truth sets of $4x = 3$ and $5x = 13$ are the empty set, $\{\ \}$. There are no integers that make these equations true.

On the other hand, the truth sets of $8 - 5x = -5x + 8$ and $9x - 12 = 3(3x - 4)$ are the set of integers. Every integer makes each equation a true statement.

Truth Sets of Inequalities

Using the set of integers as the replacement set, the truth set of $x > -2$ is $\{-1,0,1,2,\ldots\}$.

Similarly, the truth set of $-3x \geq 33$ can be determined by inspection to be $\{-11,-12,-13,\ldots\}$. Remember that when the coefficient of x is negative the sign of inequality needs to be reversed.

Focus on truth sets of inequalities

With respect to the set of integers:

the truth set of $x < 14$ is $\{\ldots,-3,-2,-1,0,1,2,3,\ldots,13\}$.

the truth set of $7x > -17$ is $\{-2,-1,0,1,2,\ldots\}$.

Unit 2 Exercise

Part A: Answers for all Part A problems are at the back of the book.

1. Use braces to show the set of integers.
2. If m and n are integers and $m + n = p$, then p is an _____.
3. The product of two integers is an integer because of the _____ Law of Multiplication of Integers.
4. Is the multiplication of negative integers closed?
5. Are $16 + t$ and $t + 16$ equivalent expressions?
6. According to the Commutative Law of Addition, $5y - 6z =$ _____.
7. According to the Commutative Law of Multiplication, $(4x - 3y)(x + 8) =$ _____.
8. According to the Commutative Law of Multiplication, $(4x + 7y) \cdot 6 =$ _____.
9. According to the Associative Law of Addition, $(6y - 4x) - 4z =$ _____.
10. According to the Associative Law of Multiplication, $5(-4y) =$ _____.
11. According to the Associative Law of Multiplication, $3[(x + 4)(x - 6)] =$ _____.
12. The Inverse Law of Addition states that the sum of an integer and its opposite is _____.
13. According to the Multiplication Law of One, $1(4a - 5b)$ is equivalent to _____.
14. $-y$ is equivalent to _____.
15. To make $(x + 5)(x - 2) = 0$ a true statement, at least one of the factors, $(x + 5)$ or $(x - 2)$, must be _____.

For problems 16-18, solve using the set of integers as the replacement set.

16. $5x - 13 = 3x + 19$
17. $4x + 7 = 5 - x$
18. $9x + 4 = 3x - 8$

For problems 19-20, find those integers that make the inequalities true.

19. $2x - 11 > -18$
20. $9 - 4x < 30$

System of Rational Numbers   47

Part B: Answers for odd-numbered problems of Part B are at the back of the book.

1. Is $\frac{25}{4}$ an integer?
2. What integer is neither positive nor negative?
3. By the Commutative Law of Addition $-5x + 3$ is equivalent to _____.
4. $(3x + 5y) \cdot 7$ is equivalent to _____ by the Commutative Law of Multiplication.
5. The Associative Law of Addition states that $(-4 + x) + 15 =$ _____.
6. The Associative Law of Multiplication states that $3 \cdot (5 \cdot -10) =$ _____.
7. Is $0 \cdot [(485 - 67) \cdot (-83 + 14)] = 0$ a true statement?
8. If $(x - 9)(2x + 5) = 0$, then $x - 9 = 0$ or _____.

In problems 9-19, solve the equations using the set of integers as the replacement set.

9. $x + 11 = 4$
10. $5x = -20$
11. $-7x = 42$
12. $x + 9 = 9$
13. $-3x = 1$
14. $5x - 9 = 21$
15. $-6x = -54$
16. $9x + 17 = 5x - 3$
17. $-5(2x - 3) = 15 - 10x$
18. $6y - 13 = 8 - y$
19. $4z + 12 = 9z - 23$

For problems 20-25 what integers will make the inequalities true?

20. $6x \geq 18$
21. $5x < -13$
22. $4x + 1 > 4x$
23. $5x + 4 < 8x + 13$
24. $3x - 5 \geq 11 + x$
25. $2x + 9 > 6x + 41$

Part C: No answers are given for these problems. However, each is accompanied by an ordered pair «C,U» showing the chapter and unit in which it was taught.

1. Multiply $-3a^3b^2c \cdot 9a^2bc^3$ «0,2»
2. Add $(5x^4 - 3x^2 + x - 7) + (x^4 + 3x^3 - x + 6)$ «0,2»
3. Using the set of integers, find the truth set of $5(2x - 7) - 11 = 24$.   «0,3»
4. Simplify $7x - 5 \leq 31 - 2x$ «0,4»
5. Factor $-12ab^3 + 60a^3b^3$ «0,5»
6. Factor $3x^2 - 9x - 27$ «0,5»
7. Does every division problem involving integers have an integer answer?   «1,1»
8. Is subtraction a commutative operation on the set of counting numbers?   «1,1»
9. $1 \cdot x =$ _____   «1,1»
10. $8(5 + 3x) =$ _____   «1,1»

Unit 3: THE SYSTEM OF RATIONAL NUMBERS

The Set of Rational Numbers

The set of integers is $\{\ldots, -2, -1, 0, 1, 2, \ldots\}$. The set of all meaningful quotients of integers must exclude division by zero, but accept all other combinations. This set of quotients is the set of rational numbers, $\{\frac{x}{y} \mid x \text{ and } y \text{ are integers and } y \neq 0\}$.

The word "rational" as applied to numbers refers to the "ratio" of two integers.

Focus on the set of rational numbers

Every counting number is also a rational number. Therefore, numbers such as $5 = \frac{5}{1}$, $28 = \frac{28}{1}$, and $519 = \frac{519}{1}$ are counting numbers, integers, and rational numbers.

Zero, which is not a counting number, is an integer and a rational number, $0 = \frac{0}{1}$.

Every negative integer is also a rational number. $-3 = \frac{-3}{1}$ and $-67 = \frac{-67}{1}$.

Proper fractions such as $\frac{2}{3}$ or $\frac{7}{9}$ are rational numbers. Improper fractions like $\frac{-18}{5}$ and $\frac{32}{7}$ are also rational numbers.

Definition	Variable		
A letter x or y is a **variable for the rational numbers**	**if and only if**	it represents a position where a rational number can replace the letter.	

Expressions written with one or more variables are **open expressions** which means that each variable represents a position that is open for replacement by a rational number.

The Laws of the Rational Numbers

The properties for the rational numbers include all of the properties of the counting numbers and integers, but one new property which is gained by a system containing reciprocals. The new property for the rational numbers that was not a property of the integers is shown in boldface type. The letters a, b, and c are variables for rational numbers in the following list.

1. For addition:
   a. Closure Law. a + b is a unique rational number.
   b. Commutative Law. a + b = b + a
   c. Associative Law. (a + b) + c = a + (b + c)
   d. Identity Law. a + 0 = 0 + a = a
   e. Inverse Law. a + -a = -a + a = 0

2. For multiplication:
   a. Closure Law. a • b is a unique rational number.
   b. Commutative Law. a • b = b • a
   c. Associative Law. (a • b) • c = a • (b • c)
   d. Multiplicative Law of One. a • 1 = 1 • a = a
   e. Multiplication Law of Negative One. a • -1 = -1 • a = -a
   f. Multiplication Law of Zero. a • 0 = 0 • a = 0
   g. **Inverse Law. If $a \neq 0$ then $\frac{1}{a}$ is a rational number and $a \cdot \frac{1}{a} = 1$**

3. Distributive Law of Multiplication over Addition. a(b + c) = ab + ac

4. a. If a = b, then a + c = b + c
   b. If a = b then ac = bc
   c. If ab = 0, then a = 0 or b = 0

5. a. If a > b, then a + c > b + c
   b. If a > b, then ac > bc when c > 0 and ac < bc when c < 0

> **Definition** **Reciprocals (Multiplication Inverses)**
>
> Two rational numbers are reciprocals **if and only if** their product is 1.
>
> The rational numbers $\frac{2}{3}$ and $\frac{3}{2}$ are reciprocals because $\frac{2}{3} \cdot \frac{3}{2} = 1$. Another pair of reciprocals is $\frac{-12}{5}$ and $\frac{-5}{12}$. They are reciprocals because $\frac{-12}{5} \cdot \frac{-5}{12} = 1$.

The Inverse Law of Multiplication

The set of rational numbers is more powerful than the set of integers because each rational number except zero has a reciprocal.

For example, recall that equations such as $5x = 13$ and $-8x = 5$ had no integer answers, but every such equation will have a solution in the set of rational numbers.

The root of $5x = 13$ is $\frac{13}{5}$.

The root of $-8x = 5$ is $\frac{-5}{8}$.

> **Definition** **Division**
>
> The division of x by y ($x \div y$) is z **if and only if** $x \cdot \frac{1}{y} = z$.

According to the definition of division, it is a form of multiplication where the dividend remains unchanged but the divisor is changed to its reciprocal.

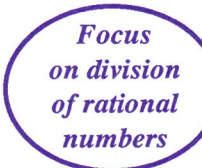
Focus on division of rational numbers

To divide $\frac{1}{4} \div \frac{5}{8}$, multiply $\frac{1}{4}$ by $\frac{8}{5}$, the reciprocal of $\frac{5}{8}$.

$$\frac{1}{4} \div \frac{5}{8} = \frac{1}{4} \cdot \frac{8}{5} = \frac{1}{\cancel{4}_1} \cdot \frac{\cancel{8}^2}{5} = \frac{2}{5}$$

Solving Equations with Rational Numbers

The rational numbers are more powerful than the integers for solving equations. The increase in power is due to the fact that every rational number except zero has a reciprocal. This improvement in equation solving ability means that every equation of the form ax + b = cx + d where a, b, c, and d are rational and a ≠ c has a unique solution in the set of rational numbers.

By adding opposites, an equation such as 7x + 8 = 4x − 15 can be written in the form kx = b where k ≠ 0. This means that the reciprocal of k exists and both sides of kx = b can be multiplied by $\frac{1}{k}$ to acquire the simple equation $x = \frac{b}{k}$.

$$
\begin{array}{rl}
7x + 8 &= 4x - 15 \\
-4x & -4x \\ \hline
3x + 8 &= 0x - 15
\end{array}
$$

$$
\begin{array}{rl}
3x + 8 &= -15 \\
-8 & -8 \\ \hline
3x + 0 &= -23
\end{array}
$$

$$
\begin{aligned}
3x &= -23 \\
\tfrac{1}{3} \cdot 3x &= \tfrac{1}{3} \cdot -23 \\
x &= \frac{-23}{3}
\end{aligned}
$$

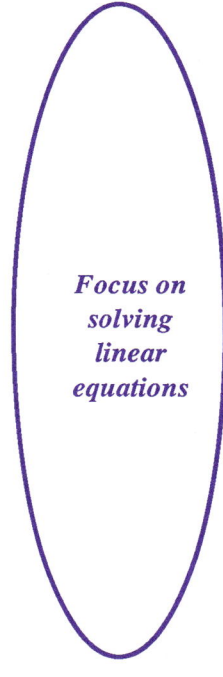

Focus on solving linear equations

The denominators of $\frac{3}{5}x + \frac{1}{3} = \frac{1}{2}x - \frac{4}{9}$ are integers and the LCM (least common multiple) of 5, 3, 2, and 9 is 90. Multiply each term on both sides of the equation by 90. Cancellation greatly simplifies the calculations.

$$\frac{3}{5}x + \frac{1}{3} = \frac{1}{2}x - \frac{4}{9}$$
$$90 \cdot \tfrac{3}{5}x + 90 \cdot \tfrac{1}{3} = 90 \cdot \tfrac{1}{2}x - 90 \cdot \tfrac{4}{9}$$
$$54x + 30 = 45x - 40$$

$$\begin{array}{r} 54x + 30 = 45x - 40 \\ -45x \quad\quad -45x \\ \hline 9x + 30 = 0x - 40 \end{array}$$

$$\begin{array}{r} 9x + 30 = -40 \\ -30 = -30 \\ \hline 9x + 0 = -70 \end{array}$$

$$9x = -70$$
$$x = \frac{-70}{9}$$

Truth Sets of Inequalities

Using the set of rational numbers as the replacement set, the truth set of $5x > 17$ can be indicated by a simpler inequality or a number line graph. The simpler inequality is $x > \frac{17}{5}$. The number line graph is shown below. The open dot (o) at $\frac{17}{5}$ indicates that $\frac{17}{5}$ is not an element of the truth set.

Similarly, the truth set of $-5x \geq 18$ can be shown as $x \leq \frac{-18}{5}$ or by the number line graph shown below. The closed dot (●) indicates that $\frac{-18}{5}$ is an element of the truth set.

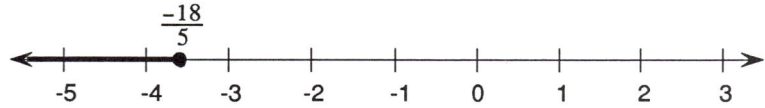

Focus on truth sets of inequalities

With respect to the set of rational numbers, the truth set of $9x < 13$ is shown as $x < \frac{13}{9}$ or as the number line graph shown below.

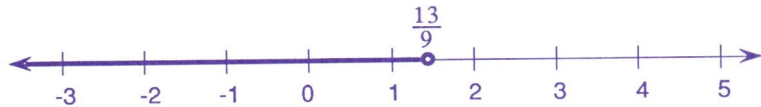

The truth set of $18 > -11x$ is shown as $x > \frac{-18}{11}$ or as the number line graph shown below.

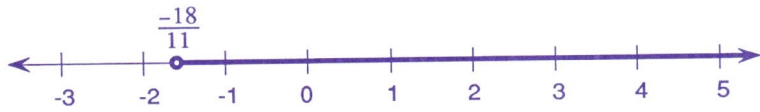

Unit 3 Exercise

Part A: Answers for all Part A problems are at the back of the book.

1. Is $\frac{4}{0}$ a rational number?
2. Is $\frac{0}{9}$ a rational number?
3. Which of the following are not rational numbers? $3, 0, \frac{7}{2}, \frac{-1}{0}, \frac{13}{-4}, \frac{0}{8}, -9, \frac{17}{1}$
4. What is the opposite of $\frac{3}{4}$?
5. What is the reciprocal of $\frac{3}{4}$?
6. What is the reciprocal of $\frac{-5}{9}$?
7. What is the reciprocal of 9?
8. What is the reciprocal of -7?
9. What is the reciprocal of $-x$ if $x \neq 0$?
10. Are $\frac{1}{2}(\frac{2}{3} - \frac{1}{6})$ and $(\frac{1}{2} \cdot \frac{2}{3}) - (\frac{1}{2} \cdot \frac{1}{6})$ equivalent expressions?
11. Use the LCM of the denominators to write an equivalent equation for $\frac{2}{3}x - \frac{3}{5} = \frac{4}{9}$.
12. Use the LCM of the denominators to write an equivalent equation for $\frac{10}{3x} + \frac{7}{4} = \frac{11}{6x}$.
13. Add -4 to each side of $5x + 4 > 23$ and complete the following inequality. $5x > $ _____
14. Multiply each side of $7x < -4$ by the reciprocal of 7. What simple inequality is found?
15. Which inequality is equivalent to $-5 > 9x$?
    $9x > -5$ or $9x < -5$
16. Add -9 to each side of $9 - 4x > 20$ and complete the following inequality. $-4x > $ _____
17. Multiply each side of $-6x < -5$ by the reciprocal of -6. What simple inequality is found?
18. Which inequality is equivalent to $-3x > 6$?
    $x > -2$ or $x < -2$

Chapter 1

Part B: Answers for odd-numbered problems of Part B are at the back of the book.

For problems 1-17, state the law that justifies each statement.

1. $\frac{3}{5}(\frac{1}{2} - \frac{7}{8}) = (\frac{3}{5} \cdot \frac{1}{2}) - (\frac{3}{5} \cdot \frac{7}{8})$
2. $\frac{3}{4} - \frac{7}{9} = \frac{-7}{9} + \frac{3}{4}$
3. $\frac{7}{3} + \frac{-7}{3} = 0$
4. $3(\frac{1}{4} + \frac{7}{2}) = (3 \cdot \frac{1}{4}) + (3 \cdot \frac{7}{2})$
5. $-6 \cdot \frac{-1}{6} = 1$
6. $(\frac{2}{3} \cdot \frac{1}{5}) \cdot \frac{3}{4} = \frac{2}{3} \cdot (\frac{1}{5} \cdot \frac{3}{4})$
7. If w and z are rational numbers and $w \cdot z = m$, then m is a unique rational number.
8. $-1 \cdot 13 = -13$
9. If a and b are rational numbers and $a + b = c$, then c is a unique rational number.
10. $\frac{3}{4}x + (\frac{-5}{8}x + \frac{1}{2}) = (\frac{3}{4}x + \frac{-5}{8}x) + \frac{1}{2}$
11. If $x \neq 0, \frac{1}{x} \cdot x = 1$
12. $-(3x - 1) = -1 \cdot (3x - 1)$
13. $(x - 3)(2x + 3) = x(2x + 3) - 3(2x + 3)$
14. $(8x) \cdot \frac{-1}{2} = \frac{-1}{2}(8x)$
15. $\frac{-3}{4}x + \frac{3}{4}x = 0$
16. $2(\frac{-1}{2} \cdot -5) = (2 \cdot \frac{-1}{2}) \cdot -5$
17. $-1 \cdot 3x = -3x$

For problems 18-25 solve.

18. $4x - 13 = 21$
19. $2x - 31 = 12 - x$
20. $6 - 5x = 14 + x$
21. $4(x - 3) - x = 3x + 7$
22. $\frac{5}{8}x - \frac{1}{6} = \frac{x}{3} + \frac{1}{4}$
23. $\frac{4}{x} - \frac{3}{5} = \frac{7}{2x}$
24. $\frac{4}{3} + \frac{3}{2x} = \frac{2}{9x} - 1$
25. $\frac{3}{5}x - 2 = x - \frac{5}{6}$

For problems 26-30 simplify each inequality and then graph the truth set on a number line.

26. $2x + 5 > 3$
27. $6 - 5x < 16$
28. $4 < 8 - 3x$
29. $2x - 6 \geq 8 + 5x$
30. $7 + 4x \leq 2x + 4$

Part C: No answers are given for these problems. However, each is accompanied by an ordered pair «C,U» showing the chapter and unit in which it was taught.

1. Add $(3x^4 - x + 7) + (x^4 - x^3 + x - 9)$ «0,2»
2. Simplify $-5x - 3 < 9 + x$ «0,4»
3. Factor $-15x^3y + 35x^2y$ «0,5»
4. Multiply $(2x - 1)(4x^2 + 2x + 1)$ «1,1»
5. Simplify $\frac{-2}{5}(15x - 4) + 6x$ «1,1»
6. Factor $2(3 - 4x) - 3x(3 - 4x)$ «1,1»
7. $(4x - 7) \cdot 3$ is equivalent to _____ by the Commutative Law of Multiplication. «1,2»
8. Is division an associative operation on the set of integers? «1,2»
9. If $(x + 3)(2x - 5) = 0$, then $x + 3 = 0$ or _____. «1,2»
10. What open expression must be added to $-5c$ to give a sum of zero? «1,2»

Unit 4: FACTORING POLYNOMIALS

Common Factor Method

Always look for a common factor when factoring. The factorization of $2x^2 + 10x + 20$ is completed as follows.

$$2x^2 + 10x + 20 = 2(x^2 + 5x + 10)$$

Never attempt any other type of factorization until any common factors other than 1 and -1 have been factored out of the polynomial.

Focus on common factors

To factor any polynomial first begin by looking for common factors other than 1 and -1. This process will never increase the difficulty of a factoring situation and frequently makes it far easier.

For example, the factorization of $14x^3 - 21x^2 + 49x$ is dependent on recognizing that each term contains $7x$.

$$14x^3 - 21x^2 + 49x$$
$$7x \cdot 2x^2 - 7x \cdot 3x + 7x \cdot 7$$
$$7x(2x^2 - 3x + 7)$$

Factoring Trinomials of the Form $x^2 + bx + c$

The trinomial $x^2 + 7x + 12$ has no common factors other than 1 and -1. Consequently, if the trinomial is to be factored its factors must be two binomials.

$$x^2 + 7x + 12 = (a + b)(c + d)$$

with x^2 from the outer terms and 12 from the inner terms.

To factor $x^2 + 7x + 12$ find the numerical factors of 12 which have a sum of 7, the coefficient of the middle term. Because 12 is positive, both of its factors must have the same sign (+ or −), but since their sum is to be 7, we are interested only in the positive factors of 12.

$$x^2 + 7x + 12 = (x + 3)(x + 4)$$

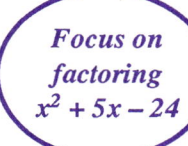

Focus on factoring $x^2 + 5x - 24$

To factor $x^2 + 5x - 24$ it is necessary to find the numerical factors of -24 which have a sum of -5. Because -24 is negative, one factor of it must be positive and the other negative. Since their sum is to be 5, we select factors of -24 where the larger one is positive.

$$x^2 + 5x - 24 = (x - 3)(x + 8)$$

Trinomials with Two Variables

To factor $x^2 - 10xy + 21y^2$, first note that the factors must be of the form $(x - \_\_\_y)(x - \_\_\_y)$ and the numbers that fill the blanks must be factors of 21, and have a sum of -10.

$$x^2 - 10xy + 21y^2 = (x - 3y)(x - 7y)$$

Focus on factoring with two variables

To factor $x^2 - 9y^2$, think of factoring $x^2 + 0xy - 9y^2$. Again, the factors must be of the form $(x + \_\_\_y)(x - \_\_\_y)$ and the numbers that fill the blanks must be factors of -9 which have a sum of zero.

$$x^2 + 0xy - 9y^2 = (x + 3y)(x - 3y)$$

Factoring by Substitution

The polynomial $(a + b)^2 + 5(a + b) + 6$, has a binomial $(a + b)$ where a single letter is normally expected. To factor the trinomial replace $(a + b)$ by an x. This replacement of $(a + b)$ by x is called **substitution**.

	$(a + b)^2 + 5(a + b) + 6$
Let $x = (a + b)$.	$x^2 + 5x + 6$
Factor the trinomial.	$(x + 2)(x + 3)$
Replace x by $(a + b)$.	$([a + b] + 2)([a + b] + 3)$
Remove the brackets.	$(a + b + 2)(a + b + 3)$

To factor $5 - 4(x + y) - (x + y)^2$ follow this process.

Focus on factoring by substitution

	$5 - 4(x + y) - (x + y)^2$
Let $a = (x + y)$.	$5 - 4a - a^2$
Factor the trinomial.	$(5 + a)(1 - a)$
Replace a by $(x + y)$.	$(5 + [x + y])(1 - [x + y])$
Remove the brackets.	$(5 + x + y)(1 - x - y)$

Pay special attention to the last line of the process. When the square brackets are removed, the signs of the terms may be changed.

Factoring Trinomials of the Form $ax^2 + bx + c$

To factor $8x^2 - 29x - 12$ it is possible to test all possible factors $(ax + b)(cx + d)$ where $ax \cdot cx = 8x^2$ and $b \cdot d = -12$. However, a direct factoring approach requiring no guessing is available. This process is illustrated below.

$$8x^2 - 29x - 12$$

1. Multiply the first and last terms.

   $8x^2 \cdot -12 = -96x^2$

2. Find factors of $-96x^2$ with a sum of $-29x$, the middle term.

   $3x \cdot -32x = -96x^2$
   and
   $3x + -32x = -29x$

3. Rewrite the trinomial so it has four terms.

   $8x^2 - 29x - 12$
   becomes
   $8x^2 + 3x - 32x - 12$

4. Factor the terms grouped in pairs.

   $(8x^2 + 3x) + (-32x - 12)$
   $x(8x + 3) - 4(8x + 3)$

5. Use $(8x + 3)$ as the common factor.

   $(8x + 3)(x - 4)$

Focus on factoring directly

The method for factoring directly also shortens the work when the trinomial is prime. This is illustrated below with the trinomial $6x^2 - 13x + 4$.

$$6x^2 - 13x + 4$$

1. Multiply the first and last terms.

   $6x^2 \cdot 4 = 24x^2$

2. Find factors of $24x^2$ with a sum of $-13x$, the middle term.

   $-1x + -24x \neq -13x$
   $-2x + -12x \neq -13x$
   $-3x + -8x \neq -13x$
   $-4x + -6x \neq -13x$

3. Since there are no factors of $24x^2$ with a sum of $-13x$, this establishes that $6x^2 - 13x + 4$ is a prime polynomial.

Factoring the Difference of Two Cubes

The last type of factoring to be shown in this unit deals with polynomials such as $a^3 - b^3$, $8x^3 + 27y^3$, and $64x^3 - y^6$ which are called the sum/difference of two cubes. Each term of the binomials is a perfect third power.

Notice that the binomial $x^3 - y^3$ is the difference of two expressions where the exponent is 3, $(x)^3$ and $(y)^3$. Expressions that have 3 as the exponent are called **perfect cubes**. x^3 is a perfect cube. y^3 is a perfect cube. $x^3 - y^3$ is called the difference of two cubes.

To factor any binomial that is the difference of two cubes use the form
$$a^3 - b^3 = (a - b)(a^2 + ab + b^2)$$

$x^3 - 8$ is the difference of two cubes, x^3 and 2^3. To factor $(x)^3 - (2)^3$, the form for factoring $a^3 - b^3$ is used.

$$a^3 - b^3 = (a - b)(a^2 + ab + b^2)$$
$$(x)^3 - (2)^3 = (x - 2)(x^2 + 2x + 4)$$

Focus on factoring the difference of two cubes

To factor $27x^3 - 8$, write $27x^3 - 8$ as $(3x)^3 - (2)^3$ and use the form $a^3 - b^3 = (a - b)(a^2 + ab + b^2)$.

$$27x^3 - 8$$
$$(3x)^3 - 2^3$$
$$(3x - 2)[(3x)^2 + 3x \cdot 2 + 2^2]$$
$$(3x - 2)(9x^2 + 6x + 4)$$

Factoring the Sum of Two Cubes

The form for factoring the sum of two cubes is
$$a^3 + b^3 = (a + b)(a^2 - ab + b^2)$$

A sum of two cubes is factored as follows.
$$x^3 + 8b^3$$
$$x^3 + (2b)^3$$
$$(x + 2b)(x^2 - 2xb + 4b^2)$$

Focus on factoring the sum of two cubes

Use the form $a^3 + b^3 = (a + b)(a^2 - ab + b^2)$ to factor the sum of two cubes.

$$8x^3 + 125$$
$$(2x)^3 + 5^3$$
$$(2x + 5)[(2x)^2 - 2x \cdot 5 + 5^2]$$
$$(2x + 5)(4x^2 - 10x + 25)$$

$$x^6 + 64y^9$$
$$(x^2)^3 + (4y^3)^3$$
$$(x^2 + 4y^3)[(x^2)^2 - x^2 \cdot 4y^3 + (4y^3)^2]$$
$$(x^2 + 4y^3)(x^4 - 4x^2y^3 + 16y^6)$$

Unit 4 Exercise

Part A: Answers for all Part A problems are at the back of the book.

1. Factor $6x^2 + 11x + 4$ by first listing all the possible ways to produce the first term $6x^2$ and the last term 4.

2. Factor $x^2 - x - 12$ by finding factors of -12 which have a sum equal to the coefficient of x.

3. Determine if $x^2 + 9xy + 10y^2$ is prime by listing all the possible ways to produce the first term x^2 and the last term $10y^2$.

4. Completely factor $5x^2 - 35x + 60$ by first finding a common factor for all three terms.

5. Factor $(x-y)^2 + 7(x-y) - 8$ by substituting k for $(x-y)$, factoring the resulting trinomial, and then substituting $(x-y)$ for k.

6. Factor $(x-5)^2 + 4(x-5)y + 3y^2$ using substitution.

7. Factor $(a+b)^2 - 4d(a+b) - 5d^2$ using substitution.

8. Factor $7 - 8(a+d) + (a+d)^2$ using substitution.

9. $x^3 - y^3$ factors as $(x-y)(x^2 + xy + y^2)$. Factor $r^3 - s^3$.

10. Factor $a^3 - 64$ using the form $x^3 - y^3 = (x-y)(x^2 + xy + y^2)$.

11. $x^3 + 1$ is the sum of two cubes, x^3 and 1^3. Factor $x^3 + 1$ using the form $a^3 + b^3 = (a+b)(a^2 - ab + b^2)$.

12. Factor $8x^3 + 125$, which is equivalent to $(2x)^3 + 5^3$.

13. $125y^3 - 8x^3 = (5y)^3 - (2x)^3$. Factor $125y^3 - 8x^3$.

14. Factor $64a^3 - 27b^3$ which is the difference of two cubes.

15. Factor $x^3 + z^3$ which is the sum of two cubes.

Part B: Answers for odd-numbered problems of Part B are at the back of the book.

Factor or state the polynomial is prime.

1. $x^2 - 7xy + 12y^2$
2. $x^3 + y^3$
3. $x^2 + 6xy + 9y^2$
4. $(x-5)^2 - 6(x-5)(y+w) + 8(y+w)^2$
5. $x^3 - 125$
6. $x^2 - 9xy + 18y^2$
7. $x^2 + 5xy - 6y^2$
8. $r^3 + s^3$
9. $x^2 + 15xy + 50y^2$
10. $x^2 - 16y^2$
11. $x^3 + 8$
12. $x^2 - 18xy + 80y^2$
13. $8x^3 - 1$
14. $x^2 - 18xy + 81y^2$
15. $(x+y)^2 + 3(x+y)(a-b) - 4(a-b)^2$
16. $x^2 + 7xy + 11y^2$
17. $x^2 - 25y^2$
18. $8x^3 + y^3$
19. $x^2 + 8xy - 33y^2$
20. $7x^2 + 28x + 28$
21. $x^3 - 1000$
22. $3x^2 - 12$
23. $4x^2 - 12x - 40$

24. $x^2 + 16x(2y + z) + 64(2y + z)^2$
25. $x^2 + 13x - 14$
26. $4 + (d - c) + (d - c)^2$
27. $1000 - 27b^3$
28. $(x + 3)^2 - 6(x + 3)y + 5y^2$
29. $y^2 - 2y(x + a) + (x + a)^2$
30. $(x + y)^2 + 3a(x + y) + 2a^2$
31. $1000y^3 - 1$
32. $(d + 3)^2 - (d + 3)b - 12b^2$
33. $1 - 5(a - r) + 4(a - r)^2$
34. $(x + 3)^2 - 11y(x + 3) - 12y^2$
35. $a^3 + 8b^3$
36. $2 + 3(r + s) + (r + s)^2$
37. $(x - 7)^2 + 10(x - 7)y + 16y^2$
38. $1 + 64r^3$
39. $(x + a)^2 - 11(x + a) + 10$
40. $x^2 - 11xy + 10y^2$
41. $x^2 + xy + y^2$
42. $(a + b)^2 - 10(a + b)(d + c) + 25(d + c)^2$
43. $x^2 + 6xy - 16y^2$
44. $5x^2 - 20x - 60$
45. $(x - y)^2 + 3(x - y)z + 2z^2$
46. $x^9 + z^{21}$
47. $D^{12} + E^6$
48. $(x - y)^2 - 7(x - y)(a - b) + 12(a - b)^2$
49. $x^3y^3 + z^3$
50. $125r^3 - 64s^3$

Part C: No answers are given for these problems. However, each is accompanied by an ordered pair «C,U» showing the chapter and unit in which it was taught.

1. Multiply $(5x - 2)(25x^2 + 10x + 4)$ «1,1»
2. Is subtraction an associative operation on the set of counting numbers?   «1,1»
3. $-5(3x - 4) = $ _____ by the Distributive Law of Multiplication.   «1,1»
4. Is multiplication an associative operation on the set of integers?   «1,2»
5. If $x(x - 3) = 0$, then $x = 0$ or _____.   «1,2»
6. Factor $4x(w - t) - 3y(w - t)$ «1,2»
7. The opposite of $\frac{-3}{5}$ is _____.   «1,3»
8. The reciprocal of $\frac{-7}{16}$ is _____.   «1,3»
9. According to the Closure Law, if p and t are rational numbers and $pt = s$, then s is a _____.   «1,3»
10. $g(h + x) = $ _____ by the Distributive Law of Multiplication.   «1,3»

Unit 5: SOLVING QUADRATIC EQUATIONS

Simple Quadratic Equations

Equations such as $x^2 = 16$ and $x^2 = 10$ are the simplest types of quadratic equations. In both cases the variable has 2 as its greatest exponent.

The equation $x^2 = 16$ asks the question: What number squared is 16? Two rational numbers, 4 and -4, are correct answers for the equation and the truth set is $\{4,-4\}$.

The equation $x^2 = 10$ asks the question: What number squared is 10? There is no rational number that correctly answers the question and the truth set, with respect to the rational numbers, is the empty set, $\{\ \}$.

Focus on solving simple quadratic equations

No procedure is necessary to solve $x^2 = 64$. Just ask the question: What number squared is 64? Both 8 and -8 are correct answers and the truth set is $\{8,-8\}$.

The situation for the equation $x^2 = -36$ is similar, but in this case there is no correct rational number answer. No rational number, when squared, is -36. The truth set is $\{\ \}$.

Solving Quadratic Equations

Equations such as $x^2 - 3x = 0$, $x^2 - 7x + 12 = 0$, and $10x^2 - 13x - 3 = 0$, are quadratic equations because the variable, x, has 2 as its largest exponent. Each of these equations is solved using the property: **If the product of two factors is zero then one of the factors must itself be zero.**

The truth set of $5x^2 - 15x = 0$ is found by factoring the binomial so that the product of the two factors is zero.

$$5x^2 - 15x = 0$$
$$5x(x - 3) = 0$$
$$5x = 0 \text{ or } x - 3 = 0$$
$$x = 0 \text{ or } x = 3$$
$$\{0,3\}$$

Both 0 and 3 are roots of $5x^2 - 15x = 0$ and the roots are checked as shown below.

If $x = 0$ then $5x^2 - 15x = 0$ becomes:
$5 \cdot 0^2 - 15 \cdot 0 = 0$ or $0 - 0 = 0$ which is true.

If $x = 3$ then $5x^2 - 15x = 0$ becomes:
$5 \cdot 3^2 - 15 \cdot 3 = 0$ or $45 - 45 = 0$ which is true.

Focus on solving a quadratic equation by factoring

The truth set of $x^2 - 7x + 12 = 0$ is found by taking the following steps.

$$x^2 - 7x + 12 = 0$$
$$(x - 3)(x - 4) = 0$$
$$x - 3 = 0 \text{ or } x - 4 = 0$$
$$x = 3 \text{ or } x = 4$$

The roots 3 and 4 need to be separately checked in the original equation to show that a true numerical statement is obtained.

Quadratic Equations in Proper Form

The equation $10x^2 - 3 = 13x$ is not in proper form to be solved by factoring because the solution depends on one side of the equation being zero. Add opposites until a zero on one side of the equation is obtained.

$$10x^2 - 3 = 13x$$
$$\underline{-13x \qquad -13x}$$
$$10x^2 - 13x - 3 = 0$$
$$(2x - 3)(5x + 1) = 0$$
$$2x - 3 = 0 \text{ or } 5x + 1 = 0$$
$$2x = 3 \text{ or } 5x = -1$$
$$x = \frac{3}{2} \text{ or } x = \frac{-1}{5}$$
$$\left\{ \frac{3}{2}, \frac{-1}{5} \right\}$$

If $x = \frac{3}{2}$ then $10x^2 - 3 = 13x$ becomes:
$10\left(\frac{3}{2}\right)^2 - 3 = 13 \cdot \frac{3}{2}$ or $10 \cdot \frac{9}{4} - 3 = \frac{39}{2}$ or $\frac{45}{2} - \frac{6}{2} = \frac{39}{2}$ which is true.

If $x = \frac{-1}{5}$ then $10x^2 - 3 = 13x$ becomes:
$10\left(\frac{-1}{5}\right)^2 - 3 = 13 \cdot \frac{-1}{5}$ or $10 \cdot \frac{1}{25} - 3 = \frac{-13}{5}$ or $\frac{2}{5} - \frac{15}{5} = \frac{-13}{5}$ which is true.

Focus on the proper form for a quadratic equation

The equation $x^2 + 5x + 6 = 8$ is not in a proper form for solving. Even though the trinomial $x^2 + 5x + 6$ can be factored there is no benefit to that because there are an infinite number of factors of 8. To solve the equation, one side must be equal to zero. Then, and only then, one of the factors must be zero.

$$x^2 + 5x + 6 = 8$$
$$\underline{ -8 \quad -8}$$
$$x^2 + 5x - 2 = 0$$

$x^2 + 5x - 2$ is a prime polynomial; it has no factors other than itself and 1. Since $x^2 + 5x - 2$ cannot be factored, $x^2 + 5x - 2 = 0$ has no rational number solution. Its truth set with respect to the set of rational numbers is { }.

Solving a Quartic Equation

The equation $x^4 - 21x^2 - 100 = 0$ is a **quartic** equation because its largest exponent is 4. To solve this equation, make a substitution of y for x^2.

When $y = x^2$ $\quad x^4 - 21x^2 - 100 = 0$
becomes $\quad\quad\quad y^2 - 21y - 100 = 0$

Now the quadratic equation $y^2 - 21y - 100 = 0$ is solved.
$$y^2 - 21y - 100 = 0$$
$$(y - 25)(y + 4) = 0$$
$$y - 25 = 0 \quad \text{or} \quad y + 4 = 0$$
$$y = 25 \quad \text{or} \quad y = -4$$

But since $y = x^2$ this means
$$x^2 = 25 \quad \text{or} \quad x^2 = -4$$

The first equation has 5 and -5 as its solutions and the second equation has no real solutions because -4 is negative. Therefore, the only rational number solutions of the original quartic equation are 5 and -5.

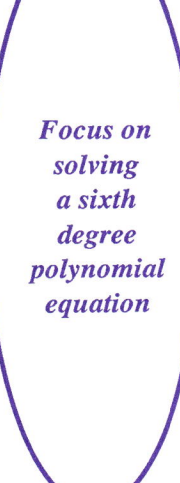

Focus on solving a sixth degree polynomial equation

The equation $x^6 - 7x^3 - 8 = 0$ is a sixth degree polynomial equation because its largest exponent is 6. To solve the equation substitute y for x^3.

When $y = x^3$ then $x^6 - 7x^3 - 8 = 0$
becomes $y^2 - 7y - 8 = 0$

Now the quadratic equation $y^2 - 7y - 8 = 0$ is solved.

$$y^2 - 7y - 8 = 0$$
$$(y - 8)(y + 1) = 0$$
$$y - 8 = 0 \text{ or } y + 1 = 0$$
$$y = 8 \text{ or } y = -1$$

But since $y = x^3$ this means
$$x^3 = 8 \text{ or } x^3 = -1$$

The first equation has 2 as its only rational number solution. $2^3 = 8$
The second equation has -1 as its only rational number solution. $(-1)^3 = -1$
The rational number solutions of the original sixth degree equation are 2 and -1.

Solving Cubic Equations

To solve $x^3 - 1000 = 0$ use the following steps.

$$x^3 - 1000 = 0$$
$$(x - 10)(x^2 + 10x + 100) = 0$$
$$x - 10 = 0 \quad x^2 + 10x + 100 = 0$$
$$x = 10$$

The polynomial $x^2 + 10x + 100$ is prime so no rational number solutions can be found for $x^2 + 10x + 100 = 0$. Consequently, the only rational number solution for $x^3 - 1000 = 0$ is 10.

Focus on solving cubic equations

To find all rational number solutions of $8x^3 + 27 = 0$ the following steps are used.

$$8x^3 + 27 = 0$$
$$(2x + 3)(4x^2 - 6x + 9) = 0$$
$$2x + 3 = 0 \qquad 4x^2 - 6x + 9 = 0$$
$$2x = -3 \qquad 4x^2 - 6x + 9 \text{ is prime.}$$
$$x = \frac{-3}{2}$$

$\frac{-3}{2}$ is the only rational number solution of $8x^3 + 27 = 0$.

Unit 5 Exercise

Part A: Answers for all Part A problems are at the back of the book.

Throughout this exercise use the set of rational numbers as replacements for the variables.

1. Solve $x^2 = 49$ by finding two numbers that can be squared to give 49.
2. Solve $3x^2 - 12x = 0$ by finding a common factor for $3x^2$ and $-12x$.
3. Solve $7x^2 + 14x = 0$ by factoring the binomial.
4. Solve $4x^2 = 5x$ by first adding $-5x$ to each side of the equation.
5. Solve $14x^2 = 21x$ by generating an equivalent equation which has one side equal to 0.
6. Why does the equation $x^2 - 8x + 10 = 0$ have $\{\ \}$ as its truth set?
7. Solve $x^2 + 7x + 4 = 0$.
8. Solve $9x^2 - 49 = 0$ by factoring the binomial as the difference of two squares.
9. Solve $x^3 + 8 = 0$ by factoring the binomial as the sum of two cubes.
10. Solve $x^4 - 17x^2 + 16 = 0$ by substituting k for x^2.

Part B: Answers for odd-numbered problems of Part B are at the back of the book.

1. $x^2 - 5x = 0$
2. $x^2 - 36 = 0$
3. $x^2 = 9x$
4. $x^2 - 16x = 0$
5. $2x^2 - 7x = 0$
6. $3x + 28 = x^2$
7. $x^2 + x = 6$
8. $x^2 - 6x + 1 = 0$
9. $2x^2 + 3 = -7x$
10. $x^2 - 6x + 8 = 0$
11. $x^2 = x + 20$
12. $x^2 + 7x + 6 = 0$
13. $x^2 = 2x - 7$
14. $x^2 - 10x = 24$
15. $2x^2 + 5x - 3 = 0$
16. $3x^2 + 10 = 13x$
17. $3x^2 - 14x + 8 = 0$
18. $x^4 - 2x^2 - 15 = 0$
19. $x^4 - 5x^2 = -4$
20. $x^6 + 26x^3 - 27 = 0$
21. $64x^3 - 125 = 0$

Part C: No answers are given for these problems. However, each is accompanied by an ordered pair «C,U» showing the chapter and unit in which it was taught.

1. Multiply: $(4x - 1)(2x^2 - 3x + 6)$ «1,1»
2. If $(3x - 5)(x - 6) = 0$, then $3x - 5 = 0$ or _____. «1,2»
3. Is division an associative operation on the set of integers? «1,2»
4. True or false? $(\frac{3}{4} \cdot \frac{1}{7}) \cdot \frac{2}{5} = \frac{3}{4} \cdot (\frac{1}{7} \cdot \frac{2}{5})$ «1,3»
5. True or false? If m and n are rational numbers and $m \cdot n = w$, then w is a unique rational number. «1,3»

For problems 6-10 factor, or state the polynomial is prime.

6. $x^2 - 3xy - 4y^2$ «1,4»
7. $a^3 - b^3$ «1,4»
8. $3x^2 - 75$ «1,4»
9. $x^2 + 5x(a - b) - 24(a - b)^2$ «1,4»
10. $27x^3 - 64w^3$ «1,4»

Unit 6: SOLVING QUADRATIC INEQUALITIES

Positive and Negative Inequalities

Inequalities which have one side as zero are positive or negative inequalities depending on the particular symbol of inequality involved. For example, the claim that $(4x - 9)$ is positive is equivalent to the linear inequality $(4x - 9) > 0$. Conversely, $5x - 3 > 0$ is true whenever $(5x - 3)$ is positive.

Similarly, the inequality $(x - 7)(2x + 5) > 0$ is a positive inequality. There are two ways that $(x - 7)(2x + 5)$ can be positive.

1. $(x - 7)$ is positive **and** $(2x + 5)$ is positive
   or
2. $(x - 7)$ is negative **and** $(2x + 5)$ is negative

Focus on negative inequalities

The inequality $7x + 3 < 0$ is a negative inequality. It becomes true only when $(7x + 3)$ is negative.

Similarly, $(4x + 3)(2x - 1) < 0$ is a negative inequality. It becomes true only when the product of $(4x + 3)$ and $(2x - 1)$ is negative. However, there are two ways for the product to be negative:

1. $(4x + 3)$ is negative **and** $(2x - 1)$ is positive
   or
2. $(4x + 3)$ is positive **and** $(2x - 1)$ is negative

Solving Quadratic Inequalities in Factored Form

The graphing of positive inequalities of the form $(x - a)(x - b) > 0$ depends upon the correct use of the words "and" and "or." When "**and**" is used to connect two inequalities, then both must be true. When "**or**" is used to connect two inequalities, then at least one must be true.

68   Chapter 1

To solve $(x - 7)(x + 5) > 0$, first note that this is a positive inequality and can become true in two ways:

1. Each factor is positive,

or

2. Each factor is negative.

Consequently, if $(x - 7)(x + 5) > 0$

$[x - 7 > 0 \text{ and } x + 5 > 0]$ or $[x - 7 < 0 \text{ and } x + 5 < 0]$
$[x > 7 \text{ and } x > -5]$ or $[x < 7 \text{ and } x < -5]$
$[x > 7]$ or $[x < -5]$

On the number line the solutions would be shown as two rays with open dots at 7 and -5.

Focus on graphing positive inequalities

Graphically, truth sets of inequalities of the form $(x - a)(x - b) > 0$ consist of two half-lines. This is because at some point on the right, both factors must become positive and remain so thereafter. Similarly, at some point on the left, both factors must become, and remain, negative.

To graph $(4x - 3)(2x + 1) > 0$, it is necessary to consider two possibilities: both factors are positive or both factors are negative. If both factors are positive, an open sentence with the word "and" is used.

Consequently, if $(4x - 3)(2x + 1) > 0$

$[4x - 3 > 0 \text{ and } 2x + 1 > 0]$ or $[4x - 3 < 0 \text{ and } 2x + 1 < 0]$
$[x > \frac{3}{4} \text{ and } x > \frac{-1}{2}]$ or $[x < \frac{3}{4} \text{ and } x < \frac{-1}{2}]$
$[x > \frac{3}{4}]$ or $[x < \frac{-1}{2}]$

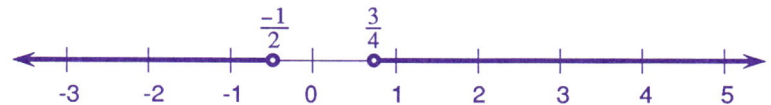

System of Rational Numbers   69

Solving Negative Inequalities

Truth sets of inequalities of the form (x − a)(x − b) < 0 will always consist of a line segment joining a and b. This is because the portion of the line between a and b represents numbers that make one of the factors positive and the other negative.

To solve (3x + 4)(5x − 1) < 0, first note that this is a negative inequality and can become true in two ways:
1. (3x + 4) is positive and (5x − 1) is negative,

    or

2. (3x + 4) is negative and (5x − 1) is positive.

Consequently, if (3x + 4)(5x − 1) < 0
$[3x + 4 > 0 \text{ and } 5x − 1 < 0]$ or $[3x + 4 < 0 \text{ and } 5x − 1 > 0]$
$[x > \frac{-4}{3} \text{ and } x < \frac{1}{5}]$ or $[x < \frac{-4}{3} \text{ and } x > \frac{1}{5}]$
$[x > \frac{-4}{3} \text{ and } x < \frac{1}{5}]$ or $[\text{No numbers make this true}]$

On the number line the solutions would be shown as a line segment with open dots at $\frac{-4}{3}$ and $\frac{1}{5}$.

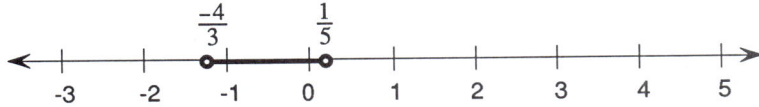

Focus on graphing inequalities with the symbol ≥

The graph of (x − 7)(x + 2) ≥ 0 is shown below. It is almost identical to the graph using the symbol > but now must include the two endpoints, 7 and -2. To show that the endpoints are included, these points are darkened.

Focus on graphing inequalities using ≤

The graph of (5x + 3)(2x − 1) ≤ 0 is shown below. It is almost identical to the graph using the symbol < but now must include the two endpoints, $\frac{-3}{5}$ and $\frac{1}{2}$. To show that the endpoints are included, these points are darkened.

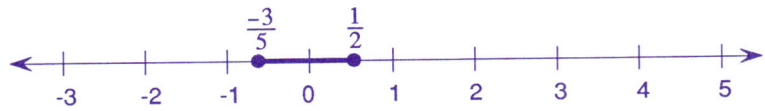

Chapter 1

Unit 6 Exercise

Part A: Answers for all Part A problems are at the back of the book.

1. If $xy > 0$, then both x and y are positive or both are _____.

2. If $(x + 10)(x - 2) > 0$ and $(x + 10) > 0$ then $(x - 2)$? 0.

3. If $(x + 3)(x - 9) > 0$ and $(x + 3) < 0$ then $(x - 9)$? 0.

4. If $(x - 3)(x + 7) > 0$, then _____ or _____.

5. If $(x - 12)(x + 5) < 0$ and $(x - 12) > 0$ then $(x + 5)$? 0.

6. If $(x + 8)(x + 3) < 0$ and $(x + 8) < 0$ then $(x + 3)$? 0.

7. If $(x - 4)(x - 9) < 0$, then _____ or _____.

8. The graph of $x(x - 7) > 0$ is:
   a. Two half-lines
   b. A line segment

9. The graph of $(x - 5)(x - 8) < 0$ is:
   a. Two half-lines
   b. A line segment

10. The difference in the graphs of $(x - 9)(x + 2) > 0$ and $(x - 9)(x + 2) \geq 0$ is _____.

11. The difference in the graphs of $(x + 5)(x - 2) \geq 0$ and $(x + 5)(x - 2) \leq 0$ is _____.

Part B: Answers for odd-numbered problems of Part B are at the back of the book.

Graph truth sets for each of the following.

1. $x(x + 6) > 0$
2. $(x - 8)(x + 3) \leq 0$
3. $(x - 7)(x - 3) \geq 0$
4. $(x + 5)(x + 2) < 0$
5. $(x + 4)(x - 7) > 0$
6. $x(x + 2) < 0$
7. $(x + 3)(x - 10) \geq 0$
8. $(x - 2)(x + 5) < 0$
9. $(x + 3)(x + 7) \leq 0$
10. $x(x - 4) \geq 0$
11. $(x - 6)(x - 1) < 0$
12. $(x + 8)(x - 8) > 0$
13. $(x - 1)(x + 5) \geq 0$
14. $(x + 4)(x - 2) > 0$
15. $(3x - 2)(x + 5) < 0$
16. $(x + 3)(4x + 7) \leq 0$
17. $3x(9x - 4) \geq 0$
18. $(x - 6)(8x - 3) < 0$
19. $(3x + 8)(x - 8) > 0$
20. $(x - 1)(6x + 5) \geq 0$

Part C: No answers are given for these problems. However, each is accompanied by an ordered pair «C,U» showing the chapter and unit in which it was taught.

1. Multiply $(3x - 1)(9x^2 + 3x + 1)$ «1,1»
2. If $(t - 5)(t + 2) = 0$, then $t - 5 = 0$ or _____. «1,2»
3. True or false? $3a + (9a + 7) = (3a + 9a) + 7$ «1,3»
4. Factor $3x^2 - 2xy - 5y^2$ «1,4»
5. Factor $8a^3 - 27b^3$ «1,4»
6. Solve $x^2 = 5x$ «1,5»
7. Solve $2x^2 - 72 = 0$ «1,5»
8. Solve $x^2 - 6x = 27$ «1,5»
9. Solve $15x^2 + 14x - 8$ «1,5»
10. Solve $x^4 - 6x^2 + 5 = 0$ «1,5»

Unit 7: APPLICATIONS: VALUE, NUMBER, AND COST

Representing Values

The formula V = np shows the relation between the value (V), the number (n), and the price (p) of objects. The value of a group of objects is the product of the number of objects and the price of each individual object.

For example, the value of 20 lawn mowers that cost $70 each is shown by the expression 20 • 70, or by the amount $1,400. If eight cars cost $2,000 each, their total value can be shown as a product, 8 • 2,000, which is equal to $16,000. Using the formula V = np, if n represents the number of $15 pairs of slacks Mr. Armour sells, then 15n represents their total value.

Focus on representing value with open expressions

The formula V = np is useful for representing the value of several collections of objects. For example, if x represents a number of 15¢ stamps and x – 10 represents a number of 25¢ stamps, then 15x shows the value of the 15¢ stamps and 25(x – 10) shows the value of the 25¢ stamps. The sum of the values, 15x + [25(x – 10)], represents the value of all the 15¢ and 25¢ stamps.

Similarly, if y represents a number of $5 books and 2 represents a number of $8 books, the total value of the books can be found by obtaining products of the number and price of each book and by then adding these products. The expression 5y + 16 represents the total value of all the books.

A Procedure for Solving Word Problems

The problems explained in this unit will be presented in five steps. The reader should adopt the five-step method for solving problems presented here. Word problems can be solved more effectively if an organized method is used for their solution. The five steps are as follows:

1. Read the problem carefully looking for relationships between quantities that can be written as open expressions and/or equations.
2. List the quantities given in the problem.
3. Use one or two variables to write open expressions for the quantities listed in step 2; use numbers where possible.
4. Use the open expressions from step 3 to write an equation (or equations) to show the number relationships given in the problem.
5. Check all answers in the original problem.

The word problem process is illustrated with the following example. The five-step process is numbered to emphasize the need for organizing the problem.

Problem: Mr. Retaw sold 10 tools for a total price of $53. One type of tool sold at $5 a piece and another at $6 a piece. How many of each did he sell?

Solution

1. The problem involves two types of numbers: (1) The numbers of tools, and (2) The numbers for the values of the tools.
2. Now the quantities of the problem are listed.

    the number of $5 tools
    the number of $6 tools
    the value of the $5 tools
    the value of the $6 tools
    the total value of all tools

3. Now variables and open expressions are assigned to the quantities listed in step 2.

    x represents the number of $5 tools
    y represents the number of $6 tools
    $5x$ represents the value of the $5 tools
    $6y$ represents the value of the $6 tools
    $5x + 6y$ represents the total value of all the tools

System of Rational Numbers

4. Two equations are written showing the number relationships in the problem.

   The total number of tools sold is 10.
   $$x + y = 10$$
   The total value of the tools is $53.
   $$5x + 6y = 53$$

5. The common solution of the two equations in step 4 is found by multiplying the first equation by -5 and by adding the result to the second equation. The common solution is (7,3). It should be verified that the sale of seven of the $5 tools and three of the $6 tools meet the conditions of the problem — that is, their combined sale will total $53.

Focus on the word problem process

The next example does not give the total number of objects sold but it does indicate the relationship between the numbers of the two objects. Again, the five-step process is applied.

Problem: A ticket agent sold $123.75 worth of tickets. He sold some for $1.75 and twice as many for $1.60. How many of each did he sell?

Solution

1. The problem involves (1) tickets, and (2) value of the tickets.
2. The quantities are listed.

   the number of $1.75 tickets
   the number of $1.60 tickets
   the value of the $1.75 tickets
   the value of the $1.60 tickets
   the total value of the tickets

3. Open expressions are written for the quantities listed in step 2.

   x is the number of $1.75 tickets
   2x is the number of $1.60 tickets
   1.75x is the value of the $1.75 tickets
   3.20x is the value of the $1.60 tickets
   1.75x + 3.20x is the total value of the tickets

4. The total value of the tickets is $123.75.
   $$1.75x + 3.20x = 123.75$$

5. The solution of the equation in step 4 is 25. The number of $1.75 tickets is 25 and the evaluation of 2x, 50, is the number of $1.60 tickets. These results should be checked against the conditions in the problem.

Mixture Problems

The next problem is a situation in which two differently priced materials are mixed with the intent of producing a mixture that will have a price between the prices of the two ingredients of the mix. Such mixtures are common in both retail and wholesale transactions. For example, if 6 gallons of paint worth $7 a gallon is mixed with 4 gallons of paint worth $5 a gallon, the resulting 10 gallons of paint will be worth $62, or $6.20 a gallon. Notice that the unit price of the mixture, $6.20, is between the unit prices of the two ingredients of the mixture, which sell for $5 and $7 a gallon.

Problem: Mrs. Snipe has to mix 1000 pounds of bird seed that will sell for 59¢ a pound from seed priced at 55¢ a pound and from seed priced at 65¢ a pound. How much of each price seed should she use.

Solution

1. The total value of the mix can be found by multiplying 59¢ by 1000. If one variable represents the weight of the 55¢ seed, subtracting the variable from 1000 (the total weight) will give an open expression that represents the weight of the 65¢ seed. An equation can be made by setting the open expression that represents the total value of the seed equal to the actual value.

2. The quantities are:
   the weight of the seed at 55¢ a pound
   the weight of the seed at 65¢ a pound
   the value of the seed at 55¢ a pound
   the value of the seed at 65¢ a pound
   the total value of the seed

3. x is the weight of the seed at 55¢ a pound
   $1000 - x$ is the weight of the seed at 65¢ a pound
   $55x$ is the value of the seed at 55¢ a pound
   $65(1000 - x)$ is the value of the seed at 65¢ a pound
   $55x + 65(1000 - x)$ is the total value of the seed

System of Rational Numbers 75

4. The sum of the values of the seeds gives the total value.

$$55x + 65(1000 - x) = 59 \cdot 1000$$

5. The solution of the equation is 600. Evaluating 1000 − 600 gives 400, the weight of the seed priced at 65¢ a pound. It should be checked that the mixture of 600 pounds of seed at 55¢ a pound and 400 pounds of seed at 65¢ a pound meets the conditions in the problem.

Focus on solving a mixture problem

Problem: Forty pounds of tea worth 30¢ a pound is to be mixed with tea worth 50¢ a pound to make a mix worth 45¢ a pound. How much of the higher priced tea is needed for the mix?

Solution

1. When a variable is used to represent the weight of the 50¢ tea, an open expression can be found to represent the total weight of the mixture by adding the variable to 40. An equation can be written by setting the sum of the values of the two teas equal to the total value of the mix.

2. The quantities are
    the weight of the tea at 50¢ a pound
    the weight of the tea at 30¢ a pound
    the total weight of tea at 45¢ a pound
    the value of the tea at 50¢ a pound
    the value of the tea at 30¢ a pound
    the total value of the mix at 45¢ a pound
    the sum of the values of the 50¢ tea and 30¢ tea

3. x is the weight of the 50¢ tea
    40 is the weight of the tea at 30¢ a pound
    $x + 40$ is the total weight of the tea
    $50x$ is the value of the tea at 50¢ a pound
    $40 \cdot 30$ or 1200, is the value of the tea at 30¢ a pound
    $45(x + 40)$ is the total value of the mix at 45¢ a pound
    $50x + 1200$ is the sum of the values of the 50¢ tea and the 30¢ tea

4. The sum of the values of the tea is the total value of the mix.

$$50x + 1200 = 45(x + 40)$$

5. The solution of the equation is 120. It should be checked that 120 pounds of tea at 50¢ a pound meets the conditions of the problem.

Unit 7 Exercise

Part A: Answers for all Part A problems are at the back of the book.

1. Represent the value of a set of books if x represents the number of books and the cost of each book is $3.

2. Richard has an order to make some tables to sell for $45 each. If y + 8 represents the number of tables that he is to make, write an open expression for the value of the tables.

3. Write an open expression to represent the total value of some nickels and dimes when n represents the number of nickels and 2n represents the number of dimes.

4. Mr. Theat sold some $3 and some $4 tickets. Write an open expression to represent the total value of the tickets when t represents the number of $3 tickets and t + 2 represents the number of $4 tickets.

5. A theater box office sold 700 tickets for $745.00. Some were $1.75 adult tickets and the rest were $.55 children tickets. How many of each were sold?

6. A grower mixed some beans worth 12¢ a pound with 10 more pounds of beans worth 15¢ a pound. The cost of the entire mix was $19.59. How many pounds of each price bean did he use?

7. A barrel contains a mixture of candy worth 60¢ a pound and candy worth 80¢ a pound. If there is 100 pounds of candy in the barrel and the total cost of the candy is $74.20, how many pounds of each price candy is in the barrel?

8. A car dealer bought some cars at $1,800 each and twice as many other models for $3,000 each. If he spent $46,800 altogether on the cars, how many of each type did he buy?

9. A chemist has to make 200 gallons of alcohol that will sell for 65¢ a gallon by mixing two types of alcohol together: one type selling for 80¢ a gallon and another that sells for 60¢ a gallon. How many gallons of each price alcohol does he need for the mixture?

10. A party mix of candy that will sell for 58¢ a pound is made by mixing candy worth 40¢ a pound with 10 pounds less of another candy that sells for 80¢ a pound. How many pounds of each type of candy is needed for the mix?

11. How much antifreeze that costs $4 a gallon should be mixed with antifreeze that costs $6 a gallon to make 1000 gallons of antifreeze that will sell for $4.50 a gallon?

12. Miss Caffy must make some coffee to sell for 58¢ a pound by mixing some that sells for 70¢ a pound with 100 pounds less of a brand that sells for 42¢ a pound. How many pounds of each brand does she need?

Part B: Answers for odd-numbered problems of Part B are at the back of the book.

1. A theater sold 800 tickets for $920. Some were $1 children's tickets and the rest were $1.50 adult tickets. How many of each were sold?

2. Mr. Pentil, a bank teller, had 25 more $50 bills than he had $5 bills. How many did he have of each denomination if their total value was $4000?

3. A distillery wants to make 900 quarts of whiskey that will sell for $4.60 a quart by blending some $4 a quart whiskey with some whiskey that is worth $5 a quart. How much of each type do they need?

4. Mr. Lettem needs to mix twice as many peas priced at 12¢ a pound with peas priced at 3¢ a pound to make a mixture that will be worth $108. How many pounds of each type of pea does he need?

5. Mrs. Cavite mixed 28 pounds of candy worth 60¢ a pound with some worth 80¢ a pound to make a mixture that will sell for 72¢ a pound. How much of the 80¢ candy does she need?

6. A hardware store ordered 900 lights for $73.20. Some were 7¢ each and the rest were 10¢ each. How many of each type did they order?

7. How many quarters and dimes does a bank teller need if he has to have $11.25 in change made up of twice as many dimes as quarters?

8. An 800-pound mix of cattle feed that is to sell for 14¢ a pound is to be made by mixing a grain worth 8¢ a pound with a grain worth 16¢ a pound. How much of each price of grain is needed?

9. Miss Sour is to make up a solution of acid that will cost 12¢ a gram by adding some acid that costs 15¢ a gram to 600 grams of another acid that costs 10¢ a gram. How many grams of 15¢ acid should she use?

10. Write an open expression for the total value of a box of rivets when x represents the number of rivets and each rivet costs 3¢.

11. If x represents a number of 8¢ stamps and $3x - 1$ represents a number of 5¢ stamps, write an open expression that represents the total value of the stamps.

12. A piggy bank contains only dimes and quarters. Find the number of each that are in the bank if there are 15 coins and their total value is $2.85.

13. Mr. Trowel has two more firebrick than red brick. If firebrick are 15¢ each and red brick are 8¢ each, and their total value is $3.75, how many of each type does Mr. Trowel have?

14. A pet shop owner wants to mix red gravel worth 50¢ a pound with blue gravel worth 20¢ a pound for use in aquariums. How many pounds does he need of each color to make 100 pounds of mix to sell for 38¢ a pound?

Chapter 1

Part C: No answers are given for these problems. However, each is accompanied by an ordered pair «C,U» showing the chapter and unit in which it was taught.

1. Multiply $(4x - 5)(3x^2 - 2x + 7)$ «1,1»
2. Factor $125x^3 - 8y^3$ «1,4»
3. Factor $a(t - s) - b(t - s)$ «1,4»
4. Solve $5x(2x + 7) = 0$ «1,5»
5. Solve $2x^2 - 15x - 17 = 0$ «1,5»
6. Solve $x^4 - 3x^2 = -2$ «1,5»

Graph truth sets for problems 7-10.

7. $x(x - 5) < 0$ «1,6»
8. $(x + 3)(x - 2) \geq 0$ «1,6»
9. $(5x - 2)(x + 4) \leq 0$ «1,6»
10. $(4x - 9)(3x + 5) > 0$ «1,6»

Chapter 1 Test

«1,U» shows the unit in which this problem was studied in this chapter.

1. Is subtraction a closed operation on the set of negative integers? «1,1»

2. What makes two open expressions equivalent? «1,1»

3. What law of addition makes $(2x + 7) + 5x$ equivalent to $5x + (2x + 7)$? «1,1»

4. Use the Distributive Law of Multiplication over Addition to write an equivalent expression to $9x^2y - 15xy$. «1,1»

5. Is $\frac{0}{5}$ a rational number? «1,3»

6. Is $\frac{8}{0}$ a rational number? «1,3»

Factor or state the polynomial is prime for problems 7-13.

7. $3x^2 - xy - 4y^2$ «1,4»

8. $x^2 - 7xy + 10y^2$ «1,4»

9. $64a^3 + 125$ «1,4»

10. $4x^2 - 8x - 60$ «1,4»

11. $3x^2 + y$ «1,4»

12. $a^2 - 3a(x + y) - 10(x + y)^2$ «1,4»

13. $(a + 4)^2 - 16b^2$ «1,4»

14. Simplify $5x - 3 \geq 4$ «1,4»

15. Simplify $6 - 4x < 7 + x$ «1,4»

Solve the equations for problems 16-22 using the set of rational numbers.

16. $\frac{5}{8}x - \frac{3}{4} = \frac{1}{6}$ «1,5»

17. $\frac{4}{x} - \frac{9}{3x} = \frac{5}{6}$ «1,5»

18. $x^2 - 7x = 0$ «1,5»

19. $3x^2 + x = 4$ «1,5»

20. $2x - 15 = x^2$ «1,5»

21. $x^4 - 13x^2 + 36 = 0$ «1,5»

22. $x^6 - 64 = 0$ «1,5»

Graph truth sets for the inequalities for problems 23-24.

23. $x(x + 3) < 0$ «1,6»

24. $(x - 5)(2x + 5) \geq 0$ «1,6»

25. How much antifreeze that costs $3.00 a gallon should be mixed with antifreeze that costs $4.00 a gallon to make 100 gallons that will sell for $3.70 a gallon? «1,7»

2
The Real Number System

Unit 1: THE SYSTEM OF REAL NUMBERS

The Set of Real Numbers

The figure shown below is the **real number line**.

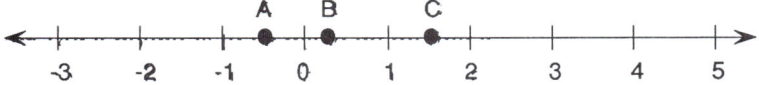

Three points have been marked on the number line and labeled with capital letters. The letter A is located at the approximate position of $\frac{-1}{2}$, B is at $\frac{1}{4}$, and C is at $\frac{3}{2}$.

Focus on real numbers indicated by radical signs

Recall that $\sqrt{4}$ is equal to 2 because 2 is positive and $2^2 = 4$. Similarly, $\sqrt{9}$ is equal to 3 because $3^2 = 9$. Since $\sqrt{4} = 2$ and $\sqrt{9} = 3$, and since 7 is between 4 and 9, it follows that $\sqrt{7}$ is greater than $\sqrt{4}$ and less than $\sqrt{9}$.

On the number line below if one of the capital letters is intended to show the location of $\sqrt{7}$, then it must be C because that is the only letter shown that is located between 2 and 3, or between $\sqrt{4}$ and $\sqrt{9}$.

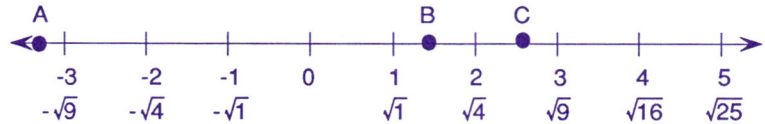

The letter A on the number line above may represent $-\sqrt{13}$ because $\sqrt{13}$ is between $\sqrt{9}$ and $\sqrt{16}$. Consequently, $-\sqrt{13}$ must be between -3 and -4. Similarly, B might represent the location of $\sqrt{2}$ because it is between $\sqrt{1}$ and $\sqrt{4}$.

Rational And Irrational Numbers

$\frac{3}{4}$ is a rational number because it is a ratio $\frac{x}{y}$ where x and y are integers and $y \neq 0$. $\sqrt{3}$ is not a rational number because it cannot be written in the form $\frac{x}{y}$ where x and y are integers and $y \neq 0$. The letter B is located at the approximate position of $\sqrt{3}$ on the real number line.

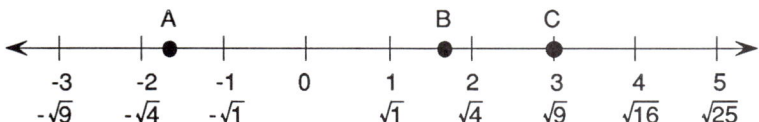

$\sqrt{3}$ is an **irrational number** because it has a position on the real number line, but it is not a rational number. Similarly, π has a position on the real number line, but π is not a rational number. Therefore, π is an irrational number.

The Real Number System   83

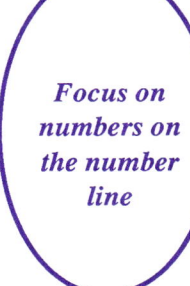

Focus on numbers on the number line

Every point of the real number line represents either a rational or an irrational number. If a point on the real number line does not represent an irrational number, then it must represent a rational number. The **set of real numbers** includes all the rational numbers and all the irrational numbers. Every rational number is a real number and every irrational number is a real number.

The number $\frac{47}{15}$ is a real number because it is a rational number. The set of real numbers includes all the rational numbers and all the irrational numbers. Every irrational number is a real number; π is irrational, therefore, π is a real number.

Real Numbers

Every real number is either a rational or an irrational number.

$\sqrt{3}$ is a real number and $\sqrt{3}$ is an irrational number.

The fraction $\frac{-3}{4}$ is a rational number, and $\frac{-3}{4}$ is also a real number.

$\sqrt{17}$ is an irrational number, and $\sqrt{17}$ is also a real number.

In unit 4 of this chapter we will look at some decimal numerals for real numbers. As you will see, both rational and irrational numbers have decimal numerals.

.25 is a decimal numeral for $\frac{1}{4}$.

.17555 is a decimal numeral for $\frac{79}{450}$.

0.15115111511115 . . . is a decimal numeral for an irrational number.

Focus on numbers that are not real numbers

In Chapter 3, we will study some numbers that are not real numbers. $\sqrt{-1}$ is the number that if multiplied by itself would give -1. Since the product of any real number with itself cannot be negative, $\sqrt{-1}$ is not a real number. Similarly, $\sqrt{-3}$ is another example of a number that is not a real number.

$\sqrt{-5}$ is not a real number because any real number multiplied by itself will not give a negative product.

> **Definition**   **Variable**
>
> A letter x or y is a **variable** for the real numbers   **if and only if**   it represents a position where a real number can replace the letter.

Expressions written with one or more variables are **open expressions** which means that each variable represents a position that is open for replacement by a real number.

The Laws of the Real Numbers

The properties for the real numbers include all of the properties of the rational numbers, and one new property, the completeness property. In the listing below, the letters a, b, c are variables for real numbers.

1. For addition:
   a. Closure Law. a + b is a unique real number.
   b. Commutative Law. a + b = b + a
   c. Associative Law. (a + b) + c = a + (b + c)
   d. Identity Law. a + 0 = 0 + a = a
   e. Inverse Law (Opposites). a + -a = -a + a = 0

2. For multiplication:
   a. Closure Law. a • b is a unique real number.
   b. Commutative Law. a • b = b • a
   c. Associative Law. (a • b) • c = a • (b • c)
   d. Multiplication Law of One. a • 1 = 1 • a = a
   e. Multiplication Law of Negative One. a • -1 = -1 • a = -a
   f. Multiplication Law of Zero. a • 0 = 0 • a = 0
   g. Inverse Law (Reciprocals). If a ≠ 0 then $\frac{1}{a}$ is a real number and a • $\frac{1}{a}$ = 1

3. Distributive Law of Multiplication over Addition. a(b + c) = ab + ac

4. If a = b, then a + c = b + c
   If a = b, then ac = bc
   If ab = 0, then a = 0 or b = 0

5. a. If a > b, then a + c > b + c
    b. If a > b, then ac > bc when c > 0 and ac < bc when c < 0

6. The Completeness Property. Every point on the number line represents a unique real number and every real number is represented by a unique point on the number line.

Completeness Property

It seems to many students that if all the rational numbers were found on a real number line then every point on it would be labeled by a rational number. That is not the case. In fact, the real number line with only the rational numbers on it would have an infinite number of holes where the irrational numbers belong. The irrational numbers are needed to "fill up" or "complete" the number line.

The Completeness Property states that every point of the number line represents a real number and every real number is represented by a point on the number line. According to the Completeness Property, every real number is represented by a point on the real number line. If you imagined a little car driving along the number line, the Completeness Property claims that any place it stops will be a real number. It might be a rational number. It might be an irrational number, but it definitely is a real number.

Focus on the Completeness Property

According to the Completeness Property, every point on the number line represents a real number. The point designated by A on the number line below must be a real number.

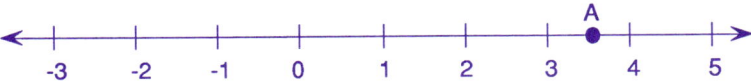

We can't tell whether A represents a rational number or an irrational number. But the Completeness Property guarantees that A represents a real number.

Addition Laws Of The Real Numbers

The addition properties of the real numbers are very similar to those for the rational numbers.

The Closure Law of Addition assumes that $2\sqrt{3} + 5\sqrt{7}$ is a real number. The sum of any two real numbers is a unique real number.

The Commutative Law of Addition assumes that $\sqrt{7} + 4$ is equivalent to $4 + \sqrt{7}$. The order of two addends may be reversed.

The Associative Law of Addition assumes that the sum of $(-3 + \pi) - \sqrt{5}$ will be equivalent to $-3 + (\pi - \sqrt{5})$. The grouping of addends may be changed.

The Identity Law of Addition assumes that $0 + \sqrt{17}$ is equivalent to $\sqrt{17}$. The sum of any real number and zero is that same real number.

The Inverse (Opposites) Law of Addition assumes that the opposite of $\sqrt{7}$ is $-\sqrt{7}$ and $\sqrt{7} + -\sqrt{7} = 0$. There are two parts to this property: First, every real number has an opposite which is also a real number, and, second, the sum of the two opposites is 0.

Focus on multiplication laws of the real numbers

The Closure Law of Multiplication assumes that $9\pi(4\sqrt{2})$ is a real number. The product of any two real numbers is a unique real number.

The Commutative Law of Multiplication assumes that $\sqrt{13} \cdot 35$ is equivalent to $35 \cdot \sqrt{13}$. The order of two factors may be reversed.

The Associative Law of Multiplication assumes that the product of $\left(-\sqrt{34} \cdot \frac{1}{2}\right) \cdot 6\pi$ will be equivalent to $-\sqrt{34} \left(\frac{1}{2} \cdot 6\pi\right)$. The grouping of factors may be changed.

The Multiplication Law of One assumes that $1 \cdot \sqrt{10}$ is equivalent to $\sqrt{10}$. The product of any real number and one is that same real number.

The Multiplication Law of Negative One assumes that $-1 \cdot \sqrt{7}$ is equivalent to $-\sqrt{7}$ and $-1 \cdot -\pi$ is equivalent to π. The product of any real number and negative one is the opposite of that same real number.

The Inverse Law (Reciprocals) of Multiplication assumes that the reciprocal of $\sqrt{7}$ is $\frac{1}{\sqrt{7}}$ and $\sqrt{7} \cdot \frac{1}{\sqrt{7}} = 1$. There are two parts to this property: First, every real number except zero has a reciprocal which is also a real number, and, second, the product of the two reciprocals is 1.

There are two important uses of zero in multiplying real numbers:
1. Zero times any real number results in a product of zero.
$$a \cdot 0 = 0 \cdot a = 0$$
2. If the product of two real numbers is zero, then at least one of the real numbers must be zero.
If $ab = 0$ then $a = 0$ or $b = 0$.

The Distributive Law of Multiplication over Addition assumes that the multiplication expression $4(\sqrt{7} - 3)$ is equivalent to the addition expression $4\sqrt{7} - 12$. It also assumes that the addition expression $2\sqrt{3} + \pi\sqrt{3}$ is equivalent to the multiplication expression $\sqrt{3}(2 + \pi)$. The distributive property makes it possible to alter the indicated order of operations for some open expressions.

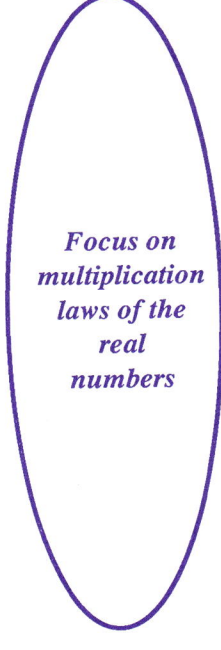

Focus on multiplication laws of the real numbers

Unit 1 Exercise

Part A: Answers for all Part A problems are at the back of the book.

1. Which capital letter is located at the approximate position of $\frac{-4}{3}$ on the real number line at the right?

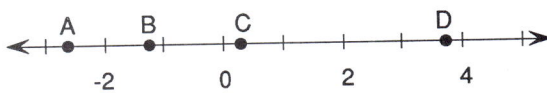

2. Which capital letter is located at the approximate position of $\frac{13}{5}$ on the real number line at the right?

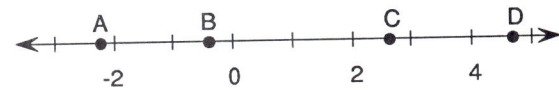

3. $-\sqrt{4} = -2$ and $-\sqrt{1} = -1$. Which capital letter is located at the approximate position of $-\sqrt{3}$ on the real number line at the right?

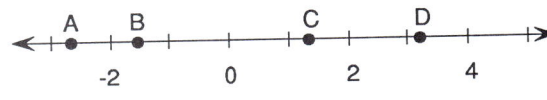

4. The number π is approximately 3.14; π represents the ratio of the circumference of a circle to its diameter. Mark the location of π on the real number line at the right.

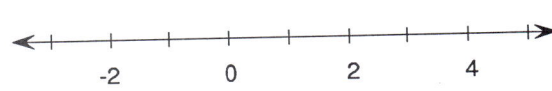

5. Every point of the real number line represents either a rational or an irrational number. Zero is a(n) _____ number.

6. Is every rational number a real number?

7. Is every irrational number a real number?

8. Is $\frac{-5}{8}$ a rational number?

9. Is $\frac{7}{8}$ a real number?

10. Is $\sqrt{15}$ a real number?

11. Is $\sqrt{-11}$ a real number?

12. $\frac{-18}{5}$ is a rational number. Is it a real number?

13. According to the Completeness Property, every point on the real number line represents a _____ number.

14. According to the Completeness Property, every real number is represented by a point on the _____ _____ _____.

15. Is $-\sqrt{913}$ represented by a point on the real number line?

16. Is $\sqrt{-473}$ represented by a point on the real number line?

17. True or False? The sum of $\frac{1}{2}$ and $\sqrt{3}$ is a real number.

18. If a, b, and c are real numbers, the Associative Law of Addition states that a + (b + c) = _____.

19. If a is a real number, the Identity Law of Addition states that 0 + a = _____.

20. The Associative Law of Multiplication states that $17 \cdot (2 \cdot \sqrt{5})$ = _____.

Part B: Answers for odd-numbered problems of Part B are at the back of the book.

1. Is $-\sqrt{4}$ a real number?

2. Is the rational number $\frac{-5}{7}$ a real number?

3. Is the irrational number $\sqrt{23}$ a real number?

4. Is zero a real number?

5. Is every rational number a real number?

6. Is every irrational number a real number?

7. $\sqrt{41}$ is an irrational number. Is it a real number?

8. Is $\sqrt{73}$ represented by a point on the real number line?

9. The Completeness Property states that every point on the number line represents a real number and every real number is represented by a point on the _____ _____ _____.

For problems 10-14 label the sentences true or false.

10. Every real number is a rational number.

11. Every irrational number is a real number.

12. Every real number is either rational or irrational.

13. The Completeness Property states that every rational number is a real number.

14. According to the Completeness Property, every real number is represented by a point on the real number line.

15. By the Closure Law of Addition of the Real Numbers, if m and n are real numbers and m + n = w, then w is a unique _____ number.

16. If a and b are real numbers, then the sum of a and b is a unique _____ number.

17. $\frac{-3}{7}$ and $\sqrt{10}$ are both real numbers. The product of $\frac{-3}{7}$ and $\sqrt{10}$ is a real number. True or False?

18. If a and b are real numbers, then the product of a and b is a unique _____ number.

19. The Commutative Law of Addition of Real Numbers states that $3 + \pi =$ _____.

20. If a and b are real numbers, the Commutative Law of Addition states that $a + b =$ _____.

21. If a and b are real numbers, the Commutative Law of Multiplication states that $a \cdot b =$ _____.

22. The Associative Law of Addition states that $\pi + (15 + \sqrt{3}) =$ _____.

23. If a, b, and c are real numbers, the Associative Law of Multiplication states that $(a \cdot b) \cdot c =$ _____.

24. The Identity Law of Addition of Real Numbers states that $\frac{-2}{3} + 0 =$ _____.

25. If x is a real number, $x + 0 =$ _____.

26. $-\pi$ is a real number. What is the opposite of $-\pi$?

27. $\sqrt{19} + -\sqrt{19} =$ _____.

28. $1 \cdot \pi =$ _____.

29. If w is a real number, the Multiplication Law of One states that $w \cdot 1 = 1 \cdot w =$ _____.

30. $-1 \cdot \frac{-1}{2} =$ _____.

31. $-1 \cdot \sqrt{68} =$ _____.

32. $-\sqrt{41} \cdot -1 =$ _____.

33. If t is a real number, the Multiplication Law of Negative One states that $t \cdot -1 = -1 \cdot t =$ _____.

34. $\frac{2}{3} \cdot$ _____ $= 1$.

35. $\frac{1}{\pi} \cdot$ _____ $= 1$.

36. If $z(x^2 - 5) = 0$, then $z = 0$ or _____.

37. If $(x - y)$ is multiplied by $(5x^2 - 3y^3)$ and $x = y$, then the product is _____.

38. $8(\sqrt{5} - 6) =$ _____.

39. $(4 + a)\sqrt{3} =$ _____.

40. By the Distributive Law, if a, b, and c are real numbers, $a \cdot (b + c) =$ _____.

Part C: No answers are given for these problems. However, each is accompanied by an ordered pair «C,U» showing the chapter and unit in which it was taught.

1. If $(9x - 7)(3x + 5) = 0$, then $9x - 7 = 0$ or _____. «1,2»
2. Factor $x^2 - 4xy + 4y^2$ «1,4»
3. Factor $a^2 - 5a(x - y) + 6(x - y)^2$ «1,4»
4. Solve $x^2 - 3x = 0$ «1,5»
5. Solve $3x^2 - 4x = 4$ «1,5»
6. Solve $2x^2 - 32 = 0$ «1,5»

For problems 7-10 graph the truth set.

7. $x(x - 5) > 0$ «1,6»
8. $(x - 7)(x + 3) \leq 0$ «1,6»
9. $(2x - 3)(x + 2) \geq 0$ «1,6»
10. $(5x + 4)(3x - 2) < 0$ «1,6»

Unit 2: SIMPLIFYING SQUARE ROOT EXPRESSIONS

Simplifying A Square Root Expression

The expression $5\sqrt{43}$ means $5 \cdot \sqrt{43}$. The number 5 preceding the radical sign is the **coefficient** and the number 43 under the radical sign is the **radicand**.

A radical expression, $\sqrt{12}$, is **simplified** when an equivalent expression, $2\sqrt{3}$, with a smaller radicand is found. The process for simplifying square root expressions depends upon finding perfect square factors of the radicand.

The counting numbers 4, 9, 16, 25, 36, 49, 64, 81, 100, 121, and 144 are the most important perfect squares and should be committed to memory. When the radicand is a perfect square integer, a square root expression names a counting number:

$\sqrt{36}$ is a counting number because $\sqrt{36} = 6$.
$\sqrt{25}$ is the counting number 5.
$\sqrt{15}$ is not a counting number because 15 is not a perfect square.

Focus on simplifying square root expressions

Square root expressions can be simplified if the radicand has a perfect square factor.

$\sqrt{20}$ is simplified as: $\quad \sqrt{20} = \sqrt{4} \cdot \sqrt{5} = 2\sqrt{5}$

$\sqrt{32}$ is simplified as: $\quad \sqrt{32} = \sqrt{16} \cdot \sqrt{2} = 4\sqrt{2}$

$\sqrt{18}$ is simplified as: $\quad \sqrt{18} = \sqrt{9} \cdot \sqrt{2} = 3\sqrt{2}$

$\sqrt{72}$ is simplified as: $\quad \sqrt{72} = \sqrt{36} \cdot \sqrt{2} = 6\sqrt{2}$

$5\sqrt{32}$ is simplified as: $\quad 5\sqrt{32} = 5\sqrt{16} \cdot \sqrt{2} = 5 \cdot 4 \cdot \sqrt{2} = 20\sqrt{2}$

$-\sqrt{40}$ is simplified as: $\quad -\sqrt{40} = -1\sqrt{40} = -1\sqrt{4} \cdot \sqrt{10} = -1 \cdot 2 \cdot \sqrt{10} = -2\sqrt{10}$

Reducing Square Root Fractions

Always try to "reduce" a square root fraction. Reduction is possible when both numerator and denominator have a common integer factor or a common square root factor.

$$\frac{10}{12} = \frac{5}{6} \text{ and } \frac{\sqrt{10}}{\sqrt{12}} = \frac{\sqrt{5}}{\sqrt{6}}$$

Similarly, $\frac{6\sqrt{24}}{9\sqrt{14}} = \frac{2\sqrt{12}}{3\sqrt{7}}$

[Note: $\frac{\sqrt{24}}{9}$ and $\frac{\sqrt{12}}{3}$ cannot be reduced.]

Focus on reducing square root fractions

Every fraction is an indicated division, and square root fractions need to be reduced whenever it is possible to reduce the integer coefficients or the square root expressions.

$$\frac{2\sqrt{30}}{8\sqrt{14}} = \frac{\sqrt{15}}{4\sqrt{7}} \text{ and } \frac{22\sqrt{35}}{14\sqrt{7}} = \frac{11\sqrt{5}}{7}$$

The square root fraction $\frac{2\sqrt{15}}{3\sqrt{22}}$ cannot be reduced because neither $\frac{2}{3}$ nor $\frac{\sqrt{15}}{\sqrt{22}}$ can be reduced.

Simplifying Square Root Fractions

The simplification of square root fractions such as $\frac{2\sqrt{3}}{\sqrt{8}}, \frac{3}{\sqrt{5}}, \frac{\sqrt{3}}{\sqrt{7}}$ requires writing a new fraction which has:

1. a positive integer as its denominator,
2. a simplified square root expression in its numerator, and
3. a completely reduced rational number coefficient.

The process can always be accomplished by any sequence of this three-step process. The sequence can and should be varied to fit the particular fraction. The simplification shown below illustrates the process.

$$\frac{8\sqrt{10}}{6\sqrt{56}} = \frac{4\sqrt{5}}{3\sqrt{28}} = \frac{4\sqrt{5}}{3 \cdot 2\sqrt{7}} = \frac{2\sqrt{5}}{3\sqrt{7}} = \frac{2\sqrt{5}}{3\sqrt{7}} \cdot \frac{\sqrt{7}}{\sqrt{7}} = \frac{2\sqrt{35}}{3\sqrt{49}} = \frac{2\sqrt{35}}{21}$$

Focus on simplifying a square root fraction

The following steps may be used to simplify $\dfrac{3\sqrt{98}}{5\sqrt{6}}$.

$$\dfrac{3\sqrt{98}}{5\sqrt{6}} = \dfrac{3\sqrt{49}}{5\sqrt{3}} = \dfrac{21}{5\sqrt{3}} = \dfrac{21}{5\sqrt{3}} \cdot \dfrac{\sqrt{3}}{\sqrt{3}} = \dfrac{21\sqrt{3}}{5 \cdot 3} = \dfrac{7\sqrt{3}}{5}$$

Simplifying Addition Expressions

The following steps are taken to simplify $3\sqrt{2} + 4\sqrt{2}$.

$$3\sqrt{2} + 4\sqrt{2} = (3 + 4)\sqrt{2} = 7\sqrt{2}$$

$4\sqrt{7} + 5\sqrt{3} + 2$ cannot be simplified because the square root expressions are different in each term.

Simplify $3\sqrt{2} + \sqrt{8}$ by first simplifying $\sqrt{8}$ to $2\sqrt{2}$. The procedure is as follows:

$$3\sqrt{2} + \sqrt{8}$$
$$3\sqrt{2} + 2\sqrt{2}$$
$$5\sqrt{2}$$

Focus on simplifying addition expressions

To simplify $5(\sqrt{3} + 2) - 2\sqrt{3}$, first remove the parentheses.

$$5(\sqrt{3} + 2) - 2\sqrt{3}$$
$$5\sqrt{3} + 10 - 2\sqrt{3}$$
$$3\sqrt{3} + 10$$

Multiplying Two-Term Square Root Expressions

The multiplication of $(3x - 5y)(7w - 4z)$ is completed as an application of the Distributive Law of Multiplication over Addition.

$$(3x - 5y)(7w - 4z)$$
$$3x(7w - 4z) - 5y(7w - 4z)$$
$$21xw - 12xz - 35yw + 20yz$$

The same process is followed to multiply $(\sqrt{3} - \sqrt{5})(\sqrt{7} - \sqrt{11})$.

$$(\sqrt{3} - \sqrt{5})(\sqrt{7} - \sqrt{11})$$
$$\sqrt{3}(\sqrt{7} - \sqrt{11}) - \sqrt{5}(\sqrt{7} - \sqrt{11})$$
$$\sqrt{21} - \sqrt{33} - \sqrt{35} + \sqrt{55}$$

Focus on multiplying binomial expressions

The multiplication and simplification of $(4 + 5\sqrt{2})(2 - \sqrt{2})$ is shown below:

$$(4 + 5\sqrt{2})(2 - \sqrt{2})$$
$$4(2 - \sqrt{2}) + 5\sqrt{2}(2 - \sqrt{2})$$
$$8 - 4\sqrt{2} + 10\sqrt{2} - 10$$
$$-2 + 6\sqrt{2}$$

Conjugates of Binomials

The **conjugate** of $(4\sqrt{7} - \sqrt{6})$ is $(4\sqrt{7} + \sqrt{6})$. Two binomials are conjugates when the first terms are identical and the second terms are opposites. An interesting product results whenever a square root expression and its conjugate are multiplied.

$$(4\sqrt{7} - \sqrt{6})(4\sqrt{7} + \sqrt{6})$$
$$4\sqrt{7}(4\sqrt{7} + \sqrt{6}) - \sqrt{6}(4\sqrt{7} + \sqrt{6})$$
$$112 + 4\sqrt{42} - 4\sqrt{42} - 6$$
$$106$$

Focus on the multiplication of conjugates

When a binomial containing square roots and its conjugate are multiplied, the product contains no square roots.

$$(\sqrt{7} - 2\sqrt{3})(\sqrt{7} + 2\sqrt{3})$$
$$\sqrt{7}(\sqrt{7} + 2\sqrt{3}) - 2\sqrt{3}(\sqrt{7} + 2\sqrt{3})$$
$$7 + 2\sqrt{21} - 2\sqrt{21} - 12$$
$$-5$$

If a rational number product is desired, then any binomial containing square roots can be multiplied by its conjugate. The result will contain the squares of the square roots and consequently there will be no square root terms in the product.

Simplifying Division Expressions

The division of 5 by $3 - 2\sqrt{7}$ is simplified using the conjugate of $3 - 2\sqrt{7}$.

$$\frac{5}{3 - 2\sqrt{7}} = \frac{5}{3 - 2\sqrt{7}} \cdot \frac{3 + 2\sqrt{7}}{3 + 2\sqrt{7}}$$

$$= \frac{15 + 10\sqrt{7}}{-19}$$

$$= \frac{-15 - 10\sqrt{7}}{19}$$

Notice that both numerator and denominator of the original fraction were multiplied by $3 + 2\sqrt{7}$. This, in effect, is equivalent to multiplying the fraction by 1.

The fraction $\frac{15 + 10\sqrt{7}}{-19}$ is equivalent to $\frac{-15 - 10\sqrt{7}}{19}$ because the sign of each term was changed. This was done to give a positive denominator.

Focus on simplifying division expressions

To simplify $\frac{3\sqrt{2} - 4}{1 + \sqrt{5}}$ multiply both the numerator and denominator of the fraction by the conjugate of $1 + \sqrt{5}$.

$$\frac{3\sqrt{2} - 4}{1 + \sqrt{5}} = \frac{3\sqrt{2} - 4}{1 + \sqrt{5}} \cdot \frac{1 - \sqrt{5}}{1 - \sqrt{5}}$$

$$= \frac{3\sqrt{2} - 3\sqrt{10} - 4 + 4\sqrt{5}}{1 - 5}$$

$$= \frac{3\sqrt{2} - 3\sqrt{10} - 4 + 4\sqrt{5}}{-4}$$

$$= \frac{-3\sqrt{2} + 3\sqrt{10} + 4 - 4\sqrt{5}}{4}$$

Unit 2 Exercise

Part A: Answers for all Part A problems are at the back of the book.

1. $\sqrt{100}$ is a(n) _____ number.

2. $\sqrt{13}$ is a(n) _____ number.

3. Complete the simplification:
   $\sqrt{24} = \sqrt{4} \cdot \sqrt{6} =$ _____.

4. Simplify $3\sqrt{8}$.

5. Simplify $\frac{\sqrt{25}}{\sqrt{16}}$ by finding the square root of the numerator and denominator of the radicand.

6. Simplify $\frac{-5\sqrt{3}}{\sqrt{5}}$ by multiplying both numerator and denominator by $\sqrt{5}$.

7. Simplify $\sqrt{\frac{2}{3}}$ by multiplying both numerator and denominator by 3.

8. Simplify $\frac{2 + 5\sqrt{3}}{\sqrt{7}}$ by multiplying both numerator and denominator by $\sqrt{7}$.

9. Simplify $5\sqrt{3} + 3\sqrt{3}$ by adding the coefficients of the radical expressions.

10. Can $7\sqrt{6} - 5\sqrt{2} + 5$ be simplified?

11. Simplify $\sqrt{18} + \sqrt{50}$ by first separately simplifying each radical expression.

12. Simplify $8(\sqrt{2} - 3) + \sqrt{2}$ by first removing the parentheses.

13. Multiply $(5 - \sqrt{3})(2 + \sqrt{7})$ using the same process as with $(2x - 3)(x + 7)$.

14. Multiply $(2\sqrt{5} - 3)(\sqrt{5} + 6)$ and simplify the result.

15. Find the conjugate of $(3\sqrt{5} - 2\sqrt{6})$ by changing the sign of the second term.

16. Multiply $(3\sqrt{5} - 2\sqrt{6})(3\sqrt{5} + 2\sqrt{6})$ and simplify the result.

17. Simplify $\frac{\sqrt{6} + 5}{3\sqrt{6} - 2}$ by multiplying both numerator and denominator by $3\sqrt{6} + 2$.

18. Simplify $\frac{5\sqrt{2} + 7}{\sqrt{2} + 3}$ by multiplying both numerator and denominator by the conjugate of the denominator.

Part B: Answers for odd-numbered problems of Part B are at the back of the book.

Simplify.
1. $\sqrt{48}$
2. $\sqrt{45}$
3. $\sqrt{75}$
4. $\sqrt{60}$
5. $\sqrt{125}$
6. $\sqrt{8}$
7. $\sqrt{\frac{9}{100}}$
8. $\sqrt{\frac{64}{25}}$
9. $17\sqrt{9}$
10. $\frac{1}{2}\sqrt{64}$
11. $3\sqrt{20}$
12. $-5\sqrt{20}$
13. $-6\sqrt{24}$
14. $-\sqrt{\frac{16}{81}}$
15. $\frac{7}{\sqrt{2}}$

16. $\dfrac{9}{\sqrt{6}}$

17. $\dfrac{5}{\sqrt{18}}$

18. $\dfrac{3}{\sqrt{5}}$

19. $\dfrac{-2}{\sqrt{8}}$

20. $\dfrac{7\sqrt{3}+2}{\sqrt{5}}$

21. $\dfrac{2-3\sqrt{5}}{\sqrt{6}}$

22. $\sqrt{\dfrac{3}{5}}$

23. $\sqrt{\dfrac{7}{3}}$

24. $\sqrt{\dfrac{1}{6}}$

25. $\sqrt{\dfrac{1}{11}}$

26. $4\sqrt{7}+8\sqrt{7}$

27. $-3\sqrt{2}+5\sqrt{2}-6\sqrt{2}$

28. $4\sqrt{7}+9+2\sqrt{7}$

29. $\sqrt{3}-5+3\sqrt{3}+7$

30. $\sqrt{27}-\sqrt{12}$

31. $-3\sqrt{5}+4\sqrt{2}+\sqrt{5}-3\sqrt{2}$

32. $\sqrt{18}-\sqrt{2}+\sqrt{16}$

33. $\sqrt{50}-\sqrt{45}-\sqrt{32}+\sqrt{80}$

34. $2(\sqrt{3}-5)-4(\sqrt{3}+1)$

35. $3(4-\sqrt{5})+2\sqrt{5}$

36. $(\sqrt{10}-\sqrt{6})(\sqrt{7}-\sqrt{13})$

37. $(\sqrt{5}+\sqrt{6})(3\sqrt{5}-\sqrt{6})$

38. $(3-4\sqrt{2})(5+6\sqrt{5})$

39. $(7\sqrt{3}-\sqrt{2})(8\sqrt{3}+5\sqrt{2})$

40. $(5\sqrt{3}-2)(5\sqrt{3}+2)$

41. $(4\sqrt{3}-\sqrt{5})(3\sqrt{3}+\sqrt{5})$

42. $(3+\sqrt{10})(3-\sqrt{10})$

43. $(6\sqrt{5}-4)(6\sqrt{5}+4)$

44. $(4-3\sqrt{5})(8+\sqrt{5})$

45. $(\sqrt{5}-\sqrt{3})(2\sqrt{5}+\sqrt{3})$

46. $(2\sqrt{3}-5)(\sqrt{7}+\sqrt{10})$

47. $\dfrac{\sqrt{5}-\sqrt{7}}{\sqrt{5}+\sqrt{7}}$

48. $\dfrac{\sqrt{6}-\sqrt{5}}{\sqrt{6}+\sqrt{5}}$

49. $\dfrac{4-3\sqrt{2}}{1+\sqrt{2}}$

50. $\dfrac{6\sqrt{5}-2}{\sqrt{5}+3}$

Part C: No answers are given for these problems. However, each is accompanied by an ordered pair «C,U» showing the chapter and unit in which it was taught.

1. Multiply $(4x-3)(16x^2+12x+9)$ «1,1»
2. Factor $x^2-7x(a-b)-18(a-b)^2$ «1,4»
3. Solve $5x^2+2x-3=0$ «1,5»
4. Solve $2a^2-5a=3$ «1,5»
5. Simplify $x(x+3)\le 0$ «1,6»
6. Simplify $(4x-3)(2x+5)>0$ «1,6»
7. Which of the following are real numbers? $\sqrt{3},\ 5,\ -\sqrt{11},\ \pi,\ \dfrac{-9}{7},\ \sqrt{-4},\ 0,\ -7$ «2,1»
8. What property states that every point on the real number line represents a real number? «2,1»
9. True or false? The sum of $\dfrac{-7}{8}+\sqrt{10}$ is a real number. «2,1»
10. The Associative Law of Addition states $(24+\sqrt{7})+\dfrac{2}{5}=$ _____. «2,1»

Unit 3: SIMPLIFYING 3rd and 4th ROOT EXPRESSIONS

Simplifying A Cube Root Expression

A radical expression, $\sqrt[3]{54}$, is **simplified** when an equivalent expression, $3\sqrt[3]{2}$, with a smaller radicand is found. The process for simplifying cube root expressions depends upon finding a perfect cube factor of the radicand.

The counting numbers 8, 27, 64, 125, 216, 343, 512, 729, and 1000 are the most important perfect cubes. When the radicand is a perfect cube integer, a cube root expression names a counting number.

$\sqrt[3]{64}$ is a rational number because $\sqrt[3]{64} = 4$, and 4 can be written $\frac{4}{1}$.

$\sqrt[3]{125}$ is another rational number.
$\sqrt[3]{36}$ is an irrational number because 36 is not a perfect cube and, consequently, $\sqrt[3]{36}$ cannot be written as a ratio $\frac{a}{b}$ where a and b are integers.

27 is a perfect cube. $\sqrt[3]{27}$ is the rational number 3.
93 is not a perfect cube. $\sqrt[3]{93}$ is an irrational number.

Focus on simplifying cube root expressions

Cube root expressions can be simplified if the radicand has a perfect cube factor.

$\sqrt[3]{54}$ is simplified as: $\quad \sqrt[3]{54} = \sqrt[3]{27} \cdot \sqrt[3]{2} = 3\sqrt[3]{2}$

$\sqrt[3]{32}$ is simplified as: $\quad \sqrt[3]{32} = \sqrt[3]{8} \cdot \sqrt[3]{4} = 2\sqrt[3]{4}$

The simplifications of cube roots with negative radicands is possible.

$\sqrt[3]{-16}$ is simplified as: $\quad \sqrt[3]{-16} = \sqrt[3]{-8} \cdot \sqrt[3]{2} = -2\sqrt[3]{2}$

$-7\sqrt[3]{-40}$ is simplified as: $\quad -7\sqrt[3]{-40} = -7 \cdot \sqrt[3]{-8} \cdot \sqrt[3]{5} = -7 \cdot -2 \cdot \sqrt[3]{5} = 14\sqrt[3]{5}$

Simplifying A Fourth Root Expression

A radical expression, $\sqrt[4]{80}$, is **simplified** when an equivalent expression, $2\sqrt[4]{5}$, with a smaller radicand is found. The process for simplifying fourth root expressions depends upon finding a perfect fourth power factor of the radicand.

The counting numbers 16, 81, 256, 625, 1296, 2401, 4096, 6561, and 10,000 are the most important perfect fourth powers. When the radicand is a perfect fourth power integer, a fourth power root expression names a counting number.

$\sqrt[4]{625}$ is a rational number because $\sqrt[4]{625} = 5$, and 5 can be written $\frac{5}{1}$.

$\sqrt[4]{256}$ is another rational number.

$\sqrt[4]{25}$ is an irrational number because 25 is not a perfect fourth power and, consequently, $\sqrt[4]{25}$ cannot be written as a ratio $\frac{a}{b}$ where a and b are integers.

16 is a perfect fourth power. $\sqrt[4]{16}$ is the rational number 2.
47 is not a perfect fourth power. $\sqrt[4]{47}$ is an irrational number.

Focus on simplifying fourth power root expressions

Fourth power root expressions can be simplified if the radicand has a perfect fourth power factor.

$\sqrt[4]{80}$ is simplified as: $\sqrt[4]{80} = \sqrt[4]{16} \cdot \sqrt[4]{5} = 2\sqrt[4]{5}$

$-6\sqrt[4]{162}$ is simplified as: $-6\sqrt[4]{162} = -6\sqrt[4]{81 \cdot 2} = -6 \cdot 3\sqrt[4]{2} = -18\sqrt[4]{2}$

Fourth power root expressions with negative radicands do not represent real numbers. $\sqrt[4]{-16}$ is not a real number because no real number can be raised to the fourth power and give a negative result.

The Real Number System

Reducing Fractions Involving Radicals

Always try to "reduce" a square root fraction. Reduction is possible when both numerator and denominator have a common integer factor or a common radicand factor.

$\frac{10}{12} = \frac{5}{6}$ and $\frac{\sqrt{10}}{\sqrt{12}}$ can be reduced to $\frac{\sqrt{5}}{\sqrt{6}}$ because numerator and denominator have $\sqrt{2}$ as a common factor.

Similarly, $\frac{8\sqrt[3]{20}}{12\sqrt[3]{15}} = \frac{2\sqrt[3]{4}}{3\sqrt[3]{3}}$ because $4\sqrt[3]{5}$ is a common factor.

Focus on reducing radical fractions

Every fraction is an indicated division and radical fractions need to be reduced whenever it is possible to reduce the integer coefficients or the radical expressions.

$\frac{-8\sqrt[4]{21}}{10\sqrt[4]{24}} = \frac{-4\sqrt[4]{7}}{5\sqrt[4]{8}}$ because $2\sqrt[4]{3}$ is a common factor that can be eliminated by division.

Simplifying Radical Fractions

The complete simplification of radical fractions such as $\frac{2\sqrt[3]{4}}{5\sqrt[3]{3}}$ requires writing a new fraction which has an integer as the denominator. The process (rationalizing the denominator) is illustrated below. Notice that the radicand of the denominator is multiplied so that it becomes a perfect third power.

$$\frac{2\sqrt[3]{4}}{5\sqrt[3]{3}} = \frac{2\sqrt[3]{4} \cdot \sqrt[3]{3^2}}{5\sqrt[3]{3^1} \cdot \sqrt[3]{3^2}} = \frac{2\sqrt[3]{36}}{5 \cdot 3} = \frac{2\sqrt[3]{36}}{15}$$

Focus on simplifying a fourth root radical fraction

The process for rationalizing the denominator of a fraction with a fourth power radical is shown below. Notice that the radicand of the denominator is multiplied so that it becomes a perfect fourth power.

$$\frac{-4\sqrt[4]{7}}{5\sqrt[4]{8}} = \frac{-4\sqrt[4]{7}}{5\sqrt[4]{2^3}} = \frac{-4\sqrt[4]{7} \cdot \sqrt[4]{2^1}}{5\sqrt[4]{2^3} \cdot \sqrt[4]{2^1}} = \frac{-4\sqrt[4]{14}}{5\sqrt[4]{2^4}} = \frac{-4\sqrt[4]{14}}{5 \cdot 2} = \frac{-4\sqrt[4]{14}}{10} = \frac{-2\sqrt[4]{14}}{5}$$

Simplifying Addition Expressions

The following steps are taken to simplify $2\sqrt[3]{36} + 9\sqrt[3]{36}$.

$$2\sqrt[3]{36} + 9\sqrt[3]{36} = (2 + 9)\sqrt[3]{36} = 11\sqrt[3]{36}$$

$4\sqrt[3]{36} + 5\sqrt[3]{22} + 7$ cannot be simplified because the radicands are different in each term.

Focus on simplifying addition expressions

To simplify an addition expression containing radicals begin by simplifying each separate term. Then add the like terms to complete the simplication.

$$4\sqrt[3]{72} + 8\sqrt[4]{72} - 5\sqrt[3]{9} - 6$$
$$4\sqrt[3]{8} \cdot \sqrt[3]{9} + 8\sqrt[4]{72} - 5\sqrt[3]{9} - 6$$
$$4 \cdot 2 \cdot \sqrt[3]{9} + 8\sqrt[4]{72} - 5\sqrt[3]{9} - 6$$
$$8\sqrt[3]{9} + 8\sqrt[4]{72} - 5\sqrt[3]{9} - 6$$
$$(8\sqrt[3]{9} - 5\sqrt[3]{9}) + 8\sqrt[4]{72} - 6$$
$$3\sqrt[3]{9} + 8\sqrt[4]{72} - 6$$

Unit 3 Exercise

Part A: Answers for all Part A problems are at the back of the book.

1. Three perfect cubes are 8, 27, and 64. What are the next three perfect cubes?

2. $\sqrt[3]{27}$ is a rational number. $\sqrt[3]{343}$ is a _____ number.

3. $\sqrt[3]{19}$ is an irrational number. $\sqrt[3]{107}$ is an _____ number.

4. $\sqrt[3]{100}$ is a(n) _____ number.

5. $\sqrt[3]{216}$ is a(n) _____ number.

6. Complete the simplication of $\sqrt[3]{16}$.
$\sqrt[3]{16} = \sqrt[3]{8 \cdot 2} = \sqrt[3]{8} \cdot \sqrt[3]{2} =$ _____.

7. Simplify $\sqrt[3]{250}$.

8. Complete the simplication of $\sqrt[3]{-81}$.
$\sqrt[3]{-81} = \sqrt[3]{-27 \cdot 3} = \sqrt[3]{-27} \cdot \sqrt[3]{3} =$ _____.

9. Simplify $\sqrt[3]{-48}$.

10. Two perfect fourth power integers are 16 and 81. What are the next two fourth power integers?

11. Complete the simplification of $\sqrt[4]{48}$.
$\sqrt[4]{48} = \sqrt[4]{16} \cdot \sqrt[4]{3} =$ _____.

12. Simplify $-5\sqrt[4]{162}$

13. Complete the simplication of $\dfrac{4\sqrt[3]{5}}{3\sqrt[3]{9}}$.

    $\dfrac{4\sqrt[3]{5}}{3\sqrt[3]{9}} = \dfrac{4\sqrt[3]{5}}{3\sqrt[3]{3^2}} \cdot \dfrac{\sqrt[3]{3}}{\sqrt[3]{3}} = \dfrac{4\sqrt[3]{15}}{3\sqrt[3]{3^3}} = $ _____.

14. Simplify $\dfrac{-6\sqrt[3]{2}}{2\sqrt[3]{16}}$

15. Simplify $\sqrt[3]{8} + \sqrt[3]{128} - 4 + \sqrt[3]{2}$ by simplifying each term first.

Part B: Answers for odd-numbered problems of Part B are at the back of the book.

1. List the first five perfect cubes.
2. List the first four perfect fourth power integers.
3. Simplify $\sqrt[3]{625}$
4. Simplify $3\sqrt[3]{108}$
5. Simplify $-8\sqrt[3]{216}$
6. Simplify $\frac{1}{2}\sqrt[3]{4000}$
7. Simplify $\sqrt[4]{32}$
8. Simplify $-20\sqrt[4]{243}$
9. Simplify $\frac{1}{2}\sqrt[4]{64}$
10. Simplify $\dfrac{-21\sqrt[3]{80}}{14\sqrt[3]{32}}$
11. Simplify $\dfrac{5\sqrt[4]{16}}{\sqrt[4]{3}}$
12. Simplify $2\sqrt[3]{54} - \sqrt[4]{81} + 3\sqrt[4]{7} - 5\sqrt[3]{3}$
13. Simplify $\sqrt[4]{60} - 2\sqrt[4]{48} + 5\sqrt[3]{64} + \sqrt[4]{243}$
14. Simplify $\sqrt[3]{16} + \sqrt[4]{32} + \sqrt[3]{54} - \sqrt[4]{162}$

Part C: No answers are given for these problems. However, each is accompanied by an ordered pair «C,U» showing the chapter and unit in which it was taught.

1. Multiply $(3x-4)(2x^2-x+7)$ «1,1»
2. Factor $4x^2-12xy+9y^2$ «1,4»
3. Solve $6x^2-9x=15$ «1,5»
4. Simplify $(2x-1)(3x+2)<0$ «1,6»
5. Is $\sqrt{-7}$ a real number? «2,1»

Simplify problems 6-10.

6. $\sqrt{54}$ «2,2»
7. $\dfrac{\sqrt{16}}{\sqrt{81}}$ «2,2»
8. $\dfrac{-7}{\sqrt{18}}$ «2,2»
9. $\sqrt{96} + \sqrt{5} - \sqrt{75} + \sqrt{24}$ «2,2»
10. $\dfrac{5-2\sqrt{3}}{1-\sqrt{3}}$ «2,2»

Unit 4: DECIMAL REPRESENTATIONS OF REAL NUMBERS

Finding A Terminating Decimal

The rational number $\frac{5}{8}$ can be represented by the decimal numeral .625 which equals $\frac{625}{1000}$. The decimal numeral .625 is a **terminating** decimal. When 5 is divided by 8 and the quotient is carried to three decimal places, the long division remainder is zero.

```
       . 6 2 5
   8 ) 5. 0 0 0
      -4 8
         2 0
        -1 6
           4 0
          -4 0
             0
```

Any rational number $\frac{x}{y}$ where y is a divisor of a power of 10 (10, 100, 1000, etc.) can be represented by a terminating decimal. This is because the division of x by y will eventually result in a zero remainder. Hence, the division terminates and the result is a terminating decimal.

Focus on finding a terminating decimal

The long division associated with the rational number $\frac{11}{16}$ is shown at the right.

Because 16 evenly divides 10,000 the division results in a zero remainder after the fourth decimal place.

The long division quotient, .6875, is the terminating decimal for $\frac{11}{16}$.

```
            . 6 8 7 5
    16 ) 1 1. 0 0 0 0
          - 9 6
            1 4 0
           -1 2 8
              1 2 0
             -1 1 2
                  8 0
                 -8 0
                    0
```

Finding A Repeating Decimal

The rational number $\frac{5}{12}$ can be represented by the decimal numeral $.41\overline{6}$ where the bar over the digit 6 indicates it will be repeated forever. When 5 is divided by 12, the remainder from the second decimal place is 8. When this remainder is repeated at the third decimal place, it means the quotient digit will repeat also.

```
      . 4 1 6 6
12 ) 5.0 0 0 0
     -4 8
        2 0
       -1 2
          8 0
         -7 2
            8 0
           -7 2
              8
```

Any rational number $\frac{x}{y}$ where y is not a divisor of a power of 10 (10, 100, 1000, etc.) can be represented by a repeating decimal. This is because the division of x by y will eventually result in a repeated remainder. The division will repeat from that point.

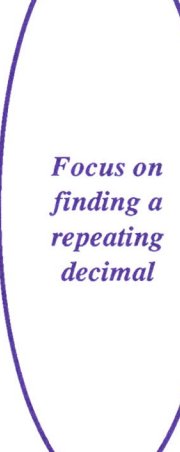

Focus on finding a repeating decimal

The long division associated with the rational number $\frac{5}{7}$ is shown at the right.

Because 7 will never evenly divide a power of ten the division never results in a zero remainder.

However, the long division will repeat itself whenever a remainder is repeated. In the division shown at the right, this occurs with the remainder 1.

The repeating decimal $.\overline{714285}$ represents the rational number $\frac{5}{7}$.

```
       . 7 1 4 2 8 5 7
7 ) 5.0 0 0 0 0 0 0
    -4 9
       1 0
      - 7
         3 0
        -2 8
           2 0
          -1 4
             6 0
            -5 6
               4 0
              -3 5
                 5 0
                -4 9
                   1
```

Chapter 2

Decimal Representations of Rational Numbers

Every rational number has either a terminating or a repeating decimal representation and every terminating or repeating decimal represents a rational number.

Each of the following equalities can be checked by long division.

$$\frac{4}{15} = .2\overline{6}$$

$$\frac{7}{32} = .21875$$

$$\frac{3}{14} = .2\overline{142857}$$

$$\frac{1049}{3300} = .31\overline{78}$$

Focus on decimal numerals for rational numbers

It is relatively easy to find the rational number represented by any terminating decimal. Just convert the decimal to its equivalent fraction and then reduce if that is possible.

$$.875 = \frac{875}{1000} \quad \text{and} \quad \frac{875}{1000} \text{ reduces to } \frac{7}{8}$$

To find the rational number of $.87\overline{35}$ the following process may be used.

1. Multiply $.87\overline{35}$ by a power of 10 that will put a single sequence of the repetition left of the decimal point.
   If $N = .87\overline{35}$ then $10{,}000N = 8735.\overline{35}$
2. Multiply $.87\overline{35}$ by a power of 10 that will put the first sequence of the repetition directly to the right of the decimal point.
   If $N = .87\overline{35}$ then $100N = 87.\overline{35}$
3. Subtract in columns 100N from 10,000N.
   $$\begin{aligned} 10{,}000N &= 8735.\overline{35} \\ -\ 100N &= -\ 87.\overline{35} \\ \hline 9900N &= 8648.00 \end{aligned}$$
4. Solve the equation to find N.
   $$N = \frac{8648}{9900} \text{ and this simplifies to } N = \frac{2162}{2475}$$

Decimal Approximations of Irrational Numbers

Both terminating and repeating decimal numerals represent rational numbers. A third and last type of decimal numeral is a non-terminating, non-repeating decimal. That is, the decimal numeral will never establish a pattern or sequence of digits. Knowing the digits in the first ten, hundred, or million decimal places gives no basis for accurately predicting the digit in the next decimal place.

The decimal approximation for $\sqrt{19}$ is 4.359. $\sqrt{19}$ is an irrational number so its decimal numeral is non-terminating and non-repeating. When we say its decimal approximation is 4.359 we mean that to the nearest one-thousandth $\sqrt{19}$ is rounded off to 4.359, but this information is not sufficient to determine the digit in a four decimal place approximation of $\sqrt{19}$.

Focus on decimal numerals for irrationals

Any irrational number will have a decimal representation that is a non-terminating, non-repeating decimal. When a decimal is used for an irrational number it will always be an approximation because it is always a rounded off numeral.

A decimal approximation for $\sqrt{11}$ is 3.317 which means that:

$$3.3165 < \sqrt{11} < 3.3175$$

Similarly, a decimal approximation for π is 3.1416 which means that:

$$3.14155 < \pi < 3.14165$$

Notice that both $\sqrt{11}$ and π which are irrational have been approximated above by terminating decimals which are rational. Rational numbers can always be represented exactly by terminating or repeating decimals. Irrational numbers can only be approximated by any decimal numeral.

Rational Number Approximations

The square root of 10, $\sqrt{10}$, can be approximated by 3.162 which is a decimal numeral for the rational number $\frac{3162}{1000}$.

The multiplication of 3.162 by itself is shown at the right. Notice that the answer is very close to 10, but is not equal to 10.

```
    3.1 6 2
  x 3.1 6 2
    6 3 2 4
   1 8 9 7 2
   3 1 6 2
   9 4 8 6
   9.9 9 8 2 4 4
```

3.162 is a three-place decimal approximation for $\sqrt{10}$. More decimal places could be used to achieve better approximations, but the number of decimal places can be increased forever without finding an exact decimal numeral for $\sqrt{10}$.

Focus on describing irrational numbers by rational numbers

$\sqrt{2}$ is an irrational number that is approximately equal to 1.414 or $\frac{1414}{1000}$ which is a rational number.

Tables have been computed to show rational approximations for many square roots. You will find such a table on page 452 of this text. The first vertical column of the table shows the counting numbers from 1 to 35. The second vertical column indicates square roots of those counting numbers. The entries in that second column are rational numbers and, in most cases, approximations which are correct to three decimal places.

A portion of the full table from page 452 is shown at the right. The square root entries for 1, 4, 9 and 16 are 1.000, 2.000, 3.000, and 4.000 because 1, 4, 9 and 16 are perfect squares. These entries are not approximations.

The square roots for the other counting numbers shown at the right are three-place, non-zero decimals. These entries are approximations. Regardless of the number of decimal places used they would be approximations.

Number	Square Root
1	1.000
2	1.414
3	1.732
4	2.000
5	2.236
6	2.449
7	2.646
8	2.828
9	3.000
10	3.162
11	3.317
12	3.464
13	3.606
14	3.742
15	3.873
16	4.000
17	4.123
18	4.243
19	4.359
20	4.472

Decimal Approximations for Other Real Numbers

The table on page 452 shows square roots for the counting numbers 1 through 100. The table can also be used to approximate $\sqrt{71.63}$ by just using the value for $\sqrt{71}$.

Approximations for numbers such as $\sqrt{1300}$ can also be determined by the table.

$$\sqrt{1300} = \sqrt{100} \cdot \sqrt{13} = 10 \cdot 3.606 = 36.06$$

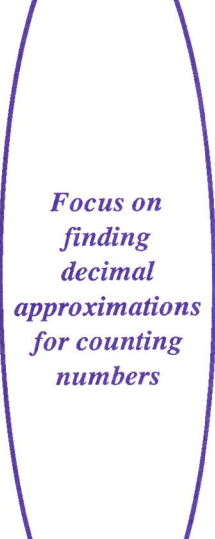

Focus on finding decimal approximations for counting numbers

A decimal approximation for the square root of any positive real number can be found by the following process.
1. Select **any** positive number as a divisor. The closer the selection is to the actual square root the faster will be the process.
2. Divide to one more decimal place than the divisor.
3. Average the divisor and quotient. This average is a first approximation. Its degree of accuracy will depend upon the original divisor selected.
4. Use the average as a divisor and repeat the process until an approximation of the desired accuracy is found.

$\sqrt{110}$ may be approximated to three decimal places by the following process:
1. Suppose 10 is selected as a divisor.
2. $110 \div 10 = 11$
3. The average of 10 and 11 is 10.5
4. $110 \div 10.5$ rounds off to 10.48
5. The average of 10.5 and 10.48 is 10.49
6. $110 \div 10.49$ rounds off to 10.486
7. The average of 10.49 and 10.486 is 10.488
8. $110 \div 10.488$ rounds off to 10.488

Because the divisor and dividend are both 10.488, this is the value for the three place decimal approximation of $\sqrt{110}$.

Unit 4 Exercise

Part A: Answers for all Part A problems are at the back of the book.

1. Find terminating decimals for:
   a. $\frac{3}{5}$ b. $\frac{17}{20}$

2. Find repeating decimals for:
   a. $\frac{1}{7}$ b. $\frac{2}{11}$

3. Find rational numbers $\frac{x}{y}$ equal to:
   a. .173   b. .195
   c. .71$\overline{4}$ d. .2$\overline{15}$

4. Use the table of square roots to find decimal approximations for:
   a. $\sqrt{57}$ b. $\sqrt{33}$
   c. $\sqrt{80}$ d. $\sqrt{96}$

5. Use simplification and the square root table to find decimal approximations for:
   a. $\sqrt{200}$ b. $\sqrt{270}$
   c. $\sqrt{80}$ d. $\sqrt{160}$

6. According to the square root table $\sqrt{11} \doteq 3.317$ and this means the smallest possible four-place decimal approximation for $\sqrt{11}$ is _____.

7. According to the square root table $\sqrt{19} \doteq 4.359$ and this means the greatest possible four-place decimal approximation for $\sqrt{19}$ is _____.

8. Find a decimal approximation for $6 + \sqrt{7}$ by adding 6 to the table entry for $\sqrt{7}$.

9. Find a decimal approximation for $5 - \sqrt{7}$ by subtracting the table entry for $\sqrt{7}$ from 5.

10. Find a decimal approximation for $\frac{6+\sqrt{7}}{11}$ by dividing the answer to problem 8 by 11 and rounding the answer off at three decimal places.

Part B: Answers for odd-numbered problems of Part B are at the back of the book.

In problems 1-10 find decimal representations.
1. $\frac{15}{16}$
2. $\frac{8}{3}$
3. $\frac{6}{7}$
4. $\frac{7}{8}$
5. $\frac{9}{15}$
6. $\frac{4}{13}$
7. $\frac{5}{9}$
8. $\frac{13}{40}$
9. $\frac{5}{6}$
10. $\frac{23}{4}$

In problems 11-15 find rational numbers $\frac{x}{y}$ equal to:
11. .832
12. .63$\overline{7}$
13. 4.3$\overline{79}$
14. -5.6$\overline{17}$
15. .$\overline{4175}$

In problems 16-30 find decimal approximations.

16. $\sqrt{41}$
17. $\sqrt{84}$
18. $\sqrt{59}$
19. $\sqrt{73}$
20. $\sqrt{800}$
21. $\sqrt{450}$
22. $\sqrt{80}$
23. $\sqrt{17} + 8$
24. $9 - \sqrt{37}$
25. $4 + 2\sqrt{11}$

26. $\frac{5 + \sqrt{5}}{2}$
27. $\frac{4 - \sqrt{5}}{3}$
28. $\frac{8 + \sqrt{5}}{2}$
29. $\frac{1 - 3\sqrt{2}}{4}$
30. $\frac{5 + \sqrt{45}}{6}$

Part C: No answers are given for these problems. However, each is accompanied by an ordered pair «C,U» showing the chapter and unit in which it was taught.

1. Factor $7a^2 + 2ab - 9b^2$ «1,4»
2. Solve $3x^2 - 7 = 4x$ «1,5»

For problems 3-10, simplify.

3. $(x - 3)(3x - 5) \geq 0$ «1,6»
4. $-\sqrt{108}$ «2,2»
5. $\sqrt{12} + \sqrt{18} - \sqrt{32} + \sqrt{48}$ «2,2»
6. $\sqrt{\frac{5}{2}}$ «2,2»
7. $\frac{3 + 2\sqrt{5}}{3 - 2\sqrt{5}}$ «2,2»
8. $\sqrt[3]{128}$ «2,3»
9. $\frac{-3\sqrt[3]{3}}{23\sqrt[3]{32}}$ «2,3»
10. $-5\sqrt[3]{8} + 2\sqrt[3]{250} - 3\sqrt[3]{54}$ «2,3»

Unit 5: SOLVING LINEAR EQUATIONS AND INEQUALITIES

Equivalent Equations

Two linear equations are equivalent when they have exactly the same solutions (truth sets). There are two primary techniques for generating equivalent linear equations.
  1. Any number or open expression may be added to both sides of the equation.
  2. Any non-zero number or open expression not equal to zero may be multiplied by both sides of the equation.

Linear equations are commonly solved using these two techniques. For example, the solution below illustrates how the equation $3x + 19 = 12$ is used to generate an equivalent equation $x = \frac{-7}{3}$ for which the solution $\frac{-7}{3}$ is obvious.

  1. Add -19 to both members of the equality.
     $$(3x + 19) + -19 = 12 + -19$$
     $$3x = -7$$

  2. Multiply both members of $3x = -7$ by $\frac{1}{3}$.
     $$\tfrac{1}{3}(3x) = \tfrac{1}{3} \cdot -7$$
     $$x = \tfrac{-7}{3}$$

$\left\{\frac{-7}{3}\right\}$ is the truth set of $3x + 19 = 12$.

The equations $3x + 19 = 12$, $3x = -7$, and $x = \frac{-7}{3}$ are equivalent. $\left\{\frac{-7}{3}\right\}$ is the truth set of each of them.

The **Additive Property of Equality** states that any number or open expression may be added to both members of an equation to produce an equivalent equation. According to the Additive Property of Equality each of the following pairs of equations is equivalent.

$$6x - 1 = 11 \text{ is equivalent to } 6x = 12.$$
$$5x = 15 \text{ is equivalent to } 5x + 2x = 15 + 2x.$$
$$2x - 9 = 5 \text{ is equivalent to } 2x = 14.$$

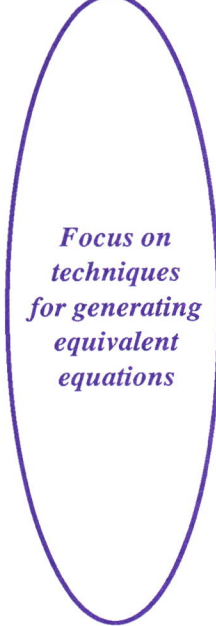

Focus on techniques for generating equivalent equations

The **Multiplication Property of Equality** states that any non-zero number or open expression not equal to zero may be multiplied by both members of an equation to produce an equivalent equation. According to the Multiplication Property of Equality each of the following pairs of equations is equivalent.

$$3x = 12 \text{ is equivalent to } x = 4.$$
$$2x = 9 \text{ is equivalent to } x = \tfrac{9}{2}.$$
$$-4x = 11 \text{ is equivalent to } x = \tfrac{-11}{4}.$$
$$\tfrac{5}{x-3} = 7 \text{ is equivalent to } 5 = 7(x-3) \text{ when } x - 3 \neq 0.$$

Solving Linear Equations

The first step in solving $\sqrt{3}x - \sqrt{7} = 6$ is to find an equivalent equation by using the Additive Property of Equality.

$$\sqrt{3}x - \sqrt{7} = 6$$
$$\sqrt{3}x = 6 + \sqrt{7}$$

The next step in finding the truth set of $\sqrt{3}x = 6 + \sqrt{7}$ is to find an equivalent equation by using the Multiplication Property of Equality.

$$\sqrt{3}x = 6 + \sqrt{7}$$

$$x = \frac{6 + \sqrt{7}}{\sqrt{3}} \quad \text{or} \quad \frac{6\sqrt{3} + \sqrt{21}}{3}$$

The root or solution of $\sqrt{3}x = 6 + \sqrt{7}$ in its simplest form is $\frac{6\sqrt{3} + \sqrt{21}}{3}$.

If the simplification is not understood, review Unit 2 of this chapter.

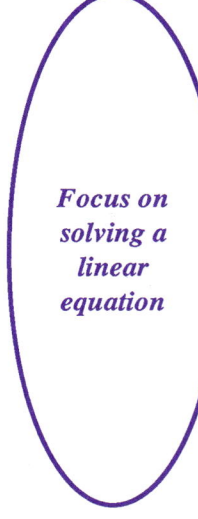

Focus on solving a linear equation

The Additive Property of Equality is used to isolate the variable terms in one member of the equation while gathering all the number terms (constants) in the other member.

$$4x - 7 = \sqrt{5} \cdot x + 3$$
$$4x - \sqrt{5} \cdot x = 3 + 7$$
$$(4 - \sqrt{5})x = 10$$

The Multiplication Property of Equality can then be used.

$$(4 - \sqrt{5})x = 10$$

$$x = \frac{10}{4 - \sqrt{5}} \text{ which simplifies to } \frac{40 + 10\sqrt{5}}{11}$$

If the simplification is not understood, review Unit 2 of this chapter.

Equivalent Linear Inequalities

Two linear inequalities are equivalent when they have exactly the same solutions (truth sets). As with equations, there are two primary techniques for generating equivalent linear inequalities but the multiplication property includes two possibilities:
1. Additive Property of Inequalities: Any number or open expression may be added to both sides of the inequality.
2. Multiplication Property of Inequalities:
   a. Any positive number or open expression assumed to be positive may be multiplied by both sides of the inequality without altering the symbol of inequality.
   b. Any negative number or open expression assumed to be negative may be multiplied by both sides of the inequality by reversing the symbol of inequality.

The **Additive Property of Inequality** states that any number or open expression may be added to both members of an inequality to produce an equivalent inequality. According to the Additive Property of Inequality each of the following pairs of inequalities is equivalent.

$8x - \sqrt{7} > 11$ is equivalent to $8x > 11 + \sqrt{7}$
$\sqrt{2}x < 9$ is equivalent to $\sqrt{2}x + \sqrt{5}x < 9 + \sqrt{5}x$
$2x - 5\sqrt{3} \leq 11\sqrt{3}$ is equivalent to $2x \leq 16\sqrt{3}$

The Real Number System

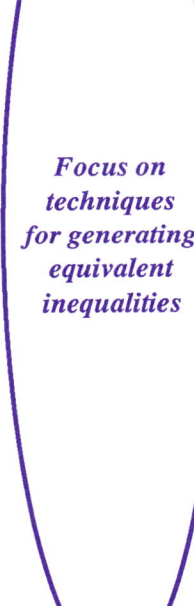

Focus on techniques for generating equivalent inequalities

The **Multiplication Property of Inequality** has two sub-parts: When any positive number or open expression assumed to be positive is multiplied by both members of an inequality the result is an equivalent inequality using the same symbol of inequality. According to this portion of the Multiplication Property of Inequality each of the following pairs of equations is equivalent.

$$3x \geq \sqrt{17} \text{ is equivalent to } x \geq \frac{\sqrt{17}}{3}$$

$$\sqrt{13}x < \sqrt{26} \text{ is equivalent to } x < \sqrt{2}$$

$$\frac{5x - 7}{3x} > 4 \text{ is equivalent to } 5x - 7 > 12x \text{ when } 3x > 0$$

The second part of the **Multiplication Property of Inequality** is: When any negative number or open expression assumed to be negative is multiplied by both members of an inequality the result is an equivalent inequality using the opposite symbol of inequality. According to this portion of the Multiplication Property of Inequality each of the following pairs of equations is equivalent.

$$-4x \geq \sqrt{11} \text{ is equivalent to } x \leq \frac{-\sqrt{11}}{4}$$

$$-\sqrt{7}x < \sqrt{42} \text{ is equivalent to } x > -\sqrt{6}$$

$$\frac{5x - 7}{3x} > 4 \text{ is equivalent to } 5x - 7 < 12x \text{ when } 3x < 0$$

Solving Linear Inequalities

The first step in solving $\sqrt{3} - 4x \geq \sqrt{7}$ is to find an equivalent inequality by using the Additive Property of Inequality.

$$\sqrt{3} - 4x \geq \sqrt{7}$$
$$-4x \geq \sqrt{7} - \sqrt{3}$$

The next step in finding the truth set of $-4x \geq \sqrt{7} - \sqrt{3}$ is to find an equivalent inequality by using the Multiplication Property of Inequality.

$$-4x \geq \sqrt{7} - \sqrt{3}$$
$$x \leq \frac{\sqrt{7} - \sqrt{3}}{-4} \text{ or } \frac{-\sqrt{7} + \sqrt{3}}{4}$$

Notice that the direction of the inequality symbol has been reversed.

Focus on solving a linear inequality

The Additive Property of Inequality is used to isolate the variable terms in one member of the inequality while gathering all the number terms (constants) in the other member.

$$7x - 4 < \sqrt{13}x + \sqrt{2}$$
$$7x - \sqrt{13}x < 4 + \sqrt{2}$$
$$(7 - \sqrt{13})x < 4 + \sqrt{2}$$

The Multiplication Property of Inequality can then be used.

$$(7 - \sqrt{13})x < 4 + \sqrt{2}$$

$$x < \frac{4 + \sqrt{2}}{7 - \sqrt{13}} \text{ which simplifies to } x < \frac{28 + 4\sqrt{13} + 7\sqrt{2} + \sqrt{26}}{36}$$

Note: It must be recognized that $7 - \sqrt{13}$ is positive to use the correct symbol of inequality in the final result.

Unit 5 Exercise

Part A: Answers for all Part A problems are at the back of the book.

1. Is $4x + 6 = 18$ equivalent to $4x = 12$?
2. Is $3x - 7 = 16$ equivalent to $3x = 23$?
3. Is $5x - 3 = 8$ equivalent to $5x = 15$?
4. Is $7x + 6 = 27$ equivalent to $7x = 21$?
5. Add $-3x$ to both members of $4x - 7 = 3x + 2$ and write the equivalent equation obtained.
6. Is $-3x - 7 = 5x + 8$ equivalent to $-7 = 8x + 8$?
7. According to the Additive Property of Equality, $4x + 11 = x - 3$ is equivalent to $4x =$ _____.
8. According to the Additive Property of Equality, $5x + 6 = 9 - 2x$ is equivalent to $7x + 6 =$ _____.
9. Is $\frac{-5x}{2} - \frac{1}{2} > \frac{2}{3}$ equivalent to $\frac{-5x}{2} > \frac{2}{3} + \frac{1}{2}$?
10. Is $\frac{-5x}{2} > \frac{1}{4}$ equivalent to $x > \frac{-2}{5} \cdot \frac{1}{4}$?
11. Is $\sqrt{7} \cdot x + 3 \geq \sqrt{11}$ equivalent to $\sqrt{7} \cdot x \geq \sqrt{11} - 3$?

12. Is $\sqrt{7} \cdot x \geq \sqrt{11} - 3$ equivalent to $x \geq \frac{\sqrt{77} - 3\sqrt{7}}{7}$?

13. Is $6x - 9\sqrt{2} < \sqrt{13} \cdot x$ equivalent to
    $6x - \sqrt{13} \cdot x < 9\sqrt{2}$?

14. Is $6x - \sqrt{13} \cdot x < 9\sqrt{2}$ equivalent to $x < \frac{9\sqrt{2}}{6 - \sqrt{13}}$?

15. Is $x < \frac{9\sqrt{2}}{6 - \sqrt{13}}$ equivalent to $x < \frac{54\sqrt{2} - 9\sqrt{26}}{23}$?

Part B: Answers for odd-numbered problems of Part B are at the back of the book.

Solve.

1. $4x = 5x - 12$
2. $9x + \sqrt{6} = x - \sqrt{5}$
3. $\sqrt{6} \cdot x - 3 = 8x + \sqrt{7}$
4. $\frac{3x}{4} = \sqrt{15}$
5. $\sqrt{21} \cdot x = 14$
6. $5x = -7\sqrt{2}$
7. $\frac{4x}{7} = \frac{-3}{8}$
8. $\frac{x}{4} = -7$
9. $5(4x + \sqrt{13}) = 2$
10. $3(\sqrt{5} \cdot x + 2) = \sqrt{6}$
11. $\frac{2x}{3} - \sqrt{3} = 5$
12. $\frac{2x}{3} + 1 = \sqrt{19}$
13. $\frac{3x}{4} - 7 = 4$
14. $\frac{2x}{3} = \frac{-1}{2}$
15. $\sqrt{3} \cdot x + \sqrt{14} = 6$
16. $\sqrt{5}x - \sqrt{3} = \sqrt{3}$
17. $x + \sqrt{7} = 6\sqrt{7} - \sqrt{3} \cdot x$
18. $4x - 7 = \sqrt{2} \cdot x + \sqrt{3}$
19. $5x - 4 = \sqrt{17} \cdot x$
20. $\sqrt{6}x - 9 = \sqrt{5}x + \sqrt{3}$
21. $\sqrt{3}x + 2 > -7$
22. $2x + \sqrt{6} < -3 + 7x$
23. $\sqrt{5} \cdot x - 4 \geq 12x + 1$
24. $x - \sqrt{5} \leq \sqrt{3} \cdot x - 8$
25. $\sqrt{26} - 4x \geq x - 4$
26. $3x - \sqrt{31} < \sqrt{2} \cdot x + 1$
27. $\sqrt{3} - 2x \geq 4 - x$
28. $\sqrt{5} \cdot x - 2 \leq 5x - 10$
29. $8x - \sqrt{7} > 3x + \sqrt{5}$
30. $x + \sqrt{2} < 6x - 2\sqrt{2}$

Part C: No answers are given for these problems. However, each is accompanied by an ordered pair «C,U» showing the chapter and unit in which it was taught.

1. Factor $2x^2 - 11xy + 9y^2$ «1,4»

Simplify problems 2-6.

2. $3x(2x - 7) \leq 0$ «1,6»
3. $\sqrt{\frac{3}{8}}$ «2,2»
4. $-\sqrt{108}$ «2,2»
5. $\sqrt[3]{162}$ «2,3»
6. $\frac{\sqrt[3]{40}}{\sqrt[3]{128}}$ «2,3»

7. Is $\frac{4}{9}$ a terminating decimal? «2,4»
8. Are all terminating or repeating decimals rational numbers? «2,4»
9. Is $\frac{2}{7}$ a repeating decimal? «2,4»
10. Is $\frac{9}{16}$ a terminating decimal? «2,4»

Unit 6: SOLVING QUADRATIC EQUATIONS

Quadratic Equations

Any equation of the form $ax^2 + bx + c = 0$ where a, b, and c are real numbers and $a \neq 0$ is a **quadratic equation** in its **standard form**.

For example, each of the following is a quadratic equation in its standard form.

$x^2 - 7x - 12 = 0$ with a = 1, b = -7, c = -12

$5x^2 + x + 8 = 0$ with a = 5, b = 1, c = 8

$x^2 - 15 = 0$ with a = 1, b = 0, c = -15

$-x^2 - 3x + \sqrt{7} = 0$ with a = -1, b = -3, c = $\sqrt{7}$

$2x^2 - x + 10 = 0$ with a = 2, b = -1, c = 10

Focus on quadratic equations not in standard form

Each of the following equations is a quadratic equation, but is not in its standard form.

$(x - 5)(x + 3) = 0$ which is equivalent to $x^2 - 2x - 15 = 0$

$4x^2 + 3 = 2x$ which is equivalent to $4x^2 - 2x + 3 = 0$

$(x - 5)^2 = 11$ which is equivalent to $x^2 - 10x + 14 = 0$

$\sqrt{5}x^2 = 9x - 4$ which is equivalent to $\sqrt{5}x^2 - 9x + 4 = 0$

$2x(x - 3) = 5$ which is equivalent to $2x^2 - 6x - 5 = 0$

Solving Quadratics By Factoring

Quadratic equations which have roots that are rational numbers are generally most easily solved by the factoring method. To solve any quadratic equation first generate an equivalent equation which has zero as one of its members and then try to factor the polynomial. If the polynomial is factorable, the equation has rational number roots.

To solve $x^2 + 12 = 7x$ by factoring, the following steps are used.

1. Add -7x to both sides of the equation so the right member becomes 0.

    $$x^2 + 12 = 7x$$
    $$\underline{ -7x\ \ -7x}$$
    $$x^2 - 7x + 12 = 0$$

2. Factor $x^2 - 7x + 12$

    $$(x - 3)(x - 4) = 0$$

3. Each factor is set equal to 0. In this case the result is two simple linear equations.

    $$x - 3 = 0 \text{ or } x - 4 = 0$$

4. Solve each linear equation.

    $$x = 3 \text{ or } x = 4$$

5. Each root is an element of the quadratic equation's truth set.

    The truth set is $\{3, 4\}$

Focus on solving a quadratic equation by factoring

$6x^2 = 5 - 13x$ is solved by factoring as shown below.

$$6x^2 = 5 - 13x$$
$$6x^2 + 13x - 5 = 0$$
$$(2x + 5)(3x - 1) = 0$$
$$2x + 5 = 0 \quad \text{or} \quad 3x - 1 = 0$$
$$2x = -5 \quad \text{or} \quad 3x = 1$$
$$x = \frac{-5}{2} \quad \text{or} \quad x = \frac{1}{3}$$

The truth set of $6x^2 = 5 - 13x$ is $\left\{\frac{-5}{2}, \frac{1}{3}\right\}$.

118   Chapter 2

Completing the Square

The process known as **completing the square** can be used on any quadratic equation including those which have roots that are not rational. The most important step in the completing the square process is writing a perfect square trinomial so the equation can be put in the form: $(x + k)^2 = m$. After that step, the equation can be solved easily because $(x + k)$ must be equal to $\sqrt{m}$ or $-\sqrt{m}$.

For example, $x^2 - 14x + 30 = 0$ is solved in the following steps. The line preceded by three asterisks is the **completing the square** step.

$$x^2 - 14x + 30 = 0$$
$$x^2 - 14x = -30$$
*** $$x^2 - 14x + 49 = -30 + 49$$
$$(x - 7)^2 = 19$$
$$(x - 7) = \sqrt{19} \text{ or } (x - 7) = -\sqrt{19}$$
$$x = 7 + \sqrt{19} \text{ or } x = 7 - \sqrt{19}$$

Notice that the roots of $(x - 7)^2 = 19$ are irrational numbers. This means that the original equation could not have been solved by factoring.

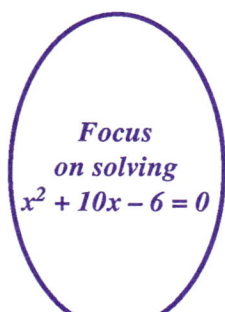

Focus on solving $x^2 + 10x - 6 = 0$

$x^2 + 10x - 6 = 0$ is solved in the following steps.

$$x^2 + 10x - 6 = 0$$
$$x^2 + 10x = 6$$
$$x^2 + 10x + 25 = 6 + 25$$
$$(x + 5)^2 = 31$$
$$(x + 5) = \sqrt{31} \text{ or } (x + 5) = -\sqrt{31}$$
$$x = -5 + \sqrt{31} \text{ or } x = -5 - \sqrt{31}$$

Again, the roots of $(x + 5)^2 = 31$ are irrational numbers and the equation could not be solved by factoring.

The Quadratic Formula

The quadratic formula, $x = \dfrac{-b \pm \sqrt{b^2 - 4ac}}{2a}$, may be applied to any equation of the form $ax^2 + bx + c = 0$, $a \neq 0$. The formula can be derived using the completing the square process. That derivation is shown in the Appendix on page 489.

The Real Number System

To solve $5x^2 - 7x + 1 = 0$, the following steps are used.

$$5x^2 - 7x + 1 = 0 \qquad a = 5, b = -7, c = 1$$

$$x = \frac{-b \pm \sqrt{b^2 - 4ac}}{2a} = \frac{-(-7) \pm \sqrt{(-7)^2 - 4 \cdot 5 \cdot 1}}{2 \cdot 5}$$

$$= \frac{7 \pm \sqrt{49 - 20}}{10}$$

$$= \frac{7 \pm \sqrt{29}}{10}$$

The two roots are $\frac{7 + \sqrt{29}}{10}$ and $\frac{7 - \sqrt{29}}{10}$.

Focus on using the quadratic formula

To solve $3x^2 + 6 = -8x$, the equation must first be written in the form $ax^2 + bx + c = 0$.

$$3x^2 + 6 = -8x \text{ becomes } 3x^2 + 8x + 6 = 0$$
$$3x^2 + 8x + 6 = 0 \quad a = 3, b = 8, c = 6$$

$$x = \frac{-b \pm \sqrt{b^2 - 4ac}}{2a} = \frac{-(8) \pm \sqrt{(8)^2 - 4 \cdot 3 \cdot 6}}{2 \cdot 3}$$

$$= \frac{-8 \pm \sqrt{64 - 72}}{6}$$

$$= \frac{-8 \pm \sqrt{-8}}{6}$$

The roots of the equation are not real numbers because there is no real number that when squared will give -8. In the next chapter complex numbers will be introduced and the equation will have two complex roots, but $3x^2 + 6 = -8x$ has { } as its truth set with respect to the set of real numbers.

Equations with Radical Signs on the Variable

To solve the equation $\sqrt{x-7} = 13$ each side of the equation can be squared. This eliminates the radical sign and results in a simple linear equation.

$$\sqrt{x-7} = 13$$
$$(\sqrt{x-7})^2 = (13)^2$$
$$x - 7 = 169$$
$$x = 176$$

A similar process can be used with $\sqrt{x} + \sqrt{x+16} = 8$, but squaring the left side of the equation does not completely eliminate the radical sign. The process must be repeated before the radical signs are eliminated.

$$\sqrt{x} + \sqrt{x+16} = 8$$
$$(\sqrt{x} + \sqrt{x+16})^2 = (8)^2$$
$$x + 2\sqrt{x(x+16)} + x + 16 = 64$$
$$2x + 2\sqrt{x(x+16)} + 16 = 64$$
$$2\sqrt{x(x+16)} = 48 - 2x$$
$$\sqrt{x(x+16)} = 24 - x$$
$$(\sqrt{x(x+16)})^2 = (24 - x)^2$$
$$x(x+16) = 576 - 48x + x^2$$
$$x^2 + 16x = 576 - 48x + x^2$$
$$16x = 576 - 48x$$
$$64x = 576$$
$$x = 9$$

The root, 9, needs to be checked in the original equation. This check is extremely important because the process of separately squaring the two sides of an equation does not guarantee the generation of an equivalent equation.

Focus on solving equations with variable radicands

To solve $\sqrt{x + 8} = -6$, first read the equation carefully because no square root of a real number can be negative, in this case -6. The truth set is { }.

Notice that if both sides of the equation are squared, as shown below, the solution seems to be 28.

$$\sqrt{x + 8} = -6$$
$$(\sqrt{x + 8})^2 = (-6)^2$$
$$x + 8 = 36$$
$$x = 28$$

When 28 is checked in the original equation, it is not a solution of the original equation. Squaring both sides of an equation is not guaranteed to generate an equivalent equation.

The equation $\sqrt{8x} - \sqrt{4x + 1} = 1$ is solved by squaring each side and, after simplifying, again squaring each side.

$$\sqrt{8x} - \sqrt{4x + 1} = 1$$
$$(\sqrt{8x} - \sqrt{4x + 1})^2 = (1)^2$$
$$8x - 2\sqrt{8x(4x + 1)} + 4x + 1 = 1$$
$$12x - 2\sqrt{8x(4x + 1)} + 1 = 1$$
$$12x - 2\sqrt{8x(4x + 1)} = 0$$
$$-2\sqrt{8x(4x + 1)} = -12x$$
$$\sqrt{8x(4x + 1)} = 6x$$
$$(\sqrt{8x(4x + 1)})^2 = (6x)^2$$
$$32x^2 + 8x = 36x^2$$
$$-4x^2 + 8x = 0$$
$$-4x(x - 2) = 0$$
$$x = 0 \text{ or } x = 2$$

Both roots, 0 and 2, must be checked in the original equation.

$\sqrt{0} - \sqrt{1} = 1$ is false. $\sqrt{16} - \sqrt{9} = 1$ is true.

The truth set of $\sqrt{8x} - \sqrt{4x + 1} = 1$ is {2}.

Interval Notation

The truth set for the inequality x > 5 is graphed on the number line below.

An open dot (º) is used to mark the point x = 5 and means that 5 is not included in the truth set. The portion of the number line to the right of x = 5 is darkened to indicate each number on that half line is in the truth set. The portion of the number line to the left of x = 5 is not darkened and that half line is not included in the truth set.

The **interval notation** method for showing the truth set of x > 5 is (5,∞). The parenthesis to the left of 5 means that 5 is the end point of a half line. The comma and **infinity symbol** (∞) indicate the half line extends to the right forever.

The truth set of x ≤ 3 is shown on the number line below. Notice that 3 is in this truth set and that is indicated by the closed dot (•) at x = 3.

The truth set of x ≤ 3 would be shown in interval notation as (-∞,3]. The square bracket to the right of 3 indicates 3 is included in the truth set. The symbol -∞ indicates that the numbers get smaller forever.

Focus on interval notation

The truth set of 5 < x ≤ 12 is that portion of the number line between 5 and 12 which excludes 5 and includes 12.

In interval notation the truth set of 5 < x ≤ 12 is (5,12].

The truth set of x < -2 or x > 3 is two half lines: one extends to the left from -2 and the other extends to the right from 3. The truth set is the union of two half lines.

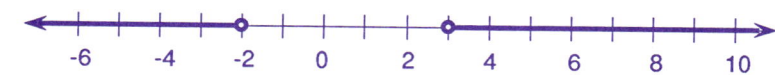

In interval notation the truth set of x < -2 or x > 3 is (-∞,-2) ∪ (3,∞).

Quadratic Inequalities

The truth set of $x^2 + 7x - 18 > 0$ is found by the following process:
1. Change the inequality to an equation and find its roots.
$$x^2 + 7x - 18 = 0$$
$$(x + 9)(x - 2) = 0$$
$$x = -9 \text{ or } x = 2$$
2. Plot the two roots on the number line.

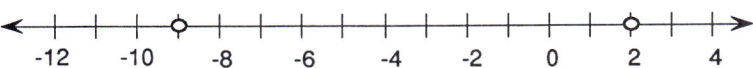

3. The number line is now divided into five parts.
   a. Two of the parts are the points: $x = -9$ and $x = 2$
   b. Three of the parts are portions of the line:
      i. The portion to the left of $x = -9$: $(-\infty, -9)$
      ii. The portion between $x = -9$ and $x = 2$: $(-9, 2)$
      iii. The portion to the right of $x = 2$: $(2, \infty)$
4. Test a number from each of these five parts to see if it makes the original inequality, $x^2 + 7x - 18 > 0$, true.
   a. When $x = -9$ or $x = 2$ they make the inequality false.
   b. Any number chosen from $(-\infty, -9)$ makes the inequality true.
   c. Any number chosen from $(-9, 2)$ makes the inequality false.
   d. Any number chosen from $(2, \infty)$ makes the inequality true.
5. The truth set of $x^2 + 7x - 18 > 0$ is the union of those parts which make it true. Therefore, the truth set is $(-\infty, -9) \cup (2, \infty)$.

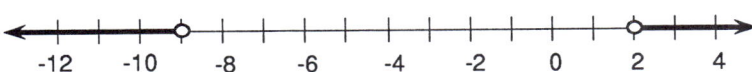

Focus on solving a quadratic inequality

To solve $x^2 - 5x - 3 \leq 0$,

1. Solve $x^2 - 5x - 3 = 0$. The quadratic formula gives the two roots $\frac{5 \pm \sqrt{37}}{2}$.

2. Plot the two roots on the number line. Use the table of square roots to approximate $\sqrt{37}$ and then estimate the sizes of the two roots.

3. Test $x = \frac{5 + \sqrt{37}}{2}$, $x = \frac{5 - \sqrt{37}}{2}$, and a number from each interval to determine where the original inequality is true.

4. The truth set is the union of the two roots and the segment between them which can be shown simply as $\left[\frac{5 - \sqrt{37}}{2}, \frac{5 + \sqrt{37}}{2}\right]$.

Unit 6 Exercise

Part A: Answers for all Part A problems are at the back of the book.

1. Solve $x^2 - 8x + 15 = 0$ by first factoring $x^2 - 8x + 15$.
2. Solve $x^2 + 8x = 0$ by first factoring $x^2 + 8x$.
3. Solve $-4x = -x^2 - 4$ by first writing the equation in the standard form $ax^2 + bx + c = 0$.
4. Solve $(x - 7)^2 = 9$ using the two linear equations $x - 7 = 3$ and $x - 7 = -3$.
5. Solve $(x - 4)^2 = 11$ using the two linear equations $x - 4 = \sqrt{11}$ and $x - 4 = -\sqrt{11}$.
6. Solve $x^2 + 8x + 16 = 25$ by first factoring $x^2 + 8x + 16$.
7. Solve $x^2 - 4x + 6 = 0$ using the quadratic formula.
8. Use interval notation to write the truth set of $x < 7$.
9. Use interval notation to write the truth set of $-3 \leq x < 8$.
10. Solve $\sqrt{x + 8} = 5$ by first squaring each side of the equation.
11. Are the equations $\sqrt{x + 2} = -7$ and $(\sqrt{x + 2})^2 = (-7)^2$ equivalent?
12. Square both sides of $\sqrt{x + 12} - \sqrt{x - 3} = 1$. After simplifying, what equation is obtained?
13. Find the root of $\sqrt{(x + 12)(x - 3)} = x + 4$ by first squaring both sides of the equation.
14. Use interval notation to write the truth set of $x^2 - 3x - 4 > 0$.
15. Use interval notation to write the truth set of $2x^2 - x - 3 \leq 0$.

Part B: Answers for odd-numbered problems of Part B are at the back of the book.

Find truth sets.

1. $x^2 - x - 12 = 0$
2. $(x + 8)^2 = 36$
3. $x^2 - 2x - 10 = 0$
4. $x^2 + 3x > 0$
5. $x^2 + 6x + 9 = 64$
6. $x^2 - 16 = 0$
7. $5x^2 - 14x - 3 \leq 0$
8. $x^2 - 7x - 6 = 0$
9. $x^2 - 3x + 5 = 0$
10. $x^2 - 5x + 3 = 0$

11. $x^2 + 3x + 7 \geq 0$
12. $x^2 + x + 1 = 0$
13. $x^2 - 7x + 5 = 0$
14. $x^2 - 5 = 11$
15. $3x^2 - 5x + 1 = 0$
16. $5x^2 - 6x = 1$
17. $x^2 - 4x - 5 < 0$
18. $x^2 + 14 = -15x$
19. $x^2 > -2x - 5$
20. $x^2 - 3x - 10 = 0$
21. $x^2 - 3x + 2 = 0$
22. $x^2 - 3x + 5 = 0$
23. $x^2 - 49 = 0$
24. $2x^2 - x - 4 = 0$
25. $\sqrt{x - 5} = 9$
26. $\sqrt{3x + 2} = 5$
27. $9 + \sqrt{x - 4} = 11$
28. $6 - \sqrt{x + 2} = 10$
29. $\sqrt{4x + 1} - \sqrt{3x - 2} = 1$
30. $\sqrt{2x + 1} - \sqrt{x + 1} = 2$
31. $x^2 + 9 = 7x$
32. $x^2 - 7x \leq 0$

Part C: No answers are given for these problems. However, each is accompanied by an ordered pair «C,U» showing the chapter and unit in which it was taught.

1. $3\sqrt{32} - 8 + \sqrt{72} + 2\sqrt{16}$ «2,2»
2. $\dfrac{-5\sqrt{24}}{\sqrt{50}}$ «2,2»
3. $\sqrt[3]{243}$ «2,3»
4. Decimals are rational numbers if they are either terminating or _____. «2,4»
5. Is $\dfrac{5}{7}$ a terminating decimal? «2,4»

For problems 6-10 find the truth set.

6. $7x + \sqrt{3} = 4x - \sqrt{5}$ «2,5»
7. $4(2x - \sqrt{7}) = 3$ «2,5»
8. $\dfrac{2x}{5} - \sqrt{7} = 1$ «2,5»
9. $x\sqrt{6} - 2 < 2x - 4$ «2,5»
10. $x + \sqrt{3} \geq 4x - 3\sqrt{3}$ «2,5»

Unit 7: FRACTIONAL EQUATIONS AND INEQUALITIES

Fractional Equations Equivalent to Linear Equations

To solve $\frac{5x}{6} - \frac{3}{4} = \frac{7}{8}$ first multiply both members of the equation by the least common multiple of 6, 4, and 8, which is 24. This is done as follows:

$$\frac{5x}{6} - \frac{3}{4} = \frac{7}{8}$$

$$24\left(\frac{5x}{6} - \frac{3}{4}\right) = 24 \cdot \frac{7}{8}$$

$$20x - 18 = 21$$

$$20x = 39$$

$$x = \frac{39}{20}$$

Focus on solving a fractional equation

The Multiplication Property of Equality states that both members of an equation may be multiplied by any nonzero real number to produce an equivalent equation. Both members of $\frac{5}{x} - \frac{3}{4} = \frac{5}{12}$ should be multiplied by 12x, but this can only be done with the assumption that $12x \neq 0$.

$$\frac{5}{x} - \frac{3}{4} = \frac{5}{12}$$

$$12x\left(\frac{5}{x} - \frac{3}{4}\right) = 12x \cdot \frac{5}{12} \text{ and } 12x \neq 0$$

$$60 - 9x = 5x$$

$$60 = 14x$$

$$\frac{30}{7} = x$$

The value $\frac{30}{7}$ does not contradict the assumption that $12x \neq 0$. Therefore, $\frac{30}{7}$ is acceptable in the truth set and $\left\{\frac{30}{7}\right\}$ is the truth set of $\frac{5}{x} - \frac{3}{4} = \frac{5}{12}$.

Solving A Fractional Inequality

The first step in solving $\frac{4}{3x} - \frac{3}{2} > \frac{5}{x}$ is to multiply both members of the equation by 6x, but this presents a problem. If $6x > 0$ then the new inequality will have the same symbol of inequality. But, if $6x < 0$, a new inequality will have the reverse symbol of inequality.

The complete solution requires two cases:

1. Suppose $6x > 0$. This means we can only accept solutions in the interval $(0, \infty)$. Under this assumption, when both members of the inequality are multiplied by 6x the symbol of inequality will remain unchanged.

$$\frac{4}{3x} - \frac{3}{2} > \frac{5}{x} \text{ becomes } 6x\left(\frac{4}{3x} - \frac{3}{2}\right) > 6x \cdot \frac{5}{x}$$
$$8 - 9x > 30$$
$$-9x > 22$$
$$x < \frac{-22}{9}$$

   Case 1 gives us the empty set because none of the solutions of $x < \frac{-22}{9}$ are in the interval $(0, \infty)$.

2. Suppose $6x < 0$. This means we can only accept solutions in the interval $(-\infty, 0)$. Under this assumption, when both members of the inequality are multiplied by 6x the symbol of inequality will be reversed.

$$\frac{4}{3x} - \frac{3}{2} > \frac{5}{x} \text{ becomes } 6x\left(\frac{4}{3x} - \frac{3}{2}\right) < 6x \cdot \frac{5}{x}$$
$$8 - 9x < 30$$
$$-9x < 22$$
$$x > \frac{-22}{9}$$

   The numbers in the interval $(-\infty, 0)$, which make $x > \frac{-22}{9}$ true are in the interval $\left(\frac{-22}{9}, 0\right)$.

The final truth set for the original inequality is the union of the two cases.

$$\emptyset \cup \left(\frac{-22}{9}, 0\right) = \left(\frac{-22}{9}, 0\right) \text{ is the truth set of } \frac{4}{3x} - \frac{3}{2} > \frac{5}{x}.$$

Note: $\emptyset$ is the empty set, { }.

Focus on solving a linear inequality

To solve $\frac{4}{x-5} \leq 3$, both sides of the inequality need to be multiplied by x – 5. Again, this requires two cases.

1. Suppose x – 5 > 0. This means we can only accept solutions in the interval (5,∞). Under this assumption, when both members of the inequality are multiplied by (x – 5) the symbol of inequality will be unchanged.

$$\frac{4}{x-5} \leq 3 \text{ becomes } (x-5)\left(\frac{4}{x-5}\right) \leq 3(x-5)$$
$$4 \leq 3x - 15$$
$$19 \leq 3x$$
$$\frac{19}{3} \leq x$$

Every solution of $x \geq \frac{19}{3}$ is in the interval (5,∞), so Case 1 gives $\left(\frac{19}{3},\infty\right)$.

2. Suppose x – 5 < 0. This means we can only accept solutions in the interval (-∞,5). Under this assumption, when both members of the inequality are multiplied by (x – 5) the symbol of inequality will be reversed.

$$\frac{4}{x-5} \leq 3 \text{ becomes } (x-5)\left(\frac{4}{x-5}\right) \geq 3(x-5)$$
$$4 \geq 3x - 15$$
$$19 \geq 3x$$
$$\frac{19}{3} \geq x$$

Since $\frac{19}{3} = 6\frac{1}{3}$, the solutions of $x \leq \frac{19}{3}$ that are also in (-∞,5) are in (-∞,5).

The truth set of $\frac{4}{x-5} \leq 3$ is the union of the two cases: $\left(\frac{19}{3},\infty\right) \cup (-\infty,5)$

Fractional Equations Equivalent to Quadratic Equations

Some fractional equations are equivalent to quadratic equations. The first step in solving $\frac{x}{5} + \frac{1}{x} = \frac{14}{5x}$ is to multiply both sides of the equation by $5x$ and the result will be a quadratic equation. Again, since we are multiplying by $5x$ it must be assumed that $5x \neq 0$.

$$\frac{x}{5} + \frac{1}{x} = \frac{14}{5x}$$
$$5x\left(\frac{x}{5} + \frac{1}{x}\right) = 5x \cdot \frac{14}{5x} \text{ and } 5x \neq 0$$
$$x^2 + 5 = 14$$
$$x^2 - 9 = 0$$
$$(x-3)(x+3) = 0$$
$$x - 3 = 0 \text{ or } x + 3 = 0$$

The roots 3 and -3 do not contradict the requirement that $5x \neq 0$. Therefore, $\{3,-3\}$ is the truth set of $\frac{x}{5} + \frac{1}{x} = \frac{14}{5x}$.

To solve $\frac{x}{3} - \frac{5}{6} = \frac{4}{3x}$, first multiply both sides by $6x$ with the assumption that $6x \neq 0$.

$$\frac{x}{3} - \frac{5}{6} = \frac{4}{3x}$$
$$6x\left(\frac{x}{3} - \frac{5}{6}\right) = 6x \cdot \frac{4}{3x} \text{ and } 6x \neq 0$$
$$2x^2 - 5x = 8$$
$$2x^2 - 5x - 8 = 0$$

The equation cannot be solved by factoring, but the quadratic formula gives the roots $\frac{5 \pm \sqrt{89}}{4}$. These roots do not contradict the requirement that $6x \neq 0$.

Equations with Polynomial Denominators

To solve $\frac{5}{x^2 + 3x + 2} - \frac{1}{x + 2} = \frac{1}{x + 1}$ each term must be multiplied by the LCM of the denominators with the assumption that the LCM is not zero.

$$\frac{5}{x^2 + 3x + 2} - \frac{1}{x + 2} = \frac{1}{x + 1}$$

$$\frac{5}{(x + 1)(x + 2)} - \frac{1}{x + 2} = \frac{1}{x + 1} \quad \text{the LCM is } (x + 2)(x + 1)$$

$$(x + 2)(x + 1) \cdot \left[\frac{5}{(x + 1)(x + 2)} - \frac{1}{x + 2}\right] = \frac{1}{x + 1} \cdot (x + 2)(x + 1)$$

The preceding line is based on the assumption $(x + 2)(x + 1) \neq 0$

$$5 - 1 \cdot (x + 1) = x + 2$$
$$5 - x - 1 = x + 2$$
$$2 = 2x$$
$$1 = x$$

When $x = 1$, the assumption is not contradicted. So $\{1\}$ is the truth set of $\frac{5}{x^2 + 3x + 2} - \frac{1}{x + 2} = \frac{1}{x + 1}$.

To solve $\frac{x + 3}{x + 1} + \frac{6}{x^2 - x - 2} = \frac{4}{x - 2}$ the following steps are used.

$$\frac{x + 3}{x + 1} + \frac{6}{x^2 - x - 2} = \frac{4}{x - 2}$$

$$\frac{x + 3}{x + 1} + \frac{6}{(x + 1)(x - 2)} = \frac{4}{x - 2}$$

$$(x + 1)(x - 2)\left[\frac{x + 3}{x + 1} + \frac{6}{(x + 1)(x - 2)}\right] = \frac{4}{x - 2} \cdot (x + 1)(x - 2)$$

The preceding line is based on the assumption $(x + 1)(x - 2) \neq 0$

$$(x - 2)(x + 3) + 6 = 4(x + 1)$$
$$x^2 + x - 6 + 6 = 4x + 4$$
$$x^2 - 3x - 4 = 0$$
$$(x + 1)(x - 4) = 0$$
$$x = -1 \text{ or } x = 4$$

Focus on multiplying by a quadratic LCM

The root -1 is unacceptable! It makes $(x + 1)(x - 2)$ equal to zero and is therefore an **extraneous** root. This means that the quadratic equation generated by multiplying both sides of the original equation by $(x + 1)(x - 2)$ was not equivalent to the original fractional equation. The only acceptable root for the original equation is 4 and $\{4\}$ is the truth set.

Inequalities Equivalent to Quadratic Inequalities

The first step in solving $\frac{3x}{x-3} > x + 8$ is to multiply both sides of the inequality by $(x - 3)$, but this requires two cases because $(x - 3)$ may be either positive or negative.

Case 1: Suppose $(x - 3) > 0$. This means $x > 3$.

$\frac{3x}{x-3} > x + 8$ becomes $(x - 3) \cdot \frac{3x}{x-3} > (x - 3)(x + 8)$
$$3x > x^2 + 5x - 24$$
$$0 > x^2 + 2x - 24$$
$$0 > (x + 6)(x - 4)$$

From our work in unit 6 of this chapter the truth set of this quadratic inequality, in interval notation, is (-6,4). With the additional requirement that $x > 3$, Case 1 gives the truth set (3,4).

Case 2: Suppose $(x - 3) < 0$. This means $x < 3$.

$\frac{3x}{x-3} > x + 8$ becomes $(x - 3) \cdot \frac{3x}{x-3} < (x - 3)(x + 8)$
$$3x < x^2 + 5x - 24$$
$$0 < x^2 + 2x - 24$$
$$0 < (x + 6)(x - 4)$$

From our work in unit 6 of this chapter the truth set of this quadratic inequality, in interval notation, is $(-\infty,-6) \cup (4,\infty)$. With the additional requirement that $x < 3$, Case 2 gives the truth set $(-\infty,-6)$.

The truth set of the original inequality is the union of the results from the two cases which is $(3,4) \cup (-\infty,-6)$.

Focus on solving a fractional inequality

The inequality $\frac{14}{x+5} \leq x$ is solved with two cases:

Case 1: When $(x + 5) > 0$ then $x > -5$.

$\frac{14}{x+5} \leq x$ becomes $(x + 5) \cdot \frac{14}{x+5} \leq x(x + 5)$
$$14 \leq x^2 + 5x$$
$$0 \leq x^2 + 5x - 14$$
$$0 \leq (x + 7)(x - 2)$$

The truth set of this quadratic inequality is $(-\infty,-7] \cup [2,\infty)$. With the additional requirement that $x > -5$, Case 1 gives the truth set $[2,\infty)$.

Focus on solving a fractional inequality

Case 2: When $(x + 5) < 0$ then $x < -5$.

$\frac{14}{x + 5} \leq x$ becomes $(x + 5) \cdot \frac{14}{x + 5} \geq x(x + 5)$
$14 \geq x^2 + 5x$
$0 \geq x^2 + 5x - 14$
$0 \geq (x + 7)(x - 2)$

The truth set of this quadratic inequality is $[-7, 2]$. With the additional requirement that $x < -5$, Case 2 gives the truth set $[-7, -5)$.

The truth set of the original inequality is the union of the results from the two cases which is $[-7, -5) \cup [2, \infty)$.

Unit 7 Exercise

Part A: Answers for all Part A problems are at the back of the book.

1. Find an equivalent equation to $\frac{5x}{8} - \frac{3}{10} = \frac{2}{5}$ by multiplying both sides of the equation by 40, the least common multiple of 8, 10, and 5.

2. Find the least positive number that contains all the factors of 3, 5, and 6.

3. Multiply both members of $\frac{2x}{3} - \frac{4}{5} = \frac{5}{6}$ by 30, which is the least common multiple of the denominators. What equivalent equation is obtained?

4. Solve $\frac{5x}{2} + \frac{2}{3} = \frac{1}{6}$ by first multiplying both members by the least common multiple of the denominator.

5. Solve $\frac{6}{5x} - \frac{3}{4} > \frac{4}{2x}$. It is necessary to consider two cases: $20x > 0$ and $20x < 0$.

6. Solve $\frac{5}{4x} - \frac{3}{7} \leq \frac{5}{x}$ by first multiplying both members of the equation by $28x$. Use two cases: $28x > 0$ and $28x < 0$.

7. Solve $\frac{3}{8x} - \frac{1}{4x} = \frac{x}{8}$ by multiplying both sides by $8x$.

8. Solve $\frac{2}{x(x - 2)} - \frac{1}{x} = \frac{2}{x - 2}$ by first multiplying both sides of the equation by $x(x - 2)$.

9. Solve $\frac{2}{x + 3} + \frac{1}{x^2 - 9} = \frac{7}{x - 3}$ by first multiplying both sides of the equation by $(x + 3)(x - 3)$.

10. Solve $\frac{1}{x} - \frac{x}{4} > \frac{-5}{4x}$. Consider two cases: $4x > 0$ and $4x < 0$.

11. Solve $\frac{9}{x + 3} < x - 5$. Consider two cases: $x + 3 > 0$ and $x + 3 < 0$.

12. Solve $\frac{18}{x - 5} \geq x + 2$. Consider two cases: $x - 5 > 0$ and $x - 5 < 0$.

Part B: Answers for odd-numbered problems of Part B are at the back of the book.

Solve.

1. $\frac{4x}{5} - \frac{3}{10} = \frac{1}{2}$
2. $\frac{3x}{8} - \frac{5}{6} = \frac{1}{3}$
3. $\frac{1x}{3} + \frac{3}{4} = \frac{1}{6}$
4. $\frac{1x}{4} + \frac{2}{3} = \frac{5x}{6} - \frac{1}{2}$
5. $\frac{3x}{4} - \frac{2}{5} = \frac{1}{2}$
6. $\frac{1x}{5} + \frac{3}{10} > \frac{1}{6}$
7. $\frac{4x}{5} - \frac{3}{10} \leq \frac{1x}{2} - \frac{3}{4}$
8. $\frac{3x}{8} + \frac{4}{5} \geq \frac{3}{10}$
9. $\frac{3x}{5} - \frac{19}{25} < 1$
10. $\frac{5x}{6} - \frac{2}{9} = \frac{3x}{2} + 2$
11. $\frac{3x}{4} + \frac{5}{12} = \frac{1}{3}$
12. $2x - \frac{2}{9} = \frac{5x}{6} - \frac{7}{3}$
13. $\frac{3x}{4} + \frac{8}{5} = \frac{1x}{5} - \frac{7}{4}$
14. $\frac{5}{3x} - \frac{7}{12} = \frac{1}{4x}$
15. $\frac{5}{6x} - \frac{3}{10} = \frac{2}{5x}$
16. $\frac{3}{4x} + \frac{8}{5} = \frac{1}{5x} - \frac{7}{4}$
17. $\frac{3}{x} - 1 \leq \frac{7}{x}$
18. $\frac{3}{2x} + \frac{1}{2} > \frac{-2}{5}$
19. $\frac{5}{6x} - \frac{2}{9} \leq \frac{3}{2x} + 2$
20. $\frac{2}{5} + \frac{3}{x} < \frac{-2}{x}$
21. $\frac{3}{x} + \frac{8}{5} \geq \frac{1}{2}$
22. $\frac{2}{x} + \frac{2}{5} < \frac{1}{3}$
23. $\frac{2x}{3} - \frac{1}{2} > \frac{1x}{5} + \frac{3}{10}$
24. $\frac{5}{x} < -3$
25. $\frac{2}{3x} + \frac{1}{6} \geq \frac{3}{8} - \frac{1}{4x}$
26. $\frac{-3}{x^2 - 2x} - \frac{1}{x - 2} = 0$
27. $\frac{3}{x - 2} - \frac{2}{(x - 2)(x - 4)} = \frac{1}{x - 4}$
28. $\frac{-2}{x - 7} + \frac{3}{x^2 + x - 56} = \frac{1}{x + 8}$
29. $\frac{x - 1}{x + 1} = \frac{6}{x + 4}$
30. $\frac{x + 2}{x - 3} = \frac{5}{x - 3}$
31. $\frac{2x}{x + 9} = \frac{2}{x + 1}$
32. $\frac{2x}{x + 1} + \frac{7}{x - 2} + \frac{12x}{x^2 - x - 2} = 0$
33. $\frac{x - 1}{x + 2} = \frac{4}{x + 7}$
34. $\frac{2}{x + 1} + \frac{3}{x + 2} = \frac{17}{x^2 + 3x + 2}$
35. $\frac{x}{x - 1} = \frac{-3}{x + 3} + \frac{2x - 6}{x^2 + 2x - 3}$
36. $\frac{3}{x - 2} > 5$
37. $\frac{1}{x + 6} < -2$
38. $\frac{5 - 4x}{x - 6} \geq 3x$
39. $\frac{5x - 7}{x - 1} \leq 1$
40. $\frac{x - 2}{x - 5} \leq \frac{5}{2}$

Part C: No answers are given for these problems. However, each is accompanied by an ordered pair «C,U» showing the chapter and unit in which it was taught.

Simplify problems 1-3.

1. $\frac{6\sqrt{24}}{\sqrt{5}}$ «2,2»
2. $\frac{-2\sqrt{98}}{3\sqrt{75}}$ «2,2»
3. $\sqrt[3]{-54}$ «2,3»
4. What decimal fraction is equal to $\frac{7}{8}$?   «2,4»

Solve problems 5-10.

5. $\frac{4x}{3} - \sqrt{5} = 8$ «2,5»
6. $x^2 - 6 = 7x$ «2,6»
7. $3x^2 - x - 3 = 0$ «2,6»
8. $9 + \sqrt{x - 4} = 11$ «2,6»
9. $\sqrt{2x - 1} - \sqrt{x - 1} = 1$ «2,6»
10. $3x^2 - 2x - 4 \leq 0$ «2,6»

Unit 8: APPLICATIONS: MIXTURES AND INGREDIENTS

Representing Percents of Quantities

This unit is concerned with situations in which mixtures containing several ingredients have their relative strength changed by either increasing or decreasing some of the ingredients. The answers to the problems depend on finding percents of quantities and this first section deals with finding and representing percents of quantities.

To find a percent of a number, multiply the decimal equivalent of the percent by the number. Following are several examples showing percent expressions and their equivalent number expressions.

3% of 80	.03 • 80 or 2.40
50% of 20	.50 • 20 or 10
11% of 42	.11 • 42 or 4.62

Percents of numbers can be shown when the number is represented by a variable. If x represents a number, 5% of the number is shown by the open expression .05x. Similarly, .01x represents 1% of x and .15y represents 15% of the number represented by y. If the open expression x + 5 represents a number, 6% of the number is shown by .06(x + 5).

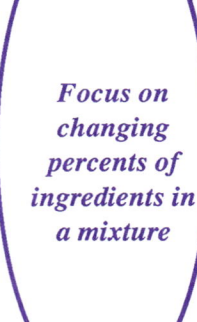

Focus on changing percents of ingredients in a mixture

If a bottle contains 10 gallons of a 20% alcohol mixture, it contains 2 gallons of alcohol (.20 • 10) and 8 gallons of other ingredients (.80 • 10). 20% of the 10 gallons is alcohol and 80% of the 10 gallons is other ingredients.

$$.20 \cdot 10 = 2 \qquad .80 \cdot 10 = 8$$

If 2 gallons of alcohol are added to the mixture, then there are 4 gallons of alcohol and 8 gallons of other ingredients, or a total of 12 gallons of the new mixture. $\frac{4}{12}$ or $33\frac{1}{3}\%$ of the new mixture is alcohol.

There are two important number relations to notice about this problem. The 2 gallons of alcohol in the original mixture were increased by 2 gallons of pure alcohol to give a total of 4 gallons in the new mixture. Also, the 10 gallons of the original mixture added to the 2 gallons of pure alcohol means that there are 12 gallons in the new mixture.

Solving an Ingredient Problem

The following problem is solved in a five-step process.

Problem: How much alcohol should be added to 20 gallons of a 10% alcohol mixture to increase the percent of alcohol in the new mixture to 28%?

Solution

1. First the problem is carefully read and analyzed. The situation might be rephrased as: The alcohol in the original mixture plus the alcohol added is the alcohol in the new mixture.

2. The quantities in the problem are listed.
    gallons of alcohol in the 10% mixture
    gallons of alcohol added
    gallons of new mixture
    gallons of alcohol in the new mixture

3. Numbers or open expressions are assigned to quantities in step 2.
    $.10 \cdot 20$ or 2 gallons of alcohol are in the 10% mixture
    x is gallons of alcohol added to the 10% mixture
    $x + 20$ is gallons in the new mixture
    $.28(x + 20)$ is gallons of alcohol in the new mixture

4. An equation is written from the number relationships in the problem.

The alcohol in the 10% mixture	plus	alcohol added	equals	alcohol in the new mixture
2	+	x	=	.28(x + 20)

    $$2 + x = .28(x + 20)$$
    $$2 + x = .28x + 5.6$$
    $$.72x = 3.6$$
    $$x = 5$$

5. The problem is checked by verifying that when 5 gallons of alcohol are added to 20 gallons of a 10% mixture, a 28% mixture is obtained.

The preceding example showed how the percent of ingredients in a mixture can be increased by adding pure ingredients. The next example shows how the percent of an ingredient in a mixture may be increased by decreasing another ingredient in the mixture. It is solved in the five-step process.

Problem: How much water would have to be evaporated from 500 pounds of milk containing 13% solids to make condensed milk containing 60% solids?

Solution

1. Since water is the only ingredient evaporated from the milk, the amount of solids in the 13% milk will be equal to the amount of solids in the 60% milk. This fact will be used to write an equation for the problem.

2. pounds of solids in the 13% milk
   pounds of water evaporated from the 13% milk
   pounds of 60% milk
   pounds of solids in the 60% milk

3. 65 pounds of solids are in the 13% milk (.13 • 500 = 65)
   x is the pounds of water evaporated from the 13% milk
   500 − x is the pounds of 60% milk
   .60(500 − x) is the pounds of solids in the 60% milk

4. The solids in the 13% milk   equal   the solids in the 60% milk.

   $$65 \quad = \quad .60(500 - x)$$
   $$65 = .60(500 - x)$$
   $$65 = 300 - .60x$$
   $$-235 = -.60x$$
   $$391.67 = x$$

5. The solution 391.67 is rounded off to the nearest hundredth. Approximately 392 pounds of water should be evaporated from the 13% milk to raise the percentage of solids to 60%. This result must be checked against the facts in the original problem.

Focus on solving an ingredient problem

Changing Mixtures by Adding Non-Pure Ingredients

In the previous problems of this unit the percentage of the ingredients in the mixtures was changed by adding a pure ingredient to the mixture. Another type of problem requires the percentage of ingredients in a mixture to be changed by adding another mixture to the original mixture.

Problem: How much 50% alcohol should be added to 8 gallons of 20% alcohol to change the percentage of alcohol to 40%?

Solution

1. Two statements can be made about the previous problem. The gallons of 50% alcohol added to the gallons of 20% alcohol will give the gallons of the 40% alcohol. In addition, the alcohol in the 50% mixture added to the alcohol in the 20% mixture gives the amount of alcohol in the 40% mixture. The last statement will provide an equation for the problem.

2. gallons of 50% mixture
   gallons of alcohol in the 50% mixture
   gallons of alcohol in the 20% mixture
   gallons of 40% mixture
   gallons of alcohol in the 40% mixture

3. x is the gallons of 50% mixture
   $.50x$ is the gallons of alcohol in the 50% mixture
   1.6 is the gallons of alcohol in the 8 gallons ($.20 \cdot 8 = 1.6$)
   $x + 8$ is the gallons of 40% mixture
   $.40(x + 8)$ is the gallons of alcohol in the 40% mixture

4. | The alcohol in the 20% mixture | added to | the alcohol in the 50% mixture | equals | the alcohol in the 40% mixture |
   |---|---|---|---|---|
   | 1.6 | + | .50x | = | .40(x + 8) |

   $$1.6 + .50x = .40x + 3.2$$
   $$.10x = 1.6$$
   $$x = 16$$

5. It needs to be confirmed that 16 gallons of 50% mixture meets the conditions of the problem.

Focus on solving a mixture problem

Problem: A chemist needs to increase the percent of salt in a mixture by removing some of the mixture and replacing it with a 90% salt mixture. How much of a 100 pound, 40% salt mixture needs to be removed and replaced by an equal amount of a 90% salt mixture to raise the percent of salt to 45%?

Solution

1. An equation can be written that shows that the salt in the original mixture, less the salt in the mixture that is removed, added to the salt in the 90% mixture, equals the salt in the 45% mixture.

2. pounds of salt in the 40% mixture
   pounds of 40% mixture that is removed
   pounds of salt in the 40% mixture that is removed
   pounds of salt in the 90% mixture that is added
   pounds of salt in the 45% mixture

3. 40 pounds of salt are in the 40% mixture (.40 • 100 = 40)
   x is the pounds of 40% mixture removed
   .40x is the pounds of salt in the 40% mixture removed
   .90x is the pounds of salt added in the 90% mixture
   45 pounds is the pounds of salt in the 100 pounds of 45% mixture

4. $\left(\begin{array}{c}\text{The salt}\\ \text{in the 40\%}\\ \text{mixture}\end{array}\right)$ minus $\left(\begin{array}{c}\text{the salt}\\ \text{removed}\end{array}\right)$ plus $\left(\begin{array}{c}\text{the salt}\\ \text{added}\end{array}\right)$ equals $\left(\begin{array}{c}\text{the salt}\\ \text{in the 45\%}\\ \text{mixture}\end{array}\right)$

   $\qquad 40 \qquad - \qquad .40x \qquad + \qquad .90x \qquad = \qquad 45$

   $$40 - .40x + .90x = 45$$
   $$.50x = 5$$
   $$x = 10$$

5. 10 pounds of 40% mixture should be removed and replaced with the 90% mixture to produce a 45% mixture. This result must be checked in the original problem.

Unit 8 Exercise

Part A: Answers for all Part A problems are at the back of the book.

1. Find 7% of 40 by multiplying .07 by 40.
2. Find 23% of 200.
3. If x represents a number, write an open expression for 10% of the number.
4. If y + 10 represents a number, write an open expression that represents 2% of the number.
5. Write an open expression to represent 5% of a number when x + 3 represents the number.
6. Mr. Lepper has to add salt to 500 pounds of a 40% brine mixture to increase the percent of salt to 70%. How much should he add?
7. How many pounds of active ingredients should be removed from 100 pounds of a mixture containing 40% active ingredients to produce a mixture containing 25% active ingredients?
8. An orange juice manufacturer has 1000 pounds of orange juice consisting of 20% solids. How many pounds of water should be evaporated from this to make condensed juice containing 60% solids? Round the answer to the nearest hundredth of a pound.
9. Miss Yancey has to increase the percent of acid in 80 gallons of a 50% acetic acid mixture by adding some 70% acetic acid mixture. How much should she add if she is to increase the acid percent to 60?
10. A milk plant wants to increase the amount of butterfat in 400 pounds of milk from 3% to 4% by adding some light cream with 10% butterfat. How much would they add?
11. A radiator contains 8 gallons of 60% antifreeze. How much mixture should be removed and replaced with an equal amount of 90% antifreeze to raise the percent of antifreeze to 70%?
12. Ms. Bulnev is instructed to remove part of a 12 gallon mixture of 4% sodium hydroxide and replace it with an equal amount of a 10% mixture of sodium hydroxide in order to increase the percent of sodium hydroxide to 5%. How much should she remove and replace with 10% mixture?

Part B: Answers for odd-numbered problems of Part B are at the back of the book.

1. How much pure acid would have to be added to 16 grams of a 25% mixture of acid to produce a 40% mixture of acid?
2. How many cubic centimeters of pure alcohol would have to be added to 25 cubic centimeters of 16% alcohol mixture to increase the alcohol percent to 58%?
3. How much water would have to be removed from 50 gallons of a mixture containing 80% water to decrease the water content to 50%?
4. A condensed milk plant wants to evaporate water from 1000 pounds of milk containing 13% active ingredients to increase the amount of active ingredients to 30%. How much water should be evaporated from the milk?

5. An engineer in an orange juice plant is to remove water from 2000 pounds of orange juice containing 11% solids to produce concentrated juice containing 20% solids. How much water should he remove from the juice?

6. A cook has to add water to 200 pounds of a lemon concentrate containing 40% active ingredients to make a drink containing 20% active ingredients. How much should he add?

7. How much of a brine mixture containing 70% salt should be added to 200 pounds of a 40% brine mixture to produce a mixture that contains 50% salt?

8. How many liters of a mixture containing 12% iodine should be added to 20 liters of a 2% iodine mixture to make a mixture containing 3% iodine?

9. Miss Eve desires to remove some of the contents of a 10 gallon bottle of medicine containing 20% active ingredients and replace it with an equal amount of medicine containing 50% active ingredients so that the resulting mixture will contain 40% active ingredients. How much should she replace?

10. Mr. Grud is to make 20 cubic centimeters of a 60% mixture of merthiolate by removing part of the contents of a 20 cubic centimeter bottle of 50% merthiolate and replacing it with an equal amount of 90% merthiolate. How much should he remove?

11. How much iodine should be added to 10 grams of a 30% mixture to increase the percent of iodine to 65%?

12. How much alcohol should be added to 20 gallons of a 30% mixture of alcohol to raise the percent of alcohol to 44%?

13. How much water should be evaporated from 1000 gallons of milk that is 87% water to reduce the percent of water to 50%?

14. How much water should be added to 100 liters mixture of 30% alcohol to lower the alcohol content to 25%?

15. How much antifreeze containing 80% active ingredients should be added to 6 gallons of antifreeze containing 20% active ingredients to raise the percent of active ingredients to 44%?

Part C: No answers are given for these problems. However, each is accompanied by an ordered pair «C,U» showing the chapter and unit in which it was taught.

1. Which of the following is not a real number?
$\sqrt{7}, -5, \frac{4}{0}, -\sqrt{10}, \frac{-19}{4}, \pi, \frac{0}{-5}, \sqrt{-25}$ «2,1»

2. Simplify $\frac{2\sqrt{32}}{-5\sqrt{48}}$ «2,2»

3. Simplify $3\sqrt{24} - \sqrt{81} + 5\sqrt{6} - \sqrt{12}$ «2,2»

4. What repeating decimal is equal to $\frac{3}{7}$? «2,4»

Solve problems 5-10.

5. $\frac{7x}{5} - \sqrt{3} = 6$ «2,5»

6. $\frac{3x}{4} - \frac{1}{5} = \frac{7}{10}$ «2,6»

7. $\frac{2}{3x} - \frac{3}{4} = \frac{1}{4x}$ «2,6»

8. $\frac{3}{x-4} - \frac{5}{2x-3} = \frac{13}{2x^2 - 11x + 12}$ «2,7»

9. $\frac{3}{x-4} < 5$ «2,7»

10. $\frac{x-4}{x-2} > x + 2$ «2,7»

Chapter 2 Test

«2,U» shows the unit in which this problem was studied in this chapter.

1. The Completeness Property states that every point on the number line represents a real number and every real number is represented by a point on the _____. «2,1»

2. If x(r² – 12) = 0, then x = 0 or _____. «2,1»

3. Simplify $\dfrac{-6\sqrt{5}}{\sqrt{6}}$ «2,2»

4. Multiply $2\sqrt{3} - 5$ by its conjugate. «2,2»

5. Simplify $\dfrac{\sqrt{3} - 2}{4\sqrt{3} + 1}$ «2,2»

6. Simplify $-2\sqrt[3]{375}$ «2,3»

7. Simplify $\sqrt[3]{64} - \sqrt[3]{54} + \sqrt[3]{250} - 4\sqrt[3]{216}$ «2,3»

8. Simplify $-5\sqrt[4]{80}$ «2,3»

Use the table of square roots to find the decimal approximations for problems 9-13.

9. $8 + \sqrt{83}$ «2,4»

10. $\dfrac{5 - \sqrt{17}}{3}$ «2,4»

11. Solve $9x + \sqrt{6} = 3x + \sqrt{5}$ «2,5»

12. Solve $\dfrac{3x}{4} - \sqrt{5} = 2\sqrt{5}$ «2,5»

13. Solve $\sqrt{3}x - 7 = \sqrt{2}x + \sqrt{5}$ «2,5»

Find truth sets for problems 14-20.

14. $x^2 + 2x - 15 = 0$ «2,6»

15. $x^2 - 11 = 7$ «2,6»

16. $5x^2 - 1 = 6x$ «2,6»

17. $x^2 + 8x = 4$ «2,6»

18. $\sqrt{x + 6} - 5 = -3$ «2,6»

19. $\dfrac{4x}{3} - \dfrac{1}{12} = \dfrac{5}{6}$ «2,7»

20. $\dfrac{x}{6} - \dfrac{7}{9} = \dfrac{x}{3} + 4$ «2,7»

Use interval notation to write truth sets for problems 21-23.

21. $x^2 + 4x - 5 > 0$ «2,6»

22. $x^2 - 9 \leq 0$ «2,6»

23. Solve $\dfrac{6}{x + 3} < x - 2$ «2,7»

24. How much water would have to be removed from 100 gallons of a mixture containing 60% water to decrease the water content to 40%? «2,7»

25. How much alcohol should be added to 40 gallons of 30% mixture to raise the percent of alcohol to 44%? «2,7»

3
System of Complex Numbers

Unit 1: IMAGINARY NUMBERS

The Number i

$\sqrt{5}$, $\sqrt{16}$, $\sqrt{73}$, and $\sqrt{913}$ are real numbers that represent points on the number line. $\sqrt{-1}$, $\sqrt{-6}$, $\sqrt{-13}$, and $4\sqrt{-7}$ are not real numbers because no real number can be multiplied by itself to give a negative product. Numbers such as $\sqrt{-1}$, $\sqrt{-6}$, $\sqrt{-13}$, and $4\sqrt{-7}$ are **imaginary** numbers. This unit is a study of imaginary numbers. Imaginary numbers are not real numbers. Therefore, they have no position on the real number line.

The square of any real number cannot be negative.
$$(5)^2 = 25$$
$$(-4)^2 = 16$$
$$0^2 = 0$$

i is an imaginary number. The letter i stands for the number that can be squared to give -1.

$$i^2 = -1$$

i is an imaginary number and $i^2 = -1$. i is not a real number, because the square of i is -1 and the square of any real number cannot be a negative number.

Focus on imaginary numbers

Any real number is represented by a point on the real number line. Real numbers have the property that when squared they cannot be negative. $\sqrt{5}$, $-\sqrt{13}$, and $\sqrt{14}$ are real numbers.

$$(\sqrt{5})^2 = 5 \qquad (-\sqrt{13})^2 = 13 \qquad (\sqrt{14})^2 = 14$$

The square root of a negative number is not a real number; it is an imaginary number. Imaginary numbers have the property that when squared they will be negative. $\sqrt{-7}$ and $-\sqrt{-6}$ are imaginary numbers.

$$(\sqrt{-7})^2 = (i\sqrt{7})^2 = i^2 \cdot (\sqrt{7})^2 = -1 \cdot 7 = -7$$
$$(-\sqrt{-6})^2 = (-1i\sqrt{6})^2 = (-1)^2 \cdot i^2 \cdot (\sqrt{6})^2 = 1 \cdot -1 \cdot 6 = -6$$

Multiplying Imaginary Numbers

The most important fact to remember for the multiplication of imaginary numbers is that $i^2 = -1$.

To multiply 3i and 2i, the following steps are used.

$$3i \cdot 2i = 6i^2 = 6 \cdot -1 = -6$$

Similarly, $-7i \cdot 5i = -35i^2 = -35 \cdot -1 = 35$

Focus on the multiplication of imaginary numbers

To multiply -5i and 3i, the following steps are used.
$$-5i \cdot 3i = -15i^2 = -15 \cdot -1 = 15$$

$i = 1i$, and $-i = -1i$. Consequently, the product of 4i and -i is:
$$4i \cdot -i = 4i \cdot -1i = -4i^2 = -4 \cdot -1 = 4$$

Similarly, $i \cdot -i = 1i \cdot -1i = -1 \cdot i^2 = -1 \cdot -1 = 1$

$i\sqrt{2} \cdot i\sqrt{2}$ is simplified as: $i\sqrt{2} \cdot i\sqrt{2} = i^2\sqrt{4} = -1 \cdot 2 = -2$

Similarly, $i\sqrt{7} \cdot i\sqrt{7} = i^2\sqrt{49} = -1 \cdot 7 = -7$

Square Roots of Real Numbers

The square root of positive 16 is 4, because 4 squared is 16.
$\sqrt{16} = 4$ and $4^2 = 16$

The square root of -3 is $i\sqrt{3}$, because $i\sqrt{3}$ squared is -3.
$\sqrt{-3} = i\sqrt{3}$ and $(i\sqrt{3})^2 = -3$

The square root of -16 is 4i, because 4i is the number that must be squared to equal -16.

$\sqrt{9} = 3$ and $3^2 = 9$
$\sqrt{-9} = 3i$ and $(3i)^2 = -9$
$\sqrt{-2} = i\sqrt{2}$ and $(i\sqrt{2})^2 = -2$
$\sqrt{-5} = i\sqrt{5}$ and $(i\sqrt{5})^2 = -5$

Focus on writing radical expressions as imaginary numbers

For any positive number x: $\sqrt{-x} = i\sqrt{x}$
$\sqrt{-19} = i\sqrt{19}$ and $\sqrt{-23} = i\sqrt{23}$

The radicand of $\sqrt{-26}$ is a negative number. $\sqrt{-26} = i\sqrt{26}$
The radicand of $\sqrt{-43}$ is a negative number. $\sqrt{-43} = i\sqrt{43}$

Whenever the number under the radical sign is negative, then the expression can be written by using the imaginary number i. $\sqrt{-11} = i\sqrt{11}$

Simplifying Imaginary Expressions

$\sqrt{-32}$ is simplified as follows.
$\sqrt{-32} = i\sqrt{32} = i\sqrt{16}\sqrt{2} = i \cdot 4\sqrt{2} = 4i\sqrt{2}$

To simplify $5\sqrt{-12}$, the following steps are used.
$5\sqrt{-12} = 5i\sqrt{12} = 5i\sqrt{4}\sqrt{3} = 5i \cdot 2\sqrt{3} = 10i\sqrt{3}$

Focus on using the Distributive Property with i numbers

The Distributive Property can be used with imaginary numbers.

$3i(4 + 2i) = 3i \cdot 4 + 3i \cdot 2i$
$= 12i + 6i^2$
$= 12i + 6 \cdot -1$
$= 12i - 6$

Simplifying Powers of i

i^3 is simplified as: $i^3 = i^2 \cdot i = -1 \cdot i = -i$

Similarly, $i^4 = i^2 \cdot i^2 = -1 \cdot -1 = 1$.

Notice that $i^2 = -1$ and $i^4 = 1$. Consequently, power expressions such as i^{15} are simplified by using i^4 and/or i^2 as factors as many times as necessary.

$$i^{15} = i^4 \cdot i^4 \cdot i^4 \cdot i^2 \cdot i = 1 \cdot 1 \cdot 1 \cdot -1 \cdot i = -i$$

Focus on simplifying power expressions

To simplify $3i^5 \cdot 4i^6$, the following steps are used.

$$\begin{aligned} 3i^5 \cdot 4i^6 &= 3 \cdot 4 \cdot i^5 \cdot i^6 \\ &= 12i^{11} \\ &= 12 \cdot i^4 \cdot i^4 \cdot i^2 \cdot i \\ &= 12 \cdot 1 \cdot 1 \cdot -1 \cdot i \\ &= -12i \end{aligned}$$

Simplifying Addition Expressions

The addition of $3i$ and $5i$ is similar to the addition of like terms.
$$3i + 5i = (3 + 5)i = 8i$$

To add $3 + 2i$ and $-4 + 5i$ the following steps are used.
$$\begin{aligned} (3 + 2i) + (-4 + 5i) &= 3 + 2i - 4 + 5i \\ &= (3 + -4) + (2i + 5i) \\ &= (3 + -4) + (2 + 5)i \\ &= -1 + 7i \end{aligned}$$

Focus on simplifying multiplication expressions

To multiply $(4 + 3i)(2 + 5i)$ use the same process followed in multiplying $(x + 7)$ and $(x - 3)$.

$$\begin{aligned} (4 + 3i)(2 + 5i) &= 4(2 + 5i) + 3i(2 + 5i) \\ &= 8 + 20i + 6i + 15i^2 \\ &= 8 + 26i + 15i^2 \\ &= 8 + 26i + 15 \cdot -1 \\ &= 8 + 26i - 15 \\ &= -7 + 26i \end{aligned}$$

System of Complex Numbers   147

Unit 1 Exercise

Part A: Answers for all Part A problems are at the back of the book.

1. Can the square of a real number be a negative number?
2. Since $i^2 = -1$, is i a real number?
3. i is an imaginary number. $i^2 =$ _____.
4. i is an imaginary number. What number can be squared to give -1?
5. $(3i)^2 = -9$. What number may be squared to give -25?
6. i is an imaginary number, and i^2 is a _____ number.
7. Since $(7i)^2 = -49$, $\sqrt{-49} =$ _____.
8. The square of any real number cannot be a _____ number.
9. The square of any real number except zero is _____. (positive, negative)
10. $\sqrt{-81} = i\sqrt{81} = 9i$ and $\sqrt{-25} = i\sqrt{25} =$ _____.
11. $\sqrt{4} = 2$ and $\sqrt{-4} = i\sqrt{4} =$ _____.
12. $\sqrt{-100} = 10i$ and $\sqrt{-64} =$ _____.
13. $\sqrt{-18} = i\sqrt{18} = i\sqrt{9}\sqrt{2} = 3i\sqrt{2}$ and $\sqrt{-27} =$ _____.
14. $-7i \cdot 2i$ is simplified by separately multiplying the coefficients and the i's.
    $-7i \cdot 2i = (-7 \cdot 2)(i \cdot i) = -14i^2 = -14 \cdot -1 = 14$.
    Simplify $5i \cdot 4i$.
15. i^7 is simplified using $i^4 = 1$ and $i^2 = -1$.
    $i^7 = i^4 \cdot i^2 \cdot i = 1 \cdot -1 \cdot i =$ _____.
16. i^{51} is simplified using $i^4 = 1$, $i^2 = -1$, and $1^{12} = 1$.
    $i^{51} = i^{48} \cdot i^2 \cdot i^1 = (i^4)^{12} \cdot i^2 \cdot i =$ _____.
17. The parentheses of $2i(3 + 5i)$ are removed by multiplying each term in the parentheses by 2i.
    $2i(3 + 5i) =$ _____.
18. The addition of $7 + 4i$ and $-9 - 7i$ is completed by adding the real numbers and separately adding the imaginary numbers.
    $(7 + 4i) + (-9 - 7i) =$ _____.
19. Multiply $(1 + 6i)$ and $(3 + 2i)$ in the same way as $(3x - 8)$ and $(x + 4)$ would be multiplied.
    $(1 + 6i)(3 + 2i) =$ _____.

Part B: Answers for odd-numbered problems of Part B are at the back of the book.

Simplify.

1. $\sqrt{-16}$
2. $\sqrt{-49}$
3. $\sqrt{-40}$
4. $7\sqrt{-18}$
5. $-3\sqrt{-24}$
6. $\sqrt{-19}$
7. $\sqrt{-85}$
8. $\sqrt{-20}$
9. $6i \cdot 3i$
10. $-4i \cdot 6i$
11. $4i \cdot -9i$
12. $-3i \cdot 8i$
13. $-6i \cdot 8i$
14. $-3i \cdot -6i$
15. $-5i \cdot -8i$
16. $-2i \cdot -5i$

17. $i\sqrt{10} \cdot i\sqrt{13}$

18. $i\sqrt{5} \cdot i\sqrt{7}$

19. $i\sqrt{15} \cdot i\sqrt{11}$

20. $(7i)^2$

21. $(9i)^2$

22. $(i\sqrt{13})^2$

23. $(i\sqrt{4})^2$

24. $(i\sqrt{19})^2$

25. $(i\sqrt{7})^2$

26. i^9

27. i^{25}

28. i^{66}

29. i^{85}

30. $5i^3 \cdot 2i^6$

31. $-2i^3 \cdot 4i^9$

32. $3i^2 \cdot -5i^3$

33. $-4i^3 \cdot 6i^2$

34. $7i^2 \cdot -3i^2$

35. $-5i^3 \cdot -3i^3$

36. $3i(6 + 5i)$

37. $2i(7 + 3i)$

38. $-5i(3 - 7i)$

39. $3i(4 - i)$

40. $5 - 3i - 7 - 2i$

41. $9 - 3i + 6 + 7i$

42. $-3 + 4i - 7 + 7i$

43. $14 - 5i - 4 - 6i$

44. $(3 - 2i)(2 + i)$

45. $(7 - 3i)(1 + i)$

46. $(3 - 2i)(2 + 3i)$

47. $(4 - 3i)(4 + 3i)$

48. $(3 + 5i)(3 - 5i)$

49. $(7 + i)(7 - i)$

50. $(2 - 5i)(2 + 5i)$

Part C: No answers are given for these problems. However, each is accompanied by an ordered pair «C,U» showing the chapter and unit in which it was taught.

1. Factor $64x^4 - 25y^2$ «1,4»
2. Solve $20x^2 + 27x = 14$ «1,5»
3. Simplify $\dfrac{-3\sqrt{24}}{2\sqrt{63}}$ «2,2»
4. Solve $2\sqrt{5} - \dfrac{3x}{2} = \dfrac{5}{6}$ «2,5»
5. Solve $\dfrac{2x}{3} - \dfrac{5}{12} = \dfrac{-1}{4}$ «2,6»
6. Simplify $\dfrac{-6}{x+2} \geq 3$ «2,7»
7. Solve $\dfrac{5}{2x-3} - \dfrac{2}{x-5} = \dfrac{-3}{2x^2 - 13x + 15}$ «2,7»

8. How much brine mixture containing 70% salt should be added to 400 quarts of 40% brine mixture to produce a mixture that contains 50% salt?   «2,8»

9. How much pure alcohol would have to be added to 32 gallons of a 25% mixture of alcohol to produce a 40% mixture of alcohol?   «2,8»

10. How much water would have to be removed from 100 gallons of a mixture containing 80% water to decrease the water content to 50%?   «2,8»

Unit 2: SOLVING QUADRATIC EQUATIONS WITHOUT REAL NUMBER SOLUTIONS

Simple Quadratic Equations

The truth set of $x^2 = 5$ contains $\sqrt{5}$ and its opposite $-\sqrt{5}$.
$$x^2 = 5$$
$$x = \sqrt{5} \text{ or } x = -\sqrt{5}$$
These two roots are checked by the equalities:
$$(\sqrt{5})^2 = 5 \text{ and } (-\sqrt{5})^2 = 5.$$

Similarly, the truth set of $x^2 = -9$ is found by:
$$x^2 = -9$$
$$x = \sqrt{-9} \text{ or } x = -\sqrt{-9}$$
$$x = i\sqrt{9} \text{ or } x = -i\sqrt{9}$$
$$x = 3i \text{ or } x = -3i$$
These two roots are checked by the equalities:
$$(3i)^2 = -9 \text{ and } (-3i)^2 = -9$$

Focus on solving equations of the form $x^2 = k$

To find the truth set of $x^2 = -17$, the following steps are used.
$$x^2 = -17$$
$$x = \sqrt{-17} \text{ or } x = -\sqrt{-17}$$
$$x = i\sqrt{17} \text{ or } x = -i\sqrt{17}$$
$$\{i\sqrt{17}, -i\sqrt{17}\}$$

To find the truth set of $x^2 = -32$, the following steps are used.
$$x^2 = -32$$
$$x = \sqrt{-32} \text{ or } x = -\sqrt{-32}$$
$$x = i\sqrt{32} \text{ or } x = -i\sqrt{32}$$
$$x = 4i\sqrt{2} \text{ or } x = -4i\sqrt{2}$$
$$\{4i\sqrt{2}, -4i\sqrt{2}\}$$

Using the Quadratic Formula

Three methods for solving second degree or quadratic equations have been learned. They are: factoring, completing the square, and the quadratic formula. When the roots of a quadratic equation are not real numbers, the easiest way to solve is normally the quadratic formula:

$$x = \frac{-b \pm \sqrt{b^2 - 4ac}}{2a}$$

where a, b, and c are determined by a quadratic equation written in standard form, $ax^2 + bx + c = 0$.

The plus-minus symbol "±" in the quadratic formula indicates that there are two numbers in the truth set. One will be obtained by using the plus sign, and the other will be obtained by using the minus sign.

To solve $x^2 + 4x + 6 = 0$ using the quadratic formula, the following steps are used.

For $x^2 + 4x + 6 = 0$, a = 1, b = 4, c = 6

$$x = \frac{-4 \pm \sqrt{16 - 24}}{2} = \frac{-4 \pm \sqrt{-8}}{2} = \frac{-4 \pm 2i\sqrt{2}}{2} = -2 \pm i\sqrt{2}$$

$\{-2 + i\sqrt{2}, -2 - i\sqrt{2}\}$ is the truth set of $x^2 + 4x + 6 = 0$.

Focus on checking $-2 \pm i\sqrt{2}$ as roots of $x^2 + 4x + 6 = 0$

To check $-2 \pm i\sqrt{2}$ as the roots of $x^2 + 4x + 6 = 0$, it is necessary to only check either $-2 + i\sqrt{2}$ or $-2 - i\sqrt{2}$.

For $-2 + i\sqrt{2}$, $x^2 + 4x + 6 = 0$ becomes:

$(-2 + i\sqrt{2})^2$	+	$4(-2 + i\sqrt{2})$	+	$6 = 0$
$(-2 + i\sqrt{2})(-2 + i\sqrt{2})$	+	$-8 + 4i\sqrt{2}$	+	$6 = 0$
$[4 - 2i\sqrt{2} - 2i\sqrt{2} + i^2(\sqrt{2})^2]$	+	$-8 + 4i\sqrt{2}$	+	$6 = 0$
$[4 - 4i\sqrt{2} - 2]$	+	$-8 + 4i\sqrt{2}$	+	$6 = 0$
$[2 - 4i\sqrt{2}]$	+	$-8 + 4i\sqrt{2}$	+	$6 = 0$
$(2 - 8 + 6)$	+	$(-4i\sqrt{2} + 4i\sqrt{2})$		$= 0$
0	+	0		$= 0$

The final sum is $0 = 0$ which proves $-2 + i\sqrt{2}$ is a root of the quadratic equation $x^2 + 4x + 6 = 0$.

Solving $x^2 - 5x + 7 = 0$

To solve $x^2 - 5x + 7 = 0$ using the quadratic formula, the following steps are used.

For $x^2 - 5x + 7 = 0$, $a = 1$, $b = -5$, $c = 7$

$$x = \frac{5 \pm \sqrt{25 - 28}}{2} = \frac{5 \pm \sqrt{-3}}{2} = \frac{5 \pm i\sqrt{3}}{2}$$

$\left\{ \frac{5 + i\sqrt{3}}{2}, \frac{5 - i\sqrt{3}}{2} \right\}$ is the truth set of $x^2 - 5x + 7 = 0$.

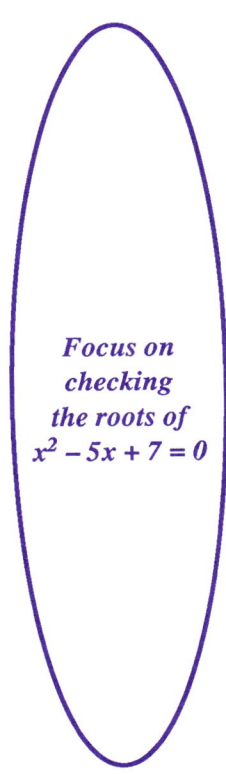

Focus on checking the roots of $x^2 - 5x + 7 = 0$

To check $\frac{5 \pm i\sqrt{3}}{2}$ as the roots of $x^2 - 5x + 7 = 0$, it is necessary to only check either $\frac{5 + i\sqrt{3}}{2}$ or $\frac{5 - i\sqrt{3}}{2}$.

For $\frac{5 - i\sqrt{3}}{2}$, $x^2 - 5x + 7 = 0$ becomes:

$\left(\frac{5 - i\sqrt{3}}{2} \right)^2$ + $-5 \left(\frac{5 - i\sqrt{3}}{2} \right)$ + $7 = 0$

$\left(\frac{25 - 10i\sqrt{3} + i^2(\sqrt{3})^2}{4} \right)$ + $\left(\frac{-25 + 5i\sqrt{3}}{2} \right)$ + $7 = 0$

$\left[\frac{25 - 10i\sqrt{3} - 3}{4} \right]$ + $\left(\frac{-25 + 5i\sqrt{3}}{2} \right)$ + $7 = 0$

$\left[\frac{22 - 10i\sqrt{3}}{4} \right]$ + $\left(\frac{-25 + 5i\sqrt{3}}{2} \right)$ + $7 = 0$

$\left[\frac{11 - 5i\sqrt{3}}{2} \right]$ + $\left(\frac{-25 + 5i\sqrt{3}}{2} \right)$ + $7 = 0$

$\frac{-14}{2}$ + $7 = 0$

-7 + $7 = 0$

The final sum is $0 = 0$ which proves $\frac{5 - i\sqrt{3}}{2}$ is a root of $x^2 - 5x + 7 = 0$.

Unit 2 Exercise

Part A: Answers for all Part A problems are at the back of the book.

1. Find the truth set of $x^2 = -16$ by solving the two linear equations $x = \sqrt{-16}$ or $x = -\sqrt{-16}$.

2. Solve $x^2 = -64$ by first writing its two linear equations.

3. Check one of the roots of $x^2 = -25$ to show that it makes the equation a true numerical statement.

4. Solve $x^2 = -19$ and check one of its roots.

5. Solve $x^2 - 2x + 2 = 0$ using the quadratic formula.

6. Simplify $(1 + i)^2$

7. Check $(1 + i)$ as a root of $x^2 - 2x + 2 = 0$ by replacing x by $(1 + i)$ and showing the numerical statement is true.

8. Solve $x^2 - 3x - 3 = 0$ using the quadratic formula.

9. Simplify $-3\left(\frac{3 + \sqrt{21}}{2}\right)$

10. Simplify $\left(\frac{3 + \sqrt{21}}{2}\right)^2$

11. Check $\left(\frac{3 + \sqrt{21}}{2}\right)$ as a root of $x^2 - 3x - 3 = 0$ by replacing x by $\left(\frac{3 + \sqrt{21}}{2}\right)$ and showing the numerical statement is true.

12. Solve $x^2 - 3x + 4 = 0$ using the quadratic formula.

13. Simplify $-3\left(\frac{3 + i\sqrt{7}}{2}\right)$

14. Simplify $\left(\frac{3 + i\sqrt{7}}{2}\right)^2$

15. Check $\left(\frac{3 + i\sqrt{7}}{2}\right)$ as a root of $x^2 - 3x + 4 = 0$ by replacing x by $\left(\frac{3 + i\sqrt{7}}{2}\right)$ and showing the numerical statement is true.

Part B: Answers for odd-numbered problems of Part B are at the back of the book.

Solve. Check one root for problems 8-13.

1. $x^2 = -100$

2. $x^2 = -23$

3. $x^2 = -7$

4. $x^2 = -81$

5. $x^2 = -8$

6. $x^2 = -20$

7. $x^2 = -1$

8. $x^2 = -37$

9. $x^2 = -45$

10. $x^2 - x + 1 = 0$

11. $x^2 + 2x + 7 = 0$

12. $x^2 + 4x + 5 = 0$

13. $x^2 - 3x + 6 = 0$

14. $2x^2 - 3x + 7 = 0$

15. $x^2 - 5x + 8 = 0$

16. $3x^2 + x + 2 = 0$

17. $x^2 - 2x + 3 = 0$

18. $x^2 + 3x + 7 = 0$

19. $x^2 - x + 1 = 0$

20. $x^2 - 2x + 7 = 0$

21. $2x^2 - x + 5 = 0$

22. $x^2 + 3x + 5 = 0$

Part C: No answers are given for these problems. However, each is accompanied by an ordered pair «C,U» showing the chapter and unit in which it was taught.

1. Factor $3x^2 - 12x + 36$ «1,4»

2. Solve $4x^2 - 20x = 96$ «1,5»

3. Solve $\frac{4x}{3} - \frac{2\sqrt{5}}{6} = \frac{1}{2}$ «2,5»

4. Solve $3 < \frac{6}{x-5}$ «2,6»

5. Solve $\frac{-3}{2x^2 - x - 3} = \frac{1}{2x-3} - \frac{2}{x+1}$ «2,7»

Simplify problems 6-10.

6. $(i\sqrt{5})^2$ «3,1»

7. $-5\sqrt{-32}$ «3,1»

8. $-2i \cdot -7i$ «3,1»

9. $13 - 5i - 9 - 6i$ «3,1»

10. $(3 - 7i)(3 + 7i)$ «3,1»

Unit 3: THE COMPLEX NUMBERS

Real, Imaginary, and Complex Numbers

$\sqrt{5}$ is a real number. -6i is an imaginary number. The sum of $\sqrt{5}$ and -6i is called a **complex number**. This unit is a study of the complex numbers.

$\frac{4}{17}$ is a real number, and 5i is an imaginary number. The sum of $\frac{4}{17}$ and 5i is a **complex** number.

Every number of the form x + yi where x and y are real numbers is called a complex number. 4 + 3i is a number of the form x + yi where 4 and 3 are real numbers. Therefore, 4 + 3i is a complex number.

Focus on the meaning of complex number

Any number of the form x + yi where x and y are real numbers is a complex number.

Since $\frac{-2}{3}$ and $-\sqrt{3}$ are real numbers, $\frac{-2}{3} - i\sqrt{3}$ is a complex number in which the coefficient of i is $-\sqrt{3}$.

Real and Imaginary Components of a Complex Number

The complex number 5 − 7i consists of the real number 5 and the imaginary number -7i. 5 is called the real number component of 5 − 7i. -7i is called the imaginary number component of 5 − 7i.

In the complex number -5 + 2i, -5 is the real component and 2i is the imaginary component.

Focus on the components of a complex number

-5 + 0i is a complex number because it consists of a real number component -5 and an imaginary component 0i. -5 + 0i is equal to -5.
-5 + 0i is both a complex number and a real number because 0i = 0.

Every real number can be written as a complex number by using 0i as the imaginary number component. 4i is an imaginary number. 0 + 4i is a complex number. $i\sqrt{3}$ is an imaginary number. $0 + i\sqrt{3}$ is a complex number.

Adding Complex Numbers

Addition of complex numbers is performed by adding the real components and separately adding the imaginary components.

$$(3 + 5i) + (2 + 3i) = (3 + 2) + (5i + 3i)$$
$$= 5 + 8i$$

Similarly,

$$(-3 + 7i) + (6 - 3i) = (-3 + 6) + (7i - 3i)$$
$$= 3 + 4i$$

Focus on adding complex numbers

The **conjugate** of $(8 - 5i)$ is $(8 + 5i)$. The signs of the second terms are different. Notice that the sum of the two conjugates is a real number.

$$(8 - 5i) + (8 + 5i) = (8 + 8) + (-5i + 5i)$$
$$= 16 + 0i$$
$$= 16$$

$(3 + 7i)$ and $(-3 - 7i)$ are **opposites**. The signs of both terms are different. The sum of two opposites is always zero.

$$(3 + 7i) + (-3 - 7i) = (3 - 3) + (7i - 7i)$$
$$= 0 + 0i$$
$$= 0$$

Multiplying Complex Numbers

Multiplication of complex numbers is performed in a manner similar to the multiplication of other binomials.

$$(3 + 2i)(4 + 5i)$$
$$3(4 + 5i) + 2i(4 + 5i)$$
$$12 + 15i + 8i + 10i^2$$
$$12 + 15i + 8i - 10$$
$$2 + 23i$$

Similarly,

$$(2 + 5i)(7 + 3i)$$
$$2(7 + 3i) + 5i(7 + 3i)$$
$$14 + 6i + 35i + 15i^2$$
$$14 + 6i + 35i - 15$$
$$-1 + 41i$$

Focus on the multiplication of conjugates

The product of a complex number and its conjugate is both interesting and important. The product of two conjugates is always a real number.

$$(5 + 3i)(5 - 3i)$$
$$5(5 - 3i) + 3i(5 - 3i)$$
$$25 - 15i + 15i - 9i^2$$
$$25 + 9$$
$$34$$

Rationalizing the Denominator

As with all radical expressions, "simplify" means to remove all radicals from the denominator of a fraction. The simplification of fractions such as $\frac{1}{2 + 3i}$, $\frac{4 - 3i}{2 - 5i}$, and $\frac{2 + i}{i}$ means to write the divisor as a rational number.

The process for rationalizing the denominator of a fraction involving complex numbers requires the use of the conjugate of the divisor. The conjugate of $(a - b)$ is $(a + b)$. The conjugate of $(\sqrt{3} - 2)$ is $(\sqrt{3} + 2)$, and the conjugate of $(4 + 3i)$ is $(4 - 3i)$.

To simplify the fraction $\frac{2 + 3i}{4 - 2i}$, the first step is to multiply the numerator and the denominator by the conjugate of the denominator as shown below.

$$\frac{2 + 3i}{4 - 2i} = \frac{(2 + 3i)(4 + 2i)}{(4 - 2i)(4 + 2i)} = \frac{2 + 16i}{20} = \frac{1 + 8i}{10}$$

Focus on rationalizing the denominator

To simplify $\frac{5 - 2i}{4 + i}$, multiply both the numerator and the denominator by the conjugate of the denominator.

$$\frac{5 - 2i}{4 + i} = \frac{(5 - 2i)(4 - i)}{(4 + i)(4 - i)} = \frac{18 - 13i}{17}$$

System of Complex Numbers

Unit 3 Exercise

Part A: Answers for all Part A problems are at the back of the book.

1. Is $\frac{3}{19}$ a real number?

2. Is 7 a real number?

3. Is $-\sqrt{5} + 6i$ a complex number?

4. $4 - 7i$ is a complex number. 4 is a real number. $-7i$ is a(n) _____ number.

5. In the complex number $7 - 5i$, 7 is the real component and $-5i$ is the _____ component.

6. What is the imaginary number component of $6 - 3i$?

7. Is $-\sqrt{11} + 0i$ both a real number and a complex number?

8. Write $-5i$ as a complex number.

9. Add the complex numbers $(2 + 7i)$ and $(3 + 2i)$ by separately adding their real and imaginary components.

10. Multiply the complex numbers $(3 + 2i)$ and $(4 + 3i)$ as with the normal multiplication of two binomials.

11. The conjugate of $(3 - 5i)$ is $(3 + 5i)$. What is the conjugate of $(8 - 5i)$?

12. What is the conjugate of $(4 - 6i)$?

13. The conjugate of $(-3 - 5i)$ is $(-3 + 5i)$. What is the conjugate of $(-4 - 10i)$?

14. What is the conjugate of $(-7 + 3i)$?

15. Simplify $\frac{2 - 7i}{5 + 3i}$ by multiplying the numerator and the denominator by the conjugate of $5 + 3i$.

Label problems 16-20 True or False.

16. $4 + 3i$ is a complex number.

17. The real number component of $5 - 3i$ is -3.

18. Every real number is also a complex number.

19. Every complex number is an imaginary number.

20. The sum of a real number and an imaginary number is a complex number.

Part B: Answers for odd-numbered problems of Part B are at the back of the book.

1. Is $\pi + i\sqrt{2}$ a complex number?

2. $-6 + i\sqrt{2}$ is a complex number consisting of the real number -6 and the imaginary number _____.

3. Write -7 as a complex number.

4. What is the real number component of $6 - 3i$?

5. $\frac{4}{17}$ is a real number. Write $\frac{4}{17}$ as a complex number.

6. Write $i\sqrt{15}$ as a complex number.

Simplify problems 7-40.

7. $(-5 + 6i) + (8 - 4i)$

8. $(5 + 4i) + (3 - 4i)$

9. $(2 + 3i) + (-2 - 3i)$

10. $(-4 + 10i) + (-1 + 2i)$

11. $(-6 + 17i) + (4 - 11i)$

12. $(7 + 2i) + (-7 + 3i)$

13. $(5 + 4i) + (-5 - 2i)$

14. $(3 + 2i) + (-3 + 4i)$

15. $(6 + 5i) + (2 - 5i)$

16. $(-4 + 2i) + (7 - 2i)$

17. $(3 - 5i) + (-3 + 5i)$

18. $(4 + 3i) + (-4 - 3i)$

19. $(12 + i) + (3 - 5i)$

20. $(2 - 17i) + (-10 + 10i)$

21. $(-3 + 5i)(4 - 2i)$

22. $(6 - 5i)(6 + 5i)$

23. $(2 - 3i)(4 - i)$

24. $(7 - 3i)(4 + i)$

25. $(3 - 7i)(2 - i)$

26. $(5 + 3i)(5 - 3i)$

27. $(1 + 2i)(1 - 2i)$

28. $(4 - 2i)(4 + 2i)$

29. $(3 - 2i)(3 + 2i)$

30. $(3 + 7i)(3 - 7i)$

31. $\dfrac{7 - i}{2 + i}$

32. $\dfrac{4 + 3i}{-5 - i}$

33. $\dfrac{7 + 3i}{4 + 2i}$

34. $\dfrac{3 - 7i}{2 + 5i}$

35. $\dfrac{3 + 6i}{3 - 6i}$

36. $\dfrac{4 + i}{-2 - i}$

37. $\dfrac{-3 + 4i}{-1 + 2i}$

38. $\dfrac{-3 + 5i}{5 + 6i}$

39. $\dfrac{7 + i}{2 - 3i}$

40. $\dfrac{3 - 3i}{5 - 2i}$

Part C: No answers are given for these problems. However, each is accompanied by an ordered pair «C,U» showing the chapter and unit in which it was taught.

1. Multiply $(2x - 3)(4x^2 + 6x + 9)$ «1,1»

2. How much light cream would have to be added to 200 pounds of milk to increase the butterfat from 3% to 4% if the light cream contains 10% butterfat? «2,8»

Simplify problems 3-5.

3. $\sqrt{-48}$ «3,1»

4. $i\sqrt{6} \cdot i\sqrt{7}$ «3,1»

5. $-3i^3 \cdot 5i^7$ «3,1»

6. Solve $x^2 = -20$ «3,2»

7. Solve $x^2 - 2x + 3 = 0$ «3,2»

8. Solve $x^2 - x + 2 = 0$ «3,2»

9. Solve $3x^2 - 2x - 4 = 0$ «3,2»

10. Solve $x^2 + 3x + 5 = 0$ «3,2»

Unit 4: SOLVING EQUATIONS WITH COMPLEX NUMBERS

> **Definition**   **Equal Complex Numbers**
>
> Two complex numbers are equal **if and only if** the real number components are equal and the imaginary number components are equal.

Equal Complex Numbers

$3 + 5i$ is not equal to $4 + 5i$ because $3 \neq 4$.

$6 - 3i$ is not equal to $6 + 3i$ because $-3i \neq 3i$.

$4 - i\sqrt{9}$ is equal to $\frac{8}{2} - 3i$ because $4 = \frac{8}{2}$ and $-i\sqrt{9} = -3i$.

According to the definition of equal complex numbers, if $(x + yi)$ is equal to $7 - 5i$ then $x = 7$ and $y = -5$.

Focus on opposites of complex numbers

To find the **opposite** of any complex number just change the signs of both the real and imaginary components. For example, the opposite of $(2 + 5i)$ is $(-2 - 5i)$ and $(2 + 5i) + (-2 - 5i) = 0 + 0i = 0$.

The requirement that $(x + yi)$ and $(w + zi)$ are opposites is the same as the requirement that x is the opposite of w and y is the opposite of z.

Adding Opposites

Every equation is solved using a set of numbers as its replacement set. In this unit, the complex numbers will be the replacement set for the variables of linear equations.

The first step in solving $(x + yi) + (2 + 3i) = (7 + 4i)$ is to add the opposite of $(2 + 3i)$ to both sides of the equation.

$$(x + yi) + (2 + 3i) = (7 + 4i)$$
$$(x + yi) + (2 + 3i) + (-2 - 3i) = (7 + 4i) + (-2 - 3i)$$
$$x + yi = 5 + i$$

Using the definition for equal complex numbers, $x = 5$ and $y = 1$. These two values constitute a solution for the equation $(x + yi) + (2 + 3i) = (7 + 4i)$.

Focus on solving using opposites

To solve $(x + yi) + (-2 + 3i) = (6 + 2i)$ for x and y, the following steps are used.

$$(x + yi) + (-2 + 3i) = (6 + 2i)$$
$$(x + yi) + (-2 + 3i) + (2 - 3i) = (6 + 2i) + (2 - 3i)$$
$$x + yi = 8 - i$$

Therefore, $x = 8$ and $y = -1$.

Multiplying by Reciprocals

The reciprocal of 5 is $\frac{1}{5}$.

The reciprocal of $(a + b)$ is $\frac{1}{(a + b)}$.

The reciprocal of $(2 + 7i)$ is $\frac{1}{(2 + 7i)}$.

System of Complex Numbers

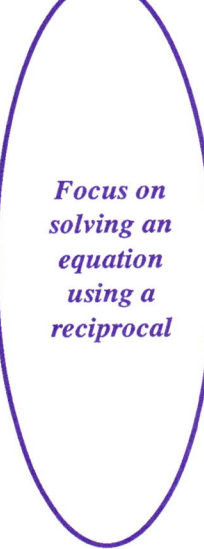

Focus on solving an equation using a reciprocal

To solve $(3 + 2i)(x + yi) = (4 - 3i)$ for x and y, the first step is to multiply both sides of the equation by the reciprocal of $3 + 2i$.

$$(3 + 2i)(x + yi) = (4 - 3i)$$

$$\frac{1}{(3 + 2i)}(3 + 2i)(x + yi) = \frac{1}{(3 + 2i)}(4 - 3i)$$

$$(x + yi) = \frac{(4 - 3i)}{(3 + 2i)}$$

To simplify the fraction $\frac{(4 - 3i)}{(3 + 2i)}$ multiply both numerator and denominator by the conjugate of $(3 + 2i)$.

$$(x + yi) = \frac{(4 - 3i)}{(3 + 2i)} \cdot \frac{(3 - 2i)}{(3 - 2i)}$$

$$(x + yi) = \frac{6 - 17i}{13}$$

$$(x + yi) = \frac{6}{13} - \frac{17i}{13}$$

Using the definition of equal complex numbers, $x = \frac{6}{13}$ and $y = \frac{-17}{13}$.

Solving Using Both Opposites and Reciprocals

To solve $(2 - i)(x + yi) - (7 - 3i) = (-3 + 6i)$, first add the opposite of $-(7 - 3i)$ to both sides of the equation.

$$\begin{array}{rl} (2 - i)(x + yi) - (7 - 3i) &= (-3 + 6i) \\ + (7 - 3i) & + (7 - 3i) \\ \hline (2 - i)(x + yi) &= (4 + 3i) \end{array}$$

To solve $(2 - i)(x + yi) = (4 + 3i)$, multiply both sides of the equation by the reciprocal of $(2 - i)$.

$$(2 - i)(x + yi) = (4 + 3i)$$

$$\frac{1}{(2 - i)}(2 - i)(x + yi) = \frac{1}{(2 - i)}(4 + 3i)$$

$$(x + yi) = \frac{(4 + 3i)}{(2 - i)}$$

$$(x + yi) = \frac{(4 + 3i)}{(2 - i)} \cdot \frac{(2 + i)}{(2 + i)}$$

$$(x + yi) = \frac{(5 + 10i)}{5}$$

$$(x + yi) = 1 + 2i$$

Therefore, $x = 1$ and $y = 2$.

Focus on solving linear equations with complex results

To solve $(3 - i)(x + yi) + (2 - 3i) = (5 + 2i)$, use the following process.

1. Add the opposite of $(2 - 3i)$ to each side of the equation.

$$\begin{aligned}(3 - i)(x + yi) + (2 - 3i) &= (5 + 2i) \\ \underline{(-2 + 3i) \qquad\qquad (-2 + 3i)} \\ (3 - i)(x + yi) + (0 + 0i) &= (3 + 5i)\end{aligned}$$

2. Multiply the reciprocal of $(3 - i)$ by each side of the equation.

$$\tfrac{1}{(3-i)}(3 - i)(x + yi) = \tfrac{1}{(3-i)}(3 + 5i)$$

3. Simplify the fraction on the right side of the equation.

$$(x + yi) = \tfrac{(3 + 5i)}{(3 - i)}$$

$$(x + yi) = \tfrac{(4 + 18i)}{10}$$

4. Simplify the x and y values if possible.

$$(x + yi) = \tfrac{(2 + 9i)}{5}$$

$$x = \tfrac{2}{5} \qquad y = \tfrac{9}{5}$$

Unit 4 Exercise

Part A: Answers for all Part A problems are at the back of the book.

1. If $(x + yi) = (7 + 4i)$, then $x = 7$ and $y =$ _____.
2. If $m + ni = p + qi$, then $m = p$ and _____.
3. If $(x + y) = 3 + 5i$, then $x = 3$ and $y =$ _____.
4. Find x and y if $x + yi = -3 - 4i$.
5. Solve $x + yi = \tfrac{5}{7}i$ by finding values for x and y that will make a true statement.
6. Solve $x + yi = \sqrt{5} + 2i$ by finding values for x and y that will make a true statement.
7. The opposite of $(-7 - 9i)$ is _____.
8. The opposite of $(-3 + 5i)$ is _____.
9. Solve $(x + yi) + (3 + 5i) = (6 + 8i)$ by adding the opposite of $(3 + 5i)$ to both sides of the equation.
10. Solve $(x + yi) + (-2 + 7i) = (-5 - 3i)$ by adding the opposite of $(-2 + 7i)$ to both sides of the equation.
11. What is the reciprocal of $(-2 - 5i)$?
12. What is the reciprocal of $(3 - 2i)$?
13. Solve $(3 - 2i)(x + yi) = (1 + 2i)$ by multiplying both sides of the equation by the reciprocal of $(3 - 2i)$.
14. Solve $(5 + i)(x + yi) = (3 + 4i)$ by multiplying both sides of the equation by the reciprocal of $(5 + i)$.
15. Solve $(2 + i)(x + yi) + (3 - 5i) = (6 + i)$ by using the opposite of $(3 - 5i)$ and then the reciprocal of $(2 + i)$.

Part B: Answers for odd-numbered problems of Part B are at the back of the book.
Solve.

1. $(x + yi) + (-5 + 4i) = (6 + 3i)$
2. $(x + yi) + (-3 - 7i) = (-4 - 19i)$
3. $(x + yi) + (-3 - 8i) = (2 + 8i)$
4. $(x + yi) + (2 + 8i) = (2 + 8i)$
5. $(1 - i)(x + yi) + (3 + 2i) = (4 - 3i)$
6. $(3 - 2i)(x + yi) = (5 - 6i)$
7. $(x + yi) + (7 - 6i) = (3 - 9i)$
8. $(2 + 3i)(x + yi) + (-3 + 4i) = (7 + i)$
9. $(3 + 5i)(x + yi) = (-2 + 3i)$
10. $(1 - i)(x + yi) = (2 - 3i)$
11. $(-4 + 3i)(x + yi) = (5 - 2i)$
12. $(2 + 3i)(x + yi) + (1 - i) = (5 - 2i)$
13. $(x + yi) + (-2 + 3i) = (7 - 2i)$
14. $(3 - 5i)(x + yi) = (1 - i)$
15. $(2 - i)(x + yi) + (4 - 3i) = (2 - 3i)$
16. $(-1 + i)(x + yi) + (-2 + 3i) = (5 - 2i)$
17. $(7 + 2i)(x + yi) + (-1 - 4i) = (6 + 5i)$
18. $(3 - 2i)(x + yi) + (6 + 2i) = (-2 - i)$
19. $(-1 - 3i)(x + yi) + (4 + 7i) = (3 + 7i)$
20. $(-5 - 4i)(x + yi) + (3 - 3i) = (5 - 4i)$

Part C: No answers are given for these problems. However, each is accompanied by an ordered pair «C,U» showing the chapter and unit in which it was taught.

1. Factor $8x^3 + 125y^3$ «1,4»
2. Simplify $i\sqrt{8} \cdot i\sqrt{6}$ «3,1»
3. Simplify $4i^8 \cdot -9i^7$ «3,1»
4. Solve $x^2 + 3 = 0$ «3,2»
5. Solve $2x^2 - x + 2 = 0$ «3,2»
6. Is $\sqrt{7} - i\sqrt{3}$ a complex number? «3,3»
7. Write $\frac{-3}{7}$ as a complex number. «3,3»

Simplify problems 8-10.

8. $(-3 + 4i) + (6 + 2i)$ «3,3»
9. $(3 - 5i)(3 + 5i)$ «3,3»
10. $\frac{6 - 3i}{2 + i}$ «3,3»

Unit 5: APPLICATIONS: SOLVING PROBLEMS USING TWO VARIABLES

Money and Digit Problems Solved Using Two Variables

When a problem requires that two quantities be found, it is frequently easier to solve by using two variables rather than one variable. The use of two variables requires that two equations be written. The common solution of the equations may be found by addition or by substitution. Following are some examples that use two variables for their solution.

Problem: 38 coins consisting of quarters and nickels are worth $6.90. How many quarters and nickels are there?

1. First, the problem should be carefully read and analyzed. The number of quarters and nickels can be represented by two variables. Two equations may be written: one about the total number of coins and the other about the value of the coins. The common solution of the two equations will give the number of quarters and nickels required by the problem.

2. Now the quantities mentioned in the problem are listed.
    the number of quarters
    the number of nickels
    the value of the quarters
    the value of the nickels
    the total number of coins
    the total value of the coins

3. A number or open expression is assigned to each quantity.
    q is the number of quarters
    n is the number of nickels
    $25q$ is the value of the quarters
    $5n$ is the value of the nickels
    $q + n$ is the total number of coins
    $25q + 5n$ is the total value of the coins

4. Two equations are written showing the number relationships in the problem.

The total number of the coins	is	38
$q + n$	=	38
The total value of the coins	is	$6.90
$25q + 5n$	=	690

5. The common solution of the two equations in step 4 is found by multiplying the first equation by -5 and adding the result to the second equation.

   The common solution is (25,13). It should be confirmed that 25 quarters and 13 nickels meet the conditions of the problem.

Focus on a problem involving digits of a number

Examine the relation that exists between the digits of a number. In the numeral 32, 3 is the tens digit and 2 is the units digit. The number 32 is 10 times the tens digit plus the units digit. $10 \cdot 3 + 2 = 32$.

If the digits in a two-digit number are represented by t for the tens digit and u for the units digit, the number itself is $10t + u$. The number with the digits reversed is $10u + t$.

Problem: Find a two-digit number such that the sum of the digits is 8, and the number decreased by the number with the digits reversed is 18.

Solution
1. The first part of the sentence will provide one equation and the second part will provide the other.

2. the ten's digit
   the unit's digit
   the number
   the number with the digits reversed

3. t is the ten's digit
   u is the unit's digit
   $10t + u$ is the number
   $10u + t$ is the number with the digits reversed

4. The sum of the digits is 8
   $$t + u = 8$$

   $\begin{pmatrix} \text{The number decreased by the} \\ \text{number with the digits reversed} \end{pmatrix}$ is 18
   $$10t + u - (10u + t) = 18$$

5. After simplifying the second equation, the u's can be eliminated by multiplying the first equation by 9 and adding. The common solution (t,u) is (5,3). 5 is the ten's digit and 3 is the unit's digit. The number is 53.

Two-Variable Rate and Mixture Problems

The speed of a boat increases when it goes downstream and decreases when it goes upstream against the current. The amount of increase or decrease will be the speed of the river. For example, if a boat that can go 10 miles per hour on still water travels downstream on a river that flows 2 miles per hour, its speed will be 12 miles per hour. But, when it returns upstream, it will slow down to 8 miles per hour (10 – 2) because of the river's flow. In 2 hours at a rate of 12 miles per hour, it could go 24 miles downstream. To go back up the river 24 miles would require 3 hours.

$$2(10 + 2) = 24$$
$$3(10 - 2) = 24$$

Following is an example that shows how to use two variables to find the boat's speed in still water and the rate of flow of the river.

Problem: When a boat traveled upstream 8 miles, it took 30 minutes. The return trip only took 20 minutes. What was the speed of the boat in still water and how fast was the river flowing?

Solution

1. After the speed of the boat and the river are represented by using two variables, the boat's speed upstream and downstream can be represented. Using the formula $D = rt$ (distance equals rate multiplied by time), two equations can be written about the distance upstream and the distance back. The time in minutes has to be converted to hours so that the rate in miles per hour can be found.

2.   the speed of the boat in still water
   the speed of the river
   the speed of the boat upstream
   the speed of the boat downstream

3.   b is the speed of the boat in still water
   r is the speed of the river
   b + r is the speed of the boat downstream
   b – r is the speed of the boat upstream

4. Distance equals rate multiplied by time. $D = rt$
       Downstream, $8 = (b + r) \cdot \frac{1}{3}$ (20 minutes = $\frac{1}{3}$ of an hour)
       Upstream, $8 = (b - r) \cdot \frac{1}{2}$ (30 minutes = $\frac{1}{2}$ of an hour)

5. The common solution of the two equations is found by multiplying the first equation by 3 and the second equation by 2 and adding the resulting equations. The common solution (b,r) is (20,4). It should be checked that 20 mph for the boat in still water and 4 mph for the river meet the conditions of the problem.

System of Complex Numbers   167

The following problem concerns mixing two alcohol mixtures to make a mixture that has a certain percent of ingredients.

Focus on an ingredients' problem

Problem: A 10% alcohol mixture and a 60% alcohol mixture are put together to produce 100 gallons of 55% alcohol. How many gallons of each type are needed?

Solution

1. Two facts can be stated from the information in the problem. The total number of gallons of the two mixtures is 100 gallons, and the amount of alcohol in the two mixtures added together will equal the alcohol in the 55% mixture. Two equations that use two variables can be written expressing the number relationships in the previous two sentences.

2. the gallons of 10% mixture
   the gallons of 60% mixture
   the alcohol in the 10% mixture
   the alcohol in the 60% mixture
   the alcohol in 100 gallons of 55% mixture

3. t is the gallons of 10% mixture
   s is the gallons of 60% mixture
   $.10t$ is the alcohol in the 10% mixture
   $.60s$ is the alcohol in the 60% mixture
   55 is the alcohol in 100 gallons of 55% mixture

4. The number of gallons of 10%
   and 60% alcohol together   equals   100 gallons
   $t + s$ =   100

   (The alcohol in the 10%   equals   (the alcohol in the
    and 60% mixtures together)   55% mixture)

   $.10t + .60s$ =   55
   $10t + 60s$ =   5500

   The second equation is often multiplied by 100 to give $10t + 60s = 5500$ and eliminate the decimals. This usually simplifies the solution.

5. The common solution of the two equations above can be found by multiplying the first equation by -10 and adding the result to the second equation. The common solution is 10 gallons of 10% mixture and 90 gallons of 60% mixture. It should be checked that these figures meet the conditions of the problem.

Unit 5 Exercise

Part A: Answers for all Part A problems are at the back of the book.

1. Mary sold 3 books and 4 magazines for $26. Later she sold 2 books and a magazine for $14. Find the price of a book and a magazine.

2. Two sides of a triangle are equal. The sum of one of the equal sides and the third side is 15. If the perimeter of the triangle is 21 inches, how long are the sides? (Hint: let x equal the length of each equal side and y equal the length of the third side.)

3. Two motorcycles sold for a total of $638. One sold for $96 more than the other. How much did each motorcycle cost?

4. The sum of the digits in a two-digit number is 12. The number with the digits reversed is 18 more than the number. Find the number.

5. A plane took 11 hours to reach its destination 5940 miles away when it was flying against the wind. On the return trip with the wind, it only took 9 hours. Find the rate of the plane in still air and the rate of the wind.

6. A pharmacist wants to dilute some 90% iodine with some 50% iodine to make 10 liters of 54% iodine. How many liters of each strength iodine should she use?

7. A racing scull took 1 hour and 15 minutes to complete the 20 mile downriver leg of a race. The 16 mile upriver leg of the race took 2 hours. Find the speed of the scull in still water and the speed of the current.

8. A distillery had some barrels of 15% alcohol and 80% alcohol. How many barrels of each type would they have to mix to make 1000 barrels of 41% alcohol?

Part B: Answers for odd-numbered problems of Part B are at the back of the book.

1. Find the length and the width of a rectangle when its perimeter is 74 inches and its width is 3 inches less than its length.

2. A banker loaned part of his $10,000 at 6%, and the rest he invested in bonds that pay 9%. Altogether he received $720 a year in interest. How much did he invest in bonds and how much in a loan?

3. The sum of the digits in a two-digit number is 6. The number with the digits reversed is 18 less than the number. Find the number.

4. The two defensive ends on a football team weigh a total of 605 pounds. If one of them weighed 45 pounds more than the other, how much did each man weigh?

5. How many gallons of gasoline containing 1% additives should be mixed with gasoline containing 3% additives to make 2000 gallons of gasoline containing 2.4% additives?

6. A plane flew with the wind 1200 miles in 3 hours and 20 minutes. On the return trip against the wind, the trip took 5 hours. Find the speed of the wind and the speed of the plane in still air.

7. One day a sports store sold 13 bats and 10 gloves for $125. The next day they sold 2 bats and 5 gloves for $40. Find the price of each item.

8. Some light cream containing 20% butterfat was mixed with some heavy cream containing 40% butterfat to make 100 pounds of cream with 31% butterfat. How many pounds of each type cream were used in the mixture?

9. A taxi company bought two cabs for a total of $6800 and paid $400 more for one of them than the other. How much did each cab cost?

10. A boat could go 2 miles downstream from its landing in 12 minutes. The return trip upstream took an hour. Find the speed of the boat in still water and the speed of the river.

11. The sum of the digits in a two-digit number is 10. The number is 54 more than the same number with the digits reversed. Find the number.

12. How many gallons of 20% alcohol and 60% alcohol should be mixed together to make 10 gallons of 48% alcohol?

13. A lady had $4000 to invest. She put part of it in a savings account paying 6% and the rest in bonds that paid 7%. Altogether she received $270 a year in interest. How much was invested at each rate?

Part C: No answers are given for these problems. However, each is accompanied by an ordered pair «C,U» showing the chapter and unit in which it was taught.

1. Solve $5x^2 - 15x - 90 = 0$ «1,5»
2. Solve $\dfrac{-2}{x-5} + \dfrac{3}{2x-1} = \dfrac{-1}{2x^2 - 11x + 5}$ «2,7»
3. Simplify $-3i^{12} \cdot 7i^9$ «3,1»
4. Simplify $i\sqrt{12} \cdot i\sqrt{6}$ «3,1»
5. Solve $5x^2 + 2x - 2 = 0$ «3,2»
6. Simplify $\dfrac{5-4i}{2+3i}$ «3,3»

Solve problems 7-10.

7. $(x + yi) + (-5 - 6i) = (3 + 2i)$ «3,4»
8. $(3 + 2i)(x + yi) = (-5 + i)$ «3,4»
9. $(3 - 2i)(x + yi) + (5 + 3i) = (-3 + 2i)$ «3,4»
10. $(1 - i)(x + yi) - (2 - 3i) = (4 - 5i)$ «3,4»

Chapter 3 Test

«3,U» shows the unit in which this problem was studied in this chapter.

1. The square of any real number cannot be a _____ number. «3,1»

2. $\sqrt{-32}$ = _____ «3,1»

3. $5i^6 \cdot 2i^5$ = _____ «3,1»

4. $(5 - 4i)(3 + 2i)$ = _____ «3,1»

5. Solve $x^2 = -20$ «3,2»

6. Solve $x^2 + x + 2 = 0$ «3,2»

7. Solve $x^2 - 2x + 5 = 0$ «3,2»

8. Solve $3x^2 - x + 2 = 0$ «3,2»

9. Simplify $(7 - 3i) + (4 + 2i)$ «3,3»

10. Simplify $(3 - 2i)(5 + 3i)$ «3,3»

11. Simplify $(5 + 2i)(5 - 2i)$ «3,3»

12. Simplify $\dfrac{2 + 4i}{3 - 2i}$ «3,3»

13. What is the opposite of $(-3 + 5i)$? «3,4»

14. What is the reciprocal of $(2 + 3i)$? «3,4»

15. Solve $(x + yi) + (-2 + 3i) = (9 - i)$ «3,4»

16. Solve $(2 + 3i)(x + yi) + (-2 + i) = (4 - 3i)$ «3,4»

17. Solve $(1 + i)(x + yi) - (4 + i) = (5 + 6i)$ «3,4»

18. The sum of the digits of a two digit number is 11. The number with the digits reversed is 27 less than the number. Find the number. «3,5»

19. The second side of a triangle is 8 inches longer than the first side, and the third side is one-half as long as the second side. What is the length of the first side if the triangle has a perimeter of 42 inches? «3,5»

20. 5 stuffed toys and 1 train set cost $210. 2 stuffed toys and 2 train sets cost $260. What is the cost of each? «3,5»

4
Absolute Value

Unit 1: THE MEANING OF ABSOLUTE VALUE

Every real number x is a complex number x + 0i. Therefore, the system of complex numbers is an extension of the system of real numbers.

The complex numbers, however, do not enjoy all of the order properties (greater than, less than) of the real numbers. Primarily for that reason, the remaining chapters in this text are discussed only in terms of the set of real numbers.

Distance Between Two Points

The distance between two distinct points on the number line is often approached by the concept of absolute value. The **distance** between two distinct points is always positive. The distance between 10 and 3 is symbolized as | 10 – 3 | or | 3 – 10 |. Both | 10 – 3 | and | 3 – 10 | stand for the distance 7.

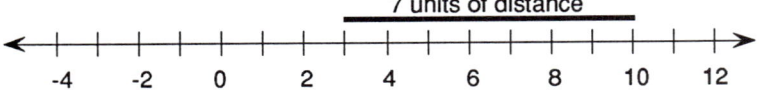

Definition	Absolute Value	
| x | is the absolute value of a real number x	if and only if	| x | is the distance that the number is from zero.

Since -8 is 8 units from 0, the absolute value of -8 is 8.

Since the distance from 9 to 0 is 9, the absolute value of 9 is 9.

Since 0 is 0 units from itself, the absolute value of 0 is 0.

The Absolute Value Symbol

The symbol | 2 | is read as "the absolute value of 2." Since the distance of 2 from 0 is 2, | 2 | = 2.

The symbol | -3 | is read as "the absolute value of -3." Since the distance of -3 from 0 is 3, | -3 | = 3.

The absolute value of a real number cannot be negative. The absolute value of a positive is a positive. The absolute value of a negative is a positive. The absolute value of zero is zero.

Absolute Value Equations

| x | = 8 is an equation involving absolute value. The truth set of
| x | = 8 consists of all real numbers that will make | x | = 8 a true statement.

The truth set of | x | = 8 is {8,-8}. Two numbers make | x | = 8 a true statement, because there are two numbers that have a distance of 8 from 0 on the real number line.

Similarly, the truth set of | x | = 5 is {5,-5}, because both 5 and -5 are numbers that are at a distance of 5 from 0.

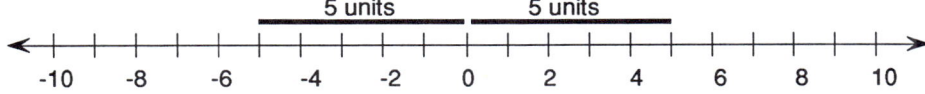

Focus on graphs of absolute value equations | x | = k

The truth set of | x | = 2 is {2,-2}, because 2 and -2 are both at a distance of 2 from 0.

| 2 | = 2 is true.
| -2 | = 2 is true.

The truth set of | x | = -5 is { }, because there is no number that has a distance of -5 from 0. Distance is never negative.

Graphing Equations and Inequalities

The truth set of $|x| = 4$ is $\{4, -4\}$. Its graph is shown below along with the graphs of $|x| < 4$ and $|x| > 4$.

The graph of $|x| = 4$ consists of two points

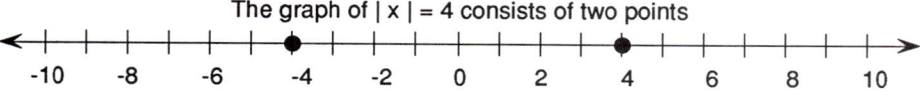

The graph of $|x| < 4$ consists of the line segment between the two points

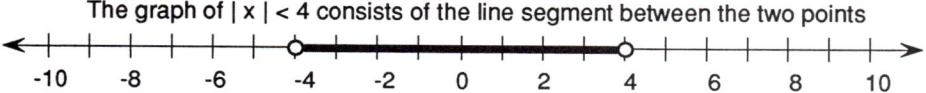

The graph of $|x| > 4$ consists of the two half-lines outside the two points

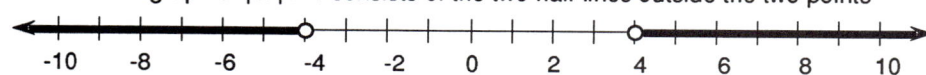

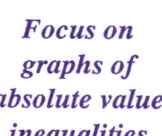

Focus on graphs of absolute value inequalities

The truth set of $|x| \geq 9$ is the set of all numbers whose distance from 0 is greater than or equal to 9. Notice that the points 9 and -9 are included in the truth set shown below.

The graph of $|x| \geq 9$ includes 9 and -9 as well as the two half-lines outside the two points

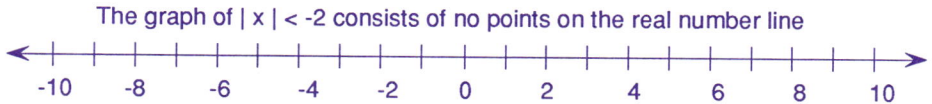

The truth set of $|x| < -2$ is the set of all numbers whose distance from 0 is less than -2. But distance is never negative and, therefore, can never be less than -2. Hence, the truth set of $|x| < -2$ is $\{\ \}$.

The graph of $|x| < -2$ consists of no points on the real number line

The truth set of $|x| > -2$ is the set of all numbers whose distance from 0 is greater than -2. Since distance is never negative, it is always greater than -2. Hence, the truth set of $|x| > -2$ is the set of all real numbers.

The graph of $|x| > -2$ consists of all points on the real number line

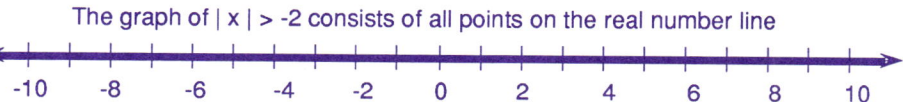

Absolute Value

Unit 1 Exercise

Part A: Answers for all Part A problems are at the back of the book.

1. On the real number line, what is the distance between -6 and 0?

2. On the real number line, what is the distance between 5 from 0?

3. The phrase "the distance between 3 and 0" can be written in math symbols as: $|3|$. Write in math symbols the phrase "the distance between -8 and 0."

4. The symbol $|-8|$ names the real number 8 because 8 is the distance between -8 and ____.

5. The symbol $|43|$ names the real number ____.

6. To find the truth set of $|x| = 6$, ask what real numbers are at distance 6 from 0. What two numbers are in the truth set?

7. To find the truth set of $|x| = -11$, ask what real numbers are at distance -11 from 0. Why are there no real numbers in the truth set?

8. To find the truth set of $|x| \geq 7$, ask what real numbers are at a distance greater than or equal to 7 from 0. Is -11 in the truth set?

9. To find the truth set of $|x| \leq 4$, ask what real numbers are at a distance less than or equal to 4 from 0. Is -9 in the truth set?

10. On the number line, graph the set of points that are at a distance greater than or equal to 6 from 0.

11. On the number line, graph the set of points that are at a distance less than or equal to 3 from 0.

Part B: Answers for odd-numbered problems of Part B are at the back of the book.

For problems 1-10, evaluate.

1. $|4| =$
2. $|13| =$
3. $|28| =$
4. $|-9| =$
5. $|-7| =$
6. $|-11| =$
7. $|-18| =$
8. $|1| =$
9. $|-27| =$
10. $|0| =$

For problems 11-20, solve the equations.

11. $|x| = 13$
12. $|x| = 0$
13. $|x| = -2$

14. $|x| = 7$

15. $|x| = 12$

16. $|x| = -14$

17. $|x| = -6$

18. $|x| = 1$

19. $|x| = 10$

20. $|x| = -3$

For problems 21-35 graph the truth sets.

21. $|x| < 10$

22. $|x| > 4$

23. $|x| \geq 6$

24. $|x| \leq 12$

25. $|x| \geq -5$

26. $|x| \geq 8$

27. $|x| < 3$

28. $|x| \leq -10$

29. $|x| \leq 11$

30. $|x| < 2$

31. $|x| > 2$

32. $|x| = 7$

33. $|x| = 0$

34. $|x| = -4$

35. $|x| = -5$

Part C: No answers are given for these problems. However, each is accompanied by an ordered pair «C,U» showing the chapter and unit in which it was taught.

1. Simplify $\dfrac{-3\sqrt{8}}{\sqrt{75}}$ «2,2»

2. Solve $-2 + \sqrt{x-4} = 4$ «2,6»

3. Simplify $\dfrac{-2\sqrt{27}}{\sqrt{48}}$ «3,1»

4. Solve $4x^2 - 3x = 2$ «3,2»

5. Simplify $\dfrac{3-3i}{2+2i}$ «3,3»

6. Solve $(x + yi) + (4 - 3i) = (7 + i)$ «3,4»

7. Solve $(4 - i)(x + yi) = (2 - 3i)$ «3,4»

8. Solve $(3 + 2i)(x + yi) + (2 + 3i) = (6 - i)$ «3,4»

9. The sum of the digits of a two digit number is 13. When the digits are reversed the new number is 27 less than the original number. Find the original number. «3,5»

10. The second angle of a triangle is 7° more than twice the first angle, and the third angle is 7° less than three times the first angle. Find the size of the first angle. «3,5»

Unit 2: SOLVING ABSOLUTE VALUE EQUATIONS

Absolute Value Bars as Grouping Symbols

The absolute value bars of $|5-1|$ act as a grouping symbol. $(5-1)$ is to be evaluated first.

$$|5-1| = |4| = 4$$

To evaluate $|2-7|$, first evaluate $(2-7)$.

$$|2-7| = |-5| = 5$$

Focus on absolute value bars as grouping symbols

To evaluate $|3 \cdot -2 - 5|$, first evaluate $(3 \cdot -2 - 5)$.

$$|3 \cdot -2 - 5| = |-6 - 5| = |-11| = 11$$

Use the absolute value bars as a grouping symbol. Always evaluate the expression within the absolute value bars first.

$$|-5 + 11| = |6| = 6$$

Expressions Using Absolute Value Symbols

$|7| - |-3|$ is an expression involving two absolute values. It is evaluated as shown below.

$$|7| - |-3| = 7 - 3 = 4$$

For the expression $|-7 + 2|$, the addition is completed before the absolute value is evaluated.

$$|-7 + 2| = |-5| = 5$$

For the expression $|-7| + |2|$, the absolute values of -7 and 2 are found first, and then the addition is completed.

$$|-7| + |2| = 7 + 2 = 9$$

Focus on evaluating opposites of absolute value expressions

Each of the following examples involves an absolute value expression preceded by a minus sign. In each case, first evaluate the absolute value expression and then find its opposite.

$$-|3-4| = -|-1| = -(1) = -1$$
$$-|8+5| = -|13| = -(13) = -13$$
$$-|-6-3| = -|-9| = -(9) = -9$$
$$-|-5+7| = -|2| = -(2) = -2$$

The Distance Between Two Numbers

For any real number replacement for x, $|x|$ can be interpreted as the distance between x and zero. The distance between any two real numbers x and y can be shown as $|x - y|$ or $|y - x|$.

The distance between 6 and 11 is the same as the evaluation of $|6 - 11|$ or $|11 - 6|$. In both cases the evaluation is 5.

The distance between -8 and -1 is the same as the evaluation of $|-8 - (-1)|$ or $|-1 - (-8)|$. In both cases the evaluation is 7.

The distance between -2 and 9 is the same as the evaluation of $|-2 - 9|$ or $|9 - (-2)|$. In both cases the evaluation is 11.

Focus on absolute value expressions as distances

The distance between any two real numbers is the absolute value of their difference. If a and b are real numbers, the distance between a and b can be written as $|a - b|$ or $|b - a|$.

The distance between -3 and 9 is found as follows.

$$|9 - (-3)| = |12| = 12 \text{ or } |-3 - 9| = |-12| = 12$$

The distance from -14 to 37 is found as follows.

$$|37 - (-14)| = |51| = 51 \text{ or } |-14 - 37| = |-51| = 51$$

(4 + 6) is equivalent to (4 − [-6]). $|4 + 6|$ is the distance between 4 and -6. (5 + 8) is equivalent to (5 − [-8]). $|5 + 8|$ is the distance between 5 and -8.

Absolute Value 179

The Distance Between Two Numbers

The open expression $|x - 2|$ can be used to determine the distance between any real number replacement for x and 2. The open expression $|x - 9|$ indicates the distance between x and 9.

The open expression $|x + 3|$ or $|x - (-3)|$ indicates the distance between x and -3.

Focus on absolute value as distance

The open expression $|x - y|$ can be used to determine the distance between any two real number replacements for x and y.

If $x = 7$ and $y = -3$, the open expression $|x - y|$ becomes $|7 - (-3)|$ or $|7 + 3|$.

Since $|7 - (-3)| = 10$, the distance between 7 and -3 is 10.

Truth Sets of Equations $|x - a| = b$

The truth set of $|x - 3| = 4$ consists of two real numbers whose distance from 3 is 4. On the number line, -1 and 7 are 4 units from 3.

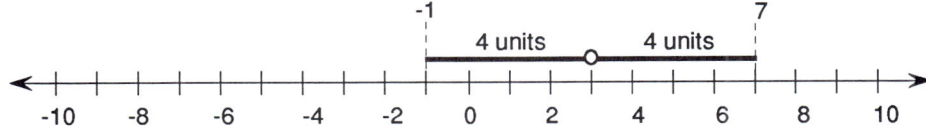

The truth set of $|x - 7| = 4$ is {3,11}, because 3 and 11 are 4 units from 7.

Focus on solving absolute value equations

The truth set of $|x + 5| = 2$ consists of two real numbers that are 2 units from -5. -7 and -3 are at a distance 2 from the point -5.

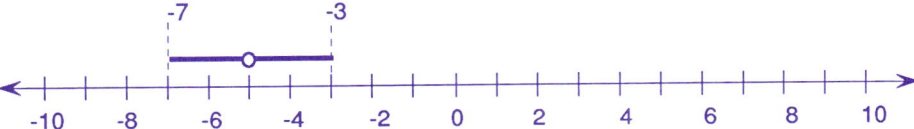

$|x - (-4)|$ is equivalent to $|x + 4|$. The truth set of $|x + 4| = 6$ consists of 2 and -10. Each is 6 units from -4.

The truth set of $|x - 3| = -5$ is { }, because there is no number at a distance -5 from 3. This is because distance or absolute value is never negative.

The Two Cases of Absolute Value

An absolute value equation can be written as an "or" sentence and solved accordingly. This is because there are two cases possible in an absolute value equation.

$$|x| = x \text{ if } x \geq 0 \text{ and}$$
$$|x| = -x \text{ if } x < 0$$

The set of real numbers may be separated into two sets.
1. the nonnegatives, which includes all the positives and zero, and
2. the negatives.

If a real number is neither positive nor zero, then it is negative.

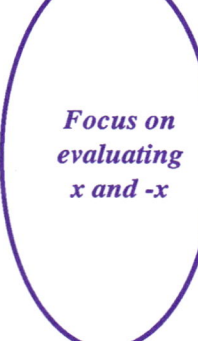

Focus on evaluating x and -x

-x is an open expression, which is equivalent to -1x. If x is replaced by 5, -x has an evaluation of -5. If x is replaced by -7, -x has an evaluation of 7.

If x is replaced by a positive number, then -x will be a negative number. If x is replaced by a negative number, then -x will be a positive number.

$|x| = x$ will become a true statement for an nonnegative replacement of x. If x is replaced by -3 in $|x| = x$, the numerical statement $|-3| = -3$ is obtained. $|-3| = -3$ is a false statement.

$|x| = x$ becomes a true statement only when x is nonnegative. The absolute value of a negative number is its opposite. $|x| = -x$ whenever $x < 0$.

Solving Equations of the Form | ax − b | = c

In dealing with the absolute value of an open expression, two possibilities must be considered.

$$|x| = x \text{ if } x \geq 0 \text{ and}$$
$$|x| = -x \text{ if } x < 0$$

When dealing with the absolute value of an open expression, two cases must be considered.
1. If $(3x - 5) \geq 0$, $|3x - 5| = (3x - 5)$, and
2. If $(3x - 5) < 0$, $|3x - 5| = -(3x - 5)$.

Absolute Value   181

To solve | 3x − 5 | = 7 consider both cases.

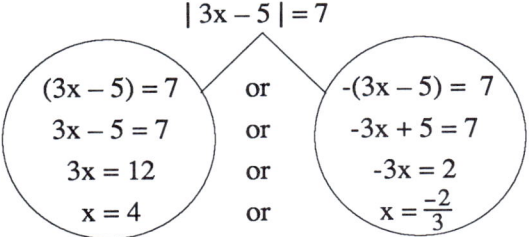

A check of these results will show that when x = 4 then (3x − 5) > 0 and when x = $\frac{-2}{3}$ then (3x − 5) < 0.

Focus on solving absolute value equations

The two cases of absolute value are used to write "or" sentences.

| x − 6 | = 2 is equivalent to: x − 6 = 2 or -(x − 6) = 2. {8} is the truth set of x − 6 = 2. {4} is the truth set of -(x − 6) = 2.

To solve | 4x − 3 | = 9, an equivalent "or" sentence is written.
| 4x − 3 | = 9 is equivalent to: 4x − 3 = 9 or -(4x − 3) = 9

Each equation of the "or" sentence is solved.

$$| 4x - 3 | = 9$$

$$\begin{array}{ccc} 4x - 3 = 9 & \text{or} & -(4x - 3) = 9 \\ 4x = 12 & \text{or} & -4x + 3 = 9 \\ x = 3 & \text{or} & -4x = 6 \\ x = 3 & \text{or} & x = \frac{-3}{2} \end{array}$$

{3, $\frac{-3}{2}$} is the truth set of | 4x − 3 | = 9.

The truth set of | 3x + 2 | = -5 is the empty set, { }, because the absolute value of a real number cannot be negative.

Unit 2 Exercise

Part A: Answers for all Part A problems are at the back of the book.

1. The absolute value expressions $|-5 - 3|$ and $|3 - (-5)|$ represent the distance on the number line between -5 and 3. Write two absolute value expressions that represent the distance between 13 and 2.

2. The absolute value expressions $|-8 - (-9)|$ and $|-9 - (-8)|$ represent the distance on the number line between -8 and -9. Write two absolute value expressions that represent the distance between -5 and 7.

3. The evaluation of $|-5 + 3|$ is the distance between -5 and _____.

4. The evaluation of $|7 - 10|$ is the distance between 7 and _____.

5. To solve $|x - 3| = 5$ ask what real numbers are at distance 5 from 3. What two real numbers are in the truth set?

6. To solve $|x + 4| = 7$ ask what real numbers are at distance 7 from -4. What two real numbers are in the truth set?

7. Will -x be a negative number for all real number replacements of x?

8. When will -x be a positive number?

9. What type of number is in the truth set of $|x| = -x$?

10. What type of number is in the truth set of $|-x| = -x$?

11. What type of number is in the truth set of $|-x| = x$?

12. To solve $|3x - 4| = 5$ two cases are considered. They are: $3x - 4 = 5$ or $-(3x - 4) = 5$. Solve the two equations and find the two roots of $|3x - 4| = 5$.

13. To solve $|-5x - 7| = 2$ two cases are considered. They are: $-5x - 7 = 2$ or $-(-5x - 7) = 2$. Solve the two equations and find the two roots of $|-5x - 7| = 2$.

14. $|x - 8| = 2$ is equivalent to the "or" sentence "$x - 8 = 2$ or _____."

15. Write the "or" sentence needed to find the two roots of $|7x + 2| = 8$.

16. Find the truth set of $|4x - 3| = 11$ by first writing an equivalent "or" sentence.

17. True or False? $|-7| + |2| = |-7 + 2|$

18. True or False? $|2 - 9| = |2| - |9|$

Part B: Answers for odd-numbered problems of Part B are at the back of the book.

Evaluate.

1. $|-6-3|$
2. $|2-8|$
3. $|6 \cdot -5+7|$
4. $|4 \cdot 3+1|$
5. $|9|-|-7|$
6. $-|-3|$
7. $|-5+3|$
8. $|-8+2|$
9. $|-6+-7|$
10. $|-8+-2|$
11. $|8+-2|+|-3|$
12. $|-3+-2|+|0|$
13. $|13-15|+|2|$
14. $|-7-2|+|-3-8|$
15. $|28+-37|+|-6+14|$

Solve.

16. $|x-1|=5$
17. $|x-8|=5$
18. $|x-5|=10$
19. $|x+9|=1$
20. $|x+2|=5$
21. $|x+8|=8$
22. $|x-3|=4$
23. $|x-5|=2$
24. $|x-6|=-4$
25. $|x+6|=2$
26. $|x+1|=5$
27. $|x+7|=-1$
28. $|x-7|=6$
29. $|x+6|=4$
30. $|x+7|=-2$
31. $|3x-5|=7$
32. $|5x+6|=4$
33. $|4x+9|=1$
34. $|3+2x|=5$
35. $|5x-3|=-4$
36. $|6x-2|=-3$
37. $|6x-7|=5$
38. $|x+2|=6$
39. $|2x+5|=7$
40. $|3x-2|=8$

Part C: No answers are given for these problems. However, each is accompanied by an ordered pair «C,U» showing the chapter and unit in which it was taught.

1. What is the conjugate of $5 - 3\sqrt{7}$? «2,2»
2. Simplify $\frac{5-2\sqrt{3}}{2-\sqrt{3}}$ «2,2»
3. Solve $(3+i)(x+yi) = (2-5i)$ «3,4»
4. Solve $(2+3i)(x+yi) - (1-i) = (3+4i)$ «3,4»
5. One day 11 bats and 25 balls sold for $257. The next day 10 bats and 15 balls sold for $195. Find the price of each item. «3,5»
6. The sum of the digits of a two digit number is 10. The number is 72 more than the number when the digits are reversed. What is the original number? «3,5»
7. Solve $|x| = 12$ «4,1»
8. Solve $|x| = -2$ «4,1»
9. Solve $|x| \leq -10$ «4,1»
10. Graph the truth set. $|x| \leq 4$ «4,1»

Unit 3: SOLVING ABSOLUTE VALUE INEQUALITIES

Graphing Absolute Value Open Sentences

The graph of the truth set of $|x - 5| = 2$ consists of the two points 7 and 3, because each point is at distance 2 from 5.

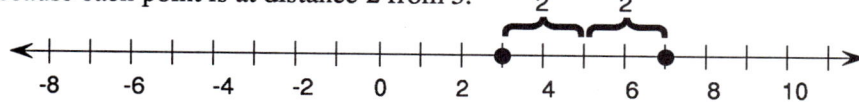

Similarly, the truth set of $|x - 4| = 7$ consists of the two points 11 and -3, because 11 and -3 are 7 units from 4. The truth set of $|x - 4| < 7$ will consist of all points whose distance from 4 is less than 7.

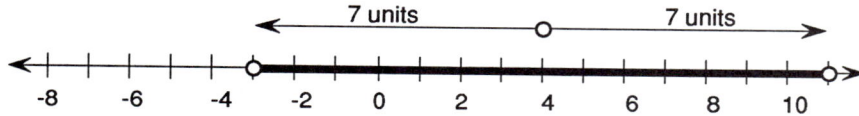

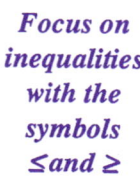

Focus on inequalities with the symbols $\leq$ and $\geq$

The truth set of $|x + 5| \leq 4$ consists of all real numbers whose distance from -5 is less than or equal to 4.

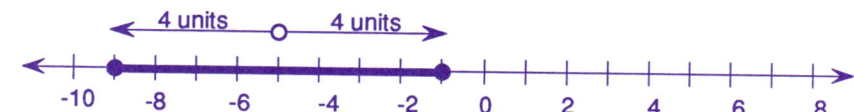

The graph above shows the truth set of $|x + 5| \leq 4$. Notice that the points -9 and -1 are indicated by closed dots (•) because they are included in the truth set.

The truth set of $|x - 4| \geq 3$ consists of all those numbers whose distance from 4 is greater than or equal to 3.

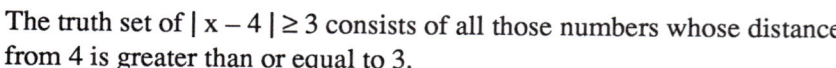
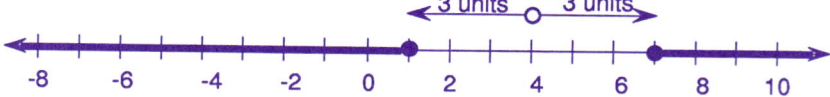

The graph above shows the truth set of $|x - 4| \geq 3$. Notice that the graph consists of two rays with the points 1 and 7 indicated by closed dots (•) because they are included in the truth set.

Open Segments as Truth Sets

The truth set of $|x - 6| = 8$ consists of exactly two numbers, 14 and -2. Any equation of the form $|x - a| = b$ where $b > 0$ will have a truth set with exactly two elements.

The graph of the truth set of $|x - 3| < 4$ is shown below. It includes all the points between -1 and 7 but not the points -1 and 7. The graph of the truth set of $|x - 3| < 4$ is an **open segment**.

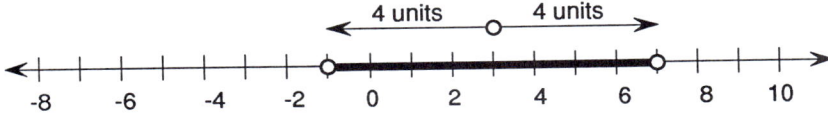

Every inequality of the form $|x - a| < b$ where $b > 0$ will have a truth set that is an open segment.

Focus on line segments as truth sets

The graph of the truth set of $|x + 5| \leq 2$ includes all the points between -7 and -3 and also the points -7 and -3. The graph of the truth set of $|x + 5| \leq 2$ is a **line segment**.

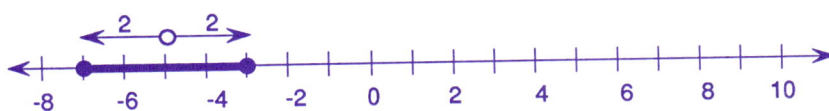

Half-Lines as Truth Sets

The graph of the truth set of $|x - 5| > 3$ consists of all points to the right of 8 or to the left of 2. The points 8 and 2 are not included. The graph of the truth set of $|x - 5| > 3$ consists of two **half-lines**.

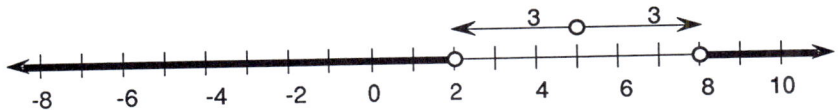

The graph of every inequality of the form $|x - a| > b$ where $b > 0$ consists of two half-lines.

186   Chapter 4

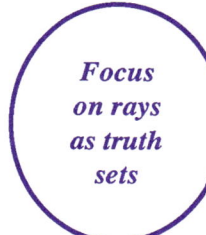
Focus on rays as truth sets

The graph of the truth set of $|x+1| \geq 4$ consists of all points to the right of 3 or to the left of -5, and also the points 3 and -5. The graph of the truth set of $|x+1| \geq 4$ consists of two **rays**.

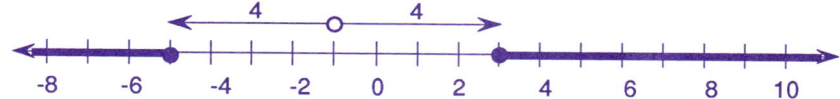

Absolute Value Inequalities and "And" Sentences

The inequality $3 < x < 7$ is a shorthand expression for writing the "and" sentence "$x > 3$ and $x < 7$." The inequality has two conditions connected by the word "and" and its truth set consists of those replacements for x that make **both** inequalities true statements.

Absolute value inequalities of the form $|x - a| < b$ where $b > 0$ are always equivalent to "and" sentences of the form "$c < x < d$." The truth set of $|x - 7| < 4$ is the same as the truth set of "$3 < x < 11$."

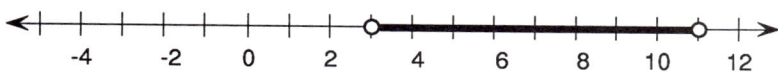

The graph above shows the truth set of the absolute value inequality $|x - 7| < 4$ which is equivalent to $3 < x < 11$.

Focus on graphs of inequalities of the form $c \leq x \leq d$

The inequality $-2 \leq x \leq 10$ is a shorthand way of writing the open sentence $x \geq -2$ and $x \leq 10$ and is equivalent to the truth set of $|x - 4| \leq 6$.

The graph above shows the truth set of $|x - 4| \leq 6$ which is equivalent to $-2 \leq x \leq 10$.

Absolute value inequalities of the form $|x - a| \leq b$ where $b > 0$ are always equivalent to "and" sentences of the form "$c \leq x \leq d$."

Absolute Value Inequalities and "Or" Sentences

"x < 2 or x > 9" has two conditions connected by the word "or." The truth set of "x < 2 or x > 9" consists of those replacements for x that make at least one of the inequalities a true statement. The result of this "or" sentence is a truth set with two half-lines.

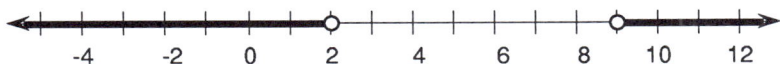

Absolute value inequalities of the form | x – a | > b where b > 0 are always equivalent to "or" sentences of the form "x < c or x > d." The truth set of | x – 3 | > 7 is the same as the truth set of "x < -4 or x > 10."

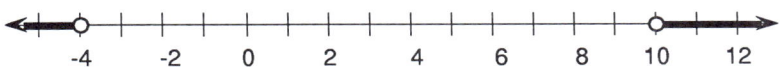

The graph above shows the truth set of the absolute value inequality | x – 3 | > 7 which is two half-lines with endpoints -4 and 10.

Focus on graphs of inequalities of the form x ≤ c or x ≥ d

The inequality x ≤ -7 or x ≥ 3 is equivalent to the truth set of | x + 2 | ≥ 5.

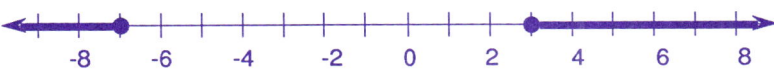

The graph above shows the truth set of | x + 2 | ≥ 5 which is two rays with endpoints -7 and 3.

Absolute value inequalities of the form | x – a | ≥ b where b > 0 are always equivalent to "or" sentences of the form "x ≤ c or x ≥ d."

Simplifying Absolute Value Inequalities

There are two cases of absolute value that must be considered in solving any open sentence.

$|x| = x$, if $x \geq 0$ and
$|x| = -x$, if $x < 0$

Using these two cases, $|x - a| < b$ can be written as:
$x - a < b$ and $-(x - a) < b$

When the two cases of absolute value are used, $|x - 5| < 6$ is solved as:

$$|x - 5| < 6$$

$x - 5 < 6$	and	$-(x - 5) < 6$
$x < 11$	and	$-x + 5 < 6$
$x < 11$	and	$-x < 1$
$x < 11$	and	$x > -1$

The "and" sentence is written as $-1 < x < 11$.

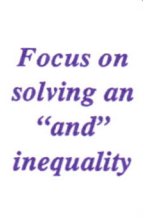

Focus on solving an "and" inequality

Because of the two cases of absolute value, $|5x + 2| < 3$ has the same truth set as "$5x + 2 < 3$ and $-(5x + 2) < 3$."

$$|5x + 2| < 3$$

$5x + 2 < 3$	and	$-(5x + 2) < 3$
$5x < 1$	and	$-5x - 2 < 3$
$5x < 1$	and	$-5x < 5$
$x < \frac{1}{5}$	and	$x > -1$

The final result may be shown as $-1 < x < \frac{1}{5}$.

Solving Absolute Value Inequalities

Always consider two cases when solving absolute value open sentences. Inequalities such as $|x + 7| > 9$ are equivalent to "or" sentences.

$$|x + 7| > 9$$

$x + 7 > 9$	or	$-(x + 7) > 9$
$x > 2$	or	$-x - 7 > 9$
$x > 2$	or	$-x > 16$
$x > 2$	or	$x < -16$

Absolute Value   189

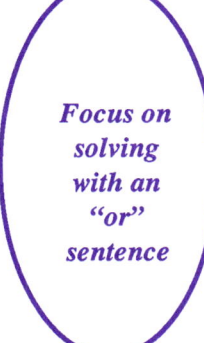

Focus on solving with an "or" sentence

The two cases of absolute value are used to solve $|3x - 2| > 4$ which has the same truth set as "$3x - 2 > 4$ or $-(3x - 2) > 4$."

The inequalities of the "or" sentences are simplified as follows:

$$|3x - 2| > 4$$

$(3x - 2) > 4$	or	$-(3x - 2) > 4$
$3x - 2 > 4$	or	$-3x + 2 > 4$
$3x > 6$	or	$-3x > 2$
$x > 2$	or	$x < \frac{-2}{3}$

Using Interval Notation for Truth Sets

$|x - 5| < 4$ has the same truth set as $1 < x < 9$. In **interval notation** the truth set is shown as (1,9) where the parentheses indicate that 1 and 9 are not part of the set.

Similarly, $|x + 4| \leq 2$ has the same truth set as $-6 \leq x \leq -2$. In **interval notation** the truth set is shown as [-6,-2] where the square brackets indicate that -6 and -2 are part of the set.

The truth set of $|x - 6| < -2$ is the empty set, { }, because no distance can be less than a negative number.

Focus on truth sets shown by interval notation

$|x - 6| > 5$ has the same truth set as $x < 1$ or $x > 11$. In interval notation the truth set is shown as a union, $(-\infty,1) \cup (11,\infty)$ where the negative **infinity symbol** ($-\infty$) indicates the set of numbers continues forever to get smaller and the positive **infinity symbol** (∞) indicates the set of numbers continues forever to get larger.

Similarly, $|x + 8| \geq 9$ has the same truth set as $x \leq -17$ or $x \geq 1$. In interval notation the truth set is shown as $(-\infty,-17) \cup (1,\infty)$.

The truth set of $|x - 3| > -5$ is the set of all real numbers, $(-\infty,\infty)$, because any distance is greater than a negative number.

Unit 3 Exercise

Part A: Answers for all Part A problems are at the back of the book.

1. The truth set of $|x - 5| < 2$ is described by the inequalities "$x > 3$ and $x < 7$." Describe the truth set of $|x - 9| < 4$ using inequalities.

2. The truth set of $|x + 7| > 3$ is described by the inequalities "$x > -4$ or $x < -10$." Describe the truth set of $|x + 5| > 8$ using inequalities.

3. The truth set of $|x - 4| \geq 8$ is described using interval notation as $(\infty, -4] \cup [12, \infty)$. Describe the truth set of $|x - 9| > 4$ using interval notation.

4. The truth set of $|x + 11| \leq 6$ is described using interval notation as $[-17, -5]$. Describe the truth set of $|x - 1| \leq 5$ using interval notation.

For problems 5-9, identify each of the open sentences by one of the following descriptions.
    a. two points
    b. an open segment
    c. two half-lines
    d. no points
    e. the entire number line

5. $|x - 9| = 6$
6. $|x - 1| = -4$
7. $|x + 2| < 4$
8. $|x - 1| > -2$
9. $|x + 5| > 3$

For problems 10-16, describe each set using inequalities and interval notation.

10. The set of real numbers between 8 and 13 that also includes 8 and 13.

11. The set of real numbers between -8 and 14 that includes 14 but not -8.

12. The set of real numbers between -6 and 2.

13. The set of real numbers between -8 and -5.

14. The set of real numbers between -3 and 11.

15. The set of real numbers between 9 and 0 that includes 9 and 0.

16. The set of real numbers between -13 and -8 that includes both -13 and -8.

Part B: Answers for odd-numbered problems of Part B are at the back of the book.

For problems 1-10, graph the truth sets.

1. $|x + 5| = 4$
2. $|x - 7| \leq 2$
3. $|x - 2| \geq 5$
4. $|x + 4| < 2$
5. $|x - 7| < -3$
6. $|7x - 4| \leq 3$
7. $|3x - 10| \geq -5$
8. $|5x + 6| \leq 1$
9. $|-3x + 5| \geq 2$
10. $|5x - 3| > 7$

For problems 11-20, use inequalities to describe the truth sets.

11. $|x + 1| > 5$
12. $|x - 7| > -6$
13. $|x + 1| \leq -3$
14. $|x + 9| \geq 1$
15. $|x - 8| > 6$
16. $|4x - 3| \geq 5$
17. $|3x + 5| \leq 8$
18. $|-2x - 1| \geq 10$
19. $|3x + 2| > 1$
20. $|6x - 2| < 1$

For problems 21-30, describe truth sets using interval notation.

21. $|x + 5| > 2$
22. $|x + 7| > 2$
23. $|x - 7| \leq 6$
24. $|x - 3| \geq 4$
25. $|x + 4| > -7$
26. $|3x - 4| \leq 6$
27. $|4x + 7| < 11$
28. $|-2x - 3| \geq -1$
29. $|6x + 5| \leq -3$
30. $|5x - 3| > 7$

Part C: No answers are given for these problems. However, each is accompanied by an ordered pair «C,U» showing the chapter and unit in which it was taught.

1. Simplify $\dfrac{3 - 5\sqrt{7}}{1 - \sqrt{7}}$ «2,2»

2. Solve $(2 + i)(x + yi) + (1 + 3i) = (1 + 3i)$ «3,4»

3. John had 40 coins that consisted of dimes and quarters. If the total value of the coins is $7.45, how many of each coin did he have? «3,5»

4. Solve $|x| = \sqrt{7}$ «4,1»

5. Graph the truth set. $|x - 2| \leq 3$ «4,1»

6. Evaluate $|7 \cdot 4 + 5|$ «4,2»

7. Evaluate $|12| - |-5|$ «4,2»

8. Solve $|x - 5| = 2$ «4,2»

9. Solve $|3x - 5| = -2$ «4,2»

10. Solve $|4x - 7| = 3$ «4,2»

Unit 4: APPLICATIONS: ABSOLUTE VALUE, DISTANCE, AND NEIGHBORHOODS

Distance and Absolute Value

On the real number line there are two numbers that are a distance 2 from 0. They are 2 and -2. The truth set of the equation $|x| = 2$ is $\{2,-2\}$. Consequently, $|x| = 2$ expresses the number relationships in the sentence "x represents a number that is 2 from 0." Similarly, the equation $|x| = 4$ shows the number relationships in the sentence "x represents a number whose distance from 0 is 4."

Focus on absolute value expressions as distances

The truth set of $|x - 1| = 4$ is $\{-3,5\}$, because the distance between both -3 and 1 and between 5 and 1 is 4.

The equation $|x - 1| = 4$ indicates the numbers that are a distance of 4 from 1.

Similarly, the equation $|x + 2| = 1$ indicates the numbers that are a distance of 1 from -2.

Verbal Problems Solved Using Absolute Value Equations

Some problems expressed in words can be solved by writing an absolute value equation that represents the number relationships in the problem. Following is an example of a problem solved using an absolute value open sentence.

Problem: A window that is 4 feet wide is to be placed in a wall of a house so that its center is 10 feet from the corner. How far are the sides of the window from the corner?

Solution

1. The window is 4 feet wide. Consequently, the sides of the window are a distance of 2 from the center of the window, which is 10 feet from the corner of the house. The problem will be solved by first writing an absolute value equation that represents the number relationship in the previous sentence.

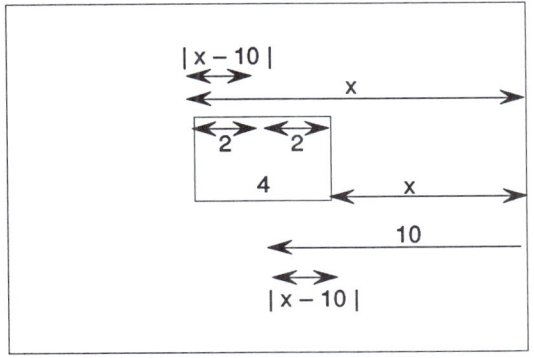

Corner of House

2. The quantities in the problem are listed.

    the distance between the sides of the window and the corner of the house

    the distance between the sides and the center of the window.

3. Open expressions are assigned to the quantities listed in step 2.

    Let x represent the distance between the sides of the window and the corner of the house.

    Then $|x - 10|$ represents the distance between the center and the sides of the window.

4. The distance between the center
    and sides of the window   is   2
    $|x - 10|$ =   2

5. $$|x - 10| = 2$$
    $x - 10 = 2$ or $-(x - 10) = 2$
    $x = 12$ or $-x + 10 = 2$
    $x = 12$ or $-x = -8$
    $x = 12$ or $x = 8$

    The sides of the window are 8 feet and 12 feet from the corner of the house.

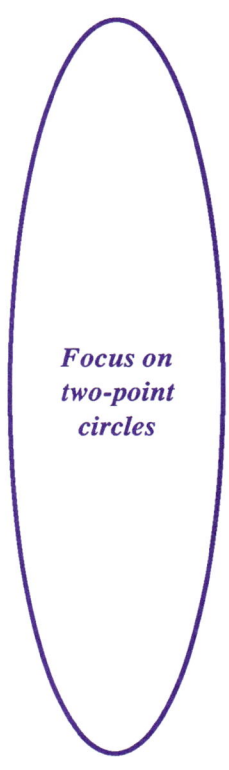

Focus on two-point circles

If a circle is placed on a number line so that the center is at 5 and every point on the circle is a distance 1 from the center, then the circle would intersect the number line at 4 and 6, because 4 and 6 are a distance 1 from the center of the circle 5. Such a circle is called a **two-point circle**. 4 and 6 constitute a two-point circle whose center is 5. The equation describing this two-point circle is $|x - 5| = 1$.

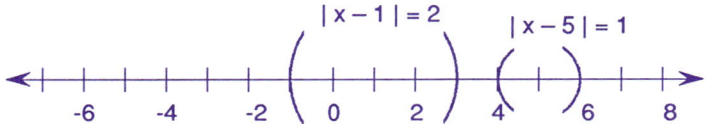

-1 and 3 constitute a two-point circle whose center is at 1. The distance between -1 and 1 is 2, and the distance between 3 and 1 is 2. The equation describing this two-point circle is $|x - 1| = 2$.

-5 is 3 from -2, and 1 is 3 from -2. Consequently, {-5,1} is a two-point circle whose center is at -2. The equation of the two-point circle {-5,1} is $|x + 2| = 3$.

To find the two-point circle whose equation is $|x - 2| = 4$, it is necessary to find the numbers that are a distance 4 from 2 on the real number line. $|x - 2| = 4$ is the equation of the two-point circle {-2,6}.

Absolute Value and Neighborhoods

The truth set of $|x - 2| = 1$ contains elements that are a distance 1 from 2. The truth set of $|x - 2| < 1$ contains elements that are a distance less than 1 from 2. It is an infinite set, because there is an unlimited amount of numbers that are a distance of less than 1 from 2. The graph below shows the truth set of $|x - 2| < 1$.

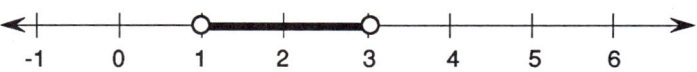

The shaded area of the graph is a **neighborhood** of 2 that consists of all points that are less than 1 from 2.

1.01, $\frac{3}{2}$, 1.99, 2.6, and 2.999 are in the neighborhood.
0, .5, 1, 3, $\frac{1}{2}$, and 4.7 are not in the neighborhood.

Absolute Value

The absolute value inequality $|x - 3| < \frac{1}{2}$ indicates a neighborhood that consists of the shaded area of the graph below.

The neighborhood consists of every point that is less than $\frac{1}{2}$ from 3. Which of the following numbers are in the above neighborhood?

$$\frac{7}{4}, 2, 2\frac{1}{2}, 2\frac{7}{10}, 3, 3\frac{2}{10}, 3\frac{1}{2}, 4, 5\frac{9}{10}$$

The previous question can be answered by finding the numbers that are less than $\frac{1}{2}$ from 3.

$2\frac{7}{10}, 3, 3\frac{2}{10}$ are the numbers that are less than $\frac{1}{2}$ from 3.

In general, if b > 0, the neighborhood that consists of all points less than b from a is indicated by the absolute value inequality $|x - a| < b$. The neighborhood of points that are less than .01 from 1 is indicated by the inequality $|x - 1| < .01$. The neighborhood around -3 that consists of all points less than .2 from -3 is indicated by the open sentence $|x + 3| < .2$.

Focus on neighborhoods on the number line

Unit 4 Exercise

Part A: Answers for all Part A problems are at the back of the book.

For problems 1-3, write an absolute value equation that describes it.

1. The distance between a number and 2 is 1.
2. The distance between a number and 5 is 3.
3. The distance between a number and -1 is 3.
4. Find any numbers that are a distance of 4 from 2 on the real number line.
5. $|x + 5| = 3$ is the equation of a two-point circle whose center is at -5 and whose radius is 3. Write the absolute value equation of the two-point circle whose center is at 3 and whose radius is 7.
6. $|x - 4| = 1$ is the equation of the two-point circle {_____, _____}.
7. $|x + 1| = 5$ is the equation of the two-point circle {_____, _____}.

8. Write the equation of the two-point circle $\{-2, 6\}$.

9. Graph the neighborhood indicated by $|x - 4| < 1$.

10. Which of the following numbers are elements of the neighborhood indicated by $|x - 3| < .2$?
    0, .5, .8, 1.8, 1.9, 2.8, 2.9, 3, 3.19, 3.2, 3.8, 3.9

11. $|x - 1| < \frac{1}{10}$ indicates a neighborhood of points less than _____ from _____.

12. Write an absolute value inequality that indicates the neighborhood around 5 containing points that are less than .03 from 5.

Part B: Answers for odd-numbered problems of Part B are at the back of the book.

1. Write an absolute value equation to indicate that the distance between x and 3 is 8.

2. Write an absolute value equation to indicate that the distance between x and -2 is 3.

3. Find any numbers that are a distance 2 from the number 1 on the number line.

4. Find any numbers that are 7 from -3 on the number line.

5. There are two numbers such that the distance between 4 times either number and 1 is 5. Find the numbers.

6. There are two numbers such that the distance between 2 times either number and 3 is 7. Find the numbers.

7. $|x - 3| = 4$ is the equation of the two-point circle _____, _____.

8. Write the equation of the two-point circle $\{1, 5\}$.

9. For the neighborhood indicated by $|x - 3| < 2$, every point is less than _____ from _____.

10. Write an absolute value inequality for the neighborhood around -7 that contains every point less than .02 from -7.

11. Is 4.3 in the neighborhood shown by $|x - 5| < .8$?

12. Is .8 in the neighborhood shown by $|x - 1| < .1$?

13. Is $\frac{5}{4}$ in the neighborhood of $|x - 1| < \frac{3}{8}$?

14. Write an open sentence that indicates that the distance between x and 2 is 5.

15. The open sentence $|x + 1| = 3$ indicates that the distance between x and _____ is _____.

16. Write an absolute value open sentence whose solution is the answer to the following problem. A window 8 feet wide has to be located so that its center is 11 feet from a door in the wall. How far are the sides of the window from the door?

17. The inequality $|x - 4| < .5$ indicates a neighborhood of points containing all points less than _____ from _____.

18. Is $\frac{-9}{10}$ in the neighborhood expressed by the absolute value inequality $|x + 1| < \frac{1}{10}$?

19. A straight stick is laid along a number line so that each end of the stick is 18 units from 5. How far are the ends of the stick from 0?

20. The center of a 10 foot wide door is located 12 feet from a garage corner. How far are the sides of the door from the corner?

21. A stick 6 units long is placed along a number line so that its center is at -1. At what points on the number line will the ends of the stick fall?

22. A compass is placed at 2 on a number line and used to make a mark on the number line 3 units on either side of 2. Where will the marks fall on the number line?

23. Mr. Hamner has to cut a window in a wall so that its center will be 5 feet from a door and the window is to be 4 feet wide. How far should each side of the window be from the door?

24. If a door 8 feet wide is placed in a wall so that its center is 5 feet from a corner, how far are the sides of the door from the corner?

Part C: No answers are given for these problems. However, each is accompanied by an ordered pair «C,U» showing the chapter and unit in which it was taught.

1. Factor $7x^2 + 29xy + 4y^2$ «1,4»
2. Simplify $5x(4x - 3) \leq 0$ «1,6»
3. Simplify $\sqrt{\frac{2}{5}}$ «2,2»
4. Solve $|7x - 2| = 12$ «4,2»
5. Solve $|x - 3| > 5$ «4,3»
6. Solve $|7x - 3| < -3$ «4,3»
7. Use inequalities to show the truth set. $|-3x - 1| \geq 8$ «4,3»
8. Use inequalities to show the truth set. $|x - 5| < 3$ «4,3»
9. Graph $|3x + 5| \leq 4$ «4,3»
10. Graph $|x + 4| = 3$ «4,3»

Chapter 4 Test

«4,U» shows the unit in which this problem was studied in this chapter.

1. Is the distance between -5 and 7 the same as the distance from 7 to -5? «4,1»

2. Do | 9 | and | -9 | have the same evaluation? «4,1»

Graph problems 3-4 on the real number line.

3. $|x| \leq 4$ «4,1»

4. $|x| > 3$ «4,1»

5. True or false? $|3 - 8| = |3| - |8|$ «4,2»

6. Evaluate $|7 - 12| - |-5|$ «4,2»

7. Solve $|x - 5| = 3$ «4,2»

8. Solve $|2x - 3| = 9$ «4,2»

9. Solve $|x - 7| = -4$ «4,2»

Graph the truth sets for problems 10-14.

10. $|x + 4| = 2$ «4,3»

11. $|x - 5| > 3$ «4,3»

12. $|x + 3| \leq 2$ «4,3»

13. $|3x - 4| < -5$ «4,3»

14. $|2x - 7| \leq 3$ «4,3»

For problems 15-18, use interval notation to describe the truth sets.

15. $|2x - 1| > 5$ «4,3»

16. $|4x + 3| \geq -5$ «4,4»

17. $|5x + 7| < 15$ «4,4»

18. $|13x - 1| \leq 7$ «4,4»

19. Write an open sentence that indicates the distance between the number represented by x and -3 is 4. «4,4»

20. Is $\frac{1}{2}$ in the neighborhood of $|x - 1| < \frac{5}{8}$? «4,4»

5
Relations and Functions

Unit 1: SETS OF ORDERED PAIRS

Definition	Equal Sets	
Two sets are equal	**if and only if**	they have exactly the same elements.

{5,7,3} is equal to {7,5,3}, because both sets have exactly the same elements. {1,2} = {2,1} is a true statement.

{4,7} ≠ {(4,7)} because {4,7} has two elements and {(4,7)} has only one element which is an ordered pair.

{2,4,6,...,100} ≠ {2,4,6, ... } because the three dots followed by 100 in the first set indicate that 100 is the largest element, but the three dots in the second set are not followed by any other element and indicate that the elements continue forever. The set is **infinite**.

Sets and Elements

Any collection of numbers, objects, or ideas may be called a set. The collection 6, 13, 19, 27, and 41 is a set of counting numbers and may be shown by the **roster method** as {6,13,19,27,41}. The set has 5 elements. The symbol for "is an element of" is "∈" and the symbol for "is not an element of" is "∉."

$19 \in \{6,13,19,27,41\}$ is a true statement.
$23 \notin \{6,13,19,27,41\}$ is another true statement.

Similarly, the set of ordered pairs {(1,7),(4,5),(-3,7),(6,-9)} has four elements.

$(4,5) \in \{(1,7),(4,5),(-3,7),(6,-9)\}$ is a true statement.
$7 \notin \{(1,7),(4,5),(-3,7),(6,-9)\}$ is another true statement.

Definition	Equal Ordered Pairs	
The ordered pairs (x,y) and (z,w) are equal	if and only if	$x = z$ and $y = w$

$(-3,7) \neq (7,-3)$ because $-3 \neq 7$
$(.5, .8) = \left(\frac{1}{2}, \frac{4}{5}\right)$ because $.5 = \frac{1}{2}$ and $.8 = \frac{4}{5}$

Relations and Functions 201

Focus on ordered pairs

The elements of {(2,7),(8,5),(9,7)} are ordered pairs.
    (2,7) ∈ {(2,7),(8,5),(9,7)} is a true statement.
    7 ∈ {(2,7),(8,5),(9,7)} is a false statement.

(6,1) ∈ {(1,6),(4,9),(7,6)} is a false statement. The ordered pair (1,6) is an element but (6,1) is not an element.

(4,6) ∉ {(1,3),(5,2),(6,4)} is true because (4,6) is not the same ordered pair as (6,4).

The set {6,7,12,19} has four elements. The set {(5,7),(6,13)} has two elements, each of which is an ordered pair.

Matching Elements of Two Sets

The figure at the right shows a pairing between two sets, {4,7,9,10} and {2,3,7,11}. This matching might be described by the set of ordered pairs {(4,2),(7,3),(9,7),(10,11)}.

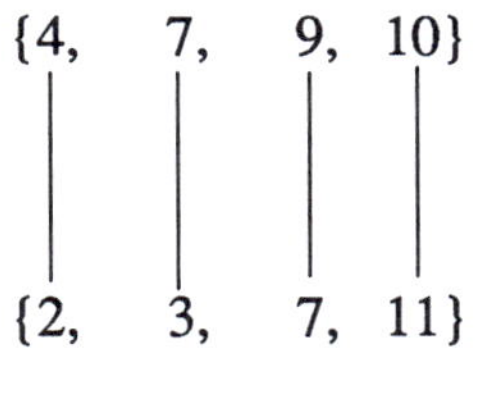

{4, 7, 9, 10}
 | | | |
{2, 3, 7, 11}

It is not possible to have a pairing between the sets {3,5,9} and {5,6,7,8} because there is always one element of {5,6,7,8} that is unpaired.

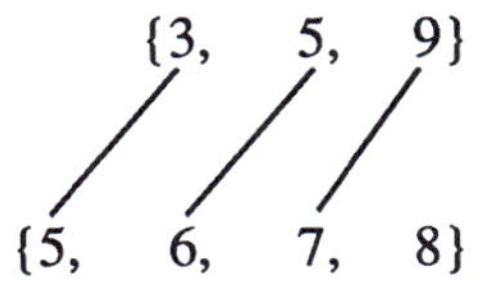

{3, 5, 9}
 /   /   /
{5, 6, 7, 8}

Focus on pairings between two sets

The sets {1,2,3,...,105} and {3,6,9,...,315} can be paired by matching each element of {1,2,3,...,105} with the number three times its size in the second set.

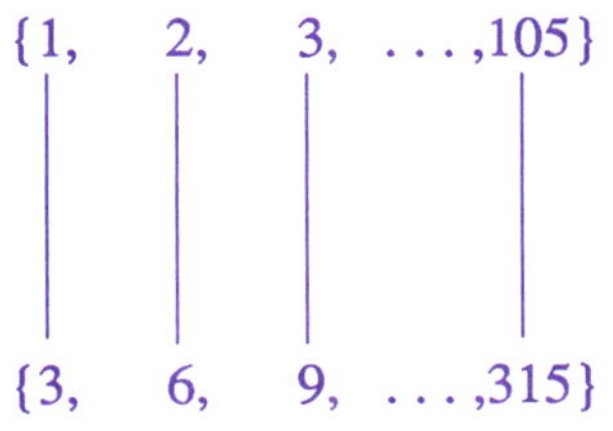

{1, 2, 3, ...,105}
 | | | |
{3, 6, 9, ...,315}

The pairing might be described by the set of ordered pairs {(1,3),(2,6),(3,9), . . . ,(105,315)}.

Definitions		Finite and Infinite Sets
A set is a finite set	if and only if	it is the empty set or it is possible to pair it with a set $\{1,2,3,\ldots,n\}$ where n is a specific counting number.
A set is an infinite set	if and only if	it is not finite.

Showing Matching of Elements Algebraically

The matching of two infinite sets is shown at the right. This matching could be described by the **rule of correspondence** $x \Rightarrow 5x$ which indicates that each element, x, of the first set is matched with 5x in the second set.

$$\{1, \quad 2, \quad 3, \quad \ldots, x\}$$
$$\{5, \quad 10, \quad 15, \ldots, 5x\}$$

Focus on rules of correspondence

The rule of correspondence $x \Rightarrow x + 8$ could describe the matching of elements for the two infinite sets shown at the right.

$$\{1, \quad 2, \quad 3, \quad \ldots \quad x, \quad \ldots\}$$
$$\{9, \quad 10, \quad 11, \quad \ldots \quad x + 8, \quad \ldots\}$$

The same rule of correspondence, $x \Rightarrow x + 8$, could describe the set of ordered pairs $\{(1,9),(2,10),(3,11),\ldots,(n,n+8),\ldots\}$.

Unit 1 Exercise

Part A: Answers for all Part A problems are at the back of the book.

1. Use braces to show the set containing 9, 51, 96, and 103.
2. Is 29 an element of {4,13,21,96}?
3. True or false? 65 ∈ {18,51,60,65,91,106}
4. True or false? 22 ∉ {4,6,15,25,36}
5. Two sets are equal if they contain exactly the same elements. Is {1,3,5, . . .} equal to {2,4,6, . . }?
6. Two ordered pairs are equal if their first components are equal and their second components are equal. Is (6,-5) equal to (-5,6)?
7. True or false? 6 ∈ (6,7)
8. How many elements are in {(3,6),(4,8),(5,10)}?
9. Is 6813 an element of {1,2,3, . . .}?
10. How many elements are in {1,3,5, . . .,79}?
11. How many elements are in {5,6,7, . . . ,53}?
12. How many elements are in {7,14,21, . . . ,84}?
13. How many elements are in {70,80,90, . . . ,240}?
14. Is the set of rational numbers between $\frac{1}{2}$ and $\frac{3}{4}$ a finite set?
15. Find a rule of correspondence that will match the elements of {1,2,3, . . .} with the elements of {3,6,9, . . .}.
16. Find a rule of correspondence that will match the elements of {1,2,3, . . .} with the elements of {4,7,10, . . .}.
17. Is {1,2,3, . . . ,210} a finite set?
18. Is {1,2,3, . . .} a finite set?
19. Is the set of integers {. . . ,-2,-1,0,1,2, . . .} finite or infinite?
20. {2,4,6, . . .} is the set of counting numbers that can be divided evenly by 2. {2,4,6, . . .} is the set of even counting numbers. Is the set finite or infinite?

Part B: Answers for odd-numbered problems of Part B are at the back of the book.

For problems 1-20, label the statements true or false.

1. {3,4,7} = {3,5,7}
2. (-11,-1) = (-1,-11)
3. {2,4,7,9} and {(2,3),(4,7)} have the same number of elements.
4. 43 ∈ {19,27,43,89}
5. 16 ∈ {2,9,13,18,27}
6. 83 ∉ {1,45,86,94,106}
7. 37 ∉ {2,28,31,37,46}
8. 7 ∈ {(2,9),(7,13),(6,1)}
9. 68 ∈ {1,2,3, . . . ,52}
10. 16 ∈ {1,2,3, . . . ,25}
11. {3,7,18} = {1,2,3}
12. {92,13,2} = {13,92,2}

13. $\{10,2,7\} = \{10,7,3\}$

14. $\{(3,5),(6,8)\} = \{(6,8),(3,5)\}$

15. $\{(7,10)\} = \{(10,7)\}$

16. $\{15,23\} = \{(15,23)\}$

17. $\{2,5\} = \{(2,5)\}$

18. $\{1,2,3,\ldots,10\}$ is an infinite set.

19. $\{(3,4)\} = \{3,4\}$

20. $\{(2,7),(3,8)\} = \{(3,8),(2,7)\}$

For problems 21-30, find a rule of correspondence that would match the elements of $\{1,2,3,\ldots\}$ with the given set.

21. $\{58,59,60,\ldots\}$

22. $\{9,18,27,\ldots\}$

23. $\{1,\frac{1}{2},\frac{1}{3},\ldots\}$

24. $\{1,4,9,16,\ldots\}$

25. $\{8,9,10,\ldots\}$

26. $\{7,9,11,\ldots\}$

27. $\{-2,-4,-6,\ldots\}$

28. $\{4,9,14,\ldots\}$

29. $\{\frac{3}{5},1,\frac{7}{5},\ldots\}$

30. $\{\frac{-1}{5},\frac{-2}{5},\frac{-3}{5},\ldots\}$

Part C: No answers are given for these problems. However, each is accompanied by an ordered pair «C,U» showing the chapter and unit in which it was taught.

1. Solve $\frac{-2}{x-3} + \frac{-2}{x^2-2x-3} = \frac{5}{x+1}$ «2,7»

2. Simplify $\frac{3-2\sqrt{5}}{2-\sqrt{5}}$ «2,2»

3. Solve $(4-3i)(x+yi) = (2-5i)$ «3,4»

4. Use interval notation to describe the truth set. $|2x-1| < 3$ «4,3»

5. Graph $|x+7| = 2$ «4,3»

6. $|x-5| = 1$ is the equation of the two-point circle $\{\_\_\_,\_\_\_\}$. «4,4»

7. Write the equation of the two point circle $\{-1,4\}$. «4,4»

8. Find the numbers that are 5 from -1 on the number line. «4,4»

9. Is 4.5 in the neighborhood shown by $|x-4| < .7$? «4,4»

10. The open sentence $|x+3| = 5$ indicates that the distance between x and ____ is ____. «4,4»

Unit 2: THE SET SELECTOR METHOD OF SHOWING SETS OF ORDERED PAIRS

Set Descriptions

Sets can be described by:
1. The roster method
    The set containing 4, 7, and 10 is shown as {4,7,10}.

2. The three dot method
    The set of counting numbers between 1 and 9 is {2,3,4, . . . ,8}.

3. The set selector method
    {2,4,6, . . .} is {x | x is an even counting number}.

Focus on sets described by the set selector method

{x | x is an even counting number} is read as "the set of all replacements for x that will make 'x is an even counting number' a true statement." The condition required for membership in the set is shown by the open sentence "x is an even counting number."

$\frac{5}{4} \in$ {x | x is a rational number} is a true statement, because the condition for membership is that the replacement for x be a rational number.

{x | x is less than 10 or greater than 15}. This is a set that has two conditions for membership.
    1. x is less than 10.
    2. x is greater than 15.
The word "or" connects these two conditions. In mathematics "or" means that either condition is sufficient for membership.

The claim that 7 ∈ {x | x is less than 10 or greater than 15} is true, because 7 meets at least one of the conditions: 7 is less than 10.

The claim that 15 ∉ {x | x is even or x is greater than 21} is true, because 15 does not satisfy either of the required conditions.

The word "or" and set operation union are related.
The union of two sets A and B can be shown by {x | x ∈ A or x ∈ B}.
For any two sets A and B, A ∪ B = {x | x ∈ A or x ∈ B}.

The Use of the Word "And"

{x | x is less than 25 and greater than 10.} This is a set that has two conditions for membership.
1. x is less than 25.
2. x is greater than 10.

The word "and" connects these two conditions. In mathematics "and" means that both conditions must be true for membership in the set.

The claim that $13 \in$ {x | x is less than 25 and x is greater that 10} is true because it satisfies both conditions for membership in the set. 13 is less than 25. Also, 13 is greater than 10.

The claim that $9 \in$ {x | x is less than 25 and x is greater than 10} is false because it does not satisfy both conditions required for membership. Specifically, 9 is not greater than 10.

Focus on the use of "and" in open sentences

The word "and" and the set operation "intersection" are related. The intersection of two sets consists of all those elements common to both sets. Two conditions connected by the word "and" create a situation that is true only when both conditions are satisfied.

The intersection of two sets A and B can be shown by
{x | x $\in$ A and x $\in$ B}.

For any two sets A and B, $A \cap B$ = {x | x $\in$ A and x $\in$ B}.

Describing Sets by the Set Selector Method

{x | x + 9 = 1} is a set described by the set selector method. To be an element of the set, a replacement for x must make x + 9 = 11 a true statement. {2} and {x | x + 9 = 11} are two methods for describing the same set.

There are two elements in {x | (x – 6)(x + 2) = 0}, because either 6 or -2 makes (x – 6)(x + 2) = 0 a true statement.

{x | (x – 6)(x + 2) = 0} = {6,-2}

Relations and Functions   207

To find the set equal to $\{x \mid x^2 - 2x - 24 = 0\}$, first factor $x^2 - 2x - 24$.

$$\{x \mid x^2 - 2x - 24 = 0\} = \{x \mid (x - 6)(x + 4) = 0\}$$
$$= \{6, -4\}$$

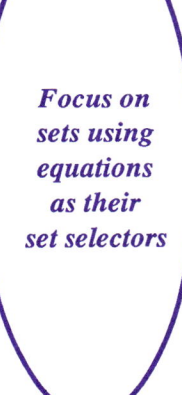

Focus on sets using equations as their set selectors

$\{x \mid x + 5 = 9 \text{ or } x + 6 = 13\}$ has two elements, 4 and 7, because 4 makes the first equation true and 7 makes the second equation true. The use of the word "or" connecting the two equations allows 4 and 7 to be elements.

$$\{x \mid x + 5 = 9 \text{ or } x + 6 = 13\} = \{4, 7\}$$

$\{x \mid x + 3 = 5 \text{ and } x - 6 = 11\}$ is the empty set $\{\ \}$. The use of "and" connecting the two equations requires that both conditions be satisfied for the same replacement of x. There is no replacement for x that will make both conditions true. Therefore, the set has no elements.

Ordered Pairs as Set Elements

The set selector sentence of $\{(x, y) \mid y = 2x - 3\}$ is an equation with two variables, x and y, and its elements are indicated to be ordered pairs (x,y).

$\{(x, y) \mid y = 2x - 3\}$ is a set of ordered pairs. To decide whether (5,7) is an element of the set, replace x by 5 and y by 7 to test whether $y = 2x - 3$ is a true statement. (5,7) is an element of $\{(x, y) \mid y = 2x - 3\}$ because $7 = 2 \cdot 5 - 3$ is a true statement.

$\{(x, y) \mid x^2 + 2y^2 = 25\}$ is another set of ordered pairs. (5,0), (-5,0), ($\sqrt{7}$,3), (-$\sqrt{7}$,3), ($\sqrt{7}$,-3) and (-$\sqrt{7}$,-3) are six elements of the set because the components of the ordered pairs, when used as replacements for x and y, result in true statements.

Focus on ordered pairs as elements

$\{(x, y) \mid x^2 = y\}$ is a set of ordered pairs. (-6,36) is an element of the set because $(-6)^2 = 36$ is true. (4,8) is not an element because $4^2 = 8$ is false.

$\{(x, y) \mid y = \sqrt{x + 1}\}$ is a set of ordered pairs. (5,$\sqrt{6}$) is an element of the set because $\sqrt{6} = \sqrt{5 + 1}$ is true. (3,6) is not an element because $6 = \sqrt{3 + 1}$ is false.

Completing Ordered Pairs as Elements

The ordered pair (7,___) can be completed as an element of $\{(x, y) \mid y = 2x - 3\}$ by replacing x with a 7 in the equation $y = 2x - 3$. When the equation is solved for y, the root will be the second component of the ordered pair.

(7,___) and $y = 2x - 3$ gives:
$$y = 2 \cdot 7 - 3 \text{ or } y = 11$$

Therefore, $(7,11) \in \{(x, y) \mid y = 2x - 3\}$.

Similarly, (4,___) can be completed as an element of $\{(x, y) \mid x^2 + 2y^2 = 90\}$.

(4,___) and $x^2 + 2y^2 = 90$ gives:
$$4^2 + 2y^2 = 90$$
$$16 + 2y^2 = 90$$
$$2y^2 = 74$$
$$y^2 = 37$$
$$y = \sqrt{37} \text{ or } y = -\sqrt{37}$$

Therefore, both $(4,\sqrt{37})$ and $(4,-\sqrt{37})$ are elements of $\{(x, y) \mid x^2 + 2y^2 = 90\}$.

Focus on completing ordered pairs as elements

To complete (5,___) as an element of $\{(x, y) \mid y = \sqrt{x + 7}\}$:

(5,___) and $y = \sqrt{x + 7}$ gives:
$$y = \sqrt{5 + 7}$$
$$y = \sqrt{12} \text{ or } y = 2\sqrt{3}$$

Therefore, $(5, 2\sqrt{3}) \in \{(x, y) \mid y = \sqrt{x + 7}\}$.

An attempt to complete (6,___) as an element of $\{(x, y) \mid y = \sqrt{4 - x}\}$ leads to the following.

(6,___) and $y = \sqrt{4 - x}$ gives:
$$y = \sqrt{4 - 6}$$
$$y = \sqrt{-2}$$

Since $\sqrt{-2}$ is not a real number, there is no element of $\{(x, y) \mid y = \sqrt{4 - x}\}$ with first component 6.

Relations and Functions 209

Unit 2 Exercise

Part A: Answers for all Part A problems are at the back of the book.

1. True or false?
   a. $7 \in \{ x \mid x$ is an even number less than $12\}$
   b. 7 is an even number less than 12.

2. True or false?
   a. $-9 \in \{ x \mid -2x + 7 = 25\}$
   b. $-2 \cdot -9 + 7 = 25$

3. The set $\{ x \mid x$ is an integer and $10 < x^2 < 40\}$ is equal to $\{4,5,6,-4,-5,-6\}$. Find the set equal to $\{ x \mid x$ is an integer and $x^2 < 20\}$.

4. $\{x \mid x$ is a counting number less than 6 or $x^2 = 5\}$ is equal to $\{1,2,3,4,5,\sqrt{5},-\sqrt{5}\}$. Find the set equal to $\{x \mid x$ an integer between -6 and -2 or $x^2 = 14\}$.

5. Every element of $\{(x, y) \mid y - 3 = \sqrt{x - 1}\}$ is an ordered pair (x,y). Are there any real number elements of the set?

6. $(17,7) \in \{(x, y) \mid y - 3 = \sqrt{x - 1}\}$ because $7 - 3 = \sqrt{17 - 1}$ is a true statement. Complete the ordered pair (50,___) as a set element.

7. Find the 8 elements of $\{(x, y) \mid x$ and y integers and $\mid x \mid + \mid y \mid = 2\}$.

8. Find $\{x \mid x \in A$ or $x \in B\}$ when $A = \{7,15,19\}$ and $B = \{15,20,22\}$.

9. Describe the set $\{x \mid x + 4 = 19\}$ in a simpler way.

10. $\{x \mid (x + 9)(x - 3) = 0\} =$ _____

11. $\{x \mid x^2 + 11x + 28 = 0\} =$ _____

12. $\{x \mid x^2 + 7x - 3 = 0\} =$ _____

13. $\{x \mid x - 7 = 1$ or $x + 4 = 15\} =$ _____

14. $\{x \mid x + 9 = 4$ and $x + 3 = 15\} =$ _____

15. $\{x \mid 5x + 1 = 2x - 8$ and $x + 5 = 2\} =$ _____

16. Is (-3,2) an element of $\{(x, y) \mid x + y + 1 = 0\}$?

17. Is (0,0) an element of $\{(x, y) \mid 3x + y = 0\}$?

18. True or false? $(2,-4) \in \{(x, y) \mid y = -x^2\}$

19. True or false? $(4,3) \in \{(x, y) \mid y = \sqrt{x} + 1\}$

Part B: Answers for odd-numbered problems of Part B are at the back of the book.

True or False?

1. $11 \in \{x \mid x$ is a counting number$\}$
2. $-7 \in \{x \mid x$ is an integer$\}$
3. $-5 \notin \{x \mid x$ is an integer$\}$
4. $13 \in \{x \mid x$ is an odd counting number$\}$
5. $12 \notin \{x \mid x$ is an odd counting number$\}$
6. $14 \in \{x \mid x$ is an even counting number$\}$
7. $\frac{5}{3} \in \{x \mid x$ is a counting number$\}$
8. $29 \in \{x \mid x$ is an integer$\}$
9. $12 \in \{x \mid x$ is a counting number less than 10$\}$
10. $5 \in \{x \mid x$ is a counting number greater than 7$\}$
11. $\frac{2}{3} \in \{x \mid x$ is a rational number greater than 0$\}$
12. $1.36 \in \{x \mid x$ is rational and greater than 2$\}$

13. $-\sqrt{9} \in \{x \mid x \text{ is an integer}\}$
14. $\frac{-3}{7} \notin \{x \mid x \text{ is a rational number}\}$
15. $6 \notin \{x \mid x \text{ is a rational number}\}$
16. $14 \in \{x \mid x \text{ is less than 10 or greater than 15}\}$
17. $45 \in \{x \mid x \text{ is even or } x \text{ is greater than 21}\}$
18. $30 \in \{x \mid x < 25 \text{ and } x > 10\}$
19. $12 \in \{x \mid x \text{ is odd and } x < 19\}$
20. $(-1,9) \in \{(x,y) \mid y = 7x - 2\}$
21. $(-2,-2) \in \{(x,y) \mid y = -x\}$
22. $(1,-1) \in \{(x,y) \mid 2x - 5y = 7\}$
23. $(-1,5) \in \{(x,y) \mid y = -2x + 3\}$
24. $(5,-2) \in \{(x,y) \mid x - y = 3\}$
25. $(1, \frac{2}{3}) \in \{(x,y) \mid \frac{-1}{3}x + y = 1\}$
26. $(-5,5) \in \{(x,y) \mid y = -x\}$
27. $(12,4) \in \{(x,y) \mid y = \sqrt{x+4}\}$
28. $(3,-15) \in \{(x,y) \mid y = -5\}$
29. $(9,3) \in \{(x,y) \mid y = x^2\}$
30. $(-1,2) \in \{(x,y) \mid y = \sqrt{5+2x}\}$
31. $(1,3) \in \{(x,y) \mid y = x^2 - 2x + 3\}$
32. $(-5,14) \in \{(x,y) \mid y = x^2 + 3x + 4\}$
33. $(-5,12) \in \{(x,y) \mid x^2 + y^2 = 169\}$
34. $(4,-7) \in \{(x,y) \mid (x-4)^2 + (y+3)^2 = 49\}$
35. $(2, 3\sqrt{5} - 3) \in \{(x,y) \mid (x-4)^2 + (y+3)^2 = 49\}$
36. $(2,3) \in \{(x,y) \mid y = \sqrt{7-x}\}$

For problems 37-43 complete the ordered pairs.

37. $(-4, \underline{}) \in \{(x,y) \mid y = 7x - 2\}$
38. $(5, \underline{}) \in \{(x,y) \mid 2x - 5y = 7\}$
39. $(-3, \underline{}) \in \{(x,y) \mid y = -2x + 3\}$
40. $(-8, \underline{}) \in \{(x,y) \mid x - y = 3\}$
41. $(6, \underline{}) \in \{(x,y) \mid \frac{-1}{3}x + y = 1\}$
42. $(-3, \underline{}) \in \{(x,y) \mid y = x^2 - 3\}$
43. $(-9, \underline{}) \in \{(x,y) \mid y = \sqrt{7-x}\}$

Part C: No answers are given for these problems. However, each is accompanied by an ordered pair «C,U» showing the chapter and unit in which it was taught.

1. Use interval notation to describe the truth set of $|x + 4| < 2$ «4,3»
2. Graph $|x - 3| = 5$ «4,3»
3. $|x - 3| = 2$ is the equation of the two-point circle {_____, _____}. «4,4»
4. Find the numbers that are 7 from -2 on the number line. «4,4»

True or false? (Problems 5-8)

5. $7 \in \{(1,7),(7,1),(7,7)\}$ «5,1»
6. $\{(4,8)\} = \{(8,4)\}$ «5,1»
7. $(3,7) \in \{(1,2),(2,5),(3,7),(7,9)\}$ «5,1»
8. $\{3,9\} = \{(3,9)\}$ «5,1»

For problems 9-10, find a rule of correspondence that would match elements of $\{1,2,3,\ldots\}$ with:

9. $\{34,35,36,\ldots\}$ «5,1»
10. $\{1, \frac{1}{4}, \frac{1}{9}, \ldots\}$ «5,1»

Unit 3: RELATIONS

> **Definition** **Relation**
>
> A set W is   if and   every element of W
> a relation   only if   is an ordered pair.

Relations as Sets of Ordered Pairs

A relation is a set of ordered pairs. Each of the following sets are relations shown by the roster or three dot method.

$\{(4,-2),(-5,5),(9,4),(6,-3),(15,-2)\}$
$\{(2,5)\}$
$\{(5,\frac{1}{2}),(7,\frac{1}{4}),(9,\frac{2}{3}),(\frac{3}{4},4)\}$
$\{(1,1),(2,4),(3,9),(4,16), \ldots ,(10,100)\}$

Any set of ordered pairs is a relation. To be a relation, every element of a set must be an ordered pair. Every element of each of the following is an ordered pair. Consequently, each is a relation.

$\{(x, y) \mid x + 7 = y\}$
$\{(x, y) \mid 4x + 3y = 7\}$
$\{(x, y) \mid y = x^2 - 7\}$
$\{(x, y) \mid x^2 + y^2 = 25\}$

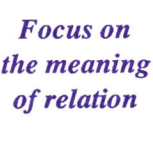

Focus on the meaning of relation

Any set of ordered pairs is a relation. Therefore, to determine whether a set is a relation determine if all elements are ordered pairs.

The elements of $\{x \mid x + 5 = 4\}$ are not ordered pairs. Consequently, the set is not a relation.

The elements of $\{(x, x + 3) \mid x \text{ is a counting number}\}$ are ordered pairs. Consequently, the set is a relation. It could be shown by the three dot method as $\{(1,4),(2,5),(3,6), \ldots ,(x,x + 3), \ldots \}$.

Showing Relations by Equations

Every equation with two variables implicitly describes a set of ordered pairs involving those two variables. Such equations, by themselves, are often treated as relations.

For example, the equation $x + 5y = 10$ is an equation with an infinite number of ordered pairs (x,y) that would result in a true numerical statement. By definition, the set containing all of these ordered pair solutions is a relation and by common practice the equation itself is treated as the same relation.

$\{(x, y) \mid x + 5y = 10\}$ is the same relation as that described by the equation $x + 5y = 10$.

Focus on relations described by rules of correspondence

A rule of correspondence such as $x \Rightarrow 5x - 3$ describes ordered pairs of real numbers $(x, 5x - 3)$ and the set of all such ordered pairs is, by definition, a relation. Consequently, a rule of correspondence is often treated as a relation.

$\{(x, 5x - 3) \mid x \text{ is a real number}\}$ is the same relation as that described by the rule of correspondence $x \Rightarrow 5x - 3$.

In the remainder of this text, all equations with two variables and rules of correspondences are to be considered relations unless specifically stated otherwise.

Domain of a Relation

The set of all first components of a relation is called the **domain** of the relation. The domain of $\{(3,5),(4,6),(11,13),(17,20)\}$ is $\{3,4,11,17\}$. The domain of $\{(2,5),(9,1),(-6,3)\}$ is $\{2,9,-6\}$. The domain of any relation described by listing the elements of a set is restricted to the first components of the ordered pairs.

The domain of a relation described by an equation or rule of correspondence is understood to be all real numbers that result in meaningful numerical expressions. For example, the domain of $\{(x, y) \mid 3x - 2 = y\}$ is the set of real numbers, because any real number can replace x and the expression $(3x - 2)$ can be calculated.

The domain of the relation described by the equation $y = \frac{3}{x-5}$ consists of all real numbers except 5. If x is replaced by 5, the result is $\frac{3}{5-5} = \frac{3}{0}$ and $\frac{3}{0}$ is not a meaningful numerical expression.

Focus on the domain of a relation

For the relation $\{(x, y) \mid \sqrt{x} = y\}$, x cannot be replaced by a negative number. This is because the square root of a negative number is not a real number (complex numbers are not used here in the study of relations).

3 is not in the domain of $y = \sqrt{x - 4}$ because $\sqrt{3-4} = \sqrt{-1}$ and $\sqrt{-1}$ is not a real number.

For the relation $x \Rightarrow \sqrt{x + 2}$, all real numbers less than -2 are excluded from the domain, because all real numbers less than -2 will give a negative evaluation to the expression $x + 2$.

Range of a Relation

The **range** of a relation is the set of all second components of the ordered pairs. For examples:

The range of $\{(4,-3),(2,6),(5,0),(4,7)\}$ is $\{-3,6,0,7\}$.
The range of $\{(9,-9),(5,-5),(-4,4),(-2,2),(0,0)\}$ is $\{-9,-5,4,2,0\}$.

The range of a relation described by an equation or rule of correspondence is understood to be those real numbers that will result from acceptable domain replacements. For example, the range of the relation $2x + 7 = y$ is the set of real numbers, because any real number may replace y.

Consider the relation $y = \frac{4}{x-7}$. If the form of the equation is changed to $x - 7 = \frac{4}{y}$ it can be seen that y cannot be replaced by 0. Therefore, 0 is not in the range of $y = \frac{4}{x-7}$.

The relation $y = x^2 + 7$ consists of ordered pairs $(x, x^2 + 7)$. Since x^2 is always greater than or equal to 0, $x^2 + 7$ can never be less than 7. Therefore, the range of $y = x^2 + 7$ is the set of real numbers greater than or equal to 7.

Focus on restrictions for domains and ranges

There are four important cases where some real number(s) must be excluded from the domain or range of a relation.

1. Replacements cannot be used if they result in a zero denominator.
   -4 is not in the domain of the relation $x \Rightarrow \frac{7}{x+4}$ because if x is replaced by -4, there would be a zero denominator.

2. Replacements cannot be used if they result in a negative radicand.
   The domain of the relation $\sqrt{x-1} = y$ excludes all real numbers less than 1, because all real numbers less than 1 will result in a negative number under the radical.

3. Replacements cannot be used if they result in a square root that is negative.
   The range of the relation $\sqrt{x-1} = y$ excludes all negative numbers, because the square root expression $\sqrt{x-1}$ cannot be negative.

4. Replacements cannot be used if they result in the square of a number being negative.
   The range of $\{(x, y) \mid y = (x-7)^2 + 1\}$ excludes all real numbers less than 1. For all replacements of x, the value of $(x-7)^2 + 1$ must be greater than or equal to 1.

Unit 3 Exercise

Part A: Answers for all Part A problems are at the back of the book.

1. Is the set $\{(4,3),(7,2),(6,-2),(5,4),(1,-3)\}$ a relation?

2. Is the set $\{6,5,4,(2,1)\}$ a relation?

3. Which of the following sets is not a relation?
   a. $\{(5,1),(4,7),(9,6),3\}$
   b. $\{(4,13)\}$
   c. $\{(1,2),(4,3),(7,1)\}$
   d. $\{1,2,4,3,7,1\}$

4. Which of the following is not a relation?
   a. $\{(x, y) \mid 3x + 2 = y\}$
   b. $\{(4,9),(5,10),(6,11),(7,12)\}$
   c. $\{x \mid 3x = 19\}$
   d. $\{(x, y) \mid x^2 + 3x - 2 = y\}$

5. Which of the following is not a relation?
   a. $\{(x, y) \mid 3x + 8 = y\}$
   b. $\{(x, y) \mid \frac{5}{x-2} = y\}$
   c. $\{4,2\}$
   d. $\{(4,2)\}$

6. What is the domain of {(5,-2),(4,1),(7,3),(-5,0)}?

7. What is the range of {(5,2),(4,3),(2,4),(1,5),(0,6)}?

8. What is the domain of $\{(x, y) \mid 2x + 5 = y\}$?

9. What is the range of $\{(x, y) \mid 2x + 7 = y\}$?

10. What replacement for x cannot be used in the equation $y = \frac{7}{x-3}$?

11. What replacement for y cannot be used in the equation $y = \frac{7}{x-3}$?

12. What replacements for x cannot be used in the equation $y = \sqrt{x-6}$?

13. What replacements for y cannot be used in the equation $y = \sqrt{x-6}$?

14. What replacement for x cannot be used in the equation $x - 3 = \frac{9}{y+2}$?

15. What replacement for y cannot be used in the equation $x - 3 = \frac{9}{y+2}$?

16. What replacements for x cannot be used in the equation $y = (x-2)^2 - 6$?

17. What replacements for y cannot be used in the equation $y = (x-2)^2 - 6$?

Part B: Answers for odd-numbered problems of Part B are at the back of the book.

Each of the problems contains a set, an equation or a rule of correspondence. State whether it is a relation and if so find its domain and range.

1. {(2,3),(4,3)}
2. {(1,7),(2,2),(0,0),(7,14)}
3. {1,2,3,4}
4. {2,3}
5. {(2,3),(2,5)}
6. {(3,7),2}
7. {(1,3),(2,0),(8,3)}
8. {(0,0),(2,8),(2,1),(8,4)}
9. $\{x \mid x + 2 = 7\}$
10. $\{(x, y) \mid x + y = 3\}$
11. $2x = 8$
12. $y = 3x - 2$
13. $\{(x, y) \mid x^2 = y\}$
14. $\{(x, y) \mid y = \frac{9}{x-6}\}$
15. $\{(x, y) \mid y - 2 = \frac{5}{x+10}\}$
16. $\{(x, y) \mid \sqrt{x} = y\}$
17. $\{(x, y) \mid y = \sqrt{x-4}\}$
18. $\{(x, y) \mid y + 3 = \sqrt{x-1}\}$
19. $x \Rightarrow \sqrt{x-9}$
20. $y = \sqrt{x-2}$
21. $\{(x, y) \mid \frac{6}{x+2} = y\}$
22. $y = \sqrt{x-10}$
23. $y = \sqrt{5-x}$
24. $\{(x, y) \mid \frac{5}{x-7} = y + 5\}$
25. $\{(x, y) \mid y = (x+5)^2\}$
26. $y - 7 = (x-6)^2\}$
27. $\{(x, y) \mid y = x^2 + 1\}$
28. $x \Rightarrow (x+4)^2 + 1$
29. $x = \sqrt{y+11}$
30. $x \Rightarrow \sqrt{2+x}$

31. $y + 8 = x^2 + 4$

32. $6 - y = (x - 7)^2$

33. $\{(x, y) \mid 3x - 4 = \frac{2}{6-y}\}$

34. $\{(x, y) \mid 7x + 2 = \frac{2}{4y-5}\}$

35. $\{(x, y) \mid y - 1 = (x + 2)^2 + 3\}$

36. $\{(x, y) \mid y = (3 + x)^2 + 2\}$

37. $\{(x, y) \mid y = 6 - x^2\}$

38. $y - 4 = \sqrt{8 - x}$

39. $y + 3 = \sqrt{9 - x^2}$

40. $x \Rightarrow \frac{7}{2-x}$

41. $y = \frac{1}{x^2 - 1}$

42. $\{(x, y) \mid 5x - 3 = y\}$

43. $\{(x, y) \mid x = \frac{4}{7+y}\}$

44. $\{(x, y) \mid y = 7 - (x + 4)^2\}$

45. $y = (x - 5)^2 + 8$

46. $9 - y = \frac{1}{5x + 2}$

47. $x^2 + y^2 = 25$

48. $x^2 + y^2 = 64$

49. $(x - 4)^2 + y^2 = 49$

50. $y = (x - 2)^2 + 3$

Part C: No answers are given for these problems. However, each is accompanied by an ordered pair «C,U» showing the chapter and unit in which it was taught.

1. Solve $(3 - 2i)(x + yi) = (3 + 2y)$ «3,4»

2. Graph $|2x + 1| = 7$ «4,3»

3. $|x - 5| = 3$ is the equation of the two-point circle {_____, _____}. «4,4»

Find the rule of correspondence that would match elements of $\{1, 3, 5, \ldots\}$ with the given set for problems 4 and 5.

4. $\{22, 24, 26, \ldots\}$ «5,1»

5. $\{-1, \frac{-1}{2}, \frac{-1}{4}, \ldots\}$ «5,1»

6. $\{x \mid (x - 7)(x + 3) = 0\} = $ _____ «5,2»

7. $\{x \mid 4x - 3 = x + 9 \text{ and } x - 1 = 3\} = $ _____ «5,2»

8. True or false? $(3, -9) \notin \{(x, y) \mid y = -x^2\}$ «5,2»

9. True or false?
$12 \notin \{x \mid x \text{ is greater than 10 and less than 15}\}$ «5,2»

10. True or false?
$(1, -6) \notin \{x \mid x^2 - 5x + 6\}$ «5,2»

Unit 4: FUNCTIONS

> **Definition**   **Function**
>
> A relation W   if and   whenever both (x,y) and
> is a **function**   only if   (x,z) are elements of W
>   then y = z.

{(4,8),(5,7),(6,8),(9,10)} is a function. It is a relation in which each first component is paired with exactly one second component. For example, if 6 is chosen as a first component then the one and only range element paired with it is 8. Choose any first component and there is exactly one second component to go with it.

{(4,8),(5,7),(6,8),(4,10)} is not a function. If 4 is chosen as a first component then it is paired with both 8 and 10. At least one first component is paired with more than one second component and this means the relation is not a function.

Functions as Special Relations

Every set of ordered pairs is a relation. Every function is a relation, and, therefore, is a set of ordered pairs. Some relations are not functions. A function is a special type of relation.

There are four elements in the relation {(4,1),(6,-5),(2,$\sqrt{2}$),(5,-9)} and four elements in the domain {4,6,2,5}. Each domain element is used as a first component exactly one time. The relation is a function.

{(6,3),(7,1),(7,9),(1,3)} is a relation with four elements, but its domain {6,7,1} has only three elements. This is because 7 is the first component of two ordered pairs (7,1) and (7,9). One element of the domain is used more than once as a first component. The relation is not a function.

(4,2) and (4,-2) are elements of the relation $x = y^2$. A domain element is used more than once as a first component. The relation is not a function.

218   Chapter 5

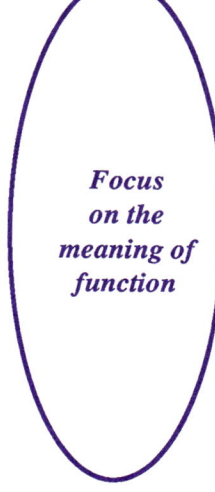

Focus on the meaning of function

A function is a relation in which each domain element is used exactly one time as a first component. {(1,3),(2,5),(3,7),(1,8)} is not a function, because the ordered pairs (1,3) and (1,8) have the same first component.

When a relation has a domain element that is used more than once as a first component, then the relation is not a function.

The relation y = 5x – 2 is a function. For any replacement of x, there will be exactly one replacement for y that will complete the solution. For example, the ordered pair (4, __) can be completed only as (4,18) to provide an element of {(x, y) | y = 5x – 2}.

{ x | x = | y |} is not a function. When x is replaced by a positive number such as 7 then y can be 7 or -7. Since both (7,7) and (7,-7) are in the relation, it is not a function.

Function Notation

When a relation is a function a different symbolism may be used in describing it.

The function described by y = 4x + 7 may be written as f(x) = 4x + 7. The symbol "f(x)" is read as "function of x" or, more simply, "f of x."

When f(x) is read as "function of x" it emphasizes the relationship between x and y in each ordered pair because it states clearly that the value of y is dependent upon the value of x. Notice that y is the same as f(x) which means y and f(x) can be substituted for each other without altering the meaning.

The function y = 3x² + 7x – 5 contains ordered pairs (x,y). It may be shown in function notation as f(x) = 3x² + 7x – 5. The function contains ordered pairs (x,f(x)). Notice that y and f(x) are interchangeable, y = f(x).

Relations and Functions 219

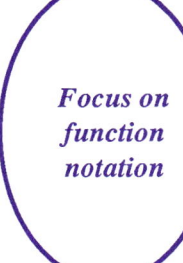

Focus on function notation

Every relation described by an equation of the form y = mx + b where m and b are real numbers, m ≠ 0, describes a **linear function.** Graphs of linear functions which are studied in Chapter 6 are straight lines.

$f(x) = 5x - 4$ is a linear function
$f(x) = \frac{1}{4}x + 9$ is a linear function

Relations described by polynomials in x are **polynomial functions.**
$f(x) = 5x^3 - 2x^2 + 4x + 1$ is a polynomial function.

Evaluating Functions of x

The linear function $f(x) = 9x - 7$ contains ordered pairs $(x, f(x))$ which means that f(4) is the second component of the ordered pair when the first component is 4. Hence, f(4) is found as:

$f(x) = 9x - 7$ and $f(4) = 9 \cdot 4 - 7$
$= 36 - 7$
$= 29$

Hence, f(4) = 29

To find f(-3) for the polynomial function $y = x^2 - x + 7$,

$y = x^2 - x + 7$ and $f(-3) = (-3)^2 - (-3) + 7$
$= 9 + 3 + 7$
$= 19$

Therefore, f(-3) = 19

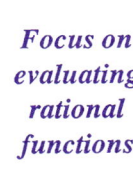

Focus on evaluating rational functions

The relation $y = \frac{5x + 4}{2x - 7}$ is a **rational function** and its domain must exclude $\frac{7}{2}$ because that replacement for x would result in a zero denominator.

To find f(-2) the following steps are used.

$y = \frac{5x + 4}{2x - 7}$ and $f(-2) = \frac{5 \cdot -2 + 4}{2 \cdot -2 - 7}$
$= \frac{-10 + 4}{-4 - 7}$
$= \frac{-6}{-11}$
$= \frac{6}{11}$

The rational function $y = \frac{x^2 - 4}{x^2 + 1}$ has no domain restrictions because no replacement for x can make the denominator 0. f(3) is found as:

$y = \frac{x^2 - 4}{x^2 + 1}$ becomes $f(3) = \frac{3^2 - 4}{3^2 + 1} = \frac{9 - 4}{9 + 1} = \frac{5}{10} = \frac{1}{2}$

Unit 4 Exercise

Part A: Answers for all Part A problems are at the back of the book.

1. In {(5,7),(6,3),(4,7),(5,3),(9,6)}, what number is used as a first component in more than one ordered pair?

2. Is there any domain element used more than once as a first component of the ordered pairs in {(5,0),(4,-2),(8,3),(4,5),(1,-6)}?

3. Is there any domain element used more than once as a first component in {(5,-2),(4,9),(8,-1),(6,4),(2,7),(1,-3)}?

4. For a relation to be a function, it is necessary for each domain element to be used exactly _____ as a first component.

5. Is {(4,7),(3,6),(5,15),(1,2)} a function?

6. Is {(1,4),(2,6),(3,9),(4,10),(5,7)} a function?

7. $(-3,7) \in \{(x, y) \mid 5x - y = -22\}$ is true. Is there any other element with -3 as its first component?

8. Every real number is in the domain of $\{(x, y) \mid 5x - y = -22\}$. Regardless of the real number chosen to replace x in (x,y) how many real numbers can correctly replace y?

9. Is $\{(x, y) \mid 5x - y = -22\}$ a function?

10. $(-3,9) \in \{(x, y) \mid y = x^2\}$ is true. Is there any other ordered pair with -3 as a first component?

11. Every real number is in the domain of $\{(x, y) \mid y = x^2\}$. Regardless of the real number chosen to replace x in (x,y) how many real numbers can correctly replace y?

12. Is $\{(x, y) \mid y = x^2\}$ a function?

13. $(9,-3) \in \{(x, y) \mid x = y^2\}$ is true. Is there any other ordered pair with 9 as a first component?

14. Is $\{(x, y) \mid x = y^2\}$ a function?

15. $(7,12) \in \{(x, y) \mid y = \mid x + 5 \mid\}$ is true. Is there any other ordered pair with 7 as a first component?

16. Is $\{(x, y) \mid y = \mid x + 5 \mid\}$ a function?

17. $(6,-4) \in \{(x, y) \mid x = \mid y - 2 \mid\}$ is true. Is there any other ordered pair with 6 as a first component?

18. Is $\{(x, y) \mid y = \mid x + 5 \mid\}$ a function?

19. If a relation has 2 ordered pairs with the same first component, then the relation _____ (is, is not) a function.

20. Which of the following statements is true?
    a. Every relation is a function.
    b. Every function is a relation.

21. For the linear function $f(x) = 7x + 6$, the value of $f(-2)$ is found by evaluating $7x + 6$ when $x = -2$. Find $f(-2)$.

22. For the polynomial function $g(x) = x^2 - 3x + 5$, the value of $g(-1)$ is found by evaluating $x^2 - 3x + 5$ when $x = -1$. Find $g(-1)$.

Part B: Answers for odd-numbered problems of Part B are at the back of the book.

1. In the relation $\{(6,3),(4,-1),(4,9),(1,3)\}$, what element of the domain is used as a first component of more than one ordered pair?

2. Is any domain element used more than once as a first component of the ordered pairs in $\{(5,9),(4,-2),(9,3),(7,-2),(1,4)\}$?

For problems 3-20 state if the set is a function and, if so, find its domain.

3. $\{(6,5),(4,7),(5,6),(4,9)\}$
4. $\{(9,3),(5,4),(-2,7),(4,6)\}$
5. $\{(2,7),(3,7),(5,7),(8,7)\}$
6. $\{(x, y) \mid y = 9x - 2\}$
7. $\{(x, y) \mid y = 5x - 6\}$
8. $\{(x, y) \mid x = 7 + y^2\}$
9. $\{(x, y) \mid y = 4 - x^2\}$
10. $\left\{(x, y) \mid y = \frac{3 - x}{3x + 5}\right\}$
11. $\left\{(x, y) \mid y = \frac{x - 4}{3 - 2x}\right\}$
12. $\{x \mid x + 8 = 10\}$
13. $\{(x, y) \mid y = 2x\}$
14. $\{(x, y) \mid y = x^2\}$
15. $\{(x, y) \mid x = y^2 + 3\}$
16. $\{(x, y) \mid y = \sqrt{x^2 - 3}\}$
17. $\{(x, y) \mid \sqrt{y + 7} = x\}$
18. $\{(x, y) \mid x^2 + y^2 = 25\}$
19. $\{(x, y) \mid |x| + |y| = 10\}$
20. $\{(x, y) \mid y = x^4\}$

21. Find f(-4) when $f(x) = 2x - 5$
22. Find f(5) when $f(x) = 3x - 2$
23. Find f(-2) when $f(x) = x^2$
24. Find f(3) when $f(x) = \frac{9}{x - 6}$
25. Find f(3) when $f(x) = \frac{5}{x + 10}$
26. Find f(7) when $f(x) = \sqrt{x}$
27. Find f(12) when $f(x) = \sqrt{x - 4}$
28. Find f(17) when $f(x) = \sqrt{x - 1}$
29. Find f(-2) when $f(x) = \sqrt{x - 9}$
30. Find f(-4) when $f(x) = \sqrt{x^2 - 4}$
31. Find f(-4) when $f(x) = \frac{x}{x - 2}$
32. Find f(-4) when $f(x) = \sqrt{5 - x}$
33. Find f(5) when $f(x) = \frac{3x + 8}{2x - 7}$
34. Find f(-4) when $f(x) = x^2 + x - 3$
35. Find f(-2) when $f(x) = (x - 6)^2$
36. Find f(5) when $f(x) = x^2 + 1$
37. Find f(-2) when $f(x) = x^2 - 3x + 1$
38. Find f(-1) when $f(x) = \sqrt{2 + x}$
39. Find f(-4) when $f(x) = -x^2 + 2x + 4$
40. Find f(2) when $f(x) = \frac{5x}{6 - x}$
41. Find f(-4) when $f(x) = \frac{2}{x - 3}$
42. Find f(-6) when $f(x) = (x + 2)^2 + 3$
43. Find f(4) when $f(x) = (3 + x)^2 + 2$
44. Find f(-3) when $f(x) = 6 - x^2$

45. Find f(-5) when $f(x) = \sqrt{8-x}$

46. Find f(-2) when $f(x) = \sqrt{9-x^2}$

47. Find f(7) when $f(x) = \dfrac{7}{2-x}$

48. Find f(-5) when $f(x) = \dfrac{1}{x^2-1}$

49. Find f(8) when $f(x) = \dfrac{3x+2}{5x}$

50. Find f(-4) when $f(x) = (x-2)^2 + 3$

Part C: No answers are given for these problems. However, each is accompanied by an ordered pair «C,U» showing the chapter and unit in which it was taught.

1. What is the conjugate of $-7 + 3\sqrt{5}$? «2,2»
2. Simplify $\sqrt{\dfrac{4}{5}}$ «2,2»
3. Use interval notation to describe the truth set $|x+2| < 4$ «4,3»
4. Graph $|2x-3| = 7$ «4,3»

Each of the problems 5-10 contains a set, an equation or rule of correspondence. State whether it is a relation and if so find its domain and range.

5. $\{(2,-2),(6,-2)\}$ «5,3»
6. $\{(1,4),(3,9),(6,3)\}$ «5,3»
7. $x \Rightarrow \sqrt{x-3}$ «5,3»
8. $(x-3)^2 = (y+2)^2 - 4$ «5,3»
9. $y = (x-1)^2 - 3$ «5,3»
10. $y = \sqrt{x-3}$ «5,3»

Unit 5: APPLICATIONS: VENN DIAGRAMS

Venn Diagrams Illustrating Sets and Set Operations

In earlier chapters of this text, the set operations of union and intersection have been briefly reviewed. To help visualize these operations, this unit uses Venn diagrams to show the effect of these operations and the set operation complementation.

For example, $A \cap B$ is the set consisting of elements that are in both set A and set B. The figure at the right is a **Venn diagram** for $A \cap B$. The shaded area in the Venn diagram is the set $A \cap B$, because the shaded area is in both A and B.

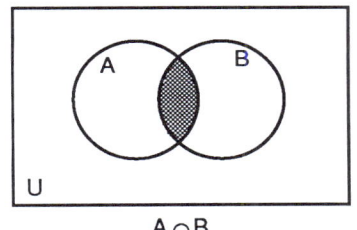
$A \cap B$

In the Venn diagram above, each set is represented by a circle. The rectangle represents the **universal set** U. The two sets A and B are subsets of U.

Focus on a Venn diagram for union of sets

$A \cup B$ represents the set whose elements are either in set A or in set B. Consequently, the Venn diagram for $A \cup B$ is made by shading the interiors of both A and B.

The interior of both sets is shaded, because all the elements of A and B are either in A or in B.

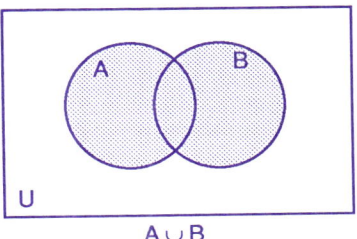
$A \cup B$

The Complementation of Two Sets

The **complement** of a set A (shown as $\bar{A}$) is the set of all elements in the universal set except for the elements of A.

The Venn diagram for $\bar{A}$ is made by shading every part of U that is not in A.

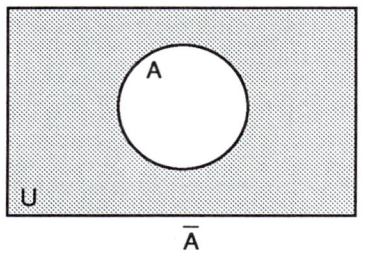

Focus on Venn diagrams involving more than one operation

The expression $A \cap \bar{B}$ involves two set operations: complement and intersection.

To make the Venn diagram of $A \cap \bar{B}$ first mark the complement of B as shown in the top figure at right. Then intersect the circle of A with the marked figure for $\bar{B}$.

The final diagram includes those areas that are common to the circle of A and the area marked for $\bar{B}$.

The bottom figure shown at the right is the final result for $A \cap \bar{B}$.

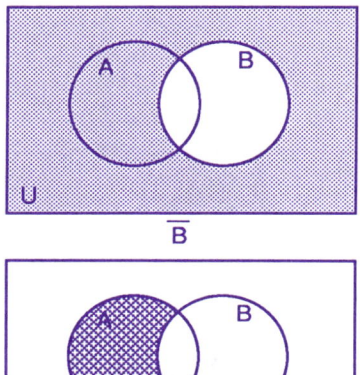

Solving Problems Using Venn Diagrams

The following example shows how a Venn diagram can be used for problem solving.

Problem: Of 74 boys in a private school, the numbers out for a sport or sports were as follows: football, 48; basketball, 20; soccer, 30; football and soccer, 10; basketball and football, 11; soccer and basketball, 8; all three sports, 3. How many boys were not out for any sport?

To solve the problem with a Venn diagram, begin with a separate circle to represent the sets of boys involved in each of the three sports. The circles are drawn so they overlap to allow for those boys that participate in more than one sport.

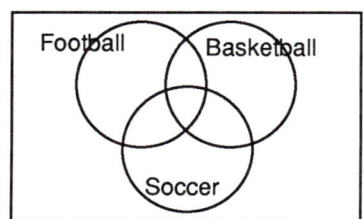

U is set of 74 boys

The three circles separate the rectangle into eight non-overlapping areas now represented by the letters m, n, p, q, x, y, z, and w. Each number clue given in the statement of the problem identifies one or more area of the Venn diagram. For example, the fact that the football circle has 48 boys means that the four separate areas of the football circle contain a total of 48. This fact gives the equation $n + p + x + y = 48$.

Write a similar addition equation for each clue given in the problem.

Focus on completing the Venn diagram

The fact that there are 74 boys in school can be written as:
$$m + n + p + q + x + y + z + w = 74$$

The fact that 20 boys play basketball can be written as: $p + q + y + z = 20$

The fact that 30 boys play soccer can be written as: $x + y + z + w = 30$

The fact that 10 play football and soccer can be written as: $x + y = 10$

The fact that 11 play football and basketball can be written as: $p + y = 11$

The fact that 8 play soccer and basketball can be written as: $y + z = 8$

The fact that 3 play all three sports can be written as: $y = 3$

But when it is known that $y = 3$, each of the preceding equations can be solved to find values for the other letters.

Each variable can be replaced by a number to provide the Venn diagram shown at the right.

Since the original question referred to boys that were not out for any sport, that answer is $m = 2$.

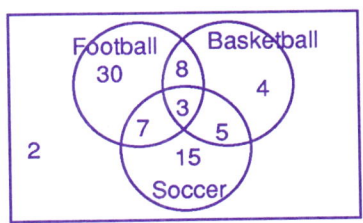
U is set of 74 boys

Unit 5 Exercise

Part A: Answers for all Part A problems are at the back of the book.

Draw a Venn diagram for each problem.

1. $B \cap A$
2. $\overline{B}$
3. $(A \cup B) \cap C$
4. $A \cup \overline{B}$

Part B: Answers for odd-numbered problems of Part B are at the back of the book.

Draw a Venn diagram for problems 1-17.

1. $A \cup B$
2. $(A \cup B) \cup C$
3. $\overline{A} \cap (B \cap C)$
4. $(A \cap B) \cap C$
5. $(A \cap B) \cup C$
6. $A \cup (B \cap C)$
7. $A \cup (B \cup C)$
8. $\overline{C}$
9. $\overline{A} \cap \overline{B}$
10. $A \cup \overline{B}$
11. $(\overline{A} \cap \overline{B}) \cap C$
12. $\overline{A} \cup \overline{B}$
13. $A \cup \overline{B}$
14. $(A \cap \overline{B}) \cap C$
15. $(A \cap \overline{B}) \cap \overline{C}$
16. $(A \cap B) \cup (A \cap C) \cup (B \cap C)$
17. $A \cap (\overline{B} \cup \overline{C})$

18. In an institution the following tabulation was made of the patients: alcoholic, 27; diabetic, 20; psychotic, 12; all three, 2; alcoholic and diabetic, 5; psychotic and alcoholic only, 7; psychotic only, 3.
    a. How many are psychotic and diabetic?
    b. How many are alcoholic and psychotic?
    c. How many are diabetic only?
    d. How many patients were studied?

19. In a survey of the executives of a company, the following tabulation was made of certain qualities: married, 12; effective, 13; female, 5; married and effective but not female, 4; married and female but not effective, 1; married and female, 2; effective but not married or female, 6.
    a. How many are married, effective, and female?
    b. How many executives were interviewed altogether?
    c. How many were female but not married or effective?

20. A survey in a small town revealed the following facts about the number of people in certain classifications: males, 35; in business, 7; rich, 8; male business people, 4; male and rich, 5; rich and in business, 2; rich males in business, 1.
   a. How many males were rich but not in business?
   b. How many people were surveyed altogether?

Part C: No answers are given for these problems. However, each is accompanied by an ordered pair «C,U» showing the chapter and unit in which it was taught.

1. Factor $27b^3 - 64a^3$ «1,4»

2. Solve $x^4 - 4x^2 + 3 = 0$ «1,5»

For problems 3-4, state the domain and range for the relations.

3. $\{(1,7),(3,4),(1,5)\}$ «5,3»

4. $(x + 1)^2 + (y - 2)^2 = 5$ «5,3»

For problems 5-7, state if the set is a function and, if so, find its domain.

5. $\{(3,5),(7,12),(9,6),(3,10)\}$ «5,4»

6. $\{(x, y) \mid y = 3x - 5\}$ «5,4»

7. $\{(x, y) \mid x^2 + y^2 = 9\}$ «5,4»

8. Find $f(4)$ when $f(x) = 9x - 3$ «5,4»

9. Find $f(-2)$ when $f(x) = x^2 - 5x + 7$ «5,4»

10. True or false? Every function is a relation and every relation is a function. «5,4»

Chapter 5 Test

«5,U» shows the unit in which this problem was studied in this chapter.

1. Is {1,2,3,...,895} a finite set?   «5,1»

2. Is {2,4,6,...} a finite set?   «5,1»

For problems 3-5, find a rule of correspondence that would match the elements of {1,2,3,...} with the given set.

3. {3,6,9,...}   «5,1»

4. {5,6,7,...}   «5,1»

5. $\{-1, \frac{-1}{4}, \frac{-1}{9}\}$ «5,1»

For problems 6-7, true or false?

6. $(12,4) \in \{(x, y) \mid y = \sqrt{x + 4}\}$ «5,2»

7. $(-1,3) \in \{(x, y) \mid y = x^2 - 2x + 3\}$ «5,2»

Complete the ordered pair for problems 8-10.

8. $(-3, \underline{}) \in \{(x, y) \mid y = x^2 - 3x + 1\}$ «5,2»

9. $(5, \underline{}) \in \{(x, y) \mid y = 2x^2 - 7y - 15\}$ «5,2»

10. $(-2, \underline{}) \in \{(x, y) \mid y = \sqrt{5 - 2x}\}$ «5,2»

11. What is the domain of {(3,8),(-1,4),(5,2),(1,6)}?   «5,3»

12. What is the range of {(1,-3),(2,5),(7,-3),(5,-2)}?   «5,3»

13. What is the domain of $y - 7 = \frac{4x + 3}{2x - 7}$?   «5,3»

14. What is the range of $y = 4 - \sqrt{x - 5}$?   «5,3»

15. Find the domain and range of $y = (2x - 5)^2 - 4$ «5,3»

16. Is $y = x^2$ a function?   «5,4»

17. Is $x^2 + y^2 = 9$ a function?   «5,4»

18. Find f(-3) when $f(x) = \frac{9}{x - 6}$ «5,4»

Draw a Venn diagram for problems 19-20.

19. $A \cup \bar{B}$ «5,5»

20. $\bar{A} \cap \bar{B}$ «5,5»

6
Graphing Linear Functions

Unit 1: EQUATIONS WITH TWO VARIABLES

Definition	Linear Equation	
An equation is a **linear equation with two variables**	**if and only if**	it is equivalent to the form $ax + by = c$ where x and y are variables, a, b, c are real numbers, and $a^2 + b^2 \neq 0$.

Notice that the requirement that $a^2 + b^2 \neq 0$ is equivalent to the claim that the coefficient of at least one of the variables is not zero. Another method for stating the same requirement is $|a| + |b| \neq 0$. Notice also that when the coefficient of y is not 0 then a linear equation with two variables is a linear function as introduced in Chapter 5.

Linear Equations With Two Variables

Equations such as $2x - y = 6$, $y = \frac{2}{3}x + 1$, and $x + 5y = 10$ are linear equations with two variables. Every such linear equation has a straight line as its graph on the x,y plane.

Other equations such as $\frac{x}{2} - \frac{7}{3} = \sqrt{2} \cdot y$ and $\frac{y+5}{x-4} = \frac{2}{3}$ are also linear equations with two variables, but are not in the form $ax + by = c$.

For example, if each term of $\frac{x}{2} - \frac{7}{3} = y\sqrt{2}$ is multiplied by 6 (the least common multiple of the denominators) the following equivalent equations are found.

$\frac{x}{2} - \frac{7}{3} = y\sqrt{2}$ becomes $\quad 6 \cdot \frac{x}{2} - 6 \cdot \frac{7}{3} = 6 \cdot y\sqrt{2}$
$$3x - 14 = 6y\sqrt{2}$$
$$3x - 6y\sqrt{2} = 14$$

Note: The coefficient of y is $6\sqrt{2}$.

Similarly, if each term of $\frac{y+5}{x-4} = \frac{2}{3}$ is multiplied by $3(x - 4)$, again the common denominator, the following equivalent equations are found.

$\frac{y+5}{x-4} = \frac{2}{3}$ becomes $\quad 3(y + 5) = 2(x - 4)$
$$3y + 15 = 2x - 8$$
$$2x - 3y = 23$$

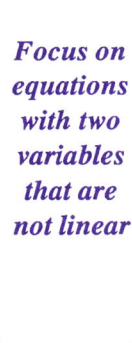

Focus on equations with two variables that are not linear

Linear equations are frequently called **first degree** equations because each variable term has only one variable and its exponent is 1.

The equation $5x^2 + 3y = 7$ is a **second degree** equation because the variable x has exponent 2.

The equation $4x - 3xy = 5$ is another **second degree** equation. In this case the sum of the variable exponents in the term $-3xy$ is 2.

To show that $\frac{5}{x} - \frac{4}{y} = \frac{1}{3}$ is not a linear equation find an equivalent form by first multiplying by the common denominator, $3xy$.

$\frac{5}{x} - \frac{4}{y} = \frac{1}{3}$ becomes $3xy \cdot \frac{5}{x} - 3xy \cdot \frac{4}{y} = 3xy \cdot \frac{1}{3}$

$15y - 12x = xy$

$15y - 12x = xy$ is a second degree equation because of the xy term and therefore is not a linear equation.

Definition	**Solution for a Linear Equation**	
The ordered pair (w,z) is a solution for a linear equation $ax + by = c$	**if and only if**	$aw + bz = c$ is a true numerical statement.

Solutions of Linear Equations

(0,2) is a solution (x,y) of $y = 5x + 2$, because if x is replaced by 0 and y is replaced by 2, $y = 5x + 2$ becomes the true statement $2 = 5 \cdot 0 + 2$.

(-6,-1) is not a solution of $y = 3x - 7$, because $-1 = 3 \cdot -6 - 7$ is a false statement.

(4,7) is a solution for $4x - y = 9$ because $4 \cdot 4 - 7 = 9$ is a true statement.

(1,-5) is another solution for $4x - y = 9$ because $4 \cdot 1 + 5 = 9$ is a true statement.

Focus on finding solutions for linear equations

To find a solution (x,y) for y = 5x – 3, **any** real number may be used as a replacement for x. If x = 2, then the equation y = 5x – 3 becomes y = 5 • 2 – 3 or y = 7. Hence, (2,7) is a solution of y = 5x – 3.

The preceding example contains a most valuable statement: any real number may be used as a replacement for x. The word "any" needs to be emphasized.

Because any real number may be used as a replacement for x, there is an infinite number of ordered pair solutions to each linear equation containing two variables.

When x = 4, the equation y = 5x – 3 becomes y = 5 • 4 – 3 or y = 17. Hence, (4,17) is a solution of y = 5x – 3.

Graphing Linear Equations

To graph y = 3x – 1, make a table of solutions like the one shown at the right. Any three numbers may be chosen for either of the variables. In this example, 0, 2, and -2 were chosen as replacements for x.

x	y
0	_
2	_
-2	_

To complete the table, find the solutions (0,__), (2,__), and (-2,__).
If x = 0, y = 3x – 1 becomes y = -1
If x = 2, y = 3x – 1 becomes y = 5
If x = -2, y = 3x – 1 becomes y = -7

x	y
0	-1
2	5
-2	-7

The three points (0,-1),(2,5), and (-2,-7) are then plotted on the x,y plane as shown at the right.

Since the three points lie on a single straight line, that entire line represents all the solutions of y = 3x – 1.

If the three points found did not lie on a single straight line, it would mean a mistake had been made and at least one of the ordered pairs found is incorrect.

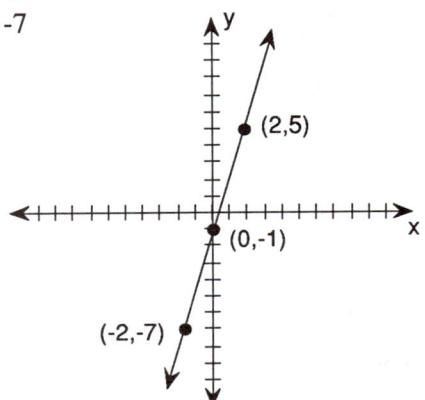

Focus on graphing a linear equation

To graph $3x + 2y = 4$,

1. Make a table of at least three solutions.

x	y
0	—
—	0
2	—

2. Complete the table.
   $3 \cdot 0 + 2y = 4$ then $y = 2$
   $3x + 2 \cdot 0 = 4$ then $x = \frac{4}{3}$
   $3 \cdot 2 + 2y = 4$ then $y = -1$

x	y
0	2
$\frac{4}{3}$	0
2	-1

3. Plot the points $(0,2)$, $\left(\frac{4}{3},0\right)$ and $(2,-1)$ on the x,y plane and draw the straight line that contains them.

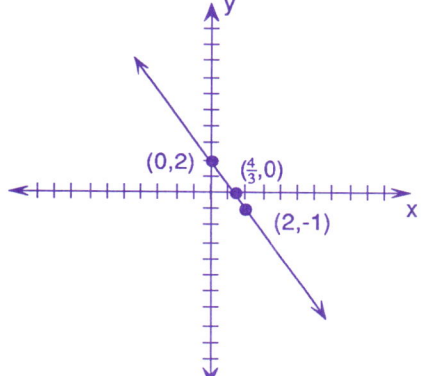

Each point on the line is a solution of the equation. Each point not on the line is not a solution.

Finding Solutions and Graphing

To graph $y = \frac{3}{5}x - 6$, at least two (preferably three) solutions need to be found. Since the coefficient of x is $\frac{3}{5}$, multiples of 5 are relatively easy replacements for x. 0, 5, and 10 are multiples of 5 and are used below.

If $x = 0$ then $y = \frac{3}{5}x - 6$ becomes
$y = \frac{3}{5} \cdot 0 - 6 = 0 - 6 = -6$ which gives $(0,-6)$.

If $x = 5$ then $y = \frac{3}{5}x - 6$ becomes
$y = \frac{3}{5} \cdot 5 - 6 = 3 - 6 = -3$ which gives $(5,-3)$.

If $x = 10$ then $y = \frac{3}{5}x - 6$ becomes
$y = \frac{3}{5} \cdot 10 - 6 = 6 - 6 = 0$ which gives $(10,0)$.

234   Chapter 6

To graph $y = \frac{3}{5}x - 6$, the solutions (5,-3), (0,-6) and (10,0) are plotted. The points (0,-6) and (10,0) are called **intercept** points because they are the points where the straight line crosses the x and y axes.

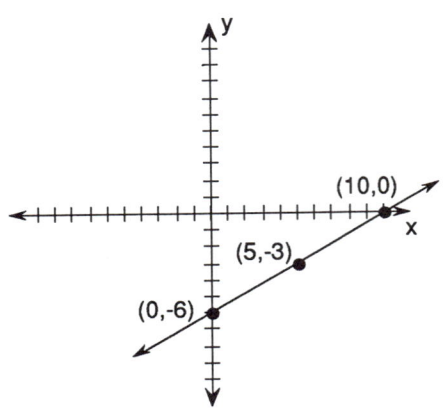

Focus on linear equations with horizontal or vertical lines

$y = 2$ is a linear equation because it is equivalent to $0x + y = 2$. Three of its solutions are (0,2), (5,2) and (-3,2). The graph of $y = 2$ is shown at the right.

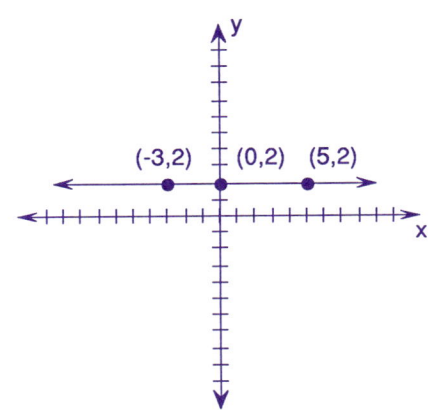

Every equation of the form $y = k$ has a graph which is the horizontal line through the solution (0,k).

$x = -3$ is another linear equation that is equivalent to $x + 0y = -3$. Three of its solutions are (-3,0), (-3,2) (-3,-5). The graph of $x = -3$ is shown at the right.

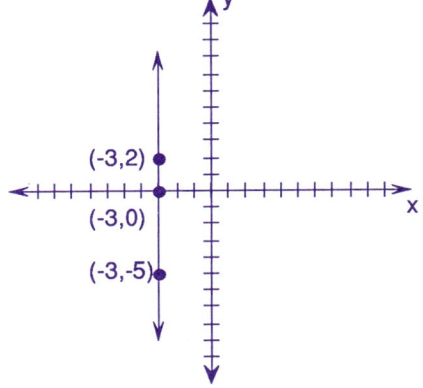

Every equation of the form $x = r$ is represented by a vertical line through the solution (r,0).

Unit 1 Exercise

Part A: Answers for all Part A problems are at the back of the book.

1. Any equation equivalent to $ax + by = c$ where not both a and b are 0 is a linear equation. Is $y = \frac{4}{5}x - 7$ a linear equation?

2. Is $\frac{x+5}{y-6} = 7$ a linear equation?

3. Is $\frac{x}{5} + \frac{y}{2} = \frac{1}{x}$ a linear equation?

4. The ordered pair (-1,-3) is a solution of $y = 5x + 2$ because $-3 = 5(-1) + 2$ is true. Is (3,17) a solution (x,y) of $y = 5x + 2$?

5. Is (4,3) a solution of $y = 3x - 7$?

6. The solution (2,___) for $y = 5x - 3$ is completed by replacing x by 2 and solving for y. Complete the solution (2,___) for $y = 5x - 3$.

7. Complete the solution (1,___) for $y = 5x - 3$.

8. Complete the table of solutions for $y = -2x - 4$.

x	y
-3	__
0	__
3	__

9. Plot the three solutions for $y = -2x - 4$ found in problem 8 on the x,y plane. Draw the straight line that connects them. What is true for every point on the line?

10. Complete the table of solutions for $y = \frac{1}{2}x + 3$.

x	y
-4	__
0	__
4	__

11. Plot the three solutions for $y = \frac{1}{2}x + 3$ found in problem 10. Draw the straight line that connects them. What is true for every point on the line?

12. Find any three ordered pair solutions of $y = 3x + 2$, locate the points on the x,y plane, and draw a straight line through them.

13. Find any three ordered pair solutions of $y = \frac{-2}{3}x + 5$, locate the points on the x,y plane, and draw a straight line through them.

14. Is the equation $y = 5$ equivalent to $0x + y = 5$?

15. Find any three solutions of $y = 5$ and draw its graph.

Chapter 6

Part B: Answers for odd-numbered problems of Part B are at the back of the book.

For problems 1-10, determine if the ordered pair is a solution of the equation.

1. (3,2) $y = 3x - 7$
2. (0,13) $y = 5x + 13$
3. (-1,6) $y = -5x + 1$
4. (3,-4) $y = -x - 1$
5. (-6,2) $y = 2x$
6. (-2,-1) $y = \frac{1}{2}x$
7. (5,3) $y = \frac{3}{5}x$
8. (5,3) $y = 5$
9. (-2,7) $x = -2$
10. $(3, \frac{-2}{13})$ $y = \frac{-5}{2}x + 1$

For problems 11-36, graph the given equation.

11. $y = 4x + 3$
12. $y = -3x + 2$
13. $y = \frac{1}{2}x$
14. $y = 3$
15. $x = 5$
16. $y = \frac{-2}{3}x - 1$
17. $y = -1$
18. $y = -5x - 2$
19. $y = -3x + 4$
20. $y = 2x - 3$
21. $y = -2x + 5$
22. $y = 5x$
23. $y = 4$
24. $y = -5x + 5$
25. $x = 6$
26. $y = -2x - 7$
27. $y = -3x - 1$
28. $x = -1$
29. $y = 5x + 7$
30. $y = x$
31. $y = -4x + 2$
32. $y = -x + 1$
33. $y = -6x + 4$
34. $y = 4$
35. $y = -x$
36. $x = -2$

Part C: No answers are given for these problems. However, each is accompanied by an ordered pair «C,U» showing the chapter and unit in which it was taught.

1. Factor $27a^3 + 64b^3$. «1,4»
2. Find the truth set for $6x^2 + 1 = 7x$. «2,6»
3. Solve $3x^2 + x + 2 = 0$. «3,2»
4. Graph the truth set for $|4x - 9| > 3$. «4,3»
5. Complete the ordered pair for $(-5, \underline{}) \in \{(x, y) \mid y = \sqrt{16 + 3x}\}$ «5,2»
6. What is the domain of $y = 17 - \sqrt{x - 3}$? «5,3»
7. What is the domain of $y = \frac{5x-1}{3x-2}$? «5,3»
8. Is $x^2 + y^2 = 25$ a function? «5,4»
9. Find $f(-5)$ when $f(x) = \frac{27}{3x + 6}$. «5,4»
10. Draw a Venn diagram for $\bar{A} \cap B$. «5,5»

Unit 2: LINEAR FUNCTIONS

> **Definition** **Linear Function**
>
> A relation is a **linear function** if and only if it is described by an equation equivalent to $y = mx + b$ where x and y are variables and m and b are real numbers.

Linear Functions

Linear equations with two variables are equivalent to equations of the form $ax + by = c$ where not both a and b are 0. The set of all linear equations can be divided into two parts:
1. Equations equivalent to $ax + 0y = c$ which are not functions, and
2. Equations equivalent to $ax + by = c$ where $b \neq 0$ which are functions.

In the remainder of this chapter we are primarily interested in those linear equations equivalent to $ax + by = c$ where $b \neq 0$. These equations are of special importance in mathematics because they are **linear functions**.

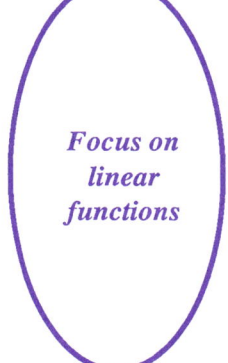

Focus on linear functions

The linear equation $5x - 3y = 4$ can be solved for y. The result is an equivalent equation of the form $y = mx + b$. This means that $5x - 3y = 4$ is a linear function.

$$5x - 3y = 4$$
$$-3y = -5x + 4$$
$$\tfrac{-1}{3} \cdot -3y = \tfrac{-1}{3} \cdot (-5x + 4)$$
$$y = \tfrac{5}{3}x - \tfrac{4}{3}$$

The linear equation $3x + 0y = 10$ cannot be written in the form $y = mx + b$ and therefore is not a linear function. Its graph is the vertical line through $(\tfrac{10}{3}, 0)$.

The Sign of the Slope of a Linear Function

Linear functions $y = mx + b$ contain two variables, y and x, and two constants, m and b. $y = \frac{5}{7}x - \frac{3}{4}$ is such a linear equation with variables y and x; its constants are the coefficient of x, $\frac{5}{7}$, and the addend, $\frac{-3}{4}$. The coefficient of x, $\frac{5}{7}$, is the **slope** of the line, and $(0, \frac{-3}{4})$ is the **y-intercept** of the line.

When the sign of the slope of a linear function is **positive** it means that as replacements for x increase in size then the matching replacements for y will also increase. In terms of the graph of the line, this means that as a point travels on the line of $y = \frac{5}{7}x - \frac{3}{4}$ it goes upward as it travels from left to right.

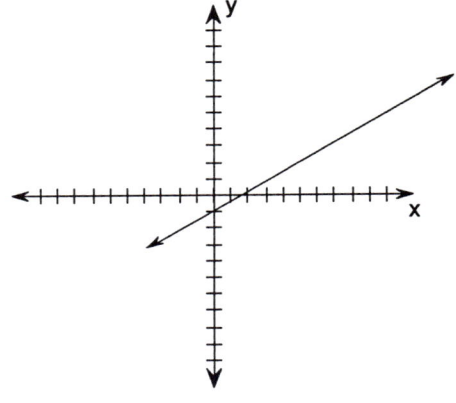

When the sign of the slope of a linear function is **negative** it means that as replacements for x increase in size then the matching replacements for y will decrease. In terms of the graph of the line, this means that as a point travels on the line of $y = \frac{-1}{2}x - \frac{3}{4}$ it goes downward as it travels from left to right.

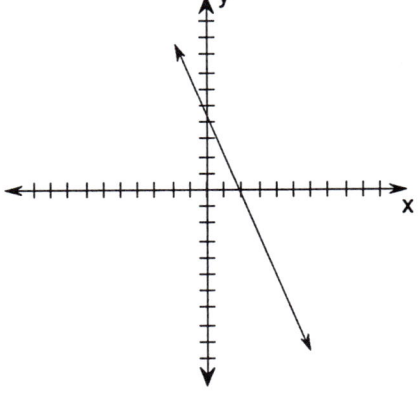

Graphing Linear Functions

Focus on the sign of the slope

The sign of the slope of a negative function can be determined by looking at the line of the function.

A line that goes from upper left to lower right has a **negative** slope because it shows that y decreases as x increases.

A line that goes from lower left to upper right has a **positive** slope because it shows that y increases as x increases.

In the graph at the right line r has a negative slope and line w has a positive slope.

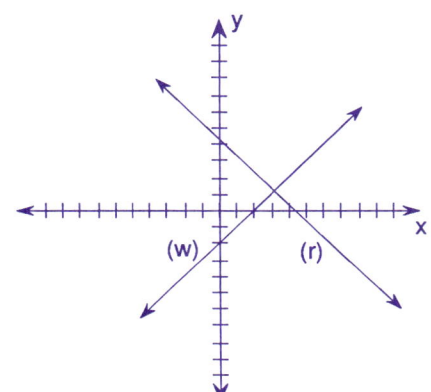

Y-Intercepts, (0,b)

Every linear function of the form y = mx + b crosses the y-axis (vertical axis) at exactly one point. That point on the y-axis is the **y-intercept** of the equation. Each point on the y-axis has a first component of zero.

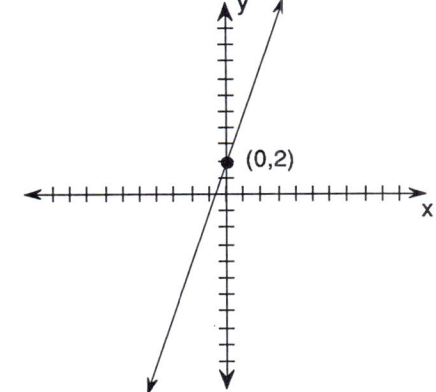

Since every ordered pair representing a point on the y-axis has a first component of zero, the point where y = 3x + 2 crosses the y-axis is the solution when x = 0. (0,2) is a solution of y = 3x + 2.

240   Chapter 6

Focus on y-intercepts of linear functions

Whenever an equation is of the form y = mx + b, then (0,b) is the y-intercept. If x = 0, then y = mx + b becomes y = m • 0 + b, or more simply y = b.

The intercept of y = 5x + 8 is (0,8). If x = 0, then y = 5 • 0 + 8 or y = 8.

To find the y-intercept of 3x − 2y = 8, the equation should first be solved for y.

$$3x - 2y = 8$$
$$-2y = -3x + 8$$
$$y = \tfrac{3}{2}x - 4$$

Since the equation $y = \tfrac{3}{2}x - 4$ is now in the form y = mx + b, the y-intercept is easily determined as (0,-4).

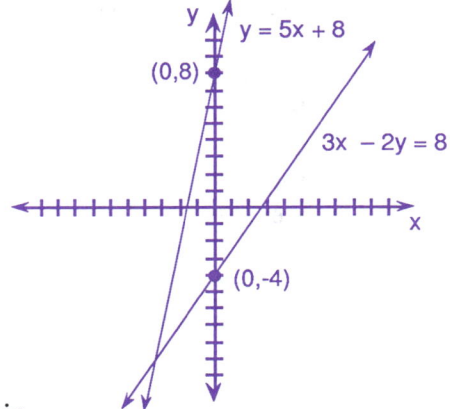

Unit 2 Exercise

Part A: Answers for all Part A problems are at the back of the book.

1. Any equation equivalent to y = mx + b is a linear function. Is y = -7x + 3 a linear function?
2. Is y = 7 a linear function?
3. Is x = 3 a linear function?
4. Is $\tfrac{x-2}{y-4} = 3$ a linear function?

Problems 5-8 depend on the graph shown at the right. For each equation, determine whether the slope is positive or negative.

5. y = 7x − 10
6. $y = \tfrac{-3}{4}x + 2$
7. y = -2x − 6
8. $y = \tfrac{1}{2}x$

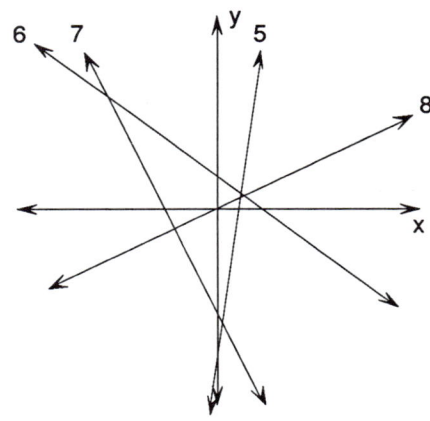

9. The y-intercept of y = 4x – 7 is the solution when x = 0. (0, ___) is the y-intercept of y = 4x – 7.

10. The y-intercept of y = $\frac{-3}{5}$x – 1 is the solution when x = 0. (0, ___) is the y-intercept of y = $\frac{-3}{5}$x – 1.

11. Solve 2x – y = 5 for y and find the y-intercept.

12. Solve 4x – 2y = -12 for y and find its y-intercept.

Part B: Answers for odd-numbered problems of Part B are at the back of the book.

Each problem contains an equation. Determine whether it is a linear function. If it is, find its y-intercept, graph its line, and determine whether its slope is positive, negative, or zero.

1. y = 1x + 3
2. y = $\frac{-2}{3}$x + 4
3. y = $\frac{5}{3}$x – 5
4. 3x – 2y = 8
5. 4x + 3y = 5
6. $\frac{y+3}{x}$ = 4
7. y = -4
8. $\frac{-2y}{x}$ = -5
9. $\frac{-3}{4}$x + $\frac{6}{y}$ = 3
10. $\frac{1}{3}$x – $\frac{5}{6}$y = -2
11. $\frac{2}{x}$ – $\frac{5}{y}$ = $\frac{6}{x}$
12. x = 7
13. y = $\frac{3}{4}$x – 7
14. 3x – 4y = 8
15. y = 2x + 5
16. y = -2x + 5
17. y = $\frac{1}{3}$x – 6
18. y = $\frac{-1}{2}$x – 4
19. y = 5x – 10
20. y = -x + $\frac{7}{5}$
21. y = 7x + 5
22. y = $\frac{-2}{3}$x – $\frac{1}{4}$
23. y = $\frac{-5}{9}$x + 1
24. y = -5x
25. x + 3y = 6
26. 4x – 3y = 12
27. 2x + y = 0
28. 3x – 2y = 5

Part C: No answers are given for these problems. However, each is accompanied by an ordered pair «C,U» showing the chapter and unit in which it was taught.

1. Simplify 7x – 13 ≤ -7 «1,4»
2. Solve 3x² + x = 4 «1,5»
3. Multiply 2√3 – √5 by its conjugate. «2,2»
4. Find the truth set of √(x + 3) – 7 = -3. «2,6»
5. Multiply (3 – 4i)(7 + 3i) «3,1»
6. Solve | 3x – 2 | = 10 «4,2»
7. Graph the truth set of | 5x – 2 | > 3 «4,4»
8. Factor 3x² – 8x – 8 «1,4»
9. Draw a Venn diagram for $\overline{A} \cup B$. «5,5»
10. Graph y = $\frac{-3}{4}$x – 3. «6,1»

Unit 3: SLOPES OF LINEAR FUNCTIONS

Ratios of the Legs of Right Triangles

The figure at the right shows a right triangle. Its legs are the vertical and horizontal sides with respective lengths of 7 and 9. The ratio of these two lengths is $\frac{7}{9}$. A hiker going from point A to B might find it a challenging, but possible climb.

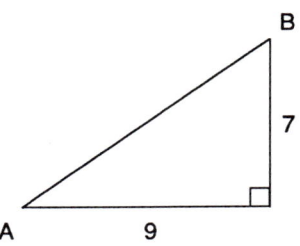

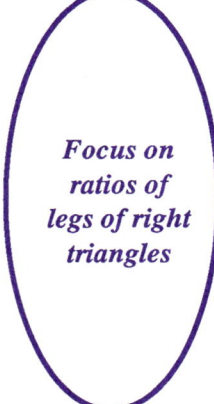

Focus on ratios of legs of right triangles

For the right triangle shown at the right, the ratio of the vertical length leg to the horizontal is $\frac{8}{2}$ or $\frac{4}{1}$. A hiker going from point A to B might find this a very difficult climb.

For the hiker going from point A to B, the bottom triangle offers a more gradual climb which can be described by the ratio $\frac{3}{15}$ or $\frac{1}{5}$.

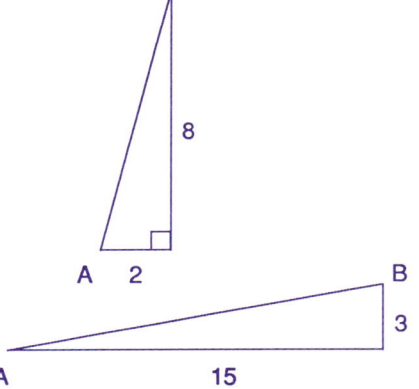

Rise Over Run

The fraction $\frac{y}{x}$ describes the steepness of a climb for a hiker attempting to go from point A to B for any of the right triangles shown at the right.

The smaller the fraction, the more gradual the slope. The larger the fraction, the steeper the slope.

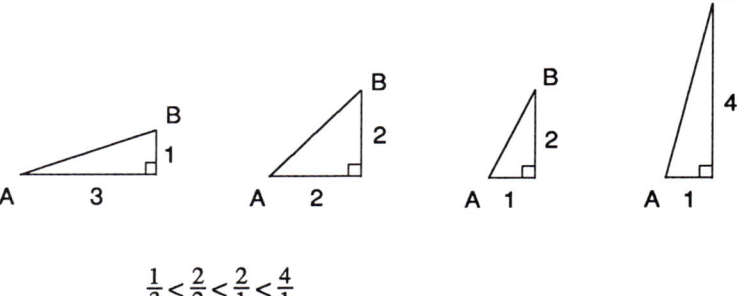

$\frac{1}{3} < \frac{2}{2} < \frac{2}{1} < \frac{4}{1}$

Graphing Linear Functions 243

The relative size of a fraction will indicate degree of steepness.

Focus on the ratio of rise over run

If we agree to use negative fractions to show descents and positive fractions to show ascents, then the ratio $\frac{1}{2}$ can be visualized as the right triangle shown at the right. The ratio $\frac{1}{2}$ represents a situation where the hiker is climbing a slope with rise of 1 over run of 2.

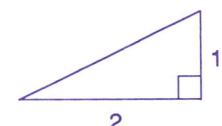

Similarly, the ratio $\frac{-2}{3}$ can be visualized as the right triangle shown at the right. The ratio $\frac{-2}{3}$ represents a situation where the hiker is on a slope with rise of -2 over run of 3.

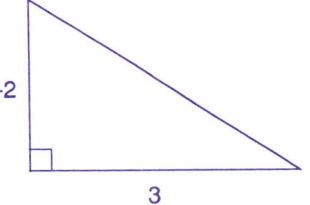

Describing the Slope Concept

The use of right triangles and ratios of rise over run can be applied to lines on the x,y plane. To write the equation for the linear function shown at the right,

1. The y-intercept is (0,-3) because the line intersects the y-axis at that point.

2. The rise over run ratio is $\frac{1}{2}$. To see that, draw the right triangle with vertices (-2,-4) to (6,0).

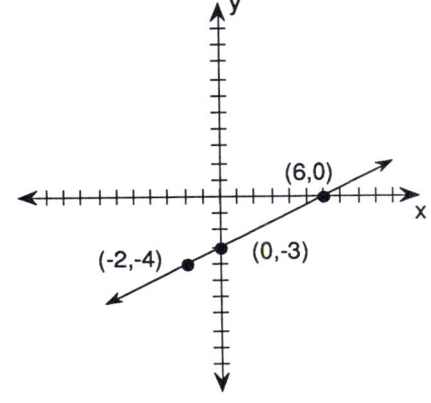

Therefore, the graph at the right shows the line of $y = \frac{1}{2}x - 3$.

Focus on writing a linear function from its graph

The line on the graph at the right is for a linear function y = mx + b. To find numbers for m and b,

1. The value of b is the y component of the y-intercept. Since the line crosses the y-axis at (0,5), b = 5.

2. The value of m is the rise over run ratio. If a right triangle is formed using (0,5) and (4,-3) as vertices, the vertical leg is 8, the horizontal leg is 4, and the line descends as it goes from right to left. $m = \frac{-8}{4}$ or m = -2.

Therefore, the linear function is:
y = -2x + 5

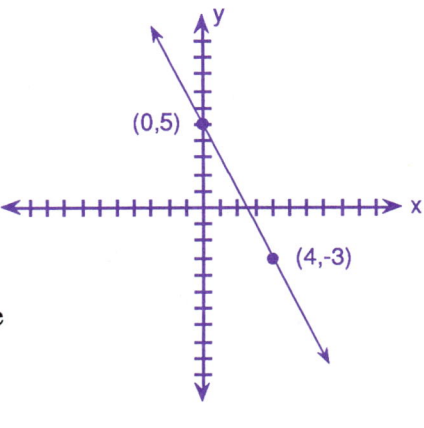

The preceding work of this unit shows the concept of a linear function's slope in terms of an informal approach to right triangles and the ratio of a vertical leg to the horizontal leg, $\frac{rise}{run}$.

This informal approach does make clear that the slope of a linear function is a comparison, by division, of the rate of change of values for y as compared to the rate of change for x. Now, it is time to turn to a more formal approach to the slope concept that will state exactly what it means and which can be applied without drawing or studying right triangles.

Definition   **Slope of a Linear Function**

The number m is the slope of a linear function   if and only if $m = \frac{f(x_1) - f(x_2)}{x_1 - x_2}$

where $x_1 \neq x_2$

The Slope as a Ratio of Differences

The slope of a linear function is found by computing the ratio $\frac{f(x_1) - f(x_2)}{x_1 - x_2}$ for two points $(x_1, f(x_1))$ and $(x_2, f(x_2))$.

For example, the slope of the linear function containing (3,7) and (2,1) is found by:

1. Noticing that (3,7) means $x = 3$ and $f(3) = 7$. Also, (2,1) means $x = 2$ and $f(2) = 1$.
2. Then the slope $m = \frac{f(3) - f(2)}{3 - 2} = \frac{7 - 1}{3 - 2} = \frac{6}{1}$.

The reader is invited to plot the points (3,7) and (2,1) on a graph, draw the right triangle containing them as vertices, and confirm that the rise over run ratio is $\frac{6}{1}$.

Similarly, the slope of the linear function containing (-2,7) and (-5,8) is found as:

$$m = \frac{f(-2) - f(-5)}{-2 - -5} = \frac{7 - 8}{-2 + 5} = \frac{-1}{3}.$$

Again, plot the points (3,7) and (2,1) on a graph and confirm that the rise over run ratio is $\frac{-1}{3}$.

Focus on finding the slope of a linear function

It is reasonable to question whether the slope of a linear function depends on the particular points chosen to compute it. For example, the linear function $y = \frac{3}{4}x - 5$ has an infinite number of solutions. Will its slope m be different for different pairs of solutions? Suppose $(a, f(a))$ and $(b, f(b))$ are two of its solutions. Then its slope is

$$m = \frac{f(a) - f(b)}{a - b}$$

$$m = \frac{(\frac{3}{4}a - 5) - (\frac{3}{4}b - 5)}{a - b}$$

$$m = \frac{\frac{3}{4}a - 5 - \frac{3}{4}b + 5}{a - b}$$

$$m = \frac{\frac{3}{4}a - \frac{3}{4}b}{a - b}$$

$$m = \frac{\frac{3}{4}(a - b)}{a - b}$$

$$m = \frac{3}{4}$$

Notice that regardless of the two solutions chosen, the slope is $\frac{3}{4}$.

Chapter 6

Using the Slope Formula

The formula for the slope of a line between two points is shown below. If (x_1, y_1) and (x_2, y_2) are the two points, and the line is a linear function then $f(x_1) = y_1$ and $f(x_2) = y_2$.

$$m = \frac{y_1 - y_2}{x_1 - x_2} \quad \text{or} \quad m = \frac{y_2 - y_1}{x_2 - x_1}$$

"x_1" is read as "x sub one." Similarly, "x_2" is read as "x sub two."

Using the formula, the slope of the line through (-8,-3) and (0,2) is found by evaluating either $\frac{2--3}{0--8}$ or $\frac{-3-2}{-8-0}$. Notice that both ratios are equal to $\frac{5}{8}$.

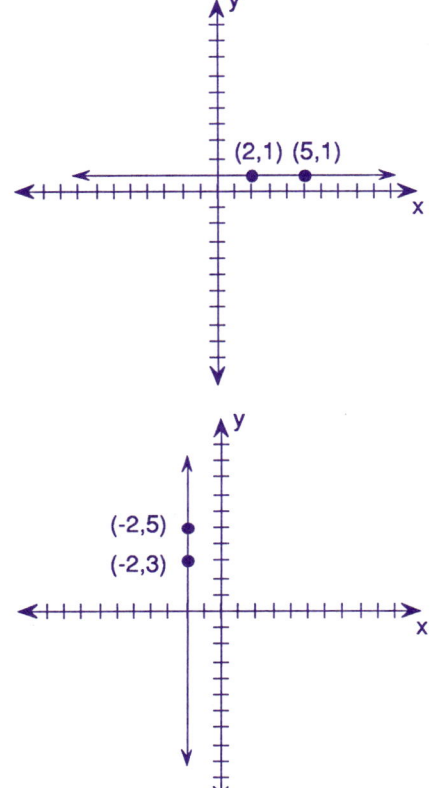

Focus on the use of the slope formula for horizontal and vertical lines

The graph at the right shows the horizontal line through (2,1) and (5,1).

The horizontal line has slope $\frac{1-1}{2-5}$ or $\frac{0}{-3}$ that simplifies to $\frac{0}{1}$. Every horizontal line has a slope of zero.

The graph at the right shows the vertical line through (-2,3) and (-2,5). Its slope is **undefined**, because $\frac{5-3}{-2--2}$ simplifies to $\frac{2}{0}$, which is not defined as a number because of the zero denominator.

Whenever two points lie on the same vertical line, the slope through the points is undefined, because the formula results in a zero denominator. In such cases the line has **no** slope.

Vertical lines have no slope and horizontal lines have slope 0.

Unit 3 Exercise

Part A: Answers for all Part A problems are at the back of the book.

1. (0,-1) and (3,5) are two solutions of $y = 2x - 1$.

   (0,-1) — ? — (3,5), with 3 between them.

   As the value of x increases by 3, the value of y _____ by _____.

2. (0,2) and (4,0) are two solutions of $y = \frac{-1}{2}x + 2$.

   (0,2) — ? — (4,0), with 4 between them.

   As the value of x increases by 4, the value of y _____ by _____.

3. (1,6) and (2,8) are solutions of $y = 2x + 4$.

   (1,6) — ? — (2,8), with 1 between them.

   As the value of x increases by 1, the value of y _____ by _____.

4. (3,17) and (4,22) are solutions of $y = 5x + 2$.

   (3,17) — ? — (4,22), with 1 between them.

   As the value of x increases by 1, the value of y _____ by _____.

5. The slope through two points is found using the formula $\frac{y_2 - y_1}{x_2 - x_1}$. Applying the formula to the points (1,2) and (5,4) gives $\frac{4-2}{5-1} =$ _____.

6. The slope through two points is found using the formula $\frac{y_2 - y_1}{x_2 - x_1}$. Applying the formula to the points (3,4) to (4,7) gives $\frac{7-4}{4-3} =$ _____.

7. Applying the slope formula to the points (3,6) to (4,7) gives $\frac{7-6}{4-3} =$ _____.

8. Applying the slope formula to the points (2,-1) and (-7,3) gives $\frac{3-(-1)}{-7-2} =$ _____.

9. The slope through (-3,-6) and (-4,8) is $\frac{8-(-6)}{-4-(-3)} =$ _____.

10. The line through the two points (1,2) and (1,7) is a vertical line parallel to the ____-axis.

11. When a line has a slope of zero, it is a horizontal line and is parallel to the _____ (x-axis, y-axis).

12. When two different ordered pairs have the same first component, the line through the two points is a vertical line and is parallel to the ____-axis.

13. When the line through two points is a vertical line parallel to the y-axis, the slope is undefined. Since the line through (-2,3) and (-2,7) is a vertical line, its slope is _____.

14. If a line is parallel to the x-axis, its slope is zero. If a line is parallel to the y-axis its slope is _____.

Chapter 6

Part B: Answers for odd-numbered problems of Part B are at the back of the book.

Find the slope through each pair of points.

1. (2,0) and (8,6)
2. (7,8) and (15,3)
3. (0,7) and (1,8)
4. (2,7) and (-3,7)
5. (3,-2) and (3,2)
6. (1,5) and (2,7)
7. (3,5) and (6,-2)
8. (-6,4) and (-1,5)
9. (2,0) and (8,-3)
10. (-3,2) and (4,-7)
11. (8,2) and (-2,12)
12. (0,1) and (3,0)
13. (-6,-2) and (-1,-8)
14. (-2,-7) and (-3,-1)
15. (0,-3) and (7,-8)
16. (2,7) and (-3,0)
17. (-1,4) and (3,-13)
18. (2,1) and (5,1)
19. (1,2) and (1,7)
20. (0,2) and (0,11)
21. (-2,3) and (-11,3)
22. (-1,1) and (0,0)
23. (3,7) and (2,1)
24. (-3,2) and (-3,-6)
25. (2,-3) and (2,7)
26. (-1,1) and (0,4)
27. (8,-2) and (1,-2)
28. (-4,2) and (-6,1)
29. (0,2) and (0,4)
30. (4,0) and (0,4)

Part C: No answers are given for these problems. However, each is accompanied by an ordered pair «C,U» showing the chapter and unit in which it was taught.

1. Solve $\frac{3}{x} - \frac{7}{5x} = \frac{-1}{10}$ «1,5»

2. Use interval notation to write the truth set for $2x^2 + x - 3 \leq 0$.   «2,6»

3. Solve $(2-i)(x+yi) - (3+i) = (8-5i)$ «3,4»

4. The sum of the digits of a two digit number is 8. The number with the digits reversed is 18 more than the number. Find the number.   «3,5»

5. Graph $|2x - 3| < -3$.   «4,3»

6. Is $\frac{3}{4}$ in the neighborhood of $|x - 1| < \frac{3}{4}$?   «4,4»

7. True or false?
   $(-5,3) \in \{(x, y \mid y = x^2 - 2x - 15\}$?   «5,2»

8. What is the range of $\{(1,-7),(2,-7),(3,-5)\}$?   «5,3»

9. Is the ordered pair (-5,7) a solution for the equation $x = -5$?   «6,1»

10. Find the slope and y intercept for the linear equation $x = -2$.   «6,2»

Unit 4: FINDING SLOPES AND Y-INTERCEPTS

Linear Functions y = mx + b Have Slope m

The slope of any linear function $y = mx + b$ represents a change in the value of y divided by the associated change in the value of x. This is the substance of the formula for the slope m of a line through two points, (x_1,y_1) and (x_2,y_2).

$$m = \frac{y_1 - y_2}{x_1 - x_2}$$

For example, (2,14) and (3,21) are two solutions of $y = 7x$.

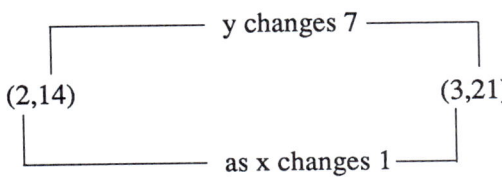

Since the slope of $y = 7x$ is the change in y divided by the change in x, the slope of $y = 7x$ is $\frac{7}{1}$.

Focus on the slope as the ratio of two changes: y to x

(5,-3) and (10,-6) are two solutions of $y = \frac{-3}{5}x$.

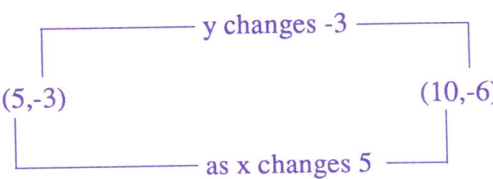

Since the slope of a line is the change in y divided by the change in x, the slope of $y = \frac{-3}{5}x$ is $\frac{-3}{5}$. Notice that the equation $y = \frac{-3}{5}x$ has a slope of $\frac{-3}{5}$, and the coefficient of x in $y = \frac{-3}{5}x$ is also $\frac{-3}{5}$. This is no coincidence. The slope of any equation of the form $y = mx$ is m, the coefficient of x. Hence, the slope of $y = -2x$ is $\frac{-2}{1}$ and the slope of $y = \frac{2}{7}x$ is $\frac{2}{7}$.

Writing Equations in Slope-Intercept Form

The y-intercept of any linear function y = mx + b is determined by b and its slope by m. For $y = \frac{-2}{3}x + \frac{5}{8}$, the y-intercept is $(0, \frac{5}{8})$ and the slope is $\frac{-2}{3}$.

For $y = \frac{4}{5}x - \frac{1}{3}$, the y-intercept is $(0, \frac{-1}{3})$ and the slope is $\frac{4}{5}$.

An equation in the form y = mx + b is in its **slope-intercept** form.

To find the **slope-intercept** form of 4x + y = 7, the equation is solved for y.

$$4x + y = 7$$
$$y = -4x + 7$$

4x + y = 7 and y = -4x + 7 are equivalent equations, because they have the same solutions. y = -4x + 7 is the slope-intercept form of the equation. From the slope-intercept form, y = -4x + 7, the slope -4 and its y-intercept (0,7) are easily determined.

Focus on slope intercept forms

The linear equation y = -13 is equivalent to the slope-intercept form y = 0x − 13. Therefore, the slope is 0, and the y-intercept is (0,-13).

The linear equation x = 5 cannot be written in the form y = mx + b, because the slope-intercept form requires y to have a coefficient of 1. x = 5 has a vertical line as its graph. A vertical line has no slope.

Equations of the form y = b have horizontal lines as their graphs and the slope of all such lines is zero. Equations of the form x = b have vertical lines as their graphs and each such line has no slope.

Linear Equations in Standard Form

y = mx + b is the slope-intercept form of a linear equation.

ax + by = c, where a, b, and c are integers and a ≥ 0, is the **standard form** of a linear equation.

Graphing Linear Functions 251

The following equations are equivalent because they have exactly the same solutions.

$y = 5x + 2$ Not standard form. Slope-intercept form
$-5x + y = 2$ Not standard form, or slope-intercept form
$5x - y = -2$ Standard form $ax + by = c$, $a \geq 0$

The following equations are equivalent because they have exactly the same solutions.

$4x - y = \frac{3}{7}$ Not standard form. $\frac{3}{7}$ is not an integer.
$7(4x - y) = 7 \cdot \frac{3}{7}$ Not standard form.
$28x - 7y = 3$ Standard form $ax + by = c$, $a \geq 0$

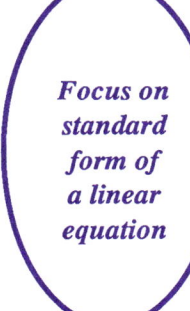

Focus on standard form of a linear equation

The following equations are equivalent. Only the last listed form is standard form $ax + by = c$, $a \geq 0$, a, b, and c integers.

$$\frac{-3}{5}x = \frac{7}{5} - y$$
$$5 \cdot \frac{-3}{5}x = 5 \cdot \frac{7}{5} - 5 \cdot y$$
$$-3x = 7 - 5y$$
$$-3x + 5y = 7$$
$$-1(-3x + 5y) = -1 \cdot 7$$
$$3x - 5y = -7$$

Slope and Y-Intercept Determine the Function

If it is known that a linear function has slope $\frac{-3}{5}$ and y-intercept $(0,-2)$, its standard form can be found in the following steps.

1. Begin with the slope-intercept form
   $y = mx + b$ with $m = \frac{-3}{5}$ and y-intercept $(0,-2)$
   $y = \frac{-3}{5}x - 2$
2. Multiply by the LCM of the denominators.
   $5 \cdot y = 5 \cdot \frac{-3}{5}x - 5 \cdot 2$
   $5y = -3x - 10$
3. Get x and y terms on the same side of the equal sign.
   $3x + 5y = -10$

Focus on writing an equation in standard form given slope and y-intercept

To write, in standard form, the linear equation with slope $\frac{5}{8}$ and y-intercept $\left(0, \frac{13}{12}\right)$,

1. Begin with the slope-intercept form

    $y = mx + b$ with $m = \frac{5}{8}$ and y-intercept $\left(0, \frac{13}{12}\right)$

    $y = \frac{5}{8}x + \frac{13}{12}$

2. Multiply by the LCM of the denominators.

    $24 \cdot y = 24 \cdot \frac{5}{8}x + 24 \cdot \frac{13}{12}$

    $24y = 15x + 26$

3. Get x and y terms on the same side of the equal sign.

    $-15x + 24y = 26$

4. Multiply by -1 to make the coefficient of x positive.

    $-1(-15x + 24y) = -1 \cdot 26$

    $15x - 24y = -26$

Slope and Any Solution Determines the Function

If it is known that a linear function has slope $\frac{3}{7}$ and contains (-2,5) as a solution, its standard form can be found in the following steps.

1. Begin with the slope formula, replace m by $\frac{3}{7}$ and (x_2, y_2) by (-2,5). Drop the subscript designations for x_1 and y_1.

    $m = \dfrac{y_1 - y_2}{x_1 - x_2}$ with $m = \frac{3}{7}$ and $(x_2, y_2) = (-2, 5)$

    $\dfrac{3}{7} = \dfrac{y - 5}{x + 2}$

2. Multiply by the LCM of the denominators.

    $7(x + 2) \cdot \dfrac{3}{7} = 7(x + 2) \cdot \dfrac{y - 5}{x + 2}$

    $3x + 6 = 7y - 35$

3. Place the x and y terms on the same side of the equal sign.

    $3x - 7y = -41$

Focus on writing an equation in standard form given slope and a solution

To write, in standard form, the linear equation with slope $\frac{-4}{9}$ and (-5,-4) as a solution,

1. Begin with the slope formula.

$$m = \frac{y_1 - y_2}{x_1 - x_2} \text{ with } m = \frac{-4}{9} \text{ and } (x_2, y_2) = (-5, -4)$$

$$\frac{-4}{9} = \frac{y + 4}{x + 5}$$

2. Multiply by the LCM of the denominators.

$$9(x + 5) \cdot \frac{-4}{9} = 9(x + 5) \cdot \frac{y + 4}{x + 5}$$

$$-4x - 20 = 9y + 36$$

3. Place the x and y terms on the same side of the equal sign.

$$-4x - 9y = 56$$

4. Multiply by -1 to make the coefficient of x positive.

$$-1(-4x - 9y) = -1 \cdot 56$$
$$4x + 9y = -56$$

Any Two Solutions Determine the Function

A straight line is determined by two points and, similarly, a linear function is completely determined by any two of its solutions. If it is known that a linear function contains (-5,-1) and (3,2) as two of its solutions, its standard form can be found in the following steps.

1. Find the slope through (-5,-1) and (3,2).

$$m = \frac{y_1 - y_2}{x_1 - x_2} = \frac{-1 - 2}{-5 - 3} = \frac{-3}{-8} = \frac{3}{8}$$

2. Use the slope and either point to write an equation.

$$\frac{3}{8} = \frac{y - 2}{x - 3} \text{ when } m = \frac{3}{8} \text{ with the solution } (3,2).$$

3. Multiply by the LCM of the denominators.

$$8(x - 3) \cdot \frac{3}{8} = 8(x - 3) \cdot \frac{y - 2}{x - 3}$$
$$3x - 9 = 8y - 16$$

4. Place the x and y terms on the same side of the equal sign.

$$3x - 8y = -7$$

Focus on writing an equation in standard form given two solutions

To write, in standard form, the linear equation with (-3,6) and (5,4) as two of its solutions,

1. Begin by finding the slope through (-3,6) and (5,4).

$$m = \frac{y_1 - y_2}{x_1 - x_2} = \frac{6 - 4}{-3 - 5} = \frac{2}{-8} = \frac{-1}{4}$$

2. Use the slope and one of the points to write an equation.

$$\frac{-1}{4} = \frac{y - 6}{x + 3} \text{ with slope } \frac{-1}{4} \text{ and solution (-3,6)}$$

3. Multiply both sides of the equation by the LCM of the denominators.

$$4(x + 3) \cdot \frac{-1}{4} = 4(x + 3) \cdot \frac{y - 6}{x + 3}$$
$$-x - 3 = 4y - 24$$

4. Place the x and y terms on the same side of the equal sign.

$$-x - 4y = -21$$

5. Multiply by -1 to make the coefficient of x positive.

$$-1(-x - 4y) = -1 \cdot -21$$
$$x + 4y = 21$$

Unit 4 Exercise

Part A: Answers for all Part A problems are at the back of the book.

1. The slope-intercept form for a linear function is y = mx + b where m is the slope and (0,b) is the y-intercept. Find the slope-intercept form of 3x − y = 5.

2. Write 2x + y = -4 in its slope-intercept form and find its slope and y-intercept.

3. The standard form for a linear function is ax + by = c where a, b, and c are integers and a ≥ 0. Change $y = \frac{3}{5}x + 2$ to its standard form.

4. Find the standard form of $\frac{y - 6}{x - 1} = \frac{3}{5}$ by first multiplying both sides of the equation by 5(x − 1).

5. Find the standard form of $\frac{y + 3}{x - 3} = \frac{9}{2}$ by first multiplying both sides of the equation by 2(x − 3).

6. If the slope and y-intercept of a linear function are known, they are sufficient to completely determine the function. Use the y-intercept form to write the equation of the linear function with slope $\frac{3}{5}$ and y-intercept (0,-4).

Graphing Linear Functions

7. If the slope and one solution of a linear function are known, they are sufficient to completely determine the function. Use the slope formula to write the standard form of the linear function with slope $\frac{-3}{4}$ and solution (5,-1).

8. If any two solutions of a linear function are known, they are sufficient to completely determine the function. Find the slope. Then use the process of problem 7 to write the standard form of the linear function with solutions (-2,4) and (5,7).

9. The slope and y-intercept of $y = 2x - 3$ are 2 and (0,-3). The slope and y-intercept of $y = \frac{-5}{3}x - 1$ are ____ and ____.

10. The slope and y-intercept of $2x - 3y = 6$ are ____ and ____.

11. The slope and y-intercept of $3y - 2x + 8 = 0$ are ____ and ____.

12. The slope and y-intercept of $\frac{1}{2}y + x = 2$ are ____ and ____.

13. The slope and y-intercept of $y = -13$ are ____ and ____.

14. Does $x = 1$ have either a slope or a y-intercept?

Part B: Answers for odd-numbered problems of Part B are at the back of the book.

For problems 1-10, write the equation in standard form.

1. $10 = y - x$
2. $y = x$
3. $y = \frac{4}{7}x - 3$
4. $\frac{y+3}{x-5} = \frac{-2}{5}$
5. $\frac{y+7}{x-6} = \frac{4}{5}$
6. $\frac{y+1}{x-4} = -3$
7. $\frac{y+1}{x-0} = \frac{3}{4}$
8. $\frac{y+8}{x-6} = \frac{-5}{2}$
9. $\frac{y+3}{x+2} = \frac{7}{3}$
10. $\frac{y-5}{x-5} = \frac{4}{3}$

For problems 11-15, write the slope-intercept form of the linear function with the given characteristics.

11. Slope -2 and y-intercept (0,7)
12. Slope 7 and y-intercept (0,-3)
13. Slope $\frac{1}{2}$ and y-intercept $(0,\frac{5}{7})$
14. Slope $\frac{-3}{4}$ and solution (-3,-2)
15. Slope $\frac{4}{3}$ and solution $(4,\frac{3}{4})$

For problems 16-20, write the standard form of the linear function with the given characteristics.

16. Slope 4 and y-intercept (0,-5)

17. Slope -3 and y-intercept $\left(0,\frac{8}{5}\right)$

18. Slope $\frac{-5}{3}$ and solution (4,-2)

19. Slope $\frac{2}{3}$ and solution $\left(\frac{6}{5},-5\right)$

20. Slope 3 and solution $\left(\frac{4}{7},\frac{1}{3}\right)$

For problems 21-25, write the slope-intercept form of the linear function with the given characteristics.

21. Solutions (-3,-8) and (4,-1)

22. Solutions (-2,1) and (3,-4)

23. Solutions (0,0) and (-1,-2)

24. Solutions (3,-2) and (-6,5)

25. Solutions (2,-1) and (3,-2)

For problems 26-30, write the standard form of the linear function with the given characteristics.

26. Solutions (5,-7) and (2,-2)

27. Solutions (-2,5) and (2,-3)

28. Solutions (-3,2) and (7,2)

29. Solutions (-2,-5) and (0,-9)

30. Solutions (3,0) and (0,6)

Part C: No answers are given for these problems. However, each is accompanied by an ordered pair «C,U» showing the chapter and unit in which it was taught.

1. Use the Distributive Law of Multiplication over Addition to write an equivalent expression to $-14x^2y + 21xy^3$. «1,1»

2. Solve $x^4 - 81$ using the set of rational numbers. «1,5»

3. Simplify $\frac{-6\sqrt{2}}{\sqrt{6}}$ «2,2»

4. Simplify $\frac{\sqrt{5}-3}{2\sqrt{5}+1}$ «2,2»

5. Multiply $(3 + 4i)(2 - 5i)$ «3,3»

6. The second side of a triangle is twice as long as the first side, and the perimeter of the triangle is 66 inches. What is the length of the third side? «3,5»

7. True or false? $|5 - 11| = |5| - |11|$ «4,2»

8. Find a rule of correspondence that would match the elements of {1,2,3,...} with the set {2,4,6,...}. «5,1»

9. Graph $\frac{-3}{5}x - 10$ «6,1»

10. Find the slope and y-intercept for the linear equation $y = -3x + 6$. «6,2»

Unit 5: GRAPHING LINEAR INEQUALITIES IN TWO VARIABLES

A Line Separates a Plane Into Three Parts

$3x - 5y = 7$ and $y = \frac{-2}{5}x - 3$ are linear functions. Their graphs are straight lines.

$2x + 3y > 6$ and $y \leq 5x + 3$ are not linear functions because many ordered pairs have the same first component. They are **linear inequalities**. This unit is a study of methods for graphing linear inequalities.

To graph the linear inequality $2x + 3y > 6$, it is necessary to first graph the linear function $2x + 3y = 6$. Three solutions of $2x + 3y = 6$ are found and the line drawn through the three points.

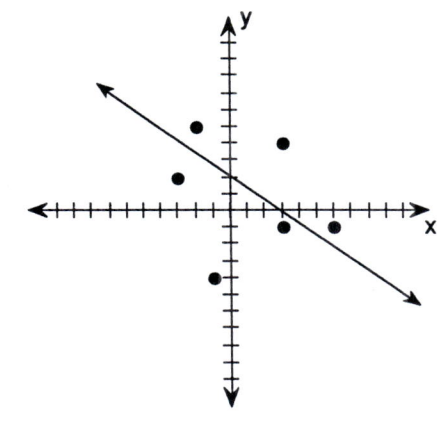

The line of $2x + 3y = 6$ is shown at right. Also six points not on the line of $2x + 3y = 6$ are shown on the graph. (-1,-4), (-4,2), (-2,5), (3,4), (3,-1), and (6,-1) are the points. Three of these points are solutions of the inequality $2x + 3y > 6$.

Every point that is to the lower left of the line of $2x + 3y = 6$ is not a solution of $2x + 3y > 6$. Every point that is to the upper right of the line $2x + 3y = 6$ is a solution of $2x + 3y > 6$.

The line of $2x + 3y = 6$ separates the plane of the graph into three parts:
1. the line itself
2. the area to the upper right of the line
3. the area to the lower left of the line

258   Chapter 6

Focus on the way a line separates the x,y plane into three parts

The graph at the right shows all the solutions of the linear inequality 2x + 3y > 6. The unmarked area to the lower left of the line is the solutions of 2x + 3y < 6. The dashed line itself is the graph of the solutions of 2x + 3y = 6.

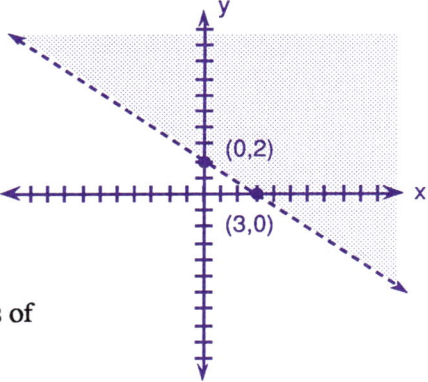

Every linear equation of the form ax + by = c separates the plane of the graph into three parts:
1. the line itself that shows the solutions of the equality ax + by = c
2. a portion on one side of the line that represents ax + by > c
3. a portion on the other side of the line that represents ax + by < c

Graphing Inequalities

To graph 4x − y > 4:

1. Graph the line of 4x − y = 4.
2. Test a point on one side of the line to see if it is a solution of 4x − y > 4.
3. If the point is not a solution of 4x − y > 4, then the graph consists of the area on the other side of the line.

The graph at the right uses a dashed line to represent 4x − y = 4. (0,0) is not a solution of 4x − y > 4.
4 • 0 − 0 > 4 is false. Therefore, the area that does not contain (0,0) is the graph of 4x − y > 4.

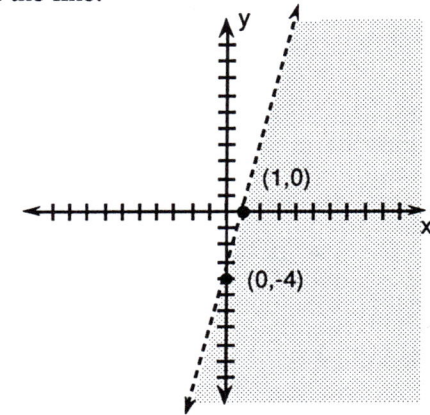

Focus on graphing a linear inequality

The inequality $y \leq \frac{-2}{5}x + 3$ uses the symbol $\leq$. The graph of $y \leq \frac{-2}{5}x + 3$ includes the line of $y = \frac{-2}{5}x + 3$ as well as the area of $y < \frac{-2}{5}x + 3$. To indicate that the solutions of $y = \frac{-2}{5}x + 3$ are included use a solid, rather than dashed, line on the graph.

To graph $y \leq \frac{-2}{5}x + 3$:

1. Graph $y = \frac{-2}{5}x + 3$.
2. Choose any point not on the line and test it as a solution for $y < \frac{-2}{5}x + 3$.
3. If the point tested is a solution, then the graph consists of the line and all points on the same side of the line.
4. If the point tested is not a solution, then the graph consists of the line and all points on the opposite side of the line.

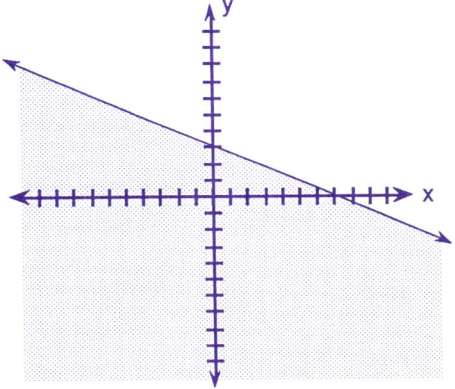

Graphing Two Inequalities on the Same Axes

The graph of
$$3x + 5y \leq 10 \text{ and } x - y > 7$$
is the intersection of the graphs obtained by separately graphing the two inequalities. This is because the word "and" requires that each ordered pair must be a solution of both inequalities.

To graph $3x + 5y \leq 10$ **and** $x - y > 7$, each inequality is first graphed separately on the same set of axes.

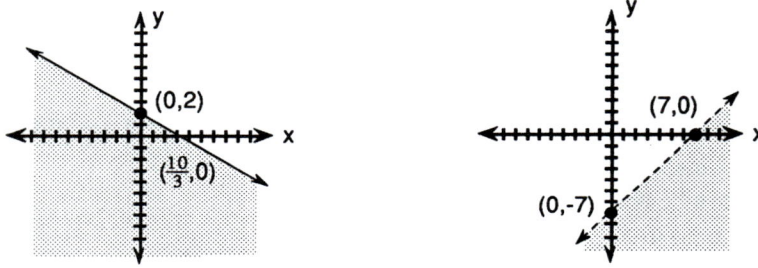

The graph of $3x + 5y \leq 10$ and $x - y > 7$ is the heavily darkened area shown at the right. The pie-shaped area has one boundary that is a portion of $3x + 5y = 10$. This boundary is a solid ray because a solid line is used to graph $3x + 5y \leq 10$. The other boundary of the pie-shaped area is a dashed ray because a dashed line is used to graph $x - y > 7$.

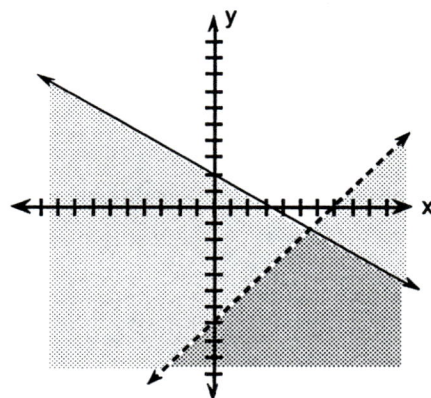

Focus on graphs that are the union of two solution sets

The graph of:
$$4x + y < 5 \text{ or } x + y \geq 7$$
is the union of the graphs obtained in graphing the two inequalities separately. This is because the word "or" allows an ordered pair to be a solution of either inequality.

The graph of:
$$x - y > 4 \text{ or } x + y \leq 3$$
is the union of the graphs of the two inequalities.

To graph
$x - y > 4$ or $x + y \leq 3$, graph the two inequalities on the same axes and accept all the points of either graph. Again note that the boundary made by $x + y \leq 3$ is a solid ray while that made by $x - y > 4$ is a dashed ray.

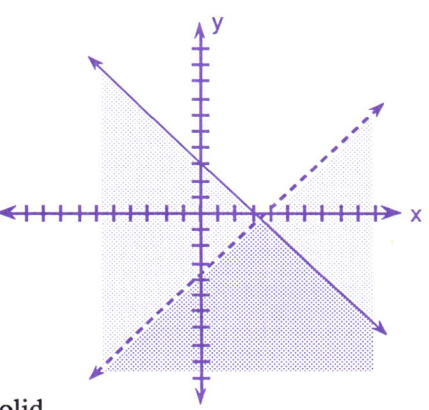

Unit 5 Exercise

Part A: Answers for all Part A problems are at the back of the book.

1. Why is $y > 5x - 3$ not a function?

2. What is the difference in the solution sets of $3x + y > 5$ and $3x + y \geq 5$?

3. The figure at right shows the line of $x + y = 3$. (0,0) is a solution of $x + y < 3$. Therefore, the area to the lower left of the line represents the solutions of _____.

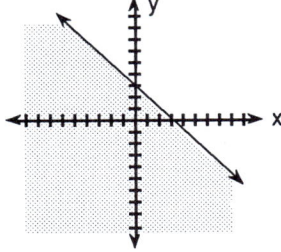

4. The line of $3x + 5y = 4$ separates the plane of the graph into three non-overlapping parts. Describe these three parts of the plane.

5. To graph $y < \frac{-2}{5}x + 3$, the line of $y = \frac{-2}{5}x + 3$ is graphed first. Why?

6. The graph of $x = -3$ is a vertical line through (-3,0). Graph the solutions of $x \geq -3$.

7. The graph of $y = -2$ is a horizontal line through (0,-2). Graph the solutions of $y \leq -2$.

8. Graph $x - y \leq -6$ and $2x + y > 5$ by graphing both inequalities on the same axes and then accepting only those points that are solutions of both inequalities.

9. The graph of $x - y \leq -6$ and $2x + y > 5$ is the (union, intersection) of the solutions of the two inequalities.

10. Graph $x + y < 3$ or $3x - y \geq 4$ by graphing both inequalities on the same axes and accepting any point graphed.

11. The graph of $x + y < 3$ or $3x - y \geq 4$ is the (union, intersection) of the solutions of the two inequalities.

12. The graph of $2x + 5y - 3 \geq 0$ or $7y < 3x - 8$ includes (all, some) of the points on the line of $2x + 5y - 3 = 0$.

13. The graph of $3x + y \leq 4$ and $9x - 4y > 7$ includes (all, some) of the points on the line of $3x + y = 4$.

Part B: Answers for odd-numbered problems of Part B are at the back of the book.

Graph solution sets.

1. $y > 2x - 3$
2. $y \geq \frac{5}{3}x - 2$
3. $y < x$
4. $x - 5y < 10$
5. $x < 2$
6. $y \leq -3$
7. $3x + 5y < 10$
8. $y \leq \frac{x}{4} - 1$
9. $x + y > 1$ and $x - y > 1$
10. $x + 2y \geq 2$ and $x - y \leq 5$
11. $x > 5$ or $x + y < 6$
12. $x + y > 1$ and $x - y > 1$
13. $y < -3$ or $y > 2x - 3$
14. $5x + 2y \leq -2$ or $x \geq -2$
15. $2x + 3y > -6$ and $y < 2x$
16. $x \geq -2$ and $-x + y \geq 1$

Part C: No answers are given for these problems. However, each is accompanied by an ordered pair «C,U» showing the chapter and unit in which it was taught.

1. Graph $(x - 5)(2x + 7) < 0$ «1,6»
2. Multiply $\sqrt[3]{5} - 8$ by its conjugate. «2,2»
3. Simply $\sqrt[3]{48} + \sqrt[3]{54} - 2\sqrt[3]{216} + \sqrt[3]{250}$ «2,3»
4. Find the truth set for $\frac{3}{4}x - \frac{1}{6} = \frac{7}{12} - 2x$. «2,7»
5. Solve $\frac{x}{x-3} > x + 3$ «2,7»
6. Multiply $3i^7 \cdot -5i^4$ «3,1»
7. Simplify $\frac{3 + 7i}{2 - 3i}$ «3,3»
8. Write an open sentence that indicates the distance between the number represented by x and -5 is 1. «4,4»
9. Find f(-5) when $f(x) = \frac{-35}{5 - 6x}$. «5,4»
10. Find the standard form of the linear equation with solutions (-3,7), (2,0). «6,4»

Chapter 6 Test

«6,U» shows the unit in which this problem was studied in this chapter.

1. If x = -5, y = _____ for the linear equation y = x − 1. «6,1»

2. If y = 4, x = _____ for the linear equation y = -3x + 7. «6,1»

3. Is the ordered pair (-4,12) a solution for the equation x = -4? «6,1»

4. Graph y = -3x + 6 «6,1»

5. Graph y = $\frac{-3}{2}$x − 3 «6,1»

6. Graph y = $\frac{5}{2}$x − 5 «6,1»

7. Find the slope for the linear equation in problem 6. «6,2»

8. Find the y-intercept for the linear equation in problem 6. «6,2»

9. Graph y = 4 «6,2»

10. Find the slope and y-intercept for the linear equation in problem 9. «6,2»

11. The line through the points (4,7) and (4,-1) is a vertical line parallel to the _____ -axis, and the slope is _____. «6,3»

Find the slope of the line through the two given points for problems 12-14.

12. (-2,5) and (1,3) «6,3»

13. (0,-3) and (4,1) «6,3»

14. (6,-3) and (6,7) «6,3»

Write the equations in standard form for problems 15-16.

15. 15 − y = 3x «6,4»

16. $\frac{x + 3}{y - 2} = \frac{-2}{5}$ «6,4»

17. Write the slope-intercept form of the linear function with the given characteristics:
    Slope $\frac{-3}{2}$ and y-intercept (0,5) «6,4»

18. Write the linear function y = $\frac{2}{3}$x − 5 in standard form. «6,4»

19. Write 3x − 4y = 12 in y-intercept form. «6,4»

20. Find the standard form of the linear equation with solutions (7,-2) and (5,6). «6,4»

21. Find the standard form of the linear equation with solutions (-3,-5) and (4,-2). «6,4»

22. Graph y < 3x − 2 «6,5»

23. Graph 2x − 3y ≥ -6 «6,5»

24. Graph x + y > 1 and x − y < 1 «6,5»

25. Graph x ≤ -3 or y > 3x + 4 «6,5»

7
Graphing Polynomial Functions and Conic Sections

Unit 1: SIMPLE TRANSFORMATIONS ON THE x,y PLANE

Moving a Point to the Origin

The point (7,2) is located 7 units to the right of the origin and 2 units above it. To move the point (7,2) to the origin a variety of paths could be taken. Three possible paths are described:

1. (7,2) can be carried horizontally 7 units to the left and then vertically 2 units down.

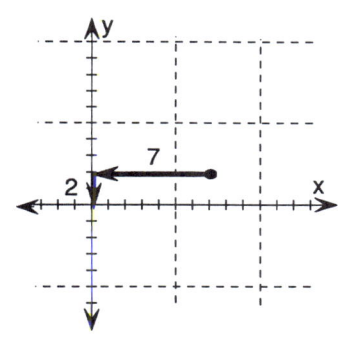

2. (7,2) can be carried vertically 5 units up and then at a 45° declination to the lower left (downwards on a slope of $\frac{1}{1}$) $7\sqrt{2}$ units.

3. (7,2) can be carried down along a slope of $\frac{2}{7}$ for $\sqrt{53}$ units. Any of these paths would move (7,2) to (0,0). The path described as number 3 is the simplest route because it requires only a direction, a slope, and a distance.

The figure above uses a single arrow or **vector** to show the movement of (7,2) to (0,0) along the path described in number 3 above. Notice that the vector has a direction, a slope, and a distance.

Focus on moving a point to the origin

The point (4,-3) can be moved to the origin in a number of ways. The simplest, most direct, movement would be:

an upward motion
along a slope of $\frac{-3}{4}$
for a distance of $\sqrt{25}$ or 5.

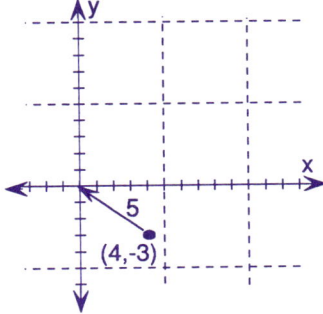

This movement of (4,-3) to (0,0) is shown in the graph at the right by a vector.

Moving Each Point of a Line with a Vector

The line $x - y = -4$ contains the point $(-8,-4)$. A vector moving that point to the origin is shown in the graph at the right. The characteristics of that vector are:

> it moves upwards
> along a slope of $\frac{1}{2}$
> for a distance of $\sqrt{80}$ or $4\sqrt{5}$.

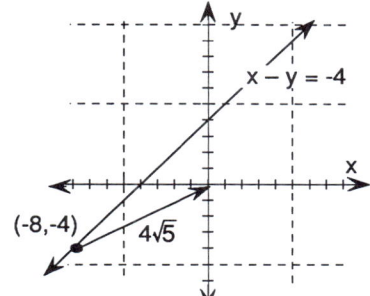

An interesting question to consider is this:

> If each other point of the line $x - y = -4$ is moved using the same vector, what will be the new shape after all points are moved?

The answer is that the shape of the line is preserved exactly. The "image" of the line after the move is a line through the origin parallel to $x - y = -4$. The image has $x - y = 0$ as its equation and is the dashed line in the figure at the right.

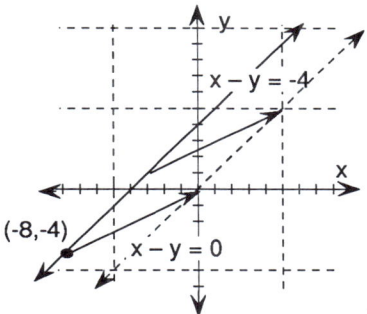

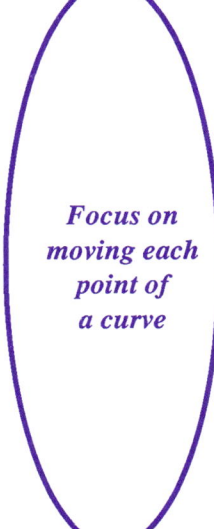

Focus on moving each point of a curve

The curve in the figure at the right contains the point $(-6,2)$. The vector that would move the point to $(0,0)$ has:

> a downwards motion
> along slope $\frac{-1}{3}$
> for distance $\sqrt{40}$ or $2\sqrt{10}$.

If that same vector is applied to every other point on the curve, what will be the shape of the "image" figure?

The answer is that the original curve and its "image" are **congruent**. They have exactly the same shape and size. The image curve is shown in the figure at the right.

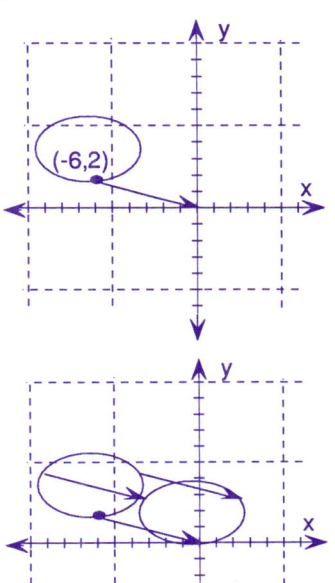

Translations on the x,y Plane

Movements on the x,y plane that can be described by a vector are called **translations**. Any translation, when applied to each point of a geometric figure will preserve shape and size. In other words, the image figure will be congruent to the original.

Any point (x,y) can be moved by a translation to its image point (x',y'). For example, the figure at the right shows a vector that moves (-4,2) to (5,-3). This translation for the entire x,y plane might be shown as
$(x',y') = (x + 9, y - 5)$ or
$(x,y) = (x' - 9, y' + 5)$.

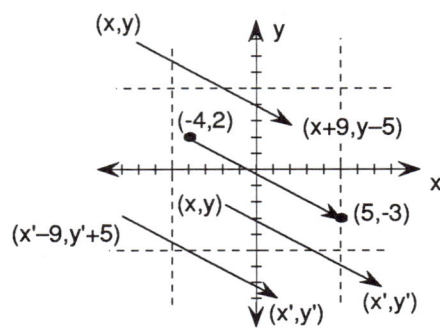

Notice: The fact that each image point (x',y') has a first component increased by 9 means that $x' = x + 9$ or, equivalently, $x = x' - 9$. Similarly, $y' = y - 5$, but $y = y' + 5$.

Focus on translations on the x,y plane

Suppose a geometric figure contains the points (7,3) and (4,-2) and a translation is applied to the plane which moves (7,3) to (-4,7). What is the image of (4,-2) under this translation?

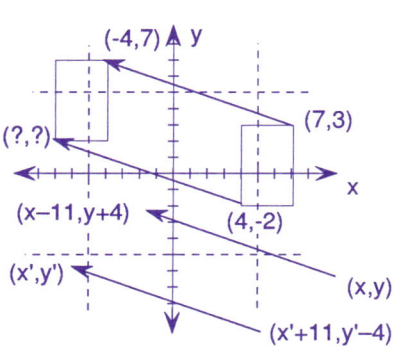

Since (7,3) has (-4,7) as its image, the translation can be described by either $(x',y') = (x - 11, y + 4)$ or $(x,y) = (x' + 11, y' - 4)$.

Applying $(x',y') = (x - 11, y + 4)$ to (4,-2) gives:
$(x',y') = (4 - 11, -2 + 4)$
or (-7,2)

Graphing Polynomial Functions and Conic Sections   269

The Image of a Line after Translation

The line $4x + y = 11$ contains the point $(2,3)$. If a translation moves $(2,3)$ to $(-6,5)$ what will be the equation of the image line?

First, the translation that takes $(2,3)$ to $(-6,5)$ can be described as
  $(x',y') = (x - 8, y + 2)$ or
  $(x,y) = (x' + 8, y' - 2)$
These equalities mean that $x = x' + 8$ and $y = y' - 2$.

Consequently, the equation for the image line can be written by substituting $(x' + 8)$ for x and $(y' - 2)$ for y.

Hence, the image line of $4x + y = 11$ is:
  $4(x' + 8) + (y' - 2) = 11$
  $4x' + 32 + y' - 2 = 11$
  $4x' + y' = -19$
Notice that $4x + y = 11$ and $4x + y = -19$ are parallel lines (same slope) which means that this is an alternative method to the one learned in chapter 6 for finding the equation of a parallel line through an exterior point.

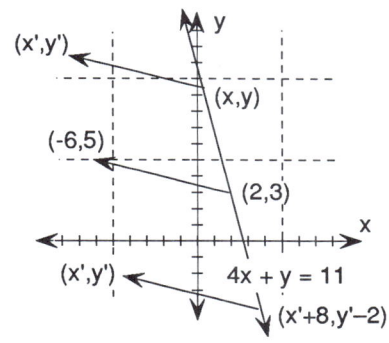

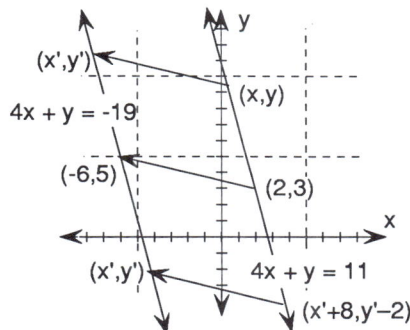

The Image of a Circle after Translation

The equation $x^2 + y^2 = 9$ has the circle shown at the right as its graph.

If each point of the circle is moved by a translation that takes it 5 units to the right and 4 units down, we know that the image figure will be a congruent circle. The translation can be described by either
  $(x',y') = (x + 5, y - 4)$ or
  $(x,y) = (x' - 5, y' + 4)$

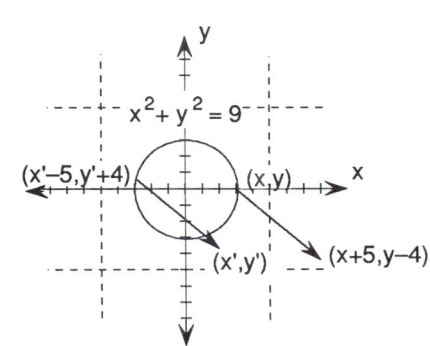

Consequently, the equation for the image circle can be found by substituting $(x' - 5)$ for x and $(y' + 4)$ for y in the original equation.

The circle of $x^2 + y^2 = 9$ has a congruent circle with equation $(x - 5)^2 + (y + 4)^2 = 9$ as its image after the translation.

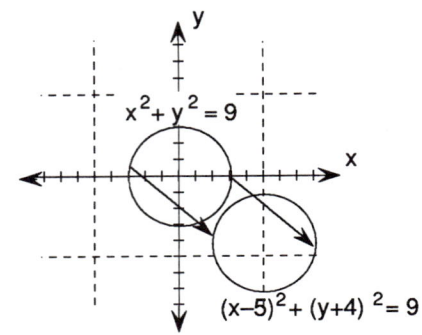

Focus on translations on the x,y plane

The graph at the right shows the curve of the function
$$y = x^2 + 6x + 14$$
which contains the point (-3,5).

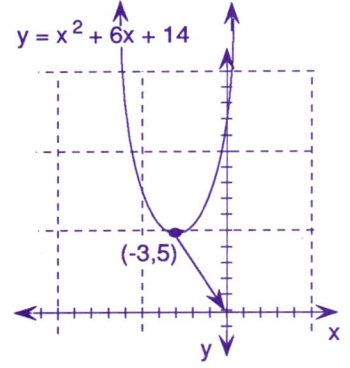

If a translation is used to move the point (-3,5) to (0,0), then the image curve will be congruent to the original, but in a new location.

To find the equation of the image first note that the translation of (-3,5) to (0,0) means that
$$(x',y') = (x + 3, y - 5) \text{ or}$$
$$(x,y) = (x' - 3, y' + 5).$$
The new equation is obtained by substituting $(x' - 3)$ for x and $(y' + 5)$ for y.

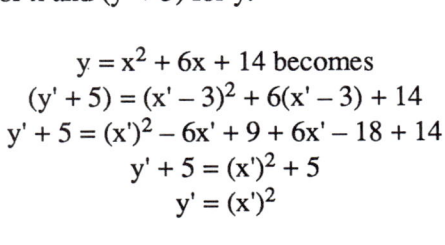

$$y = x^2 + 6x + 14 \text{ becomes}$$
$$(y' + 5) = (x' - 3)^2 + 6(x' - 3) + 14$$
$$y' + 5 = (x')^2 - 6x' + 9 + 6x' - 18 + 14$$
$$y' + 5 = (x')^2 + 5$$
$$y' = (x')^2$$

After the translation that moves (-3,5) to (0,0) the image of $y = x^2 + 6x + 14$ is the congruent curve with equation $y = x^2$.

Graphing Polynomial Functions and Conic Sections

Translations Preserve Size and Shape

By means of translations, any figure on the x,y plane can be relocated in a different location while preserving its size and shape. More specifically, any point of a figure can be translated to the origin (0,0) with the remaining points of the figure maintaining their relative positions.

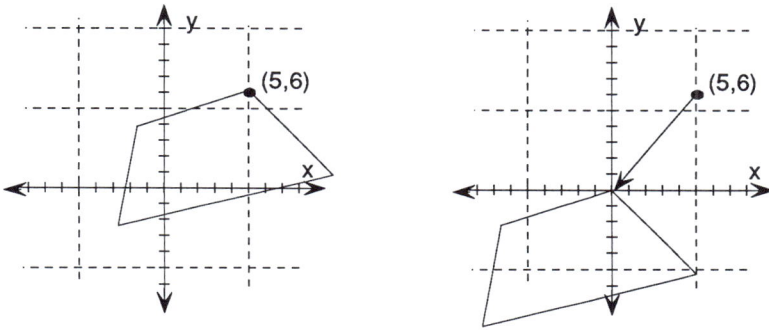

The two graphs above show a "before" and "after" situation. The figure (quadrilateral) on the left could, by a translation, be relocated so the point (5,6) would be at (0,0). The two quadrilaterals are congruent.

Focus on translations preserving size and shape

The two graphs below show the "before" and "after" of the movement of a figure with a translation that takes (-2,-5) to (0,0).

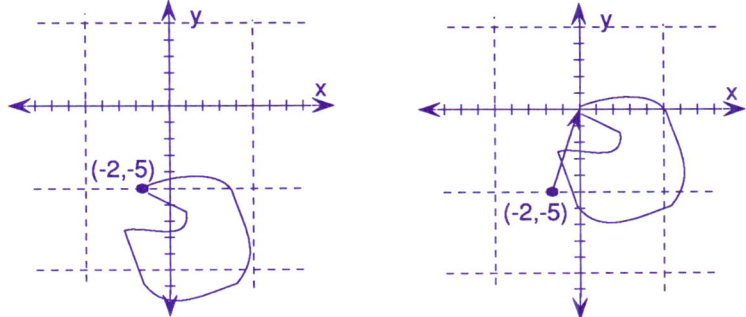

Notice that the translation preserves the original figure's shape and size. The image is congruent to the original and has been relocated to a new position on the x,y plane.

Rotations about the Origin

The curve shown at the right is "slanted" on the x,y plane. No translation could make this curve upright. However, if each point of the curve is rotated 30° counterclockwise about the origin, the curve is relocated in an upright position and neither the size nor the shape has been altered.

A circular movement about a point on the x,y plane is a **rotation**. This rotation about (0,0), like a translation, preserves the size and shape of the curve.

The figure at the right shows the results of a 30° counterclockwise rotation about the point (0,0). The image is a congruent curve.

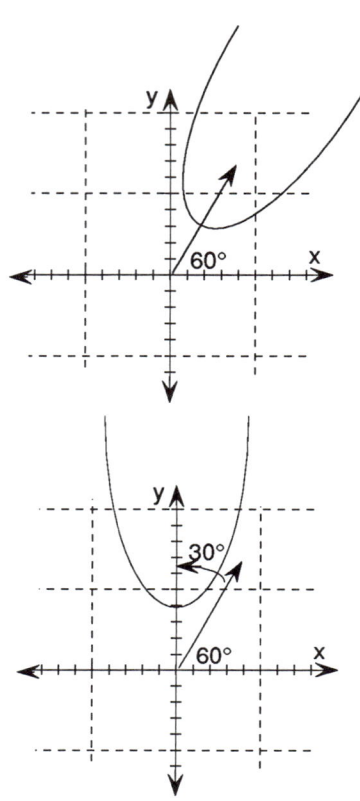

Focus on a rotation about the origin

The figure on the left below is an ellipse. The figure on the right below is the image after a 45° clockwise rotation about the origin.

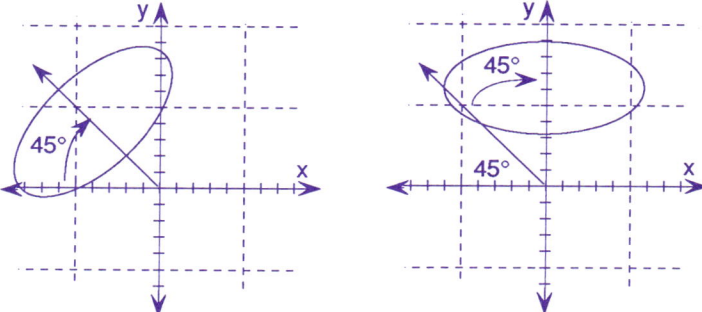

The result of the rotation is a congruent figure which is upright on the graph.

Unit 1 Exercise

Part A: Answers for all Part A problems are at the back of the book.

1. Below is shown a vector which moves the point (5,-2) to the "image point" (7,-4). Using the same vector, what would be the "image point" of (-3,5)?

   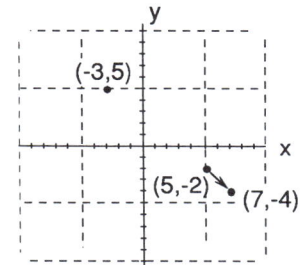

2. Below is shown a vector which moves each point of the straight line 4 units down and three units to the right. Draw the "image line."

   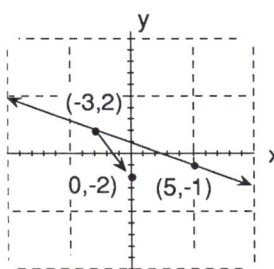

3. A vector pictorially describes a **translation** on the x,y plane. The vector that moves (6,-4) to (3,5) may be described by the equations $x' = x - 3$ and $y' = y + 9$. Write two equivalent equations describing (x,y) in terms of (x',y').

4. The line of $2x - 5y = 4$ contains (-3,-2). A translation that moves (-3,-2) to (0,0) is described by $x = x' - 3$ and $y = y' - 2$. Substitute $(x' - 3)$ for x and $(y' - 2)$ for y in $2x - 5y = 4$ and find the equation of a parallel line that contains (0,0).

5. Below is shown a circle and a vector which moves its center from (6,-2) to (-3,5). Will the "image points" of the circle form a new circle of exactly the same size?

   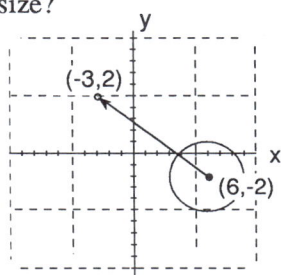

6. The circle shown in problem 5 has $(x - 6)^2 + (y + 2)^2 = 9$ as its equation. The translation in problem 5 can be described by the equations $x = x' + 9$ and $y = y' - 7$. Substitute $(x' + 9)$ for x and $(y' - 7)$ for y in the circle's equation and find the equation of the congruent circle with center (-3,5).

7. Below is shown a curve and a vector which moves one of its points from (-3,4) to (0,2). Will the "image points" of the curve form a congruent curve?

   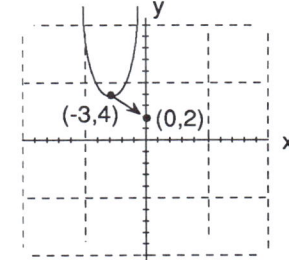

8. The curve shown in problem 7 has $y - 4 = (x + 3)^2$ as its equation. The translation in problem 7 can be described by the equations $x = x' - 3$ and $y = y' + 2$. Substitute $(x' - 3)$ for x and $(y' + 2)$ for y in the curve's equation and find the equation of a congruent curve.

9. A translation that moves (4,7) to (-3,5) can be described by the equations x = x' + 7 and y = y' – 2. These equations describe the original (x,y) in terms of the images (x',y'). Write two equations describing (x,y) in terms of their images (x',y') for the translation that moves (8,5) to (10,-3).

10. Below is shown an arc with center (0,0) that moves (7,0) to (0,7). The arc pictorially describes a rotation of the x,y plane about the point (0,0) with the amount of the rotation being 90° counterclockwise. This rotation would move (0,8) to (-8,0) and (-5,0) to _____.

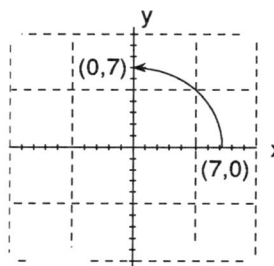

11. Below is shown a square with vertices (4,1), (7,1), (7,-2), and (4,-2). If a rotation of 90° counterclockwise using (0,0) as its center is applied, draw the image of the square and label the image points of its vertices.

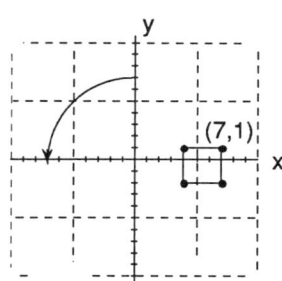

12. Below is shown a curve which is separated into two symmetric halves by the x-axis which intersects the curve at (3,0). If a rotation of 90° counterclockwise using (0,0) as its center is applied, draw the image of the curve and label the image point of (3,0).

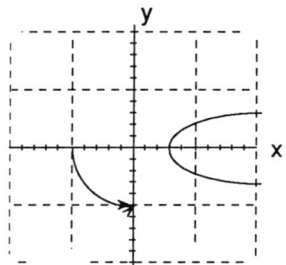

13. Below is shown a curve which is separated into two symmetric halves by an oblique line that forms a 30° angle with the x-axis. If a rotation of 60° counterclockwise using (0,0) as its center is applied, draw the image of the curve.

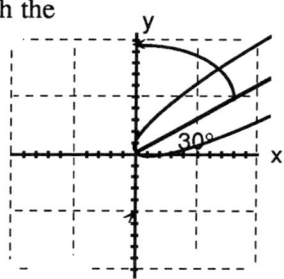

14. Below is shown a two-branched curve which is separated into two symmetric halves by an oblique line that contains (5,3) and forms a 45° angle with the line y = 3. If a rotation of 45° counterclockwise using (5,3) as its center is applied, draw the image of the curve.

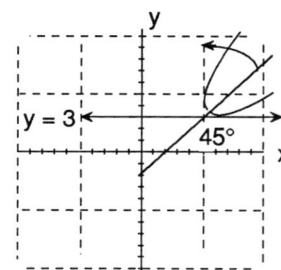

15. Are the image curves drawn in problems 12-14 congruent to their predecessors?

Part B: Answers for odd-numbered problems of Part B are at the back of the book.

For problems 1-10 write equations $x = x' \pm ?$ and $y = y' \pm ?$ for the given translations.

1. Moves a point 2 units right and 5 units down.

2. Moves a point 4 units right and 6 units up.

3. Moves a point 3 units left and 4 units up.

4. Moves a point 6 units left and 1 unit down.

5. Moves (5,3) to (7,9).

6. Moves (-2,4) to (7,3).

7. Moves (8,-1) to (6,6).

8. Moves (-3,-3) to (5,0).

9. Moves (4,-3) to (6,-2).

10. Moves (-2,2) to (4,-4).

11. The graph of $5x + y = 7$ is a straight line. Sketch the image line of $5x + y = 7$ after a translation described by $x = x' - 3$ and $y = y' + 4$.

12. The graph of $3x - 2y = 4$ is a straight line. Find the equation of the image line of $3x - 2y = 4$ after a translation described by $x = x' - 5$ and $y = y' - 3$.

13. The graph of $(x + 6)^2 + (y + 3)^2 = 7^2$ is a circle with center (-6,-3) and radius 7. Sketch its image circle after a translation described by $x = x' + 2$ and $y = y' - 4$.

14. The graph of $(x - 4)^2 + (y + 6)^2 = 4^2$ is a circle. Find the equation of its image circle after a translation described by $x = x' + 4$ and $y = y' - 3$.

15. The graph of $(x + 5)^2 + (y - 3)^2 = 2^2$ is a circle with center (-5,3) and radius 2. Sketch its image circle after a translation that moves its center to (0,0) and write the equation of the new circle.

16. The graph of $4(x - 5)^2 + 9(y + 3)^2 = 36$ is an oval which is called an ellipse. Find the equation of its image ellipse after a translation described by $x = x' - 7$ and $y = y' - 1$.

17. The graph of $6(x - 5)^2 + 3(y + 3)^2 = 18$ is an ellipse. Find the equation of its image ellipse after a translation described by $x = x' + 3$ and $y = y' + 4$.

18. The graph of $3(x - 5)^2 + 8(y + 3)^2 = 24$ is an ellipse with "center" (5,-3). Find the equation of its image ellipse after a translation that moves its "center" to (0,0).

19. The graph of $y + 5 = (x - 6)^2$ is a curve which is called a parabola. Find the equation of its image parabola after a translation described by $x = x' + 2$ and $y = y' + 1$.

20. The graph of $y - 7 = (x + 4)^2$ is a parabola. Find the equation of its image parabola after a translation described by $x = x' + 5$ and $y = y' - 2$.

21. The graph of $y - 6 = (x - 3)^2$ is a parabola which contains the point (3,6). Find the equation of its image parabola after a translation that moves (3,6) to (0,0).

22. A rotation of 90° counterclockwise has (0,0) as its center. What is the image point of (7,3)?

23. A rotation of 90° counterclockwise has (4,5) as its center. What is the image point of (7,2)?

24. The line of $x + y = 7$ goes through the point (2,5) at an angle of 45° from the line of $y = 5$. A rotation of 45° counterclockwise about the point (2,5) will result in the line with equation _____.

25. A parallelogram has vertices of (7,-7), (9,-4), (2,-1), and (4,2). Sketch the image figure after a rotation of 90° counterclockwise about (0,0).

Part C: No answers are given for these problems. However, each is accompanied by an ordered pair «C,U» showing the chapter and unit in which it was taught.

1. Factor $a^2 - 5a(x + y) - 14(x + y)^2$ «1,4»

2. Solve $\sqrt{5}x - 7 = \sqrt{3}x + \sqrt{6}$ «2,5»

3. Solve $x^2 - 2x + 5 = 0$ «3,2»

4. Simplify $\frac{5 - 2i}{3 + i}$ «3,3»

5. Solve $|3x - 5| = 10$ «4,2»

6. Use interval notation to describe the truth set of $|7x - 1| < 6$. «4,4»

7. Draw a Venn diagram for $A \cup B$. «5,5»

8. Write the equation $\frac{x - 4}{y + 1} = \frac{3}{5}$ in standard form. «6,4»

9. Write the equation $4x - 3y = -12$ in slope y-intercept form. «6,4»

10. Graph $-4x + 3y = -12$ «6,5»

Unit 2: POINT PLOTTING GRAPHS

Equations with Curves as Their Graphs

Linear equations of the form $ax + by = c$ have straight lines as their graphs. Equations such as $x^3 - 4x^2 - 32x = y$, and $x^2 + 2x - 3 = y$, and $x^2 + y^2 = 13$ have curves as their graphs. In this unit, equations associated with curves will be graphed by finding enough points (solutions) of the equations to approximate the curves.

The graphs of $x^3 - 4x^2 - 32x = y$ and $x^2 + 2x - 3 = y$ are shown below. Notice that the scales of the y-axes have been shortened to facilitate the graphing. This type of distortion is frequently useful for showing the graph in a reasonably sized space.

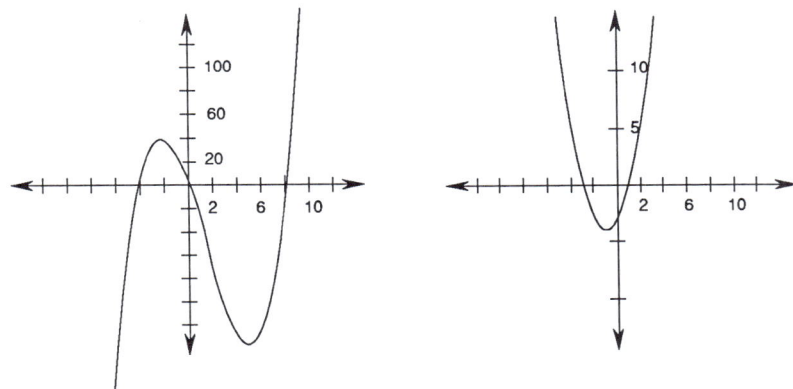

Finding Solutions of Equations

(-5,38) is a solution of $x^2 - 4x - 7 = y$. If x is replaced by -5 and y by 38, then $x^2 - 4x - 7 = y$ becomes a true statement.

For (-5,38)
$$x^2 - 4x - 7 = y$$
$$(-5)^2 - 4 \cdot -5 - 7 = 38$$
$$25 + 20 - 7 = 38$$
$$45 - 7 = 38$$

Notice: The final equality is a true statement.

278   Chapter 7

Focus on finding solutions

To find a solution for $x^2 - 4x - 7 = y$, choose any number in the domain of the relation, substitute it for x, and calculate the corresponding value for y.

For example, if x is replaced by 5 the process is:
   When x = 5, $x^2 - 4x - 7 = y$ becomes $5^2 - 20 - 7 = y$
   $25 - 27 = y$
   $-2 = y$

This means that (5,-2) is a solution for $x^2 - 4x - 7 = y$.

Similarly, if x is replaced by 0 the process is:
   When x = 0, $x^2 - 4x - 7 = y$ becomes $0^2 - 0 - 7 = y$
   $-7 = y$

(0,-7) is another solution for $x^2 - 4x - 5 = y$.

Sketching a Graph from a Set of Solutions

To graph $x^2 - 4x - 7 = y$ **enough** solutions of the equation are needed to provide a general idea of the location and shape of the curve. How many is enough? The answer depends on the degree of accuracy desired for the graph. Since $x^2 - 4x - 7$ is a second degree polynomial, we are assured that the graph will be U-shaped and the most difficult points to plot will be those where the curve is changing directions.

The graph at the right shows three solutions of $x^2 - 4x - 7 = y$. The three points are not sufficient for sketching the curve of $x^2 - 4x - 7 = y$, but line segments connecting the points from left to right will be a good clue for what other solutions are needed. For example, the figure at the right clearly indicates that a change in direction occurs between -5 and 5. This means that choices for x between -5 and 5 are likely to be productive. They will show the location of the curve where some turning occurs.

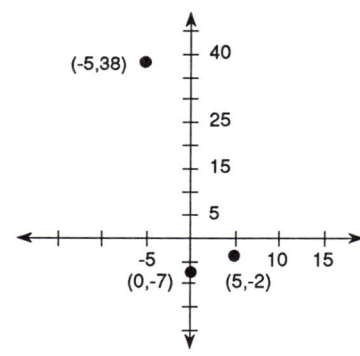

Focus on sketching the curve of $x^2 - 4x - 7 = y$

When a number of solutions of an equation are needed a table of values is one method for organizing the information.

$$x^2 - 4x - 7 = y$$

x	y
-10	133
-5	38
0	-7
5	-2
10	53

1. Begin with a table using -10, -5, 0, 5, and 10 as choices for x. This table is shown at the right. Notice: (0,-7) contains the smallest y value found.

2. From the results of step 1, choose x-values near 0.

3. Now (3,-10) contains the smallest value for y, so 2 and 4 are used as values for x.

4. Since (2,-11) contains the smallest y value, try 1 as the value for x.

x	y
-3	13
3	-10
2	-11
4	-7
1	-10

The table now can be used to plot its points. The cluster of points around x = 2 makes it possible to draw a close approximate curve for the equation in that area of the graph. All parts of the curve approximate the line segments joining the points from left to right.

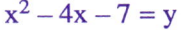

Graphing a Third Degree Polynomial

The relation $y = x^3 + 4x^2 - 7x - 10$ is a third degree (cubic) polynomial function. The curve of a third degree polynomial can be expected to change directions no more than two times. Again, the accuracy of the graph will depend upon the plotting of solutions near these "turning" points.

280   Chapter 7

To draw the graph of $y = x^3 + 4x^2 - 7x - 10$ begin with a table like the one shown below and a very rough graph connecting the solutions of the table.

$y = x^3 + 4x^2 - 7x - 10$

x	y
-10	-540
-5	0
0	-10
5	180
10	1320

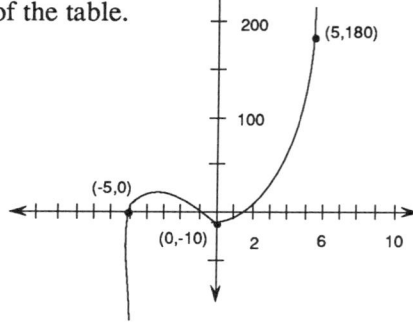

The table entries and the rough graph that accompany them indicate that -5 and 0 are possible "turning" points of the curve. A table to refine the first rough graph should include x values close to -5 and 0.

x	y
-7	-108
-3	10
3	32

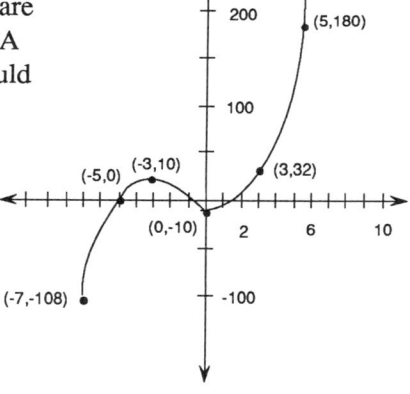

The graph above should draw our attention to seek more information about solutions near (-3,10) and (0,-10)

x	y
-4	18
-2	12
-1	0
1	-12

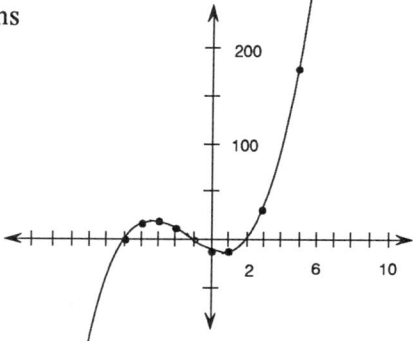

Notice that the final result is a curve that is N-shaped. It has, as expected, two changes of direction (turning points).

Focus on graphing a quartic polynomial function

$x^4 - 3x^3 - 19x^2 + 3x + 18 = y$
is a fourth degree (quartic) polynomial function. To graph it using point plotting, begin with a table like the one shown at the right. The values for x were chosen to represent a spread of points from -10 to 10.

The values of y
1. decrease as x goes from -10 to -2
2. increase as x goes from -2 to 0
3. decrease as x goes from 0 to 5
4. increase as x goes from 5 to 10

x	y
-10	11088
-7	2496
-5	528
-2	-24
0	18
2	-60
5	-192
7	480
10	5148

Plotting the points on the x,y plane and drawing the line segments that connect the points from left to right would result in a distorted W-shape.

Since we can expect a fourth degree polynomial function to have three "turning" points, the table shows clearly that the curve can be improved by finding more ordered pairs near x = -2, x = 0, and x = 5.

The figure at the right shows the curve of the quartic function $x^4 - 3x^3 - 19x^2 + 3x + 18 = y$.

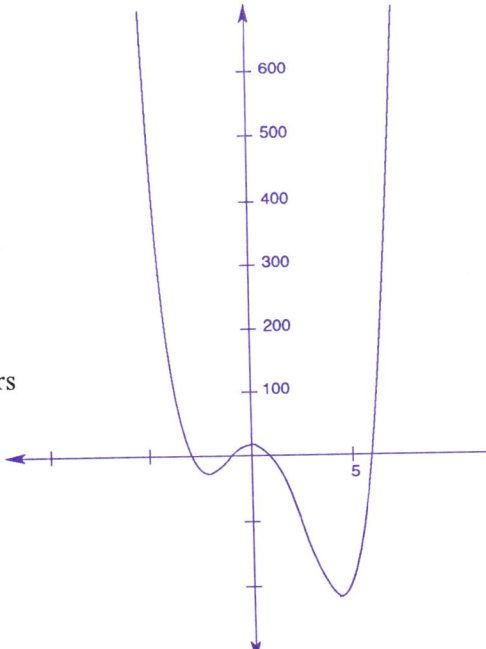

Relations with Restricted Domains

The relation $(x - 3)^2 + (y + 2)^2 = 64$, has a restricted domain. Any value of x must make $(x - 3)^2$ less than or equal to 64 because any other values will result in no matching value for y. This translates into:

$(x - 3)^2 \leq 64$
$-8 \leq (x - 3) \leq 8$
$-5 \leq x \leq 11$

Consequently, the only values for x that need to be attempted are those from -5 to 11 inclusive.

282   Chapter 7

A table of solutions for
$(x - 3)^2 + (y + 2)^2 = 64$ is
shown at the right. Notice that
some of the entries show two
values of y for a given choice
of x. This clearly shows that
$(x - 3)^2 + (y + 2)^2 = 64$
is not a function.

x	y
-5	-2
-2	$-2 \pm \sqrt{39}$
0	$-2 \pm \sqrt{55}$
3	$-2 \pm \sqrt{64}$
6	$-2 \pm \sqrt{55}$
8	$-2 \pm \sqrt{39}$
11	-2

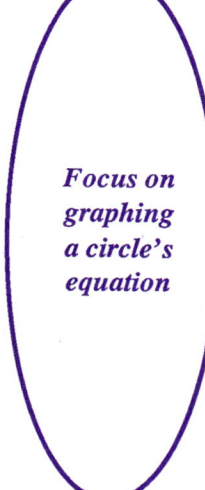

Focus on graphing a circle's equation

The entries in the table for
$(x - 3)^2 + (y + 2)^2 = 64$
have been plotted on the x,y
plane at the right. The positions
of $(-2, -2 \pm \sqrt{39})$ must be
approximated using the knowledge
that $\sqrt{39}$ is slightly more than 6.
Similarly, $-2 + \sqrt{55}$ is between
5 and 6 because $\sqrt{55}$ is between
7 and 8.

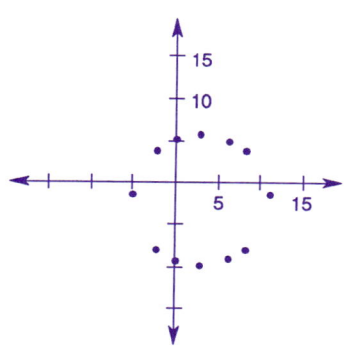

The plotted points should appear
to outline a circle because that is
the curve associated with the
relation. It is left to the reader
to sketch the circle.

Graphing Other Relations

The relation $4(x + 5)^2 + 9(y - 1)^2 = 36$, has a restricted domain. Any value of x that makes $4(x + 5)^2$ greater than 36 must be excluded from the domain because there can be no matching value of y.

$$4(x + 5)^2 \leq 36$$
$$(x + 5)^2 \leq 9$$
$$-3 \leq (x + 5) \leq 3$$
$$-8 \leq x \leq -2$$

Entries for the table at the right were found by choosing values for x between -8 and -2. Again, some of the entries show two values of y for a given choice of x. This clearly shows that

$4(x + 5)^2 + 9(y – 1)^2 = 36$

is not a function.

x	y
-8	1
-7	$\frac{3 \pm \sqrt{20}}{3}$
-6	$\frac{3 \pm \sqrt{32}}{3}$
-5	3 or -1
-4	$\frac{3 \pm \sqrt{32}}{3}$
-3	$\frac{3 \pm \sqrt{20}}{3}$
-2	-2

It is left to the reader to plot the positions of the table entries and draw a smooth curve connecting them. The curve is an oval shape, but is not a circle.

Focus on graphing a hyperbola's equation

The relation $25(x – 2)^2 – 4(y + 3)^2 = 100$, has a restricted domain, but in this case any value of x must make $25(x – 2)^2$ greater than 100.

$$25(x – 2)^2 \geq 100$$
$$(x – 2)^2 \geq 4$$
$$x – 2 \leq -2 \text{ or } x – 2 \geq 2$$
$$x \leq 0 \text{ or } x \geq 4$$

Entries for the table at the right were found by choosing values for x that are not between 0 and 4. Some of the entries show two values of y for a given choice of x. This clearly shows that $25(x – 2)^2 – 4(y + 3)^2 = 100$ is not a function.

x	y
-4	$-3 \pm 10\sqrt{2}$
-2	$-3 \pm 5\sqrt{3}$
0	-3
4	-3
6	$-3 \pm 5\sqrt{3}$
8	$-3 \pm 10\sqrt{2}$

It is left to the reader to plot the positions of the table entries and draw two separate curves for the relation. The graph will have no points in the interval of 0 < x < 4. The relation will consist of two curves: one to the left of x = 0 and the other similarly shaped to the right of x = 4.

Unit 2 Exercise

Part A: Answers for all Part A problems are at the back of the book.

1. The graph of $4x - y = 6$ is a straight line. Is the graph of $y = 3x^2 - 5x + 2$ a straight line?

2. The graph of any second degree polynomial function is always a U-shaped curve. What shape is the graph of $y = 3x^2 - 5x + 2$?

3. Use a table of values to sketch the curve of $y = 4x^2 - 8x + 2$.

4. The graph of a third degree polynomial function is expected to have two (one less than the degree of the polynomial) "turning" points. How many "turning" points are expected for the curve of $y = x^3 + 5x^2 - x + 1$?

5. Use a table of values to sketch the curve of $y = x^3 + 5x^2 - x + 1$.

6. The curve of $y = x^3$ is expected to have two "turning" points, but may have less. Plot solutions when $-5 < x < 5$ and sketch the curve of $y = x^3$.

7. The graph of a fourth degree polynomial function is expected to have three (one less than the degree of the polynomial) "turning" points. How many "turning" points are expected for the curve of $y = x^4 + x^3 + x^2 + x - 5$?

8. Use a table of values to sketch the curve of $y = x^4 + x^3 + x^2 + x - 5$.

9. The curve of $y = x^4$ is expected to have three "turning" points, but may have less. Plot solutions when $-5 < x < 5$ and sketch the curve of $y = x^4$.

10. The relation $(x - 3)^2 + (y + 2)^2 = 25$ has a circle as its graph. Solve the inequality $(x - 3)^2 \leq 25$ and find the domain of the relation.

11. Graph the circle of $(x - 3)^2 + (y + 2)^2 = 25$ by first constructing a table of values containing domain elements of the relation.

Part B: Answers for odd-numbered problems of Part B are at the back of the book.

Graph the polynomial functions of problems 1-10 by first constructing a table of solutions.

1. $y = x^2 - 6x - 7$
2. $y = x^2 + 8x + 7$
3. $y = x^2 + 2x - 3$
4. $y = x^2 + 4x + 3$
5. $y = x^2$
6. $y = 3x^2 - x - 14$
7. $y = (x - 3)^2 - 4$
8. $y = (x - 5)^2 - 9$
9. $y = x^3 - x^2 - 6x$
10. $y = x^4 + 2x^3 - 16x^2 - 2x + 15$

Graph the relations of problems 11-20 by first determining the domain.

11. $x^2 + y^2 = 36$

12. $x^2 + y^2 = 49$

13. $(x + 3)^2 + y^2 = 16$

14. $(x - 5)^2 + (y + 4)^2 = 25$

15. $y = \sqrt{64 - x^2}$

16. $x = \sqrt{9 - y^2}$

17. $4x^2 + 9y^2 = 36$

18. $9(x - 5)^2 + 4(y + 2)^2 = 36$

19. $16x^2 - 9y^2 = 144$

20. $9(x - 1)^2 - 16(y + 4)^2 = 144$

Part C: No answers are given for these problems. However, each is accompanied by an ordered pair «C,U» showing the chapter and unit in which it was taught.

1. Factor $(x - 5)^2 - 9y^2$ «1,4»

2. Simplify $-3\sqrt[3]{56}$ «2,3»

3. Solve $\frac{x}{4} - \frac{5}{12} = \frac{1}{6} - \frac{x}{3}$ «2,7»

4. Simplify $\sqrt{-48}$ «3,1»

5. $(3 - i)(x + yi) + (-3 + 2i) = (1 - i)$ «3,4»

6. What is the opposite of $(3 + 7i)$? «3,4»

7. Solve $|2x - 1| = 7$ «4,2»

8. $(-2,\underline{\quad}) \in \{(x,y) \mid y = 2x^2 + 3x - 5\}$ «5,2»

9. Graph $x \leq 2$ or $y > 2x - 1$ «6,5»

10. Find the equation of the image line of $x - 2y = 4$ after a translation described by $x = x' - 3$ and $y = y' + 1$. «7,1»

Unit 3: DISTANCE BETWEEN TWO POINTS

The Sides of a Right Triangle

The Pythagorean Theorem is a statement of the relationship between the lengths of the three sides of a right triangle. For the right triangle shown at the right, the longest side, c, is opposite the right angle. The longest side of a right triangle is called its **hypotenuse**. The two shorter sides of a right triangle (a and b in the figure) are called **legs**.

The Pythagorean Theorem states that the square of the hypotenuse is equal to the sum of the squares of the lengths of the legs. As a formula the Pythagorean Theorem can be shown as $a^2 + b^2 = c^2$ or $c = \sqrt{a^2 + b^2}$.

Focus on the sides of a right triangle

In a right triangle, the side opposite the right angle is the hypotenuse.

For triangle 1 at the right, the hypotenuse is lettered c. The hypotenuse of triangle 2 is lettered f.

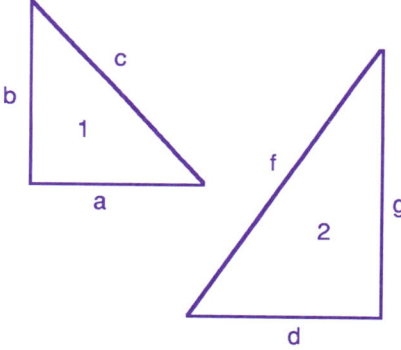

The Pythagorean Theorem

The formula showing the relationship between the three sides of a right triangle is $c = \sqrt{a^2 + b^2}$ where c is the length of the hypotenuse, and a and b are the lengths of the legs.

If a = 8 and b = 6, then the value of c can be found using the steps shown at the right.

$c = \sqrt{8^2 + 6^2}$
$c = \sqrt{64 + 36}$
$c = \sqrt{100}$
$c = 10$

If $a = 7$ and $b = 3$, then the value of c can be found using the steps shown at the right. $\sqrt{58}$ cannot be simplified, so $c = \sqrt{58}$.

$c = \sqrt{7^2 + 3^2}$
$c = \sqrt{49 + 9}$
$c = \sqrt{58}$

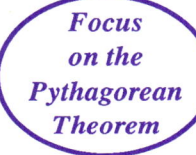

Focus on the Pythagorean Theorem

Using the formula $c = \sqrt{a^2 + b^2}$ the hypotenuse of a right triangle with legs of 3 and 6 can be found in the steps shown at the right. Notice that $\sqrt{45}$ is simplified to the final result $3\sqrt{5}$.

$c = \sqrt{3^2 + 6^2}$
$c = \sqrt{9 + 36}$
$c = \sqrt{45}$
$c = 3\sqrt{5}$

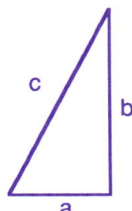

Right Triangles on the x,y Plane

In the graph at the right, a right triangle is shown. The three vertices of the triangle are (3,1), (7,1), and (3,4). The lengths of the three sides of the triangle can be determined from the ordered pairs of the vertices.

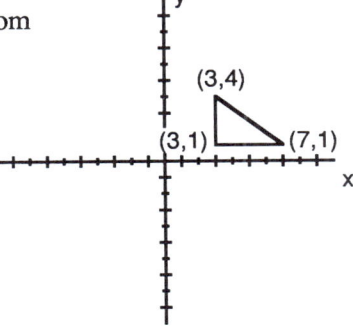

(3,1) and (7,1) are on the same horizontal line, so the distance between the two points is $|3 - 7|$ or $|7 - 3|$ or 4.

(3,4) and (3,1) are on the same vertical line, so the distance between them is $|4 - 1|$ or $|1 - 4|$ or 3. The distance between (3,4) and (7,1) can now be found using the Pythagorean Theorem.

$c = \sqrt{4^2 + 3^2}$
$c = \sqrt{16 + 9}$
$c = \sqrt{25}$
$c = 5$

Focus on triangles on the x,y plane

In the figure at the right, the lengths of the three sides of the triangle can be found using the ordered pairs of the vertices.

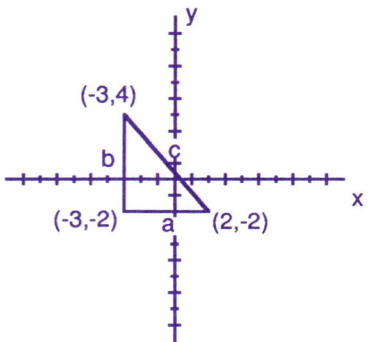

$a = |2 - -3| = |-3 - 2| = 5$
$b = |-2 - 4| = |4 - -2| = 6$

$c = \sqrt{5^2 + 6^2} = \sqrt{25 + 36} = \sqrt{61}$

Finding the Distance Between Two Points

If (x_1, y_1) and (x_2, y_2) are two points on the x,y plane, the horizontal distance between the two points is $|x_1 - x_2|$ and the vertical distance is $|y_1 - y_2|$. This means that the Pythagorean Theorem can be altered to provide the following distance formula.

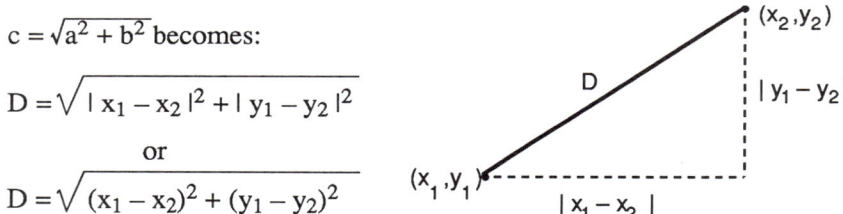

$c = \sqrt{a^2 + b^2}$ becomes:

$D = \sqrt{|x_1 - x_2|^2 + |y_1 - y_2|^2}$

or

$D = \sqrt{(x_1 - x_2)^2 + (y_1 - y_2)^2}$

To find the distance between the two points (5,1) and (2,7), it is necessary to find the hypotenuse of a right triangle with the two given points as the endpoints of that hypotenuse. Two points could serve as the third vertex of the right triangle: (2,1) or (5,7). Regardless of which point is chosen, the legs of the right triangle are 3 and 6 in length.

Therefore, the distance between (5,1) and (2,7) is:

$D = \sqrt{(5-2)^2 + (1-7)^2}$
$D = \sqrt{(3)^2 + (-6)^2}$
$D = \sqrt{9 + 36}$
$D = \sqrt{45}$
$D = 3\sqrt{5}$

vertical distance between (2,7) and (5,1) is 6.

horizontal distance between (2,7) and (5,1) is 3.

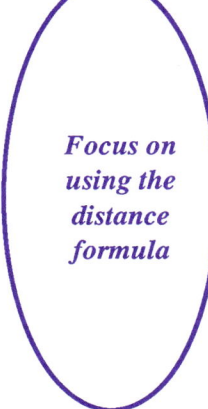

Focus on using the distance formula

To find the distance between two points use the distance formula

$$D = \sqrt{(x_1 - x_2)^2 + (y_1 - y_2)^2}$$

where $(x_1 - x_2)^2$ is the square of the horizontal distance between the points and $(y_1 - y_2)^2$ is the square of the vertical distance.

For example, the distance between (8,-3) and (-1,2) is:

$$D = \sqrt{(8 - -1)^2 + (-3 - 2)^2}$$
$$D = \sqrt{(9)^2 + (-5)^2}$$
$$D = \sqrt{81 + 25}$$
$$D = \sqrt{106}$$

vertical distance between (8,-3) and (-1,2) is 5.

horizontal distance between (8,-3) and (-1,2) is 9.

Application of the Distance Formula

Interesting problems can be written involving distance on the x,y plane. The following problem is an example.

Problem: Find a point on the y-axis that is 8 units from the point (5,-3).

Solution: Since the point is on the y-axis, its first component is 0. Hence it can be written (0,y). Also, since it is 8 units from (5,-3), the distance formula can be used with the following substitutions.

$$D = \sqrt{(x_1 - x_2)^2 + (y_1 - y_2)^2}$$
$$8 = \sqrt{(5 - 0)^2 + (-3 - y)^2}$$

Both sides of the equation are squared.

$$8^2 = (5 - 0)^2 + (-3 - y)^2$$
$$64 = 25 + 9 + 6y + y^2$$
$$y^2 + 6y - 30 = 0$$

The quadratic formula is applied to find y.

$$y = \frac{-6 \pm \sqrt{156}}{2} = \frac{-6 \pm 2\sqrt{39}}{2} = -3 \pm \sqrt{39}$$

There are two roots for the equation and two points on the y-axis that are 8 units from (5,-3). Since $\sqrt{39}$ is slightly greater than 6, the two points have y coordinates slightly greater than 3 and slightly less than -9. These are approximations for $(-3 \pm \sqrt{39})$.

290   Chapter 7

Focus on an application of the distance formula

Problem: Find the equation of the line that is the perpendicular bisector of the segment joining (4,7) and (8,9).

Solution: We are looking for an equation $ax + by = c$ that bisects (cuts into two equal parts) the segment joining (4,7) and (8,9). Because the line is to be a **perpendicular** bisector, each point (x,y) on it must be the same distance from (4,7) as it is from (8,9).

Distance from (x,y) to (4,7) is: $\sqrt{(x-4)^2 + (y-7)^2}$
Distance from (x,y) to (8,9) is: $\sqrt{(x-8)^2 + (y-9)^2}$

$$\sqrt{(x-4)^2 + (y-7)^2} = \sqrt{(x-8)^2 + (y-9)^2}$$
$$x^2 - 8x + 16 + y^2 - 14y + 49 = x^2 - 16x + 64 + y^2 - 18y + 81$$
$$-8x + 16 - 14y + 49 = -16x + 64 - 18y + 81$$
$$8x + 16 + 4y + 49 = 64 + 81$$
$$8x + 4y = 80$$

The desired line has equation $8x + 4y = 80$ or $2x + y = 20$.

Finding A Particular Point on the x,y Plane

The distance formula can be used to find particular points on the x,y plane. For example, suppose the points A = (3,0) and B = (-1,0) are given and a point C = (5,y) is sought so the sum of the distances AC and BC is equal to 10.

$$\text{Distance AC} + \text{Distance BC} = 10$$
$$\sqrt{(3-5)^2 + (0-y)^2} + \sqrt{(-1-5)^2 + (0-y)^2} = 10$$
$$\sqrt{(-2)^2 + (-1y)^2} + \sqrt{(-6)^2 + (-1y)^2} = 10$$
$$\sqrt{4 + y^2} + \sqrt{36 + y^2} = 10$$
$$(4 + y^2) + 2\sqrt{(4+y^2)(36+y^2)} + (36 + y^2) = 100$$
$$40 + 2y^2 + 2\sqrt{144 + 40y^2 + y^4} = 100$$
$$20 + y^2 + \sqrt{144 + 40y^2 + y^4} = 50$$
$$\sqrt{144 + 40y^2 + y^4} = 30 - y^2$$
$$(144 + 40y^2 + y^4) = (30 - y^2)^2$$
$$144 + 40y^2 + y^4 = 900 - 60y^2 + y^4$$
$$100y^2 = 756$$
$$y^2 = 7.56$$

$y = \sqrt{7.56}$ or $y = -\sqrt{7.56}$

Graphing Polynomial Functions and Conic Sections   291

Two points have been found. Both $(5,\sqrt{7.56})$ and $(5,\sqrt{-7.56})$ meet the requirement that the sum of their distances from points A and B is 10.

Focus on finding a particular point

The preceding example required finding a point (5,y) so that the sum of two distances would be 10. As another example, suppose the point A = (4,1) and the line y = -2 are given and a point C = (5,y) is sought so the distance AC is equal to the shortest distance from C to the line y = -2.

$$\text{Distance AC} = \text{Distance C to the point } (x,-2)$$
$$\sqrt{(4-5)^2 + (1-y)^2} = \sqrt{(5-5)^2 + (-2-y)^2}$$
$$\sqrt{(-1)^2 + (1-y)^2} = \sqrt{(0)^2 + (-2-y)^2}$$
$$\sqrt{1 + (1-y)^2} = \sqrt{(-2-y)^2}$$
$$(\sqrt{1 + (1-y)^2})^2 = (\sqrt{(-2-y)^2})^2$$
$$1 + (1-y)^2 = (-2-y)^2$$
$$1 + 1 - 2y + y^2 = 4 + 4y + y^2$$
$$2 - 2y + y^2 = 4 + 4y + y^2$$
$$-6y = 2$$
$$y = \frac{-1}{3}$$

The point $(5,\frac{-1}{3})$ is the only point that is equidistant from both (4,1) and the line y = -2.

Unit 3 Exercise

Part A: Answers for all Part A problems are at the back of the book.

1. In the figure below a right triangle is shown with legs of lengths 5 and 12. Use the Pythagorean Theorem $c = \sqrt{a^2 + b^2}$ to find the length of the hypotenuse.

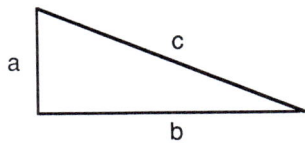

2. Use the Pythagorean Theorem to find the hypotenuse of a right triangle when one leg is 4 units long and the other is 5.

3. (5,7) and (2,-1) are endpoints of a hypotenuse. Plot the two points on the x,y plane and find the third vertex of a right triangle.

4. Plot (-5,-3) and (-1,6) on the x,y plane. Then find a third vertex so a right triangle is formed and label the lengths of the two legs.

5. Use the Pythagorean Theorem to find the length of the hypotenuse in problem 4.

6. Use the distance formula,
$$D = \sqrt{(x_2 - x_1)^2 + (y_2 - y_1)^2},$$
to find the distance between (5,7) and (2,-1).

7. Use the distance formula,
$$D = \sqrt{(x_2 - x_1)^2 + (y_2 - y_1)^2},$$
to find the distance between (-5,-3) and (-1,6).

8. Determine if (5,3), (13,8), and (9,7) are vertices of a right triangle. Hint: find the distances between the pairs of points and then apply the Pythagorean Theorem.

9. The sentence "A point (x,y) is located at distance 5 from (8,-2)" can be translated into an equation as: $\sqrt{(x-8)^2 + (y+2)^2} = 5$. Square both sides of the equation. What new equation is obtained?

10. Translate "The distance between (x,y) and (4,3) is the same as the distance between (9,-2) and (5,7)" into an equation.

Part B: Answers for odd-numbered problems of Part B are at the back of the book.

For problems 1-10 find the distance between the given pair of points.

1. (-6,2), (-1,14)
2. (-4,0), (0,3)
3. (0,0), (2,4)
4. (0,2), (0,5)
5. (0,0), (3,3)
6. (4,5), (-2,7)
7. (-2,7), (4,15)
8. (-3,-3), (1,5)
9. (0,-2), (-7,0)
10. (2,7), (2,-13)

11. Find the equation of the straight line that perpendicularly bisects the segment joining (0,5) and (-7,2).

12. Find the equation of the straight line that perpendicularly bisects the segment joining (-3,-4) and (6,3).

13. Find a point on the x-axis that is the same distance from both (5,3) and (-7,6).

14. Find a point on the y-axis that is the same distance from both (8,0) and (-5,-3).

15. A circle with center (5,-2) and radius 4 is the set of points (x,y) that are at distance 4 from (5,-2). Use the distance formula to write the equation for this circle.

16. A circle with center (-7,1) and radius 6 is the set of points (x,y) that are at distance 6 from (-7,1). Use the distance formula to write the equation for this circle.

17. Find a point on the x-axis (x,0) where the sum of its distances from (-6,3) and (2,3) is 10.

18. Find a point (x,2) so that the sum of its distances from (-1,-1) and (3,1) is $4\sqrt{5}$.

19. Find a point (10,y) so its distance from (6,1) is $\sqrt{5}$ more than its distance from (9,1).

20. Find a point (x,-2) whose distance from (2,1) is $2\sqrt{2}$ less than its distance from (-2,1).

Part C: No answers are given for these problems. However, each is accompanied by an ordered pair «C,U» showing the chapter and unit in which it was taught.

1. Solve $3x - 18 = x^2$ «1,5»

2. Graph the truth set for the inequality $(x - 3)(2x + 7) \leq 0$. «1,6»

3. Solve $\sqrt{x - 3} + 2 = 5$ «2,6»

4. Simplify $(7 + 2i)(3 - 5i)$ «3,3»

5. The sum of the digits of a two digit number is 9. The number with the digits reversed is 45 less than the number. Find the number. «3,5»

6. Solve $|2x - 5| = 9$ «4,2»

7. Graph the truth set for $|3x - 1| \leq -4$. «4,3»

8. Is $y = x^2$ a function? «5,4»

9. Write $5x - 2y = -10$ in slope intercept form. «6,4»

10. Graph the relation $(x + 3)^2 + (y - 5)^2 = 4$ «7,2»

Unit 4: Conic Sections – Circles

Conic Sections

The figure shown at the right is a double cone. The vertical line through points A and B is the **axis** of the double cone. The point A is the **vertex** of the double cone and is one of its points. The point B and all other points of the axis except A are not points of the double cone. In other words, the double cone consists of points on the cones, but not points "inside" or "outside" the cones. A straight line containing point A and another point on a cone, like point C, represents the **slant** of the double cone. Lines through A with more angle than the slant will remain entirely "inside" the cones. Lines through A with less angle than the slant will remain entirely "outside" the cones.

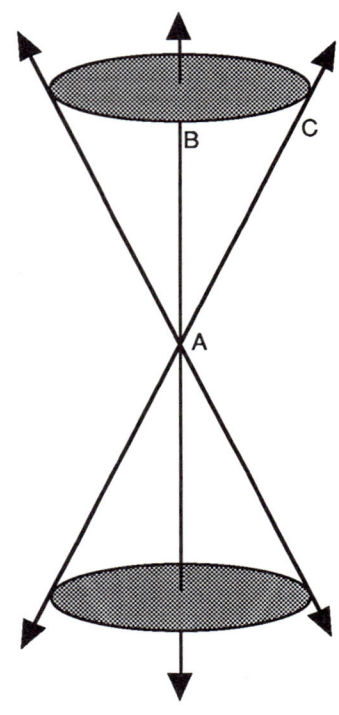

Two thousand years ago, Greek mathematicians encountered problems that required curves for their solutions. Greek mathematics was far advanced in geometry, but lagged badly in an ability to describe geometric shapes using anything like our algebra. Consequently, when the Greeks sought a family of curves for problem solving, they looked at geometric models. They became particularly interested in the different shapes that could result from the intersection of a plane and a double cone because those curves frequently led to solutions for their problems. The family of curves found by the Greeks are called conic sections because of the way they were discovered. Today, mathematicians continue to call the curves conics, but, as you will see, these curves are now described by second degree equations in two variables. To visualize these shapes as the Greeks discovered them, imagine a giant wielding a scimitar and cutting through the double cone with one mighty swing. Four shapes, called **conic sections**, can result from this intersection of a plane and the double cone. In this section, the intersection of a horizontal plane with a double cone whose axis is vertical is studied.

Focus on circles as conic sections

The intersection of a horizontal plane and a vertically held double cone is a circle. The size of the circle depends upon the distance the plane is from the point A, the double cone's vertex. If the plane is close to point A, the circle will be smaller than if the intersection occurs further from point A. The point where the horizontal plane intersects the axis of the double cone is the **center** of the circle, but the center is not itself one of the points of the circle. Only the points on both the double cone and the plane are points of the circle.

Circles as Conic Sections

The easiest conic section to visualize geometrically is the circle. In visualizing a circle, the basic idea is a curve that is everywhere equidistant from a single point — the center of the circle.

Translating a geometric concept of a circle to a set of points described algebraically is accomplished using the distance formula. We begin with a circle with (0,0) as its center.

Definition   **Circle with (0,0) as its Center**

A set of points on the x,y plane is a **circle** with (0,0) as its center   **if and only if**   each point (x,y) of the set is at a distance d (d > 0) from (0,0).

The **center** of the circle is (0,0) and its **radius** is d.

Focus on the meaning of the definition

Because the definition of circle with center (0,0) involves distance, its meaning is directly related to the distance formula:

$$D = \sqrt{(x_1 - x_2)^2 + (y_1 - y_2)^2}$$

Since one of the points is (0,0) and the distance given is d, these substitutions are made in the distance formula to give:

$$d = \sqrt{(x-0)^2 + (y-0)^2}$$
$$d = \sqrt{x^2 + y^2}$$
$$d^2 = x^2 + y^2$$

This equation will be the equation of a circle with center (0,0) whenever a positive number replaces d. For example, if d = 5 the equation becomes

$$5^2 = x^2 + y^2 \text{ or } x^2 + y^2 = 25$$

and this is the equation with center (0,0) and radius 5.

Circles with Center at the Origin

The circle with radius 9 which has the origin (0,0) as its center is described by the equation $x^2 + y^2 = 9^2$ or $x^2 + y^2 = 81$.

Every equation of the form $x^2 + y^2 = r^2$ is a circle with the center at (0,0) and radius r.

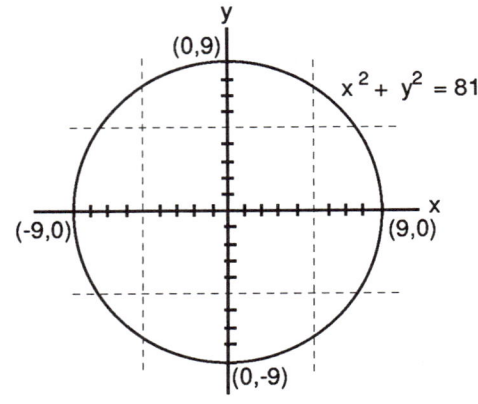

Focus on circles with centers at the origin

$x^2 + y^2 = 9$ is equivalent to $x^2 + y^2 = 3^2$ and its graph is a circle with the center at (0,0) and a radius of 3.

The graph of $x^2 + y^2 = 26$ is a circle with the center at (0,0) and a radius of $\sqrt{26}$.

The equation $x^2 + y^2 = 1$ describes a **unit circle** with center at (0,0) and radius 1.

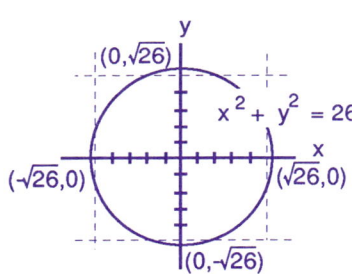

Circles with Centers Not at the Origin

Recall from Unit 1 of this chapter that a circle can be moved, using a translation, to any other location on the x,y plane.

For example, if the circle with equation $x^2 + y^2 = 36$ is to be moved so its new center is at (5,-3) the following steps can be used:
1. The translation from (0,0) to (5,-3) can be described by:
   $(x',y') = (x + 5, y - 3)$ or $(x,y) = (x' - 5, y' + 3)$
2. The equalities of step 1 mean than $x = x' - 5$ and $y = y' + 3$.
3. The expressions are substituted in the original equation.
   $x^2 + y^2 = 36$ becomes
   $(x' - 5)^2 + (y' + 3)^2 = 36$

The result means that $(x - 5)^2 + (y + 3)^2 = 36$ is the circle with center (5,-3) and radius 6.

Focus on a circle's equation after translation

The equation of a circle with center (0,0) and radius r is $x^2 + y^2 = r^2$. If an equation for a circle congruent to $x^2 + y^2 = r^2$, but with center (h,k) is sought, then the following steps are used.

1. Since (0,0) is to become (h,k), a translation to make this move is described by $(x,y) = (x' - h, y' - k)$.
2. The equality of step 1 allows the following substitutions in the original equation.

$$x^2 + y^2 = r^2 \text{ becomes}$$
$$(x' - h)^2 + (y' - k)^2 = r^2$$

The graph of any equation of the form $(x - h)^2 + (y - k)^2 = r^2$ is a circle with center (h,k) and radius r, $r > 0$. This is the basis for the definition shown below.

Definition   **Circle**

A set of points on the x,y plane is a **circle** with (h,k) as its center   **if and only if**   each point (x,y) of the set is at a distance d $(d > 0)$ from a given point (h,k).

The point (h,k) is the circle's **center** and its equation is equivalent to $(x - h)^2 + (y - k)^2 = r^2$ where r is the **radius**.

Recognizing Circles From Their Equations

Every equation $(x - h)^2 + (y - k)^2 = r^2$ is a circle with center (h,k) and radius r. This means that the graph of $(x - 1)^2 + (y - 3)^2 = 13$ is a circle with center $(1,3)$ and radius $\sqrt{13}$.

Similarly, the graph of $(x + 2)^2 + \left(y - \tfrac{1}{2}\right)^2 = 16$ is a circle with center $\left(-2, \tfrac{1}{2}\right)$ and radius 4.

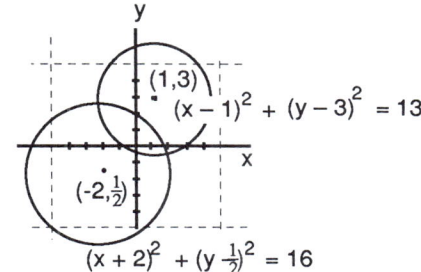

Focus on circles with centers not at the origin

The graphs of equations of the form $(x - h)^2 + (y - k)^2 = r^2$ are circles with the centers at (h,k) and radius r. The graph of $(x + 3)^2 + (y - 2)^2 = 3^2$ is a circle with center at (-3,2) and a radius of 3.

The equation of a circle with the center at (h,k) and radius r is $(x - h)^2 + (y - k)^2 = r^2$.

The equation for a circle with the center at (4,-3) and a radius of $\sqrt{3}$ is $(x - 4)^2 + (y + 3)^2 = 3$.

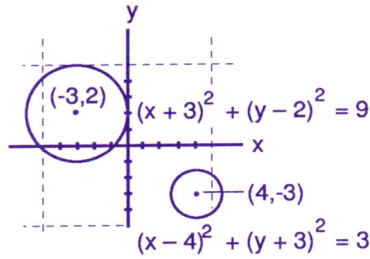

Changing the Form for a Circle's Equation

The equation $2x^2 - 8x + 14y + 2y^2 = 13$ is the equation of a circle, but its center and radius are not obvious until the form of the equation is changed as shown below.

$$2x^2 - 8x + 14y + 2y^2 = 13$$
$$x^2 - 4x + 7y + y^2 = \frac{13}{2}$$
$$(x^2 - 4x) + (y^2 + 7y) = \frac{13}{2}$$
$$(x^2 - 4x + 4) + \left(y^2 + 7y + \frac{49}{4}\right) = \frac{13}{2} + 4 + \frac{49}{4}$$
$$(x - 2)^2 + \left(y + \frac{7}{2}\right)^2 = \frac{91}{4}$$

Now the form of the equation makes it clear that it describes a circle with center $\left(2, \frac{-7}{2}\right)$ and radius $\frac{\sqrt{91}}{2}$.

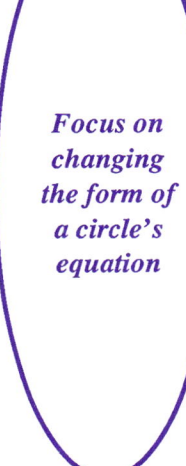

Focus on changing the form of a circle's equation

In trying to change the form of a circle's equation, begin by dividing each term of the equation by any coefficient of x^2 that is different from 1.

$$4x^2 + 4y^2 + 20x - 32y = -1$$
$$x^2 + y^2 + 5x - 8y = \frac{-1}{4}$$

Now group the terms so the same variables appear in each grouping.

$$(x^2 + 5x) + (y^2 - 8y) = \frac{-1}{4}$$

Within each grouping complete the square and add enough to the right side of the equation to balance it. Review the completing the square process if the next equation is difficult to understand.

$$\left(x^2 + 5x + \frac{25}{4}\right) + (y^2 - 8y + 16) = \frac{-1}{4} + \frac{25}{4} + 16$$

Factor each trinomial perfect square and add the terms on the right side of the equation.

$$\left(x + \frac{5}{2}\right)^2 + (y - 4)^2 = 22$$

The equation now indicates a circle with center $\left(\frac{-5}{2}, 4\right)$ and radius $\sqrt{22}$.

Unit 4 Exercise

Part A: Answers for all Part A problems are at the back of the book.

1. Draw a graph of the circle with center (0,0) and radius 2.

2. The circle with center (0,0) and radius 2 is the set of points (x,y) that are at distance 2 from (0,0). Write the equation translating the sentence "The distance from (x,y) to (0,0) is 2."

3. Square both sides of the equation
$$\sqrt{(x-0)^2 + (y-0)^2} = 2.$$
After simplifying, what is the result?

4. Draw a graph of the circle with center (7,-3) and radius 2.

5. Are the circles drawn for problems 1 and 4 congruent?

6. Use a translation that moves (0,0) to (7,-3) and the equation $x^2 + y^2 = 4$. Write the equation of the congruent circle with center (7,-3).

7. The equation of the circle with center (0,0) and radius $\sqrt{7}$ is $x^2 + y^2 = 7$. Write the equation of the circle with center (0,0) and radius 7.

8. The equation of the circle with the center (1,2) and a radius 6 is $(x-1)^2 + (y-2)^2 = 36$. Write the equation of the circle with center (-4,-2) and radius 5.

9. $(x+2)^2 + (y-5)^2 = 64$ is equivalent to $(x--2)^2 + (y-5)^2 = 8^2$. Both equations describe the circle with center _____ and radius _____.

10. The graph of $(x + 7)^2 + (y - 3)^2 = 19$ is a circle with center _____ and radius _____.

11. Why is $\left(x + \frac{2}{3}\right)^2 + (y - \sqrt{2})^2 = -9$ not the equation of a circle?

12. Why is $\left(x + \frac{1}{2}\right) + (y - \sqrt{5})^2 = 43$ not the equation of a circle?

13. Why is $\left(x + \frac{3}{4}\right)^2 + (y - \sqrt{7})^3 = 16$ not the equation of a circle?

14. Why is $\left(x + \frac{3}{8}\right)^2 - (y - \sqrt{3})^2 = 12$ not the equation of a circle?

15. $(x^2 + 6x + 9) + \left(y^2 - 7x + \frac{49}{4}\right) = 25$ is the equation of a circle. Factor the polynomials in parentheses and find the center and radius of the circle.

16. The equation $(x^2 - 8x) + (y - 3)^2 = 33$ is the equation of a circle. What number needs to be added to both sides of the equation so that the center and radius can be found?

17. The equation $(x^2 + 9x) + (y + 5)^2 = 41$ is the equation of a circle. What number needs to be added to both sides of the equation so that the center and radius can be found?

Part B: Answers for odd-numbered problems of Part B are at the back of the book.

1. Is $(-2,7)$ a solution of $(x - 1)^2 + (y - 3)^2 = 25$?
2. Is $(4,7)$ a solution of $(x - 1)^2 + (y - 3)^2 = 25$?
3. Graph $x^2 + y^2 = 36$
4. Graph $x^2 + y^2 = 10$
5. Graph $(x - 1)^2 + (y + 1)^2 = 9$
6. Graph $(x - 2)^2 + y^2 = 49$
7. Graph $x^2 + (y + 1)^2 = 15$
8. Graph $(x - 1)^2 + (y + 3)^2 = 36$
9. Graph $(x + 2)^2 + (y - 1)^2 = 8$
10. Graph $(x - 1)^2 + (y - 2)^2 = 13$

For problems 11-20 write equations for the described circles.

11. Center at $(0,0)$ and radius 5.
12. Center at $(0,3)$ and radius 7.
13. Center at $(-1,3)$ and radius 1.
14. Center at $(-2,-5)$ and radius 4.
15. Center at $(7,-2)$ and radius $\sqrt{8}$.
16. Center at $(-5,7)$ and radius $\frac{3}{4}$.
17. Center at $\left(-6,\frac{5}{8}\right)$ and radius 3.
18. Center at $(4,-5)$ and radius 6.
19. Center at $\left(2,\frac{1}{2}\right)$ and radius $\sqrt{5}$.
20. Center at $(\sqrt{3},-2)$ and radius 3.

In problems 21-25 each equation describes a circle. Use the completing the square process to change the form of the equation. Then find its center and radius.

21. $x^2 + 10x + y^2 = 24$

22. $x^2 + 2x + y^2 - 4y = 4$

23. $x^2 + 3x + y^2 - 12y = \frac{3}{4}$

24. $x^2 - 8x + y^2 - y = \frac{9}{4}$

25. $2x^2 - 14x + 2y^2 + 12y = \frac{-9}{2}$

Part C: No answers are given for these problems. However, each is accompanied by an ordered pair «C,U» showing the chapter and unit in which it was taught.

1. Factor $(a - b)^2 - 2(a - b)(x + y) - 15(x + y)^2$ «1,4»

2. If $(x - a)(3x + 5a) = 0$, then $(x - a) = 0$ or _____ $= 0$. «2,1»

3. Simplify $-12\sqrt[5]{64}$ «2,3»

4. $(3 + 5i)(4 - 3i) =$ _____ «3,1»

5. Evaluate $|13 - 27| - |-36|$ «4,2»

6. Is $\frac{1}{2}$ in the neighborhood of $|x - 3| < 4$? «4,4»

7. Find the domain and range of $y = (3x - 4) - 1$. «5,3»

8. Find the slope and y-intercept for the linear equation $3x - 4y = 12$. «6,2»

9. Graph the polynomial function $y = x^2 + 2x - 3$. «7,2»

10. Find the length of one leg of a right triangle if the other leg is 5 inches and the hypotenuse is 13 inches. «7,3»

Unit 5: Conic Sections – Ellipses

Conic Sections

Each of the four figures called **conic sections** can be visualized using a double cone like the one shown at the right. In unit 3, the circle was studied. Each circle can be visualized as the intersection of a vertically held cone with a horizontal plane.

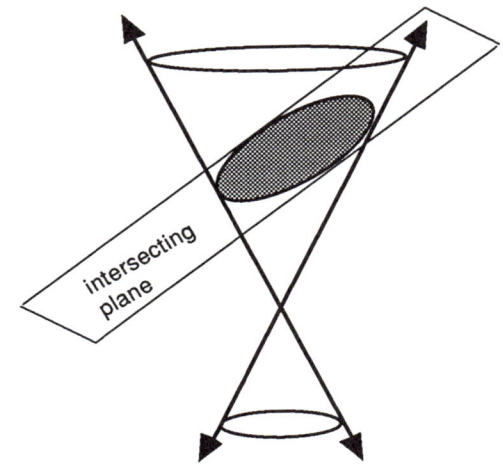

In this unit, the conic section called an **ellipse** is studied. To visualize an ellipse, use a vertically held cone and let the intersecting plane be at an angle between horizontal and the slant of the cone's edge. The resulting intersection is an oval which is the ellipse.

The shape of the orbit of the earth about the sun is elliptical. The sun is not located at the center of the ellipse, but at a point off center which is a **focus point** for the ellipse. Satellites placed in earth orbit also have elliptical-shaped orbits. Again, the earth is not at the center of the orbit; the earth is at a focus point.

An ellipse might be described as a "distorted circle" and the amount of the distortion is dependent upon the angle of the plane away from the horizontal. The more angle the greater the distortion and the less like a circle will be the ellipse.

Focus on ellipses as conic sections

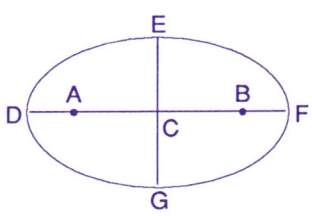

The point where the plane intersects the axis of the cone is one **focus** of the ellipse, but this focus is not at the center of its ellipse. In fact, an ellipse contains two focus points (foci) and the center of the ellipse is the midpoint for the two foci. In the figure at the left, the points A and B are the **focus points** and the point C is the **center** of the ellipse.

The points D, E, F, and G in the figure are the **vertices** of the ellipse. The line segment DF which contains the two foci and center is the **major axis** for the ellipse. The line segment EG which contains the center but not the foci is the **minor axis**.

Ellipses as Conic Sections

To describe an ellipse geometrically, the basic idea is this: A piece of plywood has two nails in it. The nails are each connected to a piece of string that is greater than the distance between the nails. The piece of string is pulled away from the nails and stretched tightly along the plywood. A pencil traces the path of this tightly stretched string. The result of the tracing will be an ellipse. If the string is close to the same length as the distance between the nails, then the ellipse will be a relatively flat oval, but as the length of the string increases the ellipse becomes more circular.

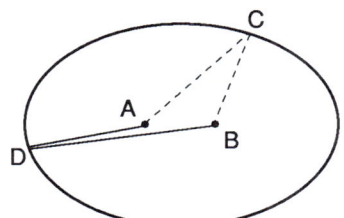

Distance from A to D to B is equal to the distance from A to C to B

Translating a plywood experiment with string into a definition of a set of points on the x,y plane is accomplished using the length of the string as a sum of the distances from the two nails (focus points).

Definition	**Ellipse**	
A set of points (x,y) on the x,y plane is an **ellipse**	if and only if	there are two distinct points A and B at distance d from each other and the sum of the distances from these two points to each point (x,y) is a constant c where c > d.

The points A and B are **focus points** of the ellipse. The midpoint for points A and B is the **center** of the ellipse.

Sketching the Graph of an Ellipse

To roughly sketch the ellipse with foci A = (3,0) and B = (-3,0) and 10 as the total distance of the points (x,y) from the foci,
1. Notice that the center must be at (0,0) because that is the mid-point of the foci.
2. The ends of the major axis are on the x-axis at points (a,0) and (-a,0) where the sum of the distances is 10. Therefore,
$$|a + 3| + |a - 3| = 10$$
We can assume a > 3 (Why?) and the equation simplifies to: a + 3 + a − 3 = 10 or 2a = 10 or a = 5.
Thus, the major axis endpoints are (5,0) and (-5,0).

3. The ends of the minor axis are on the y-axis at points (0,b) and (0,-b). Since these points are equidistant from the foci and the sum of the distances is 10, the point must be 5 units from each foci.
Therefore, $\sqrt{(3-0)^2 + (0-b)^2} = 5$
$$3^2 + b^2 = 25$$
$$b^2 = 16$$
$$b = 4 \text{ or } b = -4$$
Thus, the minor axis endpoints are (0,4) and (0,-4).

4. Using the four vertices (endpoints of the axes) the general shape of the ellipse can be sketched as shown below.

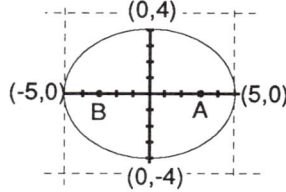

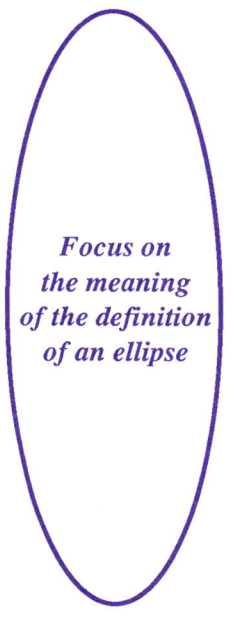

Focus on the meaning of the definition of an ellipse

In the preceding example, an ellipse was sketched. Using the same criteria for an ellipse (foci: A = (3,0), B = (-3,0) and total distance 10) the definition gives a different perspective than the sketch.

By the definition, each point C = (x,y) of the ellipse must satisfy the following equation:

Distance AC + Distance BC = total distance

For the foci A = (3,0), B = (-3,0), the equation becomes

$$\sqrt{(x-3)^2 + (y-0)^2} + \sqrt{(x+3)^2 + (y-0)^2} = 10$$

This is the equation of the desired ellipse. To convert the equation to a more usable form requires repeated squaring of both sides of the equation until all radicals are eliminated. The interesting result of that simplification is:

$$\frac{x^2}{25} + \frac{y^2}{16} = 1 \quad \text{or} \quad \frac{(x-0)^2}{5^2} + \frac{(y-0)^2}{4^2} = 1$$

where the factors (x – 0) and (y – 0) determine the center (0,0) of the ellipse and the square roots of the denominators determine the endpoints of the major and minor axes, (5,0), (-5,0), (0,4), (0,-4).

Ellipses with Center at (0,0)

Any equation of the form

$$\frac{(x-0)^2}{a^2} + \frac{(y-0)^2}{b^2} = 1 \text{ and } a^2 \neq b^2$$

has an ellipse with center (0,0) as its graph (if $a^2 = b^2$ the figure is a circle). The horizontal axis will have endpoints (a,0) and (-a,0). The vertical axis will have endpoints (0,b) and (0,-b).

For example, $\frac{x^2}{14} + \frac{y^2}{36} = 1$ is the equation of an ellipse with center (0,0) because $x^2 = (x-0)^2$ and $y^2 = (y-0)^2$. The endpoints of the horizontal axis are $(\sqrt{14},0)$ and $(-\sqrt{14},0)$. The endpoints of the vertical axis are (0,6) and (0,-6). $\sqrt{36} = 6$. Notice that the vertical axis is the major axis of the ellipse because $6 > \sqrt{14}$.

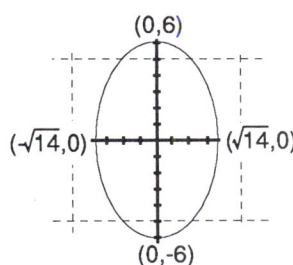

Focus on writing an equation for an ellipse with center (0,0)

To find the equation for an ellipse with center (0,0), horizontal endpoints at (8,0), (-8,0) and vertical axis endpoints at $(0,\sqrt{73})$, $(0,-\sqrt{73})$:

1. Begin with the form for the ellipse with center (0,0).
$$\frac{(x-0)^2}{a^2} + \frac{(y-0)^2}{b^2} = 1$$

2. Replace a with 8 and b with $\sqrt{73}$.
Then $a^2 = 8^2 = 64$ and $b^2 = (\sqrt{73})^2 = 73$

3. The final equation is:
$$\frac{x^2}{64} + \frac{y^2}{73} = 1$$

The vertical axis is the major axis because $\sqrt{73} > 8$.

Ellipses with Center Not at (0,0)

The definition for an ellipse does not assume that the focus points lie on the same horizontal or vertical line. When an ellipse has axes that are not horizontal or vertical, a rotation about a point can make one of the axes horizontal.

The definition also doesn't require that the center be at (0,0). A translation can be used to move any such ellipse so that its center is (0,0). Such a translation preserves the shape and size of the ellipse. This means that the center is moved, but the lengths of the two axes will remain the same.

Consequently, the equations
$$\frac{(x-0)^2}{7^2} + \frac{(y-0)^2}{5^2} = 1 \quad \text{and} \quad \frac{(x-4)^2}{7^2} + \frac{(y+9)^2}{5^2} = 1$$
have congruent ellipses as their graphs, but different locations on the x,y plane. One has its center at (0,0) and the other at (4,-9).

To graph $\frac{(x-4)^2}{7^2} + \frac{(y+9)^2}{5^2} = 1$,

1. Plot its center at (4,-9) because of the factors (x − 4) and (y + 9).
2. Use 7 as the distance between its center and the endpoints of the horizontal axis. This process gives the endpoints (11,-9) and (-3,-9).
3. Use 5 as the distance between its center and the endpoints of the vertical axis. This process gives the endpoints (4,-4) and (4,-14).
4. Sketch the graph through the four vertices (axes endpoints).

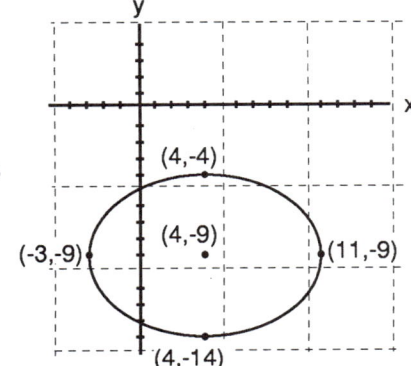

Graphing Polynomial Functions and Conic Sections   307

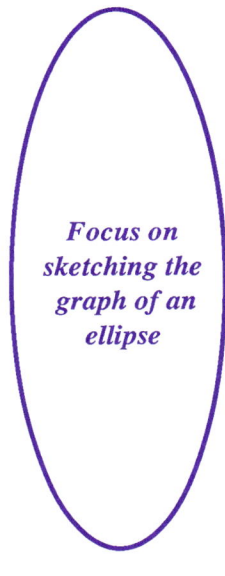

Focus on sketching the graph of an ellipse

The ellipse with equation $\frac{(x+6)^2}{9} + \frac{(y-4)^2}{14} = 1$ is graphed by:
1. Locating its center, (-6,4).
2. Locating the vertices at the ends of its horizontal axis which is the minor axis in this case because 9 is less than 14. The vertices will be 3 units right or left of the center because $3^2 = 9$.
   (-9,4) and (-3,4)
3. Locating the vertices at the ends of the major axis. These vertices will be $\sqrt{14}$ units above or below the center because $(\sqrt{14})^2 = 14$.
   $(-6, 4 + \sqrt{14})$ and $(-6, 4 - \sqrt{14})$
4. The rectangle containing the four vertices circumscribes the ellipse and allows a more accurate sketching of the curve.

Changing the Form for a Ellipse's Equation

The equation $4x^2 - 24x + 90y + 9y^2 + 225 = 0$ is the equation of an ellipse, but it is difficult to recognize and/or graph in its present form. The equation is altered as shown below.

$$4x^2 - 24x + 90y + 9y^2 + 225 = 0$$

Each term is divided by 36 which is the LCM of the coefficients of x^2 and y^2.

$$\tfrac{4}{36}x^2 - \tfrac{24}{36}x + \tfrac{90}{36}y + \tfrac{9}{36}y^2 + \tfrac{225}{36} = 0$$

$$\tfrac{1}{9}x^2 - \tfrac{2}{3}x + \tfrac{5}{2}y + \tfrac{1}{4}y^2 + \tfrac{25}{4} = 0$$

$$\left(\tfrac{1}{9}x^2 - \tfrac{2}{3}x\right) + \left(\tfrac{1}{4}y^2 + \tfrac{5}{2}y\right) = \tfrac{-25}{4}$$

$$\left(\tfrac{1}{9}x^2 - \tfrac{6}{9}x\right) + \left(\tfrac{1}{4}y^2 + \tfrac{10}{4}y\right) = \tfrac{-25}{4}$$

$$\tfrac{x^2 - 6x}{9} + \tfrac{y^2 + 10y}{4} = \tfrac{-25}{4}$$

$$\tfrac{x^2 - 6x + 9}{9} + \tfrac{y^2 + 10y + 25}{4} = \tfrac{-25}{4} + \tfrac{9}{9} + \tfrac{25}{4}$$

$$\tfrac{(x-3)^2}{3^2} + \tfrac{(y+5)^2}{2^2} = 1$$

Now the ellipse may be graphed using (3,-5) as the center point. Each endpoint on the horizontal axis will be 3 units from the center. Each endpoint on the vertical axis will be 2 units from the center.

Unit 5 Exercise

Part A: Answers for all Part A problems are at the back of the book.

1. Draw a graph of the ellipse with center (0,0), a horizontal axis of length 6, and a vertical axis of length 4. Estimate the positions of the foci points on the graph.

2. Draw a graph of the ellipse with center (0,0), a horizontal axis of length 8, and a vertical axis of length 10. Estimate the positions of the foci points on the graph.

3. Draw a graph of the ellipse with center (-3,2), a horizontal axis of length 5, and a vertical axis of length 8. Estimate the positions of the foci points on the graph.

4. Draw a graph of the ellipse with center (-2,-4), a horizontal axis of length 6, and a vertical axis of length 4. Estimate the positions of the foci points on the graph.

5. Are the ellipses drawn for problems 1 and 4 congruent?

6. The ellipse with foci (3,0), (-3,0) and total distance 10 is described as:
$$\sqrt{(x-3)^2 + (y-0)^2} + \sqrt{(x+3)^2 + (y-0)^2} = 10$$
Square both sides of the equation. After simplifying, square again to eliminate the radical signs completely. Then try to write the equation in the form $\frac{x^2}{a^2} + \frac{y^2}{b^2} = 1$

7. The ellipse described by $\frac{x^2}{3^2} + \frac{y^2}{2^2} = 1$ has center (0,0), horizontal axis of length 6 and vertical axis of length 4. (2 • 3 = 6 and 2 • 2 = 4). Write the equation of the ellipse with center (0,0) which has horizontal axis of length 8 and vertical axis of length 10.

8. The ellipse described by $\frac{(x-2)^2}{25} + \frac{(y+3)^2}{33} = 1$ has center (2,-3), horizontal axis of length 10, and vertical axis of length $2\sqrt{33}$. Write the equation of the ellipse with center (4,-1) which has horizontal axis of length 14 and vertical axis of length $2\sqrt{21}$.

9. The equations $\frac{x^2}{3^2} + \frac{y^2}{2^2} = 1$ and $\frac{x^2}{2^2} + \frac{y^2}{3^2} = 1$ describe congruent ellipses. What is the difference in their graphs?

10. The ellipse described by $\frac{(x-4)^2}{6^2} + \frac{(y-8)^2}{7^2} = 1$ has center (4,8), horizontal axis of length 12 and vertical axis of length 14. Write the equation of the ellipse with center (4,-1) which has horizontal axis of length 20 and vertical axis of length 8.

11. Why is $\frac{(x-3)^2}{5^2} + \frac{(y+6)^2}{5^2} = 1$ not the equation of an ellipse?

12. Why is $\frac{(x-9)^2}{6^2} - \frac{(y+8)^2}{3^2} = 1$ not the equation of an ellipse?

13. Why is $\frac{(x-1)^2}{47} + \frac{(y-5)^3}{13} = 1$ not the equation of an ellipse?

14. $3(x^2 + 6x + 9) + 4\left(y^2 - 7x + \frac{49}{4}\right) = 12$ is the equation of an ellipse. Divide both sides of the equation by 12 and then simplify the equation so its center and axes can be determined.

15. The equation $6(x^2 - 10x) + 8(y - 3)^2 = -102$ is the equation of an ellipse. What number needs to be added to $(x^2 - 10x)$ to complete the square inside the parentheses?

16. The equation $6(x^2 - 10x) + 8(y - 3)^2 = -102$ is the equation of an ellipse. What number needs to be added to both sides of the equation so that the center and axes can be found?

17. The equation $6(x^2 - 10x) + 8(y - 3)^2 = -102$ is the equation of an ellipse. Find its center and the lengths of its two axes.

18. If an ellipse has (2,-2) and (-2,2) as its foci then the major axis will be at a 45° angle to the x-axis. Would a rotation of 45° counterclockwise about (0,0) place the ellipse so its major axis is horizontal?

Part B: Answers for odd-numbered problems of Part B are at the back of the book.

1. Is (-2,7) a solution of $\frac{(x-4)^2}{6^2} + \frac{(y-8)^2}{7^2} = 1$?

2. Is (4,6) a solution of $\frac{(x-4)^2}{9^2} + \frac{(y-13)^2}{7^2} = 1$?

3. Graph $\frac{x^2}{4} + \frac{y^2}{9} = 1$

4. Graph $\frac{x^2}{8} + y^2 = 1$

5. Graph $\frac{(x-4)^2}{6^2} + \frac{(y-8)^2}{7^2} = 1$

6. Graph $\frac{(x-4)^2}{9^2} + \frac{(y-1)^2}{5^2} = 1$

7. Graph $\frac{(x+6)^2}{20} + \frac{(y-5)^2}{12} = 1$

8. Graph $\frac{(x-3)^2}{49} + \frac{(y+2)^2}{30} = 1$

For problems 9-12 use the distance formula to write an equation for the ellipse.

9. Foci at (4,0), (-4,0) and sum of the distances 10.

10. Foci at (0,-2), (0,2) and sum of the distances 6.

11. Foci at (7,3), (-1,3) and sum of the distances 12.

12. Foci at (-3,6), (-3,2) and sum of the distances 7.

For problems 13-20 write an equation for the ellipse.

13. Center at (0,0) with horizontal axis 6 units in length and vertical axis 14 units.

14. Center at (0,0) with vertices at (12,0) and (0,7).

15. Center at (5,-6) with horizontal axis 8 units in length and vertical axis 20 units.

16. Center at (-4,4) with horizontal axis 4 units in length and vertical axis $\sqrt{20}$ units.

17. Center at (-2,7) with vertices at (5,7) and (-2,9).

18. Center at (-5,-9) with vertices at $(-5 + \sqrt{7}, -9)$ and (-5,-6).

19. Center at (-3,0) with horizontal axis $\sqrt{24}$ units in length and vertical axis $\sqrt{32}$ units.

20. Center at (5,2) with vertices at $(5 - \sqrt{23}, 2)$ and $(5, 2 + 3\sqrt{2})$.

In problems 21-25 each equation describes an ellipse. Use the completing the square process to change the form of the equation to show its center and vertices.

21. $5x^2 + 10x + 4y^2 = 15$

22. $6x^2 + 12x + 3y^2 - 12y = 6$

23. $x^2 - 6x + 4y^2 + 16y = 11$

24. $8x^2 - 16x + 3y^2 - 24y = -32$

25. $4x^2 - 28x + 2y^2 + 12y = -63$

Part C: No answers are given for these problems. However, each is accompanied by an ordered pair «C,U» showing the chapter and unit in which it was taught.

1. Graph the truth set for $x(x - 5) > 0$. «1,6»

2. Simplify $\sqrt[3]{64} - \sqrt[3]{48} + \sqrt[3]{216} - \sqrt[3]{6}$ «2,3»

3. Find the truth set for $\sqrt{x + 4} - 1 = 5$. «2,6»

4. Simplify $\dfrac{3 - 2i}{5 + i}$ «3,3»

5. Graph the truth set for $|x - 3| < 5$. «4,3»

6. Use interval notation to describe the truth set of $|3x - 7| \geq 2$. «4,4»

7. Find the rule of correspondence that would match the elements of $\{2, 4, 6, \ldots\}$ and $\{1, \frac{1}{4}, \frac{1}{16}, \ldots\}$. «5,1»

8. What value(s) is/are not in the domain for $y + 3 = \dfrac{35 - x}{7x + 3}$? «5,3»

9. Graph $x \leq -2$ and $y = 2x + 3$ «6,5»

10. Find the distance between points (-3,5) and (2,-3). «7,3»

Unit 6: Conic Sections — Parabolas

Conic Sections

Each of the four figures called **conic sections** can be visualized using a double cone like the one shown at the right. In units 3 and 4, the circle and ellipse were studied. In this unit, the conic section called a **parabola** is studied. To visualize a parabola, use a vertically held cone and let the intersecting plane be at exactly the same angle as the slant of the cone's edge. The resulting intersection is the shape of a parabola.

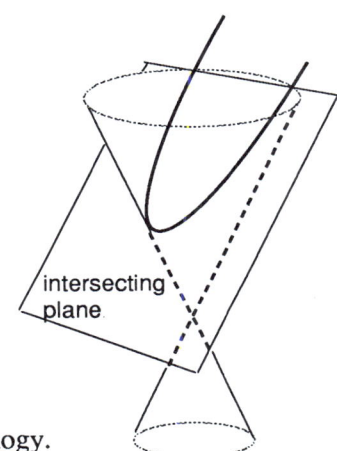

Parabolas occur frequently in nature and technology. The path of a fly ball hit in a baseball game is a parabola. The reflector for a flashlight is parabolic to direct the light in a narrow beam.

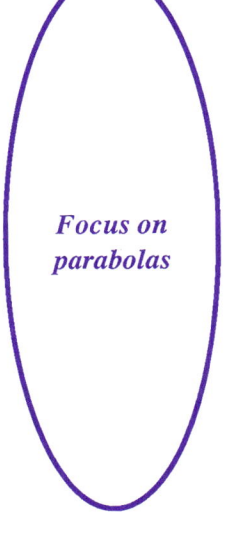

Focus on parabolas

In the figure shown at the right the parabola is accompanied by two points and two lines. The line through the points A and B is the **axis** of the parabola and separates it into two symmetric halves.

The line perpendicular to the axis is exactly the same distance from A as the distance from A to B. This perpendicular line is the **directrix** of the parabola.

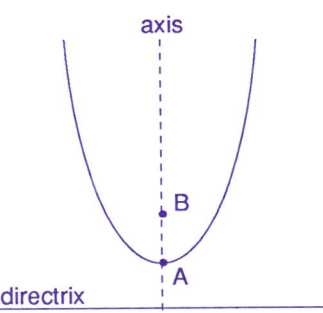

The point A which lies on the curve itself and is halfway between point B and the directrix is the **vertex** of the parabola.

The point B which is not on the curve and therefore not part of the parabola is its **focus**.

Parabolas as Conic Sections

To describe a parabola geometrically, the basic idea is this: A piece of plywood has a nail and a line not containing the nail. A piece of elastic is attached to the nail and a ball that rolls along the line. The midpoint of the elastic contains a pencil that will always be at the **same distance from the nail as it is from the line**. The pencil traces a path on the plywood. The result of the tracing will be a parabola.

Translating a plywood experiment with elastic into a definition of a set of points on the x,y plane is accomplished using the fact that the distances are the same.

> **Definition   Parabola**
>
> A set of points on the x,y plane is a parabola **if and only if** for a point B and a line L that does not contain the point the distance of (x,y) from B is equal to the distance of (x,y) from L.
>
> The point B is the **focus** of the parabola. The line L is the **directrix**.

Our definition for a parabola does not assume that the directrix is a horizontal line. In fact, it could be vertical or oblique. When that is the case, a rotation about a point could make the directrix horizontal. To simplify this unit all further discussions of parabolas is limited to those which already have a horizontal or vertical directrix (vertical or horizontal axis).

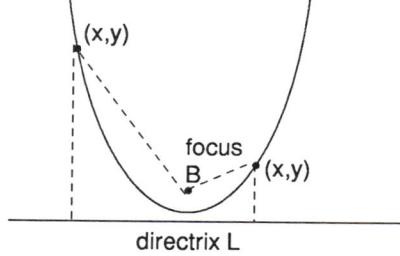

Graphing Polynomial Functions and Conic Sections 313

Sketching the Graph of a Parabola

To roughly sketch the parabola with foci A = (0,1) and directrix y = -1,

1. Notice that point (0,0) must be on the curve. It is at distance 1 from the point (0,1) and also at distance 1 from the line y = -1.

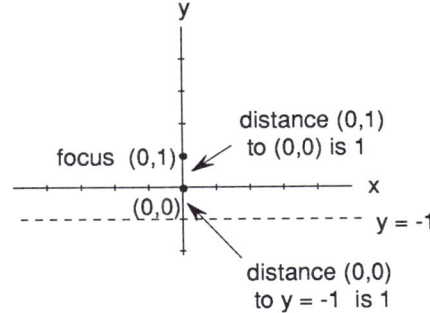

2. When y = 1, the point (x,y) will be on the same horizontal line as the focus point (0,1). This fact makes the finding of two points on the curve relatively simple.
   When y = 1, the point on the parabola is (x,1) and:
   Distance (0,1) to (x,1) = Distance (x,1) to (x,-1).
   $$\sqrt{(x-0)^2 + (1-1)^2} = \sqrt{(x-x)^2 + (1-[-1])^2}$$
   $$\sqrt{x^2 + (0)^2} = \sqrt{(0)^2 + (2)^2}$$
   $$\sqrt{x^2} = \sqrt{4}$$
   $$x^2 = 4$$
   $$x = 2 \text{ and } x = -2$$
   (2,1) and (-2,1) are on the parabola. Both are at distance 2 from the directrix and focus.

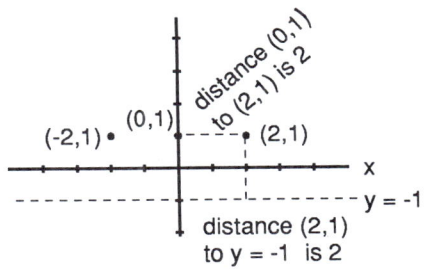

3. Two more points on the curve can be found. For example, when y = 4, the points (x,4) can be found in the following steps.
   When y = 4, the point on the parabola is (x,4) and:
   Distance (0,1) to (x,4) = Distance (x,4) to (x,-1).
   $$\sqrt{(x-0)^2 + (4-1)^2} = \sqrt{(x-x)^2 + (4-[-1])^2}$$
   $$\sqrt{x^2 + (3)^2} = \sqrt{(0)^2 + (5)^2}$$
   $$\sqrt{x^2 + 9} = \sqrt{25}$$
   $$x^2 + 9 = 25$$
   $$x^2 = 16$$
   $$x = 4 \text{ and } x = -4$$
   (4,4) and (4,-4) are on the parabola. Both are at distance 5 from the directrix and focus.

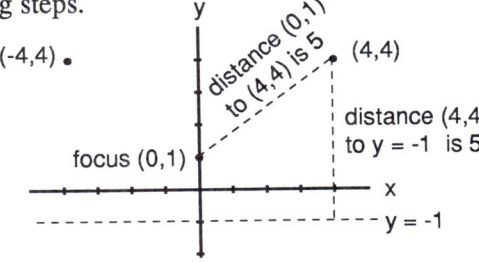

4. Using the five points, the general shape of the parabola can be sketched as shown at the right.

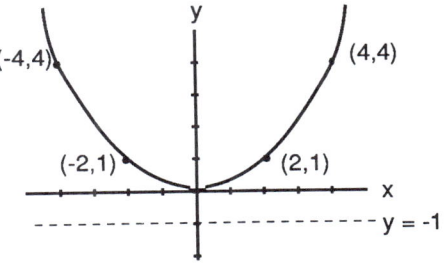

Focus on the meaning of the definition of a parabola

The equation of the parabola sketched previously can be found using its focus, A = (0,1), and directrix, the line y = -1. For any point C = (x,y) on the parabola the following equation is true.

Distance AC = Distance from C to the line with points (x,-1).

This translates as:
$$\sqrt{(x-0)^2 + (y-1)^2} = \sqrt{(x-x)^2 + (y+1)^2}$$

Squaring each side of the equation and solving for y gives:
$$(x-0)^2 + (y-1)^2 = (y+1)^2$$
$$x^2 + y^2 - 2y + 1 = y^2 + 2y + 1$$
$$x^2 = 4y$$
$$4y = x^2$$
$$y = \tfrac{1}{4}x^2 \text{ or } (y-0) = \tfrac{1}{4}(x-0)^2$$

The factors (y – 0) and (x – 0) determine the vertex (0,0) of the parabola. The coefficient $\tfrac{1}{4}$ regulates the width of the curve and will be discussed more completely later in this unit.

Notice that the parabola of $(y-0) = \tfrac{1}{4}(x-0)^2$ contains the points (0,0), (2,1), (-2,1), (4,4), and (4,-4) which were found in the preceding example.

Parabolas with Vertices at (0,0)

An equation of the form
$$y = ax^2, (a \neq 0) \text{ or } (y-0) = a(x-0)^2, (a \neq 0)$$
always is associated with a parabola that has (0,0) as its vertex and a vertical line as its axis. The numerical coefficient of x^2 or $(x-0)^2$ affects the width and direction of opening for the parabola.

$y = 3x^2$ is the equation of a parabola with vertex (0,0) because it is equivalent to $(y-0) = 3(x-0)^2$. The curve of $y = 3x^2$ opens upwards because the coefficient 3 is positive.

$y = -2x^2$ is the equation of a parabola with vertex (0,0) because it is equivalent to $(y-0) = -2(x-0)^2$. The curve of $y = -2x^2$ opens downwards because the coefficient -2 is negative.

The curve of $y = 3x^2$ is narrower than that of $y = -2x^2$. The difference in widths is caused by the relative sizes of | 3 | and | -2 |. The larger the absolute value of the coefficient, the narrower will be the parabola.

Focus on parabolas with vertices (0,0)

Any equation of the form $y = ax^2$ ($a \neq 0$) has a parabola with vertex (0,0) and a vertical line as its axis.

The coefficient of x^2 is positive when the focus point is above the directrix. This means that the parabola opens "upwards" only when the coefficient is positive. Similarly, the coefficient can be negative only when the focus point is below the directrix which means that the parabola opens "downwards." Consequently, the sign of the coefficient of x^2 completely determines the direction of the curve.

The absolute value of the size of the coefficient of x^2 determines the width of the parabola. The larger this absolute value the narrower the parabola. The size of the coefficient increases as the distance between the focus and directrix increases.

Because of the coefficient of x^2,

the parabola of $y = x^2$ opens upwards.

the parabola of $y = 6x^2$ opens upwards. It is narrower than that of $y = x^2$ because $|6| > |1|$.

the parabola of $y = -5(x - 3)^2 - 2$ opens downwards. It is narrower than $y = x^2$ and wider than $y = 6x^2$.

the parabola of $y = \frac{1}{2}x^2$ opens upwards. It is wider than $y = x^2$ and wider than $y = 6x^2$.

the parabola of $y = \frac{-3}{4}x^2$ opens downwards. It is wider than $y = x^2$ and narrower than $y = \frac{1}{2}x^2$.

Parabolas with Vertices not at (0,0)

Translations on the x,y plane can move any parabola with a vertex at (0,0) to a new position on the plane while preserving the size and shape of the figure.

For example, the parabola of $y = x^2$ with vertex (0,0) can be moved by a translation so the vertex is at (5,-4). To find the equation of the image parabola,

1. The translation can be described by:
   $(x',y') = (x + 5, y - 4)$ or $(x,y) = (x' - 5, y' + 4)$

2. The original equation was $y = x^2$ and can be used to find the image equation by substituting $(x' - 5)$ for x and $(y' + 4)$ for y.
   $y = x^2$ becomes $(y' + 4) = (x' - 5)^2$

Therefore, the parabola congruent to $y = x^2$ with vertex (5,-4) has $(y + 4) = (x - 5)^2$ as its equation.

Focus on parabolas congruent to those with vertex (0,0)

Any equation of the form $(y - k) = a(x - h)^2$ or $y = a(x - h)^2 + k$, $a \neq 0$, has a parabola as its graph. Furthermore, the parabola will be congruent to $y = ax^2$, but its vertex will be (h,k).

$y = x^2$ and $y - 5 = (x - 7)^2$ have congruent parabolas because the coefficient of both x^2 and $(x - 7)^2$ is 1. The vertices are (0,0) and (7,5).

$y = 6x^2$ and $y = 6(x - 4)^2 + 7$ have congruent parabolas because the coefficient of both x^2 and $(x - 4)^2$ is 6. The vertices are (0,0) and (4,7).

$y = \frac{1}{2}x^2$ and $y = \frac{1}{2}(x + 8)^2 + 5$ have congruent parabolas because the coefficient of both x^2 and $(x + 8)^2$ is $\frac{1}{2}$. The vertices are (0,0) and (-8,5).

$y = \frac{-3}{4}x^2$ and $y = \frac{-3}{4}(x + 9)^2 - 1$ have congruent parabolas because the coefficient of both x^2 and $(x + 9)^2$ is $\frac{-3}{4}$. The vertices are (0,0) and (-9,-1).

Recognizing Parabola Equations

The equation $4x^2 - 5y + 12x + 10 = 0$ is the equation of a parabola, but may not be recognized in its present form. The equation is altered as shown below.

$$4x^2 - 5y + 12x + 10 = 0$$
$$5y = 4x^2 + 12x + 10$$
$$5y = 4(x^2 + 3x) + 10$$
$$5y = 4\left(x^2 + 3x + \tfrac{9}{4}\right) + 10 - 4 \cdot \tfrac{9}{4}$$
$$5y = 4\left(x + \tfrac{3}{2}\right)^2 + 10 - 9$$
$$5y = 4\left(x + \tfrac{3}{2}\right)^2 + 1$$
$$y = \tfrac{4}{5}\left(x + \tfrac{3}{2}\right)^2 + \tfrac{1}{5}$$

This new form of the equation indicates that the vertex is $\left(\tfrac{-3}{2}, \tfrac{1}{5}\right)$ and the parabola opens upwards with a curve that is slightly wider than $y = x^2$.

Unit 6 Exercise

Part A: Answers for all Part A problems are at the back of the book.

1. Draw a graph of the parabola with vertex (0,0) that also contains the points (1,1), (2,4), and (3,9).

2. Draw a graph of the parabola with vertex (-2,-3) that also contains the points (-1,-2), (0,1), and (1,6).

3. Draw a graph of the parabola with vertex (0,0) that also contains the points (1,-1), (2,-4), and (3,-9).

4. Draw a graph of the parabola with vertex (5,-2) that also contains the points (6,-3), (7,-6), and (8,-11).

5. Are the parabolas drawn for problems 1-4 congruent?

6. The parabola with focus point (3,0) and directrix $y = -3$ is the set of points (x,y) where the distance from (3,0) to (x,y) is equal to the distance from (x,y) to the line $y = -3$. Using the distance formula, the equation is:
$$\sqrt{(x-3)^2 + (y-0)^2} = \sqrt{(x-x)^2 + (y+3)^2}$$
Square both sides of the equation and then write the equation in the form $(y - k) = a(x - h)^2$.

7. Use the distance formula to write the equation of a parabola with focus point (4,3) and directrix $y = 1$.

8. Use the distance formula to write the equation of a parabola with focus point $(0, \frac{1}{4})$ and directrix $y = \frac{-1}{4}$.

9. What are the focus point and directrix of the parabola $y = x^2$?

10. What is the equation of the parabola sketched in problem 1?

11. Any parabola congruent to $y = x^2$ will have an equation of the form $(y - k) = (x - h)^2$ with a vertex at (h,k). Write the equation of a parabola congruent to $y = x^2$ with vertex at (4,-1).

12. The parabolas of $y = x^2$ and $y = -x^2$ are congruent and both have (0,0) as the vertex. The difference in the curves is their directions of opening. The parabola of $y = x^2$ opens upwards and the parabola of $y = -x^2$ opens _____.

13. Why is $(y + 3)^2 = (x - 4)^2$ not the equation of a parabola?

14. The equation $x^2 + 8x + 7 = y + 4$ is the equation of a parabola. Complete the square for $(x^2 + 8x)$ and then write the equation in a form to show its vertex.

15. The equation $x = y^2$ is the equation of a parabola, but this parabola has a horizontal line as its axis. (0,0) is the vertex and (1,1), (4,2), (9,3) are three solutions of the equation. Sketch the parabola.

16. The equation $x = y^2$ is the equation of a parabola with vertex (0,0) which opens to the right side of the x,y plane. Describe the parabola of $x = -y^2$.

17. The parabolas of $y = 2x^2$, $(y - 5) = 2(x + 7)^2$, and $y = -2(x - 3)^2 + 3$ are congruent because the coefficients are equal to |2|. Find the vertex for each parabola.

18. A parabola has an axis that goes through (0,0) and (5,5). What rotation about the point (0,0) will make the axis vertical?

Part B: Answers for odd-numbered problems of Part B are at the back of the book.

1. Graph the parabola of $y = (x - 1)^2$
2. Graph the parabola of $y + 3 = (x + 2)^2$
3. Graph $y - 4 = (x - 2)^2$
4. Graph $y = 2(x + 1)^2$
5. Graph $y - 5 = -x^2$
6. Graph $y - 9 = -(x - 3)^2$
7. Graph $y = \frac{1}{4}(x + 4)^2 + 6$
8. Graph $y = \frac{-1}{2}(x + 3)^2 - 1$

For problems 9-12 use the distance formula to write an equation for the parabola.

9. Focus at (4,0) and directrix $y = 2$.
10. Focus at (3,-1) and directrix $y = 3$.
11. Focus at (-2,4) and directrix $x = 0$.
12. Focus at (-3,-4) and directrix $x = -5$.

For problems 13-16 write an equation for the parabola.

13. Vertex at (0,0), opens upwards, and contains the point (2,12).

14. Vertex at (2,-2), opens upwards, and contains the point (6,10).

15. Vertex at (0,3), opens to the right side, and contains the point (5,6).

16. Vertex at (-4,4), opens downwards, and contains the point (-5,1).

17. Graph $y = x^2$ and $y = x^2 + 3$ on the same x,y plane. Compare the curves of these two equations.

18. Graph $y = x^2$ and $y = (x - 5)^2$ on the same x,y plane. Compare the curves of these two equations.

19. Graph $y = 2x^2$ and $y = x^2$ on the same x,y plane. Compare the curves of these two equations.

20. Graph $y = 2x^2$ and $y = -2x^2$ on the same x,y plane. Compare the curves of these two equations.

In problems 21-25 each equation describes a parabola. Use the completing the square process to change the form of the equation to $(y - k) = a(x - h)^2$.

21. $5x^2 + 20x - 6y = 7$

22. $6x^2 + 5 - 3y = 30x$

23. $x^2 - 9x + 4y = -6$

24. $8x^2 - 16x + 3y = -3$

25. $2y - 20x = 4x^2 + 7$

Part C: No answers are given for these problems. However, each is accompanied by an ordered pair «C,U» showing the chapter and unit in which it was taught.

1. Which of the following is not a rational number? $\frac{-2}{3}, \frac{4}{-7}, \frac{0}{5}, \frac{3}{0}, \frac{0}{-2}$ «1,3»

2. Solve $\frac{3}{5}x - \frac{3}{4} = \frac{1}{10} - \frac{1}{2}x$ «2,7»

3. Solve $x^2 - 2x + 4 = 0$ «3,2»

4. Graph $|x + 5| = 1$ «4,3»

5. Graph the truth set for $|x - 2| \leq 2$. «4,3»

6. True or false? $(1,-3) \in \{(x,y) \mid x^2 - 2x + 3\}$ «5,2»

7. Draw a Venn diagram for $\overline{A} \cup \overline{B}$. «5,5»

8. Write $y = \frac{-1}{4}x + 1$ in standard form. «6,4»

9. Write the slope intercept form of $3x - 5y = -2$. «6,4»

10. Graph the function $y = x^3 + x^2 + 2x - 2$ «7,2»

Unit 7: Conic Sections – Hyperbolas

Conic Sections

In this unit, the conic section called a **hyperbola** is studied. To visualize a hyperbola, use a vertically held cone and let the intersecting plane be nearly vertical and certainly at a greater angle than the slant of the cone's edge. The resulting intersection is a curve with two branches – one from each cone. This curve is a hyperbola.

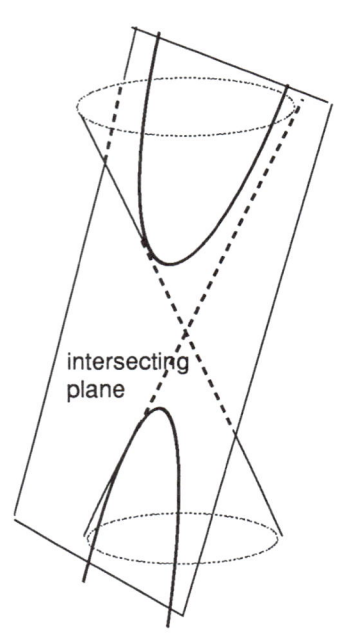

The hyperbola is sometimes mistakenly thought of as a pair of parabolas and, in fact, historically was even named in that manner. However, the hyperbola is a unique curve different from the parabola and deserving of separate treatment.

Focus on hyperbolas as conic sections

Each hyperbola has a pair of focus points that are not on either of the branches of the hyperbola. These **focus points** are shown in the figure at the right as points A and B.

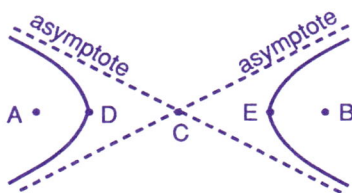

The midpoint for the two foci is the **center** point for the hyperbola. In the figure the point C is the center.

The points D and E shown on the figure are the **vertices** of the two hyperbola branches. These vertices are always on the same straight line as the foci and center.

The two straight lines of the figure which approximate the branches of the hyperbola are its **asymptotes**. The branches of the hyperbola approach but never coincide with the asymptotes.

Hyperbolas as Conic Sections

To describe a hyperbola geometrically, the basic idea is this: A piece of plywood has two nails in it. One nail is attached to a straight stick with length less than the distance between the nails. The far end of the stick and the other nail are each connected to a piece of elastic. The middle of the piece of plastic contains a pencil and traces a path on the plywood. This path will always be **the length of the stick further from its nail than the distance to the other nail**. The result of the tracing will be a hyperbola.

Translating a plywood experiment with elastic into a definition of a set of points on the x,y plane is accomplished using the fact that the distances from the two nails (focus points) has a constant difference (the length of the stick).

> **Definition**   **Hyperbola**
>
> A set of points on the x,y plane is a **hyperbola**   **if and only if**   for two distinct points A and B the absolute value of the difference of the distances from A and B to each point (x,y) of the set is d where d is a positive real number less than the distance between A and B.
>
> The points A and B are **focus points** of the hyperbola. The midpoint for points A and B is the **center** of the hyperbola.

Our definition for an hyperbola does not assume that the focus points lie on the same horizontal or vertical line. When the focus points are not on a horizontal or vertical line, a rotation could place them in one of those positions. To simplify this unit all further discussion of hyperbolas is limited to those which already have foci on the same horizontal or vertical line.

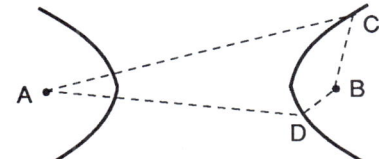

distance AC − distance CB = distance AD − distance DB

322 Chapter 7

Sketching the Graph of a Hyperbola

To roughly sketch the hyperbola with foci A = (3,0) and B = (-3,0) and 4 as the difference in the distances of points (x,y) from the foci,

1. Notice that the center must be at (0,0) because that is the midpoint of the foci.

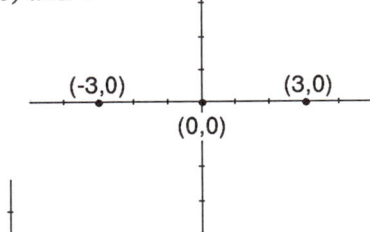

2. The vertices of the two branches of the hyperbola are on the x-axis at points (a,0) and (-a,0) where the difference of the distances is 4. Therefore,
   $|a + 3| - |a - 3| = 4$
   We can assume $0 < a < 3$ because any vertex is between the foci. Hence,
   $|a + 3| = a + 3$ and $|a - 3| = -a + 3$.
   The equation simplifies to:
   $a + 3 - (-a + 3) = 4$ or $2a = 4$ or $a = 2$.
   Thus, the vertices are (2,0) and (-2,0).

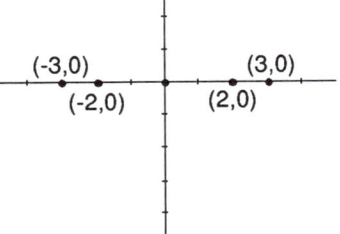

3. The asymptotes contain the center and their slopes can be found using the ratio of two distances from the center:
   $$\frac{(\text{distance: center to vertex})^2}{(\text{distance: center to focus})^2 - (\text{distance: center to vertex})^2}$$

   The distance from the center to a vertex is 2.
   The distance from the center to a focus is 3.

   One asymptote has slope $\frac{2^2}{3^2 - 2^2} = \frac{4}{5}$.
   The other asymptote has the opposite slope, $\frac{-4}{5}$.

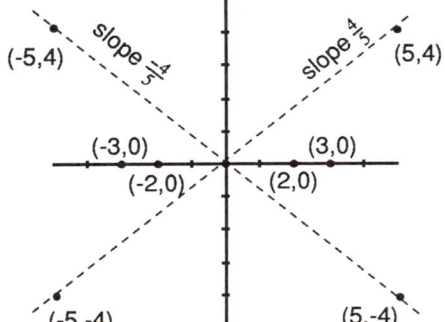

4. Using the vertices and asymptotes the general shape of the hyperbola can be sketched as shown below.

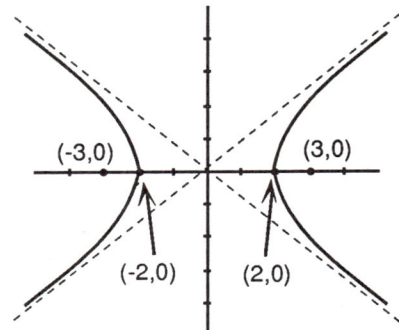

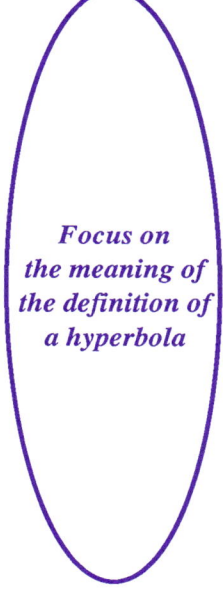

Focus on the meaning of the definition of a hyperbola

A hyperbola was sketched in the preceding example. Using the same criteria for the hyperbola (Foci : A = (3,0), B = (-3,0) and difference in distance 4) the definition gives a different perspective than the sketch.

By the definition, each point (x,y) on one branch of the hyperbola must satisfy the following equation.

$$\text{Distance AC} - \text{Distance BC} = \text{difference in distance}$$

For the foci A = (3,0), B = (-3,0) the equation becomes

$$\sqrt{(x-3)^2 + (y-0)^2} - \sqrt{(x+3)^2 + (y-0)^2} = 4$$

This is the equation of the desired hyperbola. To convert the equation to a more usable form requires repeated squaring of both sides of the equation until all radicals are eliminated. The interesting result of that simplification is:

$$\frac{x^2}{5^2} - \frac{y^2}{4^2} = 1 \text{ or } \frac{(x-0)^2}{5^2} - \frac{(y-0)^2}{4^2} = 1$$

where the factors $(x-0)$ and $(y-0)$ determine the center $(0,0)$ of the hyperbola and the ratio of the square roots of the denominators determine the slopes of the asymptotes, $\frac{4}{5}$ and $\frac{-4}{5}$.

Hyperbolas with Center at (0,0)

Any equation of the form

$$\frac{(x-0)^2}{a^2} - \frac{(y-0)^2}{b^2} = 1$$

has a hyperbola with center (0,0) and its branches opening to the sides of the x,y plane. Its vertices are (a,0) and (-a,0). Asymptotes for the branches will pass through its center with slopes $\frac{b}{a}$ and $\frac{-b}{a}$.

For example, $\frac{x^2}{49} - \frac{y^2}{30} = 1$ is the equation of a hyperbola with center (0,0) because $x^2 = (x-0)^2$ and $y^2 = (y-0)^2$. The branches will open to the sides and will have (7,0) and (-7,0) as their vertices. The slopes of the asymptotes will be $\frac{\sqrt{30}}{7}$ and $\frac{-\sqrt{30}}{7}$.

324   Chapter 7

Focus on a hyperbola with branches opening up and down

An equation of the form
$$\frac{(y-0)^2}{b^2} - \frac{(x-0)^2}{a^2} = 1$$
has a hyperbola with center (0,0) with its branches opening to the top and bottom of the x,y plane. Its vertices are (0,b) and (0,-b). Asymptotes for the branches will pass through its center and will still have slopes $\frac{b}{a}$ and $\frac{-b}{a}$.

For example, $\frac{y^2}{29} - \frac{x^2}{25} = 1$ is the equation of a hyperbola with center (0,0). The branches will open upwards and downwards and will have $(0,\sqrt{29})$ and $(0,-\sqrt{29})$ as their vertices. The slopes of the asymptotes will be $\frac{\sqrt{29}}{5}$ and $\frac{-\sqrt{29}}{5}$.

Hyperbolas with Center Not at (0,0)

The definition for a hyperbola doesn't require that the center be at (0,0). A translation can be used to move any such hyperbola so that its center is (0,0). Such a translation preserves the shape and size of the hyperbola. This means that the center is moved, but the vertices and asymptotes maintain their positions with respect to the center.

Consequently, the equations
$$\frac{(x-0)^2}{8^2} - \frac{(y-0)^2}{3^2} = 1 \quad \text{and} \quad \frac{(x+5)^2}{8^2} - \frac{(y-4)^2}{3^2} = 1$$
have congruent hyperbolas as their graphs, but their centers are located at (0,0) and (-5,4).

To graph $\frac{(x+5)^2}{8^2} - \frac{(y-4)^2}{3^2} = 1$,
1. Plot its center at (-5,4) because of the factors (x + 5) and (y − 4).
2. The two vertices are on the horizontal axis at a distance 8 from the center and therefore are located at (3,4) and (-13,4).
3. The asymptotes have slopes of $\frac{3}{8}$ and $\frac{-3}{8}$.
4. Sketch the two branches of the hyperbola using the vertices as points on the curves and the asymptotes as guidelines for approximating the curves.

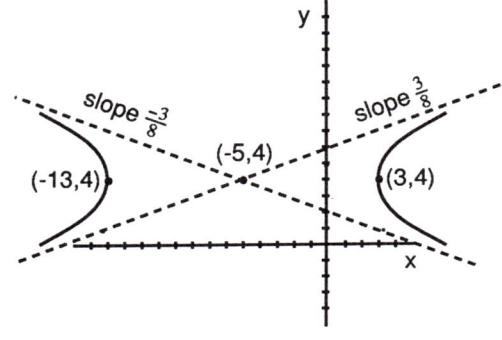

Graphing Polynomial Functions and Conic Sections   325

The hyperbola with equation $\frac{(y+9)^2}{16} - \frac{(x+2)^2}{19} = 1$, is graphed by:

1. The center is at (-2,-9) because of the factors (y + 9) and (x + 2).
2. The two vertices are on the same vertical axis and the branches open upwards and downwards. The vertices will be at distance 4 from the center and therefore are located at (-2,-5) and (-2,-13).
3. The asymptotes have slopes of $\frac{4}{\sqrt{19}} = \frac{4\sqrt{19}}{19}$ and $\frac{-4}{\sqrt{19}} = \frac{-4\sqrt{19}}{19}$.
4. Sketch the two branches of the hyperbola using the vertices as points on the curves and the asymptotes as guidelines for approximating the curves.

Focus on sketching the graph of a hyperbola

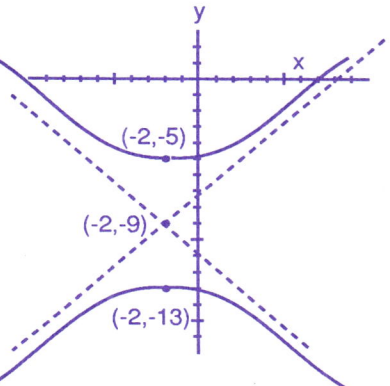

Changing the Form for a Hyperbola's Equation

The equation $5x^2 - 36y - 9y^2 - 20x = 61$ is the equation of a hyperbola, but it is difficult to recognize or graph in its present form. The equation is altered as shown below.

$$5x^2 - 36y - 9y^2 - 20x = 61$$
$$5x^2 - 20x - 9y^2 - 36y = 61$$
$$5(x^2 - 4x) - 9(y^2 + 4y) = 61$$
$$5(x^2 - 4x + 4) - 9(y^2 + 4y + 4) = 61 + 20 - 36$$
$$5(x-2)^2 - 9(y+2)^2 = 45$$
$$\frac{5(x-2)^2}{45} - \frac{9(y+2)^2}{45} = \frac{45}{45}$$
$$\frac{(x-2)^2}{9} - \frac{(y+2)^2}{5} = 1$$

Now the hyperbola may be graphed using (2,-2) as the center with each vertex 3 units from the center.
    Vertices are (5,-2) and (-1,-2)
The asymptotes will pass through (2,-2) with slopes of $\frac{\sqrt{5}}{3}$ and $\frac{-\sqrt{5}}{3}$.

Unit 7 Exercise

Part A: Answers for all Part A problems are at the back of the book.

1. Draw a graph of the hyperbola with center (0,0), a vertex at (5,0), and one asymptote with slope $\frac{3}{5}$. Estimate the positions of the foci points on the graph.

2. Draw a graph of the hyperbola with center (0,0), a vertex at (0,-4), and one asymptote with slope $\frac{7}{4}$. Estimate the positions of the foci points on the graph.

3. Draw a graph of the hyperbola with center (2,3), a vertex at (-1,3), and one asymptote with slope $\frac{-2}{3}$. Estimate the positions of the foci points on the graph.

4. Draw a graph of the hyperbola with center (2,-4), a vertex at (-3,-4), and one asymptote with slope $\frac{-3}{5}$. Estimate the positions of the foci points on the graph.

5. Are the hyperbolas drawn for problems 1 and 4 congruent?

6. The hyperbola with foci (3,0), (-3,0) and difference in the distances of 4 is described by the distance formula as:
$\sqrt{(x-3)^2 + (y-0)^2} - \sqrt{(x+3)^2 + (y-0)^2} = 4$
Square both sides of the equation. After simplifying, square again to eliminate the radical signs completely. Then try to write the equation in the form $\frac{x^2}{a^2} - \frac{y^2}{b^2} = 1$.

7. The hyperbola of $\frac{x^2}{3^2} - \frac{y^2}{2^2} = 1$ has an asymptote with slope $\frac{2}{3}$. The hyperbola of $\frac{x^2}{6^2} - \frac{y^2}{4^2} = 1$ also has an asymptote with shope $\frac{2}{3} = \frac{4}{6}$. However, the hyperbola of $\frac{x^2}{3^2} - \frac{y^2}{2^2} = 1$ has vertices at (3,0) and (-3,0) while the hyperbola of $\frac{x^2}{6^2} - \frac{y^2}{4^2} = 1$ has vertices at _____ and _____.

8. The hyperbola of $\frac{y^2}{6^2} - \frac{x^2}{5^2} = 1$ has the same center and asymptotes as the hyperbola of $\frac{y^2}{12^2} - \frac{x^2}{10^2} = 1$. However, the vertices of $\frac{y^2}{6^2} - \frac{x^2}{5^2} = 1$ are at (0,6) and (0,-6) while the vertices of $\frac{y^2}{12^2} - \frac{x^2}{10^2} = 1$ are at _____ and _____.

9. The equations $\frac{x^2}{3^2} - \frac{y^2}{2^2} = 1$ and $\frac{y^2}{3^2} - \frac{x^2}{2^2} = 1$ describe congruent hyperbolas. What is the difference in their graphs?

10. Is $\frac{(x-3)^2}{5^2} - \frac{(y+6)^2}{5^2} = 1$ the equation of a hyperbola?

11. Is $\frac{(x-9)^2}{6^2} + \frac{(y+8)^2}{3^2} = 1$ the equation of a hyperbola?

12. Is $\frac{(x-1)^2}{47} - \frac{(y-5)^3}{13} = 1$ the equation of a hyperbola?

13. $2\left(x^2 - 5x + \frac{25}{2}\right) - 5(y^2 - 10x + 25) = 10$ is the equation of a hyperbola. Divide both sides of the equation by 10 and then simplify the equation so its center and asymptotes can be determined.

14. The equation $4(x^2 - 6x) - 8(y - 3)^2 = -28$ is the equation of a hyperbola. What number needs to be added to $(x^2 - 6x)$ to complete the square inside the parentheses?

15. The equation $4(x^2 - 6x) - 8(y - 3)^2 = -28$ is the equation of a hyperbola. What number needs to be added to both sides of the equation so that the center and axes can be found?

16. The equation $4(x^2 - 6x) - 8(y - 3)^2 = -28$ is the equation of a hyperbola. Find its center and its asymptotes.

17. If a hyperbola has (-3,3) and (3,-3) as its foci then the major axis will be at a 135° angle to the x-axis. Would a rotation of 45° counter-clockwise about (0,0) place the hyperbola so its foci were on the same horizontal line?

Part B: Answers for odd-numbered problems of Part B are at the back of the book.

1. Is (12,4) a solution of $\frac{(x-4)^2}{6^2} - \frac{(y-8)^2}{7^2} = 1$?
2. Is (4,6) a solution of $\frac{(x-4)^2}{9^2} - \frac{(y-13)^2}{7^2} = 1$?
3. Graph $\frac{x^2}{4} - \frac{y^2}{9} = 1$
4. Graph $\frac{x^2}{8} - y^2 = 1$
5. Graph $\frac{(x-4)^2}{6^2} - \frac{(y-8)^2}{7^2} = 1$
6. Graph $\frac{(x-4)^2}{9^2} - \frac{(y-1)^2}{5^2} = 1$
7. Graph $\frac{(y-5)^2}{16} - \frac{(x+6)^2}{81} = 1$
8. Graph $\frac{(y+2)^2}{36} - \frac{(x-4)^2}{25} = 1$
9. Graph a hyperbola with vertices at (5,2) and (-3,2) and an asymptote with slope $\frac{3}{4}$.
10. Graph a hyperbola with foci at (3,-2) and (3,5) and an asymptote with slope $\frac{-3}{2}$.

For problems 11-14 use the distance formula to write an equation for the hyperbola.

11. Foci at (4,0), (-4,0) and difference of the distances 2.
12. Foci at (0,6), (0,-6) and difference of the distances 4.
13. Foci at (7,3), (-3,3) and difference of the distances 4.
14. Foci at (-3,6), (-3,2) and difference of the distances 6.

For problems 15-20 write an equation for a hyperbola.

15. Vertices at (9,0) and (-9,0) with an asymptote with slope $\frac{7}{9}$.
16. Vertices at (8,0) and (-8,0) and an asymptote with slope $\frac{-1}{2}$.

17. Vertices at (6,0) and (-6,0) and an asymptote with slope $\frac{2}{3}$.

18. Vertices at (6,2) and (-6,-2) and an asymptote with slope $\frac{5}{6}$.

19. Vertices at (-4,2) and (-4,8) and an asymptote with slope $\frac{-5}{4}$.

20. Vertices at (3,5) and (3,-7) and an asymptote with slope $\frac{3}{2}$.

In problems 21-25, each equation describes a hyperbola. Use the completing the square process to change the form of the equation to show its center and asymptotes.

21. $5x^2 + 20x - 4y^2 = 0$

22. $4x^2 + 32x - 3y^2 - 12y = -40$

23. $x^2 - 6x - 4y^2 + 16y = 11$

24. $3x^2 - 18x - 3y^2 - 24y = 24$

25. $2y^2 + 12y = 72 + 5x^2 - 40x$

Part C: No answers are given for these problems. However, each is accompanied by an ordered pair «C,U» showing the chapter and unit in which it was taught.

1. Solve $x^4 - 25x^2 + 144 = 0$ «1,5»

2. Simplify $\frac{2\sqrt{3} - \sqrt{7}}{\sqrt{3} + 2\sqrt{7}}$ «2,5»

3. $-8i^5 \cdot 3i^{12} = $ _____ «3,1»

4. Simplify $\frac{7 - 3i}{2 - 3i}$ «3,3»

5. Graph the truth set for $|x + 5| = 3$. «4,3»

6. Graph the truth set for $|2x + 1| \le 4$. «4,3»

7. $(5,\underline{}) \in \{(x,y) \mid y = 2x^2 - 7y - 15\}$ «5,2»

8. Find the slope of the line through the points (-3,4) and (2,6). «6,3»

9. Write the equation of a circle with a radius of 9 and its center at (3,-1). «7,4»

10. Find the center and radius for the circle $x^2 - 4x + y^2 + 6y = -4$. «7,4»

Chapter 7 Test

«7,U» shows the unit in which this problem was studied in this chapter.

1. Find the equation of the image line of $2x - 3y = 6$ after a translation described by $x = x' - 4$ and $y = y' - 2$. «7,1»

2. Graph the circle $(x - 5)^2 + (y + 4)^2 = 9$. Sketch its image circle after a translation that moves the center to $(0,0)$ and write the equation of the new circle. «7,1»

3. How many turning points would you expect to find in the graph of the polynomial function, $y = x^4 + 2x^3 - 16x^2 - 2x - 15$? «7,2»

4. Graph the relation $(x - 2)^2 + (y + 5)^2 = 16$ «7,2»

5. Graph the polynomial function $y = x^2 + 6x + 4$ «7,2»

6. Graph the relation $9x^2 + 4y^2 = 36$ «7,2»

7. Graph the function $y = x^3 + x^2 - 2x - 2$ «7,2»

8. Find the length of the hypotenuse of a right triangle if the legs of the triangle are 5 feet and 12 feet. «7,3»

9. Find the length of one side of a right triangle if the other side is 6 inches and the hypotenuse is 10 inches. «7,3»

10. Find the distance between points $(4,-5)$ and $(-2,7)$ «7,3»

11. Find the distance between points $(0,-3)$ and $(4,0)$ «7,3»

12. Find the equation of the line that is the perpendicular bisector of the segment joining $(-3,-4)$ and $(1,-6)$. «7,3»

13. Is $(-2,-2)$ a solution of $(x - 2)^2 + (y + 5)^2 = 25$? «7,4»

14. Write the equation of a circle with a radius of 3 and its center at the origin. «7,4»

15. Write the equation of a circle with a radius of 4 and its center at $(4,-1)$. «7,4»

16. Find the center and radius for the circle $x^2 + 6x + y^2 = 0$ «7,4»

17. Find the center and radius for the circle $x^2 - 4x + y^2 + 2y = 20$ «7,4»

18. Graph $\frac{x^2}{9} - \frac{y^2}{25} = 1$ «7,7»

19. Graph $\frac{x^2}{16} + \frac{y^2}{25} = 1$ «7,5»

20. Graph $y - 6 = (x - 3)^2$ «7,6»

21. Graph $\frac{(x - 2)^2}{10} + \frac{(y + 7)^2}{25} = 1$ «7,5»

22. Graph $y = \frac{1}{2}(x + 2)^2 + 5$ «7,6»

23. Graph $\frac{(y - 3)^2}{4} - \frac{(x + 5)^2}{9} = 1$ «7,7»

24. Graph $(x + 4)^2 + (y - 3)^2 = 1$ «7,4»

8
Exponents

Unit 1: COUNTING NUMBERS AS EXPONENTS

Factors in Multiplication Expression

2 • 3 is a multiplication expression. Each of the numbers, 2 and 3, is a **factor** of 2 • 3.

6 • 6 • 6 is a multiplication expression. 6 is used as a factor three times. The **exponent expression** 6^3 is equivalent to 6 • 6 • 6.

4 is used as a factor five times in the multiplication expression 4 • 4 • 4 • 4 • 4 which is equivalent to the exponent expression 4^5.

Focus on counting number exponents

6^2 is a multiplication expression in which 6 is used as a factor two times; 6^2 is equivalent to $6 \cdot 6$.

5^3 is a multiplication expression in which 5 is used as a factor three times. The multiplication expression 7^3 is equivalent to $7 \cdot 7 \cdot 7$. The exponent 3 shows that there are three factors of 7 in the expression 7^3.

The exponent 4 in the expression 11^4 shows that there are 4 factors of 11.

4^5 is a multiplication expression in which the exponent 5 shows that 4 is to be used five times as a factor.

Using Exponents to Indicate Multiplication

The expression 5^2 is equivalent to $5 \cdot 5$, which has an evaluation of 25.

$6^2 = 6 \cdot 6 = 36$

$(-6)^2$ is equivalent to $-6 \cdot -6$, which has an evaluation of 36. Similarly, the evaluation of $(-3)^2 = 9$, is equivalent to $-3 \cdot -3$.

$(-2)^3$ is evaluated as follows.

$$\begin{aligned}(-2)^3 &= -2 \cdot -2 \cdot -2 \\ &= (-2 \cdot -2) \cdot -2 \\ &= 4 \cdot -2 \\ &= -8\end{aligned}$$

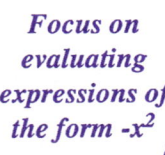
Focus on evaluating expressions of the form $-x^2$

-2^2 is read as the "opposite of (2^2)" which is equivalent to $-1 \cdot 2 \cdot 2$ and is evaluated as:
$-2^2 = -1 \cdot 2^2 = -1 \cdot 2 \cdot 2 = -4$

Similarly, -2^3 is read as the "opposite of (2^3)" which is equivalent to $-1 \cdot 2 \cdot 2 \cdot 2$ and is evaluated as:
$-2^3 = -1 \cdot 2^3 = -1 \cdot 2 \cdot 2 \cdot 2 = -8$

-2^4 is not equivalent to (-2^4). Notice the difference in their evaluations.
$(-2)^4 = -2 \cdot -2 \cdot -2 \cdot -2 = 16$
$-2^4 = -1 \cdot 2 \cdot 2 \cdot 2 \cdot 2 = -16$

The Base of an Exponent

In the expression 5^3, the 3 is an exponent and 5 is the **base**. In the expression 4^7, the base is 4 and the exponent is 7.

In the expression 6^2, 6 is the base and 2 is the exponent. In the expression 4^5, the exponent is 5 and the base is 4.

In the expression $2^1 \cdot 3^5$, the exponent 5 applies only to the number of times 3 is used as a factor. The base of the exponent in $2 \cdot 3^5$ is 3. 2 is the **coefficient** of the exponent expression, $2(3^5)$.

In the expression -4^5, the exponent applies only to the number of times 4 is used as a factor. The base of the exponent in -4^5 is 4 and -1 is the coefficient of the exponent expression. $-4^5 = -1 \cdot 4^5$

Focus on the use of variables as bases for exponents

Variables may be used to represent the bases of exponents. $4x^2$ is an open expression in which the variable x is the base of the exponent and 4 is the coefficient.

The base of the exponent in $-9x^3$ is x and -9 is the coefficient. $-9x^3$ is equivalent to -9xxx.

Similarly, the base of the exponent in $-x^5$ is x and -1 is the coefficient. $-x^5$ is equivalent to -1xxxxx.

The base of the exponent in $(3x)^2$ is 3x. $(3x)^2$ is equivalent to $3x \cdot 3x$.

The base of $19x^3$ is x, and $19x^3$ is equivalent to 19xxx. The base of the exponent 3 in the expression $(4x^2)^3$ is $4x^2$. $(4x^2)^3$ is equivalent to $4x^2 \cdot 4x^2 \cdot 4x^2$.

Unit 1 Exercise

Part A: Answers for all Part A problems are at the back of the book.

1. What are the factors of 7 • 5?

2. In the expression 7^8 the exponent is 8 and its base is 7. How many times is 7 used as a factor in the multiplication expression 7^8?

3. In the expression $(-2)^3$ the exponent is 3 and its base is -2. How many times is -2 used as a factor in the expression $(-2)^3$?

4. In the expression $-7x^5$ the exponent is 5, its base is x, and the coefficient is -7. How many times is -7 used as a factor in the expression $-7x^5$.

5. In the expression -4^6 the exponent is 6, its base is 4, and the coefficient is -1. What is the base of the exponent 3 in -5^3?

6. The expression -4^3 has -64 as its evaluation because $-1 • 4 • 4 • 4 = -64$. The expression $(-4)^3$ also has -64 as its evaluation because $-4 • -4 • -4 = -64$. Do -4^3 and $(-4)^3$ have the same meaning?

7. Is the evaluation of $(-3)^4$ equal to the evaluation of -3^4?

8. What is the base of the exponent 3 in $(-3x^5)^3$?

9. What is the base of the exponent 5 in $(-3x^5)^3$?

10. In the expression 5x the exponent for 5 is 1 and the exponent for x is also 1. $5x = 5^1 x^1$. What is the exponent for 6 in $6x^5$?

11. Simplify $(3x)^4$ using the meaning of the exponent 4 to write $(3x)(3x)(3x)(3x)$.

12. Evaluate $-4 • 2^3$ by using the expression $-4 • 2 • 2 • 2$.

Part B: Answers for odd-numbered problems of Part B are at the back of the book.

1. In the expression $(-2)^5$, the exponent is _____ and the base is _____.

2. In the expression -7^3, the exponent is _____ and the base is _____.

3. What is the exponent of the base x in $-4x^5$?

4. What is the coefficient in the expression $-x^6$?

5. What multiplication expression is equivalent to $4x^2$?

6. What multiplication expression is equivalent to $(-3x)^2$?

7. What is the exponent of 5 in $5 • 9^2$?

8. True or false? $(-6)^9 = -6^9$

9. True or false? $(-8)^6 = -8^6$

10. True or false? $(a + b)^2 = a^2 + b^2$

Evaluate.

11. $(-3)^3$
12. $(-5)^3$
13. $(-4)^3$
14. 2^4
15. $(-3)^4$
16. 3^2
17. $(-4)^2$
18. -3^2
19. -5^3
20. -2^4
21. $(-3)^2$
22. $3 \cdot (-2)^4$
23. $(-4)^4$
24. 3^4
25. -2^4
26. $(-2)^4$
27. $-3 \cdot 2^3$
28. $5(-2)^2$
29. $2 \cdot 5^3$
30. $(7-5)^5$
31. $(6-9)^4$
32. $(-1)^{37}$
33. $(-1)^{402}$
34. -1^{68}
35. -1^{961}

Part C: No answers are given for these problems. However, each is accompanied by an ordered pair «C,U» showing the chapter and unit in which it was taught.

1. Factor $8x^6 - 27y^3$ «1,4»

2. Simplify $-3\sqrt[3]{54}$ «2,3»

3. Use interval notation to write a truth set for $x^2 + 3x - 10 < 0$.   «2,6»

4. Solve $(3+i)(x+yi) + (3+2i) = (1-3i)$ «3,4»

5. Use interval notation to describe the truth set for $|4x + 1| \geq -3$.   «4,4»

6. Find the domain and the range for $y = (x-3)^2 - 2$.   «5,3»

7. Find the slope of the line through the points $(-1,4)$ and $(3,-2)$.   «6,3»

8. Find the standard form of the linear equation with solutions $(-2,4)$ and $(3,-1)$.   «6,4»

9. Graph the relation $(x-3)^2 + (y+1)^2 = 16$.   «7,2»

10. Graph the function $y = x^2 - 3x - 4 = 0$.   «7,2»

Unit 2: SIMPLIFYING POWER EXPRESSIONS

Factors of a Power Expression

An expression containing an exponent is often called a **power** expression. This unit explains the simplification of a power expression involving only counting number exponents.

A counting number exponent indicates how many times its base is to be used as a factor.

x^2 is equivalent to xx.
x^5 is equivalent to xxxxx.

2 is not a factor of x^2. 2 is an exponent and shows that x is to be used twice as a factor in the multiplication expression x^2.

3 is a factor of 3^5, but 5 is not a factor. 5 is the exponent that shows that 3 is to be used as a factor five times.

Focus on factors of a power expression

The factors of $3x^7$ are 3 and x. 3 is the coefficient and is used once as a factor, but x is the base of the exponent and is used 7 times.

The base of $(3x)^2$ is 3x and $(3x)^2$ is equivalent to $3x \cdot 3x$.

The base of $-x^4$ is x and $-x^4$ is equivalent to -1xxxx.
The base of $(-x)^4$ is -x and $(-x)^4$ is equivalent to $-x \cdot -x \cdot -x \cdot -x$.

Multiplying Like Bases

$x^5 \cdot x^4$ is equivalent to x^9. The exponent of x^5 shows that x is to be used five times as a factor. The exponent of x^4 shows that x is to be used as a factor four times. The exponents in $x^5 \cdot x^4$ show the number of times x is used as a factor. x is used as a factor nine times in $x^5 \cdot x^4$.

x^2 and x^4 have like bases. x^3 and y^3 do not have like bases. The base of x^3 is x and the base of y^3 is y.

The multiplication of like bases is accomplished by adding the exponents.
$$x^2 \cdot x^3 \text{ means } xx \cdot xxx \text{ or } x^5$$

Notice that x is equivalent to x^1. The exponent 1 is always understood whenever no other exponent appears.
$$x^3 \cdot x \text{ means } xxx \cdot x \text{ or } x^4$$

Focus on multiplying like bases

To multiply like bases, the exponents are added but the base remains the same. This means that $x^3 \cdot x^4 = x^7$ and also means that
$$2^3 \cdot 2^4 = 2^7, \ (-6)^8 \cdot (-6)^2 = (-6)^{10} \text{ and } 4^3 \cdot 4^7 = 4^{10}$$

Multiplying Monomials

To multiply $3x^2 \cdot -5x^4$ the factors are reordered and regrouped as shown below.
$$3x^2 \cdot -5x^4$$
$$(3 \cdot -5)(x^2 \cdot x^4)$$
$$-15x^6$$

Focus on simplifying multiplications of monomials

The regrouping and reordering of factors are important in simplifying $-2x^2yz^5 \cdot 4xy^7z^3$ as shown below.
$$-2x^2yz^5 \cdot 4xy^7z^3$$
$$(-2 \cdot 4)(x^2 \cdot x)(y \cdot y^7)(z^5 \cdot z^3)$$
$$-8x^3y^8z^8$$

Simplifying Expressions of the Form $(x^a)^b$

There are two exponents in the expression $(x^2)^3$. The base of the exponent 2 is x. The base of the exponent 3 is x^2. $(x^2)^3$ is equivalent to $x^2 \cdot x^2 \cdot x^2$ or x^6.

Similarly, $(x^4)^2$ is equivalent to $x^4 \cdot x^4$ or x^8.

$(x^2)^7$ is equivalent to $x^2 \cdot x^2 \cdot x^2 \cdot x^2 \cdot x^2 \cdot x^2 \cdot x^2$. The exponents are to be added. The addition of seven 2's gives the same result as $7 \cdot 2$, that is,
$$2 + 2 + 2 + 2 + 2 + 2 + 2 = 14$$
Therefore, $(x^2)^7 = x^{14}$.

338   Chapter 8

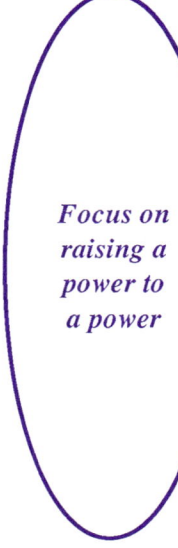

Focus on raising a power to a power

$(x^7)^3$ is equivalent to $x^7 \cdot x^7 \cdot x^7$. The exponents are added, but the sum of three 7's (7 + 7 + 7) is the same as 3 · 7.
$$(x^7)^3 = x^{21}$$
In words, x to the 7th power raised to the 3rd power is equivalent to x to the 21st power.

$(x^a)^b = x^{ab}$ becomes a true statement for any counting number replacements of a and b.
$$(x^7)^5 = x^{35}$$
There are two exponents in the expression $(3x^5)^2$. The base of the exponent 5 is x. The base of the exponent 2 is $3x^5$.
$(3x^5)^2$ is equivalent to $3x^5 \cdot 3x^5$ or $9x^{10}$.

Similarly, $(-2x^3)^2$ is equivalent to $-2x^3 \cdot -2x^3$ or $4x^6$.
$$(x^2yz^7)^3 = x^2yz^7 \cdot x^2yz^7 \cdot x^2yz^7 = x^6y^3z^{21}$$
$$(-2xy^2z^5)^3 = -2xy^2z^5 \cdot -2xy^2z^5 \cdot -2xy^2z^5 = -8x^3y^6z^{15}$$

Fractions Equal to One

When a power expression is divided by itself (with the assumption that the power expression is not equal to zero) the quotient is always equal to 1.

$$\frac{x}{x} = 1 \quad \frac{x^2}{x^2} = 1 \quad \frac{x^5}{x^5} = 1 \text{ when } x \neq 0$$

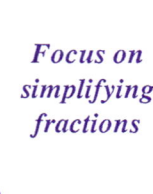

Focus on simplifying fractions

The quotient of x^{11} divided by x^5 is simplified using the fact that x^5 divided by itself equals 1 whenever $x \neq 0$.

$$\frac{x^{11}}{x^5} = \frac{x^5 x^6}{x^5} = \frac{x^6}{1} = x^6$$

Similarly, the quotient of x^7 divided by x^{12} is simplified using the division of x^7 by itself. Again, this simplification is only possible when $x \neq 0$.

$$\frac{x^7}{x^{12}} = \frac{x^7}{x^7 x^5} = \frac{1}{x^5}$$

Dividing Like Bases

All further divisions shown in this chapter are shown without the explicit claim that the denominator (divisor) is not equal to zero. Although the claim will be implicit from this point forward it is valuable for the student to constantly note it.

The numerator of $\frac{x^7}{x^4}$ has seven factors of x while the denominator has only four. The power expression with the larger exponent can always be written as a multiplication as in: $x^7 = x^4 x^3$. This makes it possible to simplify the fraction.

$$\frac{x^7}{x^4} = \frac{x^4 x^3}{x^4} = x^3$$

The numerator of $\frac{x}{x^6}$ has one factor of x while the denominator has six. The power expression with the larger exponent can always be written as a multiplication as in: $x^6 = x \cdot x^5$. This makes it possible to simplify the fraction.

$$\frac{x}{x^6} = \frac{x}{xx^5} = \frac{1}{x^5}$$

The two previous examples justify the following procedure: In dividing like bases, the exponents may be subtracted. The new power expression will have the same position in the fraction as the original power expression with the larger exponent.

Focus on dividing like bases

The division of like bases can be simplified by subtracting exponents. The division of x^3 by x^{11} is shown below. Notice that the original fraction has more x's in the denominator and the simplified fraction has the remaining x's in the denominator.

$$\frac{x^3}{x^{11}} = \frac{1}{x^{11-3}} = \frac{1}{x^8}$$

To divide like bases, subtract exponents. The division of x^8 by x^2 is shown below. Notice that the original fraction has more x's in the numerator and the simplified fraction has the remaining x's in the numerator.

$$\frac{x^8}{x^2} = \frac{x^{8-2}}{1} = x^6$$

Simplifying the Division of Power Expressions

To simplify $\frac{-8x^2y^5z^4}{-6xy^5z^7}$ four separate fractions are considered. One fraction involves the integer coefficients and the other three fractions involve the three variables of the problem.

$$\frac{-8x^2y^5z^4}{-6xy^5z^7} = \frac{-8}{-6} \cdot \frac{x^2}{x} \cdot \frac{y^5}{y^5} \cdot \frac{z^4}{z^7} = \frac{4}{3} \cdot \frac{x}{1} \cdot \frac{1}{1} \cdot \frac{1}{z^3} = \frac{4x}{3z^3}$$

Focus on simplifying the division of two power expressions

To simplify $\frac{-10x^2yz}{2xy^5}$ the following steps are used.

$$\frac{-10x^2yz}{2xy^5} = \frac{-10}{2} \cdot \frac{x^2}{x} \cdot \frac{y}{y^5} \cdot \frac{z}{1} = \frac{-5}{1} \cdot \frac{x}{1} \cdot \frac{1}{y^4} \cdot \frac{z}{1} = \frac{-5xz}{y^4}$$

Unit 2 Exercise

Part A: Answers for all Part A problems are at the back of the book.

1. Is 3 a factor in the expression x^3?
2. Is 2 a factor of 2^3?
3. Is 7 a factor of 4^7?
4. Is 2 a factor of $(3x)^2$?
5. Is 5 a factor of $3x^5$?
6. How many times is x used as a factor in x^3?
7. How many times is 7 used as a factor in $7^5 \cdot 7^4$?
8. How many times is x used as a factor in the multiplication expression $x^2 \cdot x^3$?
9. What is the total number of times that x is used as a factor in $x^2 \cdot x^4$?
10. Evaluate $3^3 \cdot 3^2$ using the meaning of the exponents.
11. Evaluate $2^3 \cdot 3^2$ using the meaning of the exponents.
12. Is $2^1 \cdot 4^2 = 8^3$ a true statement?
13. $(x^6)^3$ is equivalent to $x^6 \cdot x^6 \cdot x^6$ or x^{18}. $(x^3)^4$ is equivalent to $x^3 \cdot x^3 \cdot x^3 \cdot x^3$ or _____.
14. What is the base of the exponent 4 in $(5x^3)^4$?

15. To simplify $\frac{x^7}{x^{10}}$ use the meaning of exponents. There are 7 factors of x in the numerator and 10 in the denominator. When 7 factors of x are divided out, the quotient will have ____ factors of x remaining in the _____.

16. $\frac{x^5}{x^{12}} = \frac{x^5}{x^5 x^7} = $ ——

17. To simplify $\frac{x^{12}}{x^4}$ use the meaning of exponents. There are 12 factors of x in the numerator and 4 in the denominator. When 4 factors of x are divided out, the quotient will have ____ factors of x remaining in the _____.

18. $\frac{x^9}{x^4} = \frac{x^4 x^5}{x^4} = $ ——

Part B: Answers for odd-numbered problems of Part B are at the back of the book.

For problems 1-8 label each equality as true or false.

1. $4^2 \cdot 3^2 = 12^4$
2. $2^3 \cdot 4^2 = 8^5$
3. $2^3 \cdot 2^5 = 2^{15}$
4. $3^2 \cdot 3^4 = 3^6$
5. $2^3 \cdot 2^4 = 2^7$
6. $\frac{10^4}{5^2} = 2^2$
7. $\frac{7^4}{7^3} = 7$
8. $\frac{8^2}{6^2} = \frac{4^2}{3^2}$

Simplify.

9. $6^4 \cdot 6^5$
10. $5^4 \cdot 5^9$
11. $x^2 \cdot x^{11}$
12. $x^3 \cdot x^7$
13. $2 \cdot 2^8$
14. $x^{10} \cdot x$
15. $x^5 \cdot x^2$
16. $9x^7 \cdot -8x$
17. $3x^4 \cdot -x^3$
18. $-6x^5 \cdot -2x^3$
19. $-x^2 \cdot -6x^5$
20. $5x^2y^3z^5 \cdot -3x^4yz^2$
21. $-12xyz^2 \cdot -x^3yz^4$
22. $-3x^2y \cdot 4x \cdot y^2$
23. $-3x^2 \cdot 2yz^5$
24. $-2xyz^2 \cdot -7x^2y^4z^8$
25. $3x^4y^7 \cdot -2x^3z^7$
26. $-2x^9yz^3 \cdot x^4yz$
27. $-2x^2y \cdot 3x^3y^2 \cdot -5xy^4$
28. $5a^2b^3c^2 \cdot -2ac^2 \cdot 4a^3$
29. $-3x^2y \cdot 5xy^4$
30. $2ab \cdot -4a^2b^2$
31. $-5xz \cdot 5xy^2z$
32. $-2xy^2 \cdot 3x^2y^3 \cdot -4x^2y$
33. $2^3 \cdot -5$
34. $3^3 \cdot 3^7$
35. $3^4 \cdot 3^8$
36. $(x^9)^2$
37. $(x^3)^2$
38. $(x^4)^2$
39. $(x^5)^3$
40. $(x^9)^3$
41. $(x^5)^4$
42. $(x^6)^4$
43. $(x^5)^6$
44. $(x^4)^7$
45. $(x^8)^5$
46. $(x^9)^6$
47. $(x^{10})^4$
48. $(x^3)^4$

49. $(x^4)^5$

50. $(y^5)^9$

51. $(x^5)^6$

52. $(x^3)^5$

53. $(4x^2y)^2$

54. $(x^3y^4z)^2$

55. $(-2x^2y^3)^4$

56. $(-3x^4z)^2$

57. $(-x^3yz^4)^5$

58. $(-10y^2z^8)^3$

59. $(x^5)^7$

60. $(x^2)^5$

61. $x^2 \cdot x^5$

62. $(2x^3)^4$

63. $(-3x^2)^3$

64. $(3x^2y^5)^4$

65. $\dfrac{x^9}{x^3}$

66. $\dfrac{x^4}{x^{20}}$

67. $\dfrac{xyz}{x^2y^5z}$

68. $\dfrac{6x^5y^2}{8xy^8}$

69. $\dfrac{-12x^5y^5z^3}{-4x^4yz^2}$

70. $\dfrac{-6x^9y^2z^7}{15x^{10}y^3z^3}$

71. $\dfrac{-9xyz^5}{3xyz}$

72. $\dfrac{-5x^2y^2z^2}{-15x^2y^9z}$

73. $\dfrac{8xy^3z^7}{-6xy}$

74. $\dfrac{25x^2y^{11}z}{15x^2y^{11}z}$

75. $\dfrac{24x^2y^5z^3}{x^3y^4z^2}$

76. $\dfrac{14x^9y^2z^7}{10x^6y^5z^2}$

77. $\dfrac{-18x^9yz}{-16x^{13}y^4z^7}$

78. $\dfrac{-42y^7z}{-6y^3z}$

79. $\dfrac{9x^4y^3}{27x^7y^4}$

80. Which of the open sentences shown below will become a true statement for any counting number replacements for a and b?

   a. $x^a \cdot x^b = x^{ab}$
   b. $x^a \cdot x^b = x^{a+b}$
   c. $(x^a)^b = x^{a+b}$
   d. $(x^a)^b = x^{ab}$
   e. $\dfrac{x^a}{x^b} = x^{a-b}$
   f. $\dfrac{x^a}{x^b} = x^{b-a}$

Part C: No answers are given for these problems. However, each is accompanied by an ordered pair «C,U» showing the chapter and unit in which it was taught.

1. Solve $x^6 - 64$ using the set of rational numbers.   «1,5»

2. How much alcohol should be added to 80 gallons of 30% mixture to raise the percent of alcohol to 44%?   «2,7»

3. Solve $3x^2 - 2x + 4 = 0$ «3,2»

4. 7 lbs of flour and 2 lbs of sugar cost $18.40. 3 lbs of flour and 8 lbs of sugar cost $19.60. What is the cost per lb for flour and sugar?   «3,5»

5. Solve $|x - 3| = -4$ «4,2»

6. Find f(-6) when f(x) is $\frac{1}{2}x^2 + x - 12$.   «5,4»

7. Graph $3x - 4y = -12$ «6,1»

8. Write the equation $3x - 7y = -2$ in y-intercept form.   «6,4»

9. Find the length of the hypotenuse of a right triangle if the legs of the triangle are 4 feet and 3 feet.   «7,3»

10. Find the center of the circle $x^2 - 6x + y^2 = 41$.   «7,4»

Unit 3: INTEGER EXPONENTS

Previously in this text, the only numbers used as exponents have been counting numbers — elements of the set {1,2,3, . . .}. In this unit, the use of integers as exponents will be explained. As is frequently the case in mathematics, the use of integer exponents is defined completely in terms of the counting number exponents.

> **Definition**   **Integer Exponents**
>
> The integer k is an exponent with base x, **if and only if**
> 1. $x^k = x^k$ when $k > 0$,
> 2. $x^k = 1$ when $k = 0$ and $x \neq 0$,
> 3. $x^k = \left(\frac{1}{x}\right)^{-k}$ when $k < 0$ and $x \neq 0$.
>
> x^0 is undefined when $x = 0$.
> x^k is undefined when $x = 0$ and $k < 0$.

Integer Exponents

According to the definition of integer exponents,
1. When the integer exponent is a counting number then it continues to represent the number of times its base is used as a factor.
$$5^3 = 5 \cdot 5 \cdot 5, \quad (-4)^5 = -4 \cdot -4 \cdot -4 \cdot -4 \cdot -4$$
2. When the integer exponent is zero and the base is not zero, the power is equal to 1.
$$5^0 = 1, \quad (-4)^0 = 1, \quad \left(\tfrac{1}{2}\right)^0 = 1$$
3. When the integer exponent is negative then the reciprocal of the base becomes the new base and the opposite of the exponent is the new exponent.
$$5^{-6} = \left(\tfrac{1}{5}\right)^6, \quad \left(\tfrac{3}{4}\right)^{-2} = \left(\tfrac{4}{3}\right)^2$$

Notice that 0^k where $k \leq 0$ is undefined. The reason for leaving such power expressions undefined is directly related to the fact that division by zero is undefined.

Focus on the use of positive integer exponents

The definition of integer exponents cites three possibilities:
1. The exponent is positive.
2. The exponent is zero.
3. The exponent is negative.

The first case simply maintains the use of counting number exponents when they are considered positive integers.

$$10^7 = 10 \cdot 10 \cdot 10 \cdot 10 \cdot 10 \cdot 10 \cdot 10 = 10,000,000$$
$$\left(\tfrac{2}{3}\right)^3 = \left(\tfrac{2}{3}\right) \cdot \left(\tfrac{2}{3}\right) \cdot \left(\tfrac{2}{3}\right) = \tfrac{8}{27}$$

Zero as an Exponent

The use of zero as an exponent is defined by the following equality.

$$x^0 = 1 \text{ whenever } x \neq 0$$

$5^0 = 1$ and $(-\sqrt{2})^0 = 1$ are both true statements. $\pi^0 = 1$ and $(-17)^0 = 1$ are also true statements.

The definition for zero as an exponent proclaims that any base except zero with a zero exponent names the number 1. Many students react to this definition as they do when first confronted with the idea that the product of two negatives is positive. As is always the case in mathematics, there is a reason why the definition makes the power equal to 1. Some of the rationale is shown below.

Earlier in this chapter, it was shown that when like bases are multiplied the exponents are added. By using this property, $x^0 \cdot x^5$ is simplified as:
$$x^0 \cdot x^5 = x^{0+5} = x^5, \text{ but } 1 \cdot x^5 = x^5$$
For integers as exponents, if the property that exponents are to be added when multiplying like bases is to be preserved then x^0 must be 1.

As another example, consider the raising of a power to a power as in $(x^5)^7 = x^{35}$.
$$(x^0)^6 = x^{0 \cdot 6} = x^0 \text{ and } 1^6 = 1$$
Again, x^0 behaves like 1 and if the property that exponents are to be multiplied when raising a power to a power is to be preserved then $x^0 = 1$ maintains that property.

In dividing like bases, the exponents are subtracted as in

$$\frac{x^8}{x^3} = x^{8-3} = x^5$$

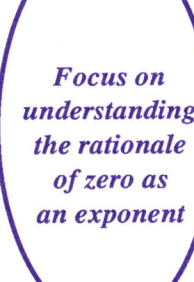

Focus on understanding the rationale of zero as an exponent

Applying this same procedure to $\frac{x^7}{x^7}$ shows why zero exponents are defined to make their powers equal to 1.

$$\frac{x^7}{x^7} = x^{7-7} = x^0, \text{ but } \frac{x^7}{x^7} = 1$$

Whether a problem involves multiplying like bases, raising a power to a power, or dividing like bases, the definition that $x^0 = 1$ whenever $x \neq 0$ maintains the properties for simplifying power expressions.

Negative Integers as Exponents

The use of negative integers as exponents is defined by the following equality.

$$x^k = \left(\frac{1}{x}\right)^{-k} \text{ when } k < 0 \text{ and } x \neq 0$$

The stipulation that $x \neq 0$ is necessary because $\frac{1}{x}$ doesn't exist when $x = 0$.

Two examples of changing negative exponents to counting numbers are:

$$4^{-3} = \left(\frac{1}{4}\right)^3 = \frac{1}{64} \qquad y^{-7} = \left(\frac{1}{y}\right)^7 = \frac{1}{y^7}$$

The definition for negative exponents has nothing to do with negative number answers. Negative integer exponents can always be changed to counting numbers and the change has no effect on the sign of the power.

As was the case with zero exponents, the reasoning which supports the definition of negative exponents is directly related to maintaining the counting number properties for exponents. It is desirable to maintain the property that when like bases are multiplied the exponents are added. Two ways of doing the same problem are shown below. Notice that the final results are the same.

$$x^{-3} \cdot x^5 = x^{-3+5} = x^2$$

$$x^{-3} \cdot x^5 = \left(\frac{1}{x}\right)^3 \cdot x^5 = \frac{1}{x^3} \cdot \frac{x^5}{1} = \frac{x^5}{x^3} = x^2$$

The previous example shows why negative exponents are defined in terms of counting number exponents. All of the exponent properties are maintained. The same properties established for counting number exponents apply equally well to negative exponents.

Focus on changing negative exponents to counting numbers

It is unnecessary and often undesirable to have negative exponents.

To change any negative exponent to a positive:
1. Change the position of the base in the fraction.
2. Change the exponent to its opposite.

$$x^{-5} = \left(\frac{x}{1}\right)^{-5} = \left(\frac{1}{x}\right)^{5} = \frac{1}{x^5}$$

— the opposite of -5
— the base moved to the denominator

$$\frac{1}{x^{-8}} = \left(\frac{1}{x}\right)^{-8} = \left(\frac{x}{1}\right)^{8} = x^8$$

— the opposite of -8
— the base moved to the numerator

Unit 3 Exercise

Part A: Answers for all Part A problems are at the back of the book.

1. A zero exponent with a nonzero base always represents the number 1. $6^0 = 1$ and $(-5)^0 =$ ____

2. Any nonzero base with a zero exponent is equal to 1. $(\sqrt{37})^0 =$ ____

3. If the base of a zero exponent is 0 then the expression is undefined. $\pi^0 = 1$ and $0^0 =$ ____

4. The power expression 6^{-2} is equal to $\left(\frac{1}{6}\right)^2$ because when the sign of an exponent is changed the base is changed to its reciprocal. $2^{-3} = \left(\frac{1}{2}\right)^3 =$ ____

5. Any negative exponent on a numerator can be changed to a positive by moving its base to the denominator. $7^{-4} = \frac{1}{7^4}$ and $8^{-3} =$ ____

6. Any negative exponent on a denominator can be changed to a positive by moving its base to the numerator. $\frac{1}{5^{-3}} = 5^3$ and $\frac{1}{9^{-7}} =$ ____

7. x^{-4} is the reciprocal of x^4 and $x^{-4} \cdot x^4 = \frac{1}{x^4} \cdot x^4 =$ ____

8. Use a negative exponent to write the reciprocal of y^8.

9. Zero cannot be the base of a negative exponent because zero has no _____.

10. Can a negative exponent have a negative base and result in a positive answer?

11. Can a negative exponent have a positive base and result in a negative answer?

Part B: Answers for odd-numbered problems of Part B are at the back of the book.

For problems 1-10 evaluate.

1. 7^{-1}
2. $(\sqrt{5})^{-2}$
3. $(\frac{4}{5})^{-3}$
4. $(\frac{3}{8})^0$
5. $(7-4)^{-4}$
6. $(1-\frac{9}{7})^{-1}$
7. $(5 + 7 \cdot \frac{32}{5})^0$
8. $(3 - \frac{1}{3} \cdot 6)^0$
9. $(8 + \frac{3}{7})^{-1}$
10. $(1 + \frac{3}{4})^{-2}$

Rewrite each expression so the exponent is positive.

11. x^{-6}
12. z^{-4}
13. y^{-7}
14. $\frac{1}{y^{-1}}$
15. $\frac{1}{x^{-8}}$
16. $\frac{1}{z^{-13}}$
17. $\frac{1}{y^{-42}}$
18. $\frac{1}{x^{-a}}$
19. $\frac{1}{y^{-3}}$
20. z^{-4}
21. y^{-8}
22. z^{-1}
23. x^{-17}
24. $\frac{1}{y^{-2}}$
25. z^{-5}
26. k^{-12}
27. $\frac{1}{z^{-15}}$
28. x^{-3}
29. y^{-71}
30. $\frac{1}{z^{-7}}$

Part C: No answers are given for these problems. However, each is accompanied by an ordered pair «C,U» showing the chapter and unit in which it was taught.

1. Using the set of rational numbers, solve $\frac{7}{8}x - \frac{1}{3} = \frac{2}{3} - \frac{1}{4}x$. «1,4»

2. Simplify $\frac{4 - 5\sqrt{3}}{2 + 3\sqrt{3}}$ «2,2»

3. Simplify $(7 - 3i) + (-2 + 4i)$ «3,3»

4. Solve $(x + yi) + (5 - 3i) = (2 + i)$ «3,4»

5. Evaluate $|3| - |5| - |2|$ «4,2»

6. Find $f(-2)$ when $f(x) = \frac{13}{2x - 1}$ «5,4»

7. Find the slope and y-intercept for $3x - 4y = 6$. «6,2»

8. Find the distance between points $(0,-3)$ and $(-2,4)$. «7,3»

9. Evaluate -2^4 «8,1»

10. Evaluate $(-3)^2$ «8,1»

Unit 4: SIMPLIFYING WITH INTEGER EXPONENTS

Multiplying Like Bases

To simplify the multiplication of x^5 and x^{-9}, either of the following processes can be used.

1. $x^5 \cdot x^{-9} = x^5 \cdot \dfrac{1}{x^9} = \dfrac{x^5}{x^9} = \dfrac{1}{x^4}$

2. $x^5 \cdot x^{-9} = x^{5 + -9} = x^{-4}$

The first process shown converted the negative exponent to a positive and then completed the problem with counting number exponents. The second process used the fact that when like bases are multiplied the exponents are added. The results are the same, but the first process emphasizes the meaning of negative exponents and the second emphasizes a quick and easy way to get the answer. Learn both processes. Answers are valuable in mathematics only when their meaning is understood.

Focus on multiplying like bases

With counting number and integer exponents the multiplication of like bases is simplified by adding exponents.

$$x^8 \cdot x^{-2} = x^6$$

$$x^{-5} \cdot x^{-3} = x^{-8} = \dfrac{1}{x^8}$$

$$6x^{-4}y^{-5} \cdot -2x^7y^{-4} = -12x^3y^{-9} = \dfrac{-12x^3}{y^9}$$

Dividing Like Bases

To simplify the division $\frac{x^5}{x^{-6}}$, either of the following processes can be used.

1. $\frac{x^5}{x^{-6}} = x^5 \cdot x^6 = x^{11}$

2. $\frac{x^5}{x^{-6}} = x^{5-(-6)} = x^{5+6} = x^{11}$

The first process shown above converted the negative exponent to a positive and then completed the problem with counting number exponents. The second process used the fact that when like bases are divided the exponents are subtracted. Again, the results are the same, but the first process emphasizes the meaning of negative exponents and the second emphasizes a quick and easy way to get the answer. Learn both processes. If you only learn how to get an answer, you'll forget it. If you learn why the shortcut works, you'll remember.

Focus on dividing like bases

With counting number and integer exponents the division of like bases is simplified by subtracting exponents.

$$\frac{x^9}{x^{-4}} = x^{9-(-4)} = x^{9+4} = x^{13}$$

$$\frac{x^{-5}}{x^3} = x^{-5-3} = x^{-8} = \frac{1}{x^8}$$

$$\frac{x^{-7}}{x^{-2}} = x^{-7-(-2)} = x^{-7+2} = x^{-5} = \frac{1}{x^5}$$

Notice that in each case the denominator exponent was subtracted from the numerator exponent and the new power was in the numerator. In each case the final result was written with only positive exponents.

$$\frac{-8x^{-3}y^4}{6x^{-7}y^{-4}} = \frac{-8}{6} \cdot \frac{x^{-3}}{x^{-7}} \cdot \frac{y^4}{y^{-4}} = \frac{-4x^4y^8}{3}$$

Numerical coefficients are not affected by the exponents. -8 and 6 are numerical coefficients and the fraction $\frac{-8}{6}$ was reduced as usual to $\frac{-4}{3}$.

Raising a Power to a Power

To simplify the power expression $(x^{-2})^5$, either of the following processes can be used.

1. $(x^{-2})^5 = (\frac{1}{x^2})^5 = \frac{1^5}{(x^2)^5} = \frac{1}{x^{10}}$

2. $(x^{-2})^5 = x^{-2 \cdot 5} = x^{-10} = \frac{1}{x^{10}}$

The first process shown above converted the negative exponent to a positive and then completed the problem with counting number exponents. The second process used the fact that when a power is raised to a power the exponents are multiplied. Again, the results are the same, but the first process emphasizes the meaning of negative exponents and the second emphasizes a quick and easy way to get the answer. Learn both processes.

Focus on raising a power to a power

With counting number and integer exponents the raising of a power to a power requires multiplying the exponents.

$$(x^4)^{-3} = x^{4 \cdot -3} = x^{-12} = \frac{1}{x^{12}}$$

$$(x^{-6})^{-4} = x^{-6 \cdot -4} = x^{24}$$

$$(-5x^6y^{-7})^{-3} = (-5)^{-3}x^{-18}y^{21} = (\frac{-1}{5})^3 x^{-18} y^{21} = \frac{-y^{21}}{125x^{18}}$$

Notice in this last example that the numerical coefficient, -5, also needs to be raised to the -3 power.

Exponents   351

Simplifying Fractions Raised to a Power

Fractions like $\left(\dfrac{-10x^{-3}y^6}{5x^{-8}y^9z^{-2}}\right)^{-3}$ can be simplified in a number of ways, but the following process is generally the easiest.

First, simplify the fraction inside the parentheses.
$$\left(\dfrac{-10x^{-3}y^6}{5x^{-8}y^9z^{-2}}\right)^{-3} = \left(\dfrac{-2x^5z^2}{y^3}\right)^{-3}$$

Invert the fraction and change the sign of the exponent.
$$= \left(\dfrac{y^3}{-2x^5z^2}\right)^3$$

Raise each factor inside the parentheses to the 3rd power.
$$= \dfrac{y^9}{-8x^{15}z^6}$$

Simplify the fraction.
$$= \dfrac{-y^9}{8x^{15}z^6}$$

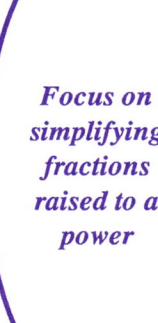

Focus on simplifying fractions raised to a power

To simplify the power expression at the right,
$$\left(\dfrac{-21x^5y^8z^{-2}}{14x^{-9}y^{-1}z}\right)^{-4}$$

1. Simplify inside the parentheses.
$$\left(\dfrac{-3x^{14}y^9}{2z^3}\right)^{-4}$$

2. Invert the fraction and change the sign of the exponent.
$$\left(\dfrac{2z^3}{-3x^{14}y^9}\right)^4$$

3. Raise each factor inside the parentheses to the 4th power.
$$\dfrac{16z^{12}}{81x^{56}y^{36}}$$

Unit 4 Exercise

Part A: Answers for all Part A problems are at the back of the book.

1. With counting number exponents, the multiplication of x^3 and x^4 is accomplished by finding the total number of x's.
$$x^3 \cdot x^4 = (xxx)(xxxx) = x^7.$$
There is no reason to memorize a rule for this process because the meaning of the exponents leads directly to the correct result. Use the meaning of the counting number exponents to simplify $x^2 \cdot x^6$.

2. If you remember the process for simplifying $x^2 \cdot x^6$, then the multiplication of $x^{-5} \cdot x^7$ will require finding the sum of the exponents.
$x^{-5} \cdot x^7 = x^{-5+7} = x^2$ and
$x^{-3} \cdot x^{-5} = x^{-3+-5} = $ ___.

3. To simplify $x^{-6} \cdot x^4$, recall what you would do with counting number exponents and then apply the same process. $x^{-6} \cdot x^4 = $ ___

4. With counting number exponents, the division of x^7 by x^4 is accomplished by finding the number of x's that are not eliminated by the cancelling process.
$$\frac{x^7}{x^4} = \frac{xxxxxxx}{xxxx} = \frac{\cancel{xxxx}xxx}{\cancel{xxxx}} = \frac{xxx}{1} = x^3$$
There is no reason to memorize a rule for this process because the meaning of the exponents leads directly to the correct result. Use the meaning of the counting number exponents to simplify the division of x^6 by x^2.

5. If you remember the process for simplifying the division of x^6 by x^2, then the division of $x^{-2} \div x^3$ will require finding the difference of the exponents.
$\frac{x^{-2}}{x^3} = x^{-2-3} = $ ___.

6. To simplify the division of x^{-7} by x^{-9}, recall what you would do with counting number exponents and then apply the same process.
$\frac{x^{-7}}{x^{-9}} = $ ___

7. With counting number exponents, the simplification of $(x^3)^4$ is accomplished by finding the total number of x's.
$$(x^3)^4 = (x^3)(x^3)(x^3)(x^3) = x^{12}.$$
There is no reason to memorize a rule for this process because the meaning of the exponents leads directly to the correct result. Use the meaning of the counting number exponents to simplify $(x^7)^3$.

8. If you remember the process for simplifying $(x^7)^3$, then the simplification of $(x^{-5})^2$ will require finding the product of the exponents. $(x^{-5})^2 = x^{-5 \cdot 2} = x^{-10}$ and $(x^3)^{-5} = x^{3 \cdot -5} = $ ___.

9. To simplify $(x^{-6})^{-4}$, recall what you would do with counting number exponents and then apply the same process. $(x^{-6})^{-4} = $ ___

10. With counting number exponents, the simplification of $(5x^2y^4)^3$ is accomplished by raising each factor inside the parentheses to the 3rd power. $(5x^2y^4)^3 = (5^1)^3(x^2)^3(y^4)^3 = 125x^6y^{12}$. Use the same process to simplify $(2x^5y^7)^4$.

11. With integer exponents, the simplification of $(-3x^6y^{-8})^{-4}$ is accomplished by raising each factor inside the parentheses to the -4 power.
$([-3]^1)^{-4}(x^6)^{-4}(y^{-8})^{-4} = [-3]^{-4}x^{-24}y^{32} = \frac{y^{32}}{81x^{24}}$.
Use the same process to simplify $(-2x^{-5}y^3)^{-5}$.

Exponents   353

12. With counting number exponents, the simplification of $(\frac{x^4}{y^3})^3$ is accomplished by raising both the numerator and denominator to the 3rd power.

$$(\frac{x^4}{y^3})^3 = \frac{(x^4)^3}{(y^3)^3} = \frac{x^{12}}{y^9}$$

Use the same process to simplify $(\frac{y^3}{x^4})^6$.

13. With integer exponents, the simplification of a fraction raised to a power is often easiest when the fraction inside the parentheses is first simplified and all exponents are converted to counting numbers. Simplify the expression below.

$$(\frac{-8x^{-4}y}{10x^{-7}y^3z^5})^{-3}$$

Part B: Answers for odd-numbered problems of Part B are at the back of the book.

Simplify. Write the final result with only positive exponents.

1. $x^{-6} \cdot x^9$
2. $\frac{x^{-3}}{x^{-7}}$
3. $(x^{-4})^{-5}$
4. $x^{-4} \cdot x^{-10}$
5. $\frac{x^{-2}}{x^{-5}}$
6. $(x^{-6})^2$
7. $x^{-7} \cdot x^{-1}$
8. $\frac{x^5}{x^{-3}}$
9. $x^{-3} \cdot x^3$
10. $(x^{-7})^{-1}$
11. $\frac{12x^{-8}}{-8x^{-4}}$
12. $-5x^{-6} \cdot 3x^7$
13. $(-3x^{-3})^{-2}$
14. $\frac{-15xy^{-3}}{-6x^{-2}y^{-7}}$
15. $-x^2y^{-6} \cdot 7x^{-5}y^4$
16. $-8x^{-5}y^{-2} \cdot -2x^7y^{-2}$
17. $(\frac{x^6y^4}{x^{-2}y^{-1}})^{-5}$
18. $(-2x^{-9}y^5)^{-3}$
19. $4x^{-7}y^7 \cdot -3x^9y^{-4}$
20. $(\frac{6x^{-4}y^3}{-2x^{-2}y^{-6}})^{-4}$
21. $(-4x^3y^{-5})^{-4}$
22. $-x^{-6}y^{-3} \cdot 2x^8y^{-5}$
23. $(\frac{-9x^{-7}y^{20}}{12x^{-5}y^{23}})^{-3}$
24. $(-2x^{-7}y^2)^3$
25. $-9x^3y^{-8} \cdot -4x^6y^{-6}$
26. $(\frac{10x^{-8}y^{-1}}{-2x^{-6}y^3})^{-3}$
27. $(-4x^{-3}y^{-6})^3$
28. $3x^{-4}y^{-9} \cdot -4x^5y^{-2}$
29. $(\frac{14x^{-9}y^{-2}}{-7x^{-10}y^2})^{-5}$
30. $(-x^7y^{-4})^{-6}$

Part C: No answers are given for these problems. However, each is accompanied by an ordered pair «C,U» showing the chapter and unit in which it was taught.

1. Factor $(x + y)^2 - 3(x + y)b - 4b^2$ «1,4»
2. Find the truth set for $x^2 + 6x = 2$ «2,6»
3. The sum of the digits of a two digit number is 10. The number with its digits reversed is 18 less than the number. Find the original number. «3,5»
4. Describe the truth set using interval notation for $|3x - 4| = 2$. «4,4»
5. Find a rule of correspondence that would match the elements of $\{1,2,3,\ldots\}$ with $\{13,14,15,\ldots\}$. «5,1»
6. Graph $y = \frac{-2}{3}x - 4$ «6,1»
7. Graph $2x - 3y \geq -12$ «6,5»
8. Find the center and radius for the circle $x^2 - 6x + y^2 = 7$. «7,4»
9. Is $(-2)^4 = -2^4$ true or false? «8,1»
10. Simplify $\frac{-10x^2y^3}{15xy}$ «8,2»

Unit 5: RATIONAL NUMBER EXPONENTS

Counting Number and Integer Exponents

In the first four units of this chapter, counting number exponents were reviewed and integer exponents were introduced. The definition of integer exponents was shown to be designed so that the exponent properties of the counting number exponents would be extended to integer exponents.

In this unit, rational numbers as exponents will be defined. Again, the definition maintains the properties for multiplying/dividing like bases and raising a power to a power.

The Index Of a Radical Expression

The real number $\sqrt{5}$ is read as "the square root of 5." The exponent of 5 is understood to be 1.
$$\sqrt{5} = \sqrt{5^1}$$

There is also an understanding that a small "2" is in the notch of the radical sign.
$$\sqrt{5^1} = \sqrt[2]{5^1}$$

The small "2" shown in the notch of the radical sign is called its **index**.

The index of $\sqrt[3]{85}$ is 3 and the expression is read as "the third root of 85." The index of $\sqrt[4]{13}$ is 4 and the expression is read as "the fourth root of 85."

Every radical expression has both an index and an exponent on the radicand.

For the radical expressions,
    $\sqrt[7]{x^4}$ The index is 7 and the exponent of the radicand is 4.
    $\sqrt[3]{x^8}$ The index is 3 and the exponent of the radicand is 8.
    $\sqrt{x^9}$ The index is 2 and the exponent of the radicand is 9.
    $\sqrt[5]{x^3}$ The index is 5 and the exponent of the radicand is 3.

Exponents   355

Focus on the index of a radical expression

Every radical expression has a number called it index, n, which indicates "the nth root of its radicand." If no index is written with the radical then it is understood to be 2.

$\sqrt[7]{5^6}$ is read as "the seventh root of 5^6"
$\sqrt[6]{8^5}$ is read as "the sixth root of 8^5"
$\sqrt{4^7}$ is read as "the second (square) root of 4^7"
$\sqrt[3]{11}$ is read as "the third root of 11^1"

Principle Roots for Real Numbers

The numeral $\sqrt[3]{x}$ represents the **principle** third root of 11 and may be positive, negative, or zero depending on whether x is positive, negative, or zero. For example, $\sqrt[3]{11}$ is positive because only a positive can be used 3 times as a factor and have 11 as its product.

$$\sqrt[3]{11} \cdot \sqrt[3]{11} \cdot \sqrt[3]{11} = 11$$

But, $\sqrt[3]{-9}$ is a negative because only a negative can be used 3 times as a factor and have -9 as its product.

$$\sqrt[3]{-9} \cdot \sqrt[3]{-9} \cdot \sqrt[3]{-9} = -9$$

The numeral $\sqrt[4]{17}$ represents the **principle** fourth root of 16 and is a **positive** real number slightly greater than 2 because $2^4 = 16$.

There is also a negative real number that can be raised to the fourth power to give 17, but the numeral $\sqrt[4]{17}$ represents the **principle** fourth root of 17. If a negative number were desired it would be shown as $-\sqrt[4]{17}$, in other words, "the opposite of the principle fourth root of 17."

Focus on principle roots of real numbers

The use of radical signs to designate real numbers is always accompanied by the concept of **principle** root.

Whenever the radical sign ($\sqrt[n]{x}$) has an odd numbered index (n) then the numeral names exactly one real number which may be positive, negative, or zero. That real number is the **principle** root.

Whenever the radical sign ($\sqrt[n]{x}$) has an even numbered index (n):
1. If the radicand is positive then the numeral names the positive nth root.
$$\sqrt{9} = 3, \text{ but } \sqrt{9} \neq -3 \text{ even though } (-3)^2 = 9$$
$$\sqrt[4]{16} = 2, \text{ but } \sqrt[4]{16} \neq -2 \text{ even though } (-2)^4 = 16$$
2. If the radicand is negative then the numeral does not name any real number and there is no principle root.
$$\sqrt{-9} \text{ is not a real number}$$
$$\sqrt[8]{(-3)^5} \text{ is not a real number (8th root of a negative number)}$$

Notice that each of the following is a real number.
$$\sqrt[8]{(-3)^6} \text{ (the 8th root of a positive number)}$$
$$\sqrt[5]{(-2)^3} \text{ (the 5th root of a negative number)}$$
$$\sqrt[4]{(-9)^{-2}} \text{ (the 4th root of a positive number)}$$

When the index is odd the radicand may be negative. When the index is even the radicand cannot be negative.

Writing Rational Exponent Expressions

Any radical expression naming a real number can be rewritten as a power expression with a **rational number exponent**.

$\sqrt[6]{(-2)^4}$ can be written as $(-2)^{4/6}$
$\sqrt[4]{9^1}$ can be written as $9^{1/4}$
$\sqrt[8]{3^6}$ can be written as $3^{6/8}$
$\sqrt[5]{(-7)^3}$ can be written as $(-7)^{3/5}$
$\sqrt{13}$ can be written as $13^{1/2}$

In each of these examples, the numerator of the rational number exponent is the exponent of the radicand and the denominator of the exponent is the index of the radical.

Exponents   357

Focus on writing rational exponent expressions

In general, the change of any radical expression to a rational exponent power is accomplished simply by using the index as the denominator and the exponent of the radicand as the numerator.

Be careful of situations where the rational number exponent can be reduced. For example, $\sqrt[4]{(-9)^2}$ can be written as $(-9)^{2/4}$ and the exponent must not be reduced.

$(-9)^{2/4}$ is a real number. It is the fourth root of a positive number.

$(-9)^{1/2}$ is not a real number. It is the second root of a negative number.

$(-9)^{2/4}$ is not equal to $(-9)^{1/2}$.

Below is an example where the exponent can be reduced. $\sqrt[6]{(-2)^4}$ can be written as $(-2)^{4/6}$ and this exponent can be reduced.

$(-2)^{4/6}$ is the sixth root of a positive number.

$(-2)^{2/3}$ is the third root of a positive number.

The radicand remained positive with the reduction.

$(-2)^{4/6}$ is equal to $(-2)^{2/3}$

Negative Rational Number Exponents

Negative rational number exponents have the same meaning as negative integer exponents. The base of the exponent is changed to its reciprocal and the exponent is changed to its opposite.

$5^{-3/8}$ is equal to $\left(\frac{1}{5}\right)^{3/8}$

$(-7)^{-5/9}$ is equal to $\left(\frac{-1}{7}\right)^{5/9}$

$\left(\frac{3}{4}\right)^{-5/3}$ is equal to $\left(\frac{4}{3}\right)^{5/3}$

$\left(\frac{1}{3}\right)^{-8/9}$ is equal to $3^{8/9}$

As with negative integer exponents, the base of a negative exponent cannot be zero. Zero has no reciprocal and the meaning of a negative exponent depends upon the reciprocal of the base.

Focus on negative exponents

Any negative exponent can be changed to a positive by altering the base to its reciprocal.

As before, caution is needed to make certain the expression names a real number. The two major difficulties are:
1. When the denominator is even and the radicand is negative.
   $(-5)^{-3/8}$ is not a real number.
2. When the base is zero the exponent cannot be 0 or negative.
   $0^{-5/7}$ is not a real number.

Unit 5 Exercise

Part A: Answers for all Part A problems are at the back of the book.

1. $\sqrt{7}$ is the "square root of 7" or the "second root of 7 to the first power." $\sqrt{7}$ is equivalent to $\sqrt[2]{7^1}$ where the index is 2 and the exponent on the radicand is 1. Write the expression for "the second root of 34."

2. $\sqrt[5]{37}$ is a numeral for the fifth root of 37 which means: $\sqrt[5]{37} \cdot \sqrt[5]{37} \cdot \sqrt[5]{37} \cdot \sqrt[5]{37} \cdot \sqrt[5]{37} = $ _____

3. $\sqrt[6]{28}$ is the sixth root of 28. This means: $\sqrt[6]{28} \cdot \sqrt[6]{28} \cdot \sqrt[6]{28} \cdot \sqrt[6]{28} \cdot \sqrt[6]{28} \cdot \sqrt[6]{28} = $ _____

4. Write a numeral for the real number that can be used 9 times as a factor to give a product of 12.

5. Write a numeral for the eighth root of 46.

6. $\sqrt[6]{7}$ is the sixth root of 7. $\sqrt[6]{7^2} \cdot \sqrt[6]{7^4} = 7$. $\sqrt[8]{5^5} \cdot$ _____ $= 5$.

7. $\sqrt[5]{9}$ is the fifth root of 9. $\sqrt[5]{9^3} \cdot$ _____ $= 9$.

8. How many different real numbers can be raised to the fourth power and have a product of 81?

9. How many different real numbers can be raised to the third power and have a product of -8?

10. How many different real numbers can be raised to the fifth power and have a product of 243?

11. How many different real numbers can be raised to the fourth power and have a product of -81?

12. A radical expression can be written using rational number exponents.
    $\sqrt[7]{5^2} = 5^{2/7}$ and $\sqrt[8]{4^6} = $ _____.

13. A power expression with a rational number exponent can be written as a radical numeral.
    $7^{4/9} = \sqrt[9]{7^4}$ and $(-5)^{3/5} = $ _____.

Exponents

Part B: Answers for odd-numbered problems of Part B are at the back of the book.

In problems 1-10, write radical numerals for:

1. The fourth root of 7 to the third power
2. The third (cube) root of 17
3. The square root of 35
4. The seventh root of -11 to the fifth power
5. The fifth root of -7 to the eighth power
6. A real number that can be used seven times in multiplication and have a product of -47
7. A real number that can be used five times in multiplication and have a product of 12
8. A real number that can be used four times in multiplication and have a product of 37
9. A number that can be multiplied by $\sqrt[7]{9^3}$ to give a product of 9
10. A number that can be multiplied by $\sqrt[9]{-6}$ to give a product of -6

In problems 11-20, write word descriptions of the following real numbers.

11. $\sqrt[3]{6^4}$
12. $\sqrt[4]{7^2}$
13. $\sqrt[8]{5^3}$
14. $\sqrt{11^6}$
15. $\sqrt[7]{6^3}$
16. $\sqrt[9]{11^8}$
17. $\sqrt[10]{8^4}$
18. $\sqrt[7]{32^2}$
19. $\sqrt[3]{8^5}$
20. $\sqrt[7]{5^2}$

In problems 21-30, write the radical expression using rational number exponents.

21. $\sqrt[3]{4^9}$
22. $\sqrt{8^7}$
23. $\sqrt[3]{x^4}$
24. $\sqrt{7^3}$
25. $\sqrt[5]{3^4}$
26. $\sqrt{3}$
27. $\sqrt[3]{5^2}$
28. $\sqrt{11^3}$
29. $\sqrt[8]{5^3}$
30. $\sqrt{5^4}$

In problems 31-40, write the power expression using a radical sign.

31. $11^{5/12}$
32. $13^{3/2}$
33. $14^{1/2}$
34. $7^{1/5}$
35. $12^{5/3}$
36. $20^{1/4}$
37. $5^{1/8}$
38. $6^{1/2}$
39. $5^{1/2}$
40. $4^{3/2}$

Part C: No answers are given for these problems. However, each is accompanied by an ordered pair «C,U» showing the chapter and unit in which it was taught.

1. Factor $4x^2 + xy - 5y^2$ «1,4»
2. Simplify $-2\sqrt[5]{64a^5b^{11}}$ «2,3»
3. The sum of the digits of a two digit number is 4. The number with the digits reversed is 18 more than the number. Find the original number. «3,5»
4. Solve $|4x - 3| = 1$ «4,2»
5. Graph the truth set for $|2x - 5| = -3$. «4,3»
6. Complete the ordered pair: $(-1, \_\_\_) \in \{(x, y) \mid 2x^2 - 5x + 3 = 0$ «5,2»
7. State the domain and range of $\{(3,5),(-2,1),(1,-6),(3,0)\}$. «5,3»
8. The line through the points (5,-2) and (-2,-2) is parallel to the ____ axis, and has a slope of _____. «6,3»
9. Graph $\frac{x^2}{9} + \frac{y^2}{16} = 1$ «7,5»
10. Simplify $5x^{-6}y^4z^{-2} \cdot -2xy^{-4}z^3$ «8,4»

Unit 6: SIMPLIFYING WITH RATIONAL NUMBER EXPONENTS

There are restrictions discussed earlier for the use of rational number exponents. In this unit variables are used for the bases of the exponents. This is done with the assumption that only those number replacements are allowed that would result in real numbers.

Multiplying Like Bases

To multiply like bases, add the exponents.

$$x^{1/7} \cdot x^{4/7} = x^{5/7}$$
$$x^{2/3} \cdot x^{1/4} = x^{8/12} \cdot x^{3/12} = x^{11/12}$$

To multiply $\sqrt[5]{x^3}$ and $\sqrt[7]{x^4}$, the following process is used.

$$\sqrt[5]{x^3} \cdot \sqrt[7]{x^4}$$

First write each factor using a rational number exponent.

$$x^{3/5} \cdot x^{4/7}$$

Write exponents with a common denominator.

$$x^{21/35 \cdot 20/35}$$

Since the bases are like, the exponents are added.

$$x^{41/35}$$

Since the exponent is greater than 1, change it to a mixed number.

$$x^{1 + 6/35}$$

$x^1 = x$ and $x^{6/35} = \sqrt[35]{x^6}$

$$x^3\sqrt[35]{x^6}$$

Focus on multiplying radical expressions

With all exponents the multiplication of **like** bases is simplified by adding exponents.

For examples

$$\sqrt[7]{3^2} \cdot \sqrt[7]{3^4} = 3^{2/7} \cdot 3^{4/7} = 3^{6/7} = \sqrt[7]{3^6}$$

$$\sqrt[4]{7^5} \cdot \sqrt[8]{7^3} = 7^{5/4} \cdot 7^{3/8} = 7^{13/8} = 7^{1+5/8} = 7\sqrt[8]{7^5}$$

$$\sqrt[9]{4} \cdot \sqrt[6]{32} = \sqrt[9]{2^2} \cdot \sqrt[6]{2^5} = 2^{2/9} \cdot 2^{5/6} = 2^{4/18} \cdot 2^{15/18} = 2^{19/18} = 2\sqrt[18]{2}$$

When the bases are **not like**, the rational number exponents must not be added. If the indices (plural of index) of two radical expressions are the same, the radicands are simply multiplied. For example,

$$\sqrt[5]{4} \cdot \sqrt[5]{7} = \sqrt[5]{28}$$

When the indices are different and there is no common base, the indices must be made identical and the radicands must be multiplied. For example,

$$\sqrt[4]{2} \cdot \sqrt[6]{5} = 2^{1/4} \cdot 5^{1/6} = 2^{3/12} \cdot 5^{2/12} = \sqrt[12]{2^3} \cdot \sqrt[12]{5^2} = \sqrt[12]{8 \cdot 25} = \sqrt[12]{200}$$

Dividing Like Bases

To divide like bases, subtract the exponents.

$$\frac{x^{7/9}}{x^{5/9}} = x^{2/9}$$

$$\frac{x^{1/2}}{x^{3/4}} = \frac{x^{2/4}}{x^{3/4}} = x^{-1/4} = \frac{1}{x^{1/4}}$$

To simplify the division $\frac{\sqrt[4]{x^3}}{\sqrt[10]{x^7}}$, each of the radical expressions is written with a rational number exponent.

First write each radical as a rational number power
$$\frac{x^{3/4}}{x^{7/10}}$$

Since the bases are like, the exponents are subtracted.
$$x^{3/4 - 7/10}$$

$$x^{15/20 - 14/20}$$

$$x^{1/20}$$

The quotient is:
$$\sqrt[20]{x}$$

Focus on dividing like bases

With all exponents the division of **like** bases is simplified by subtracting exponents.

For examples

$$\sqrt[7]{3^2} \div \sqrt[7]{3^4} = 3^{2/7 - 4/7} = 3^{-2/7} = \left(\tfrac{1}{3}\right)^{2/7} = \frac{1}{\sqrt[7]{3^2}}$$

$$\sqrt[4]{7^5} \div \sqrt[8]{7^3} = 7^{5/4 - 3/8} = 7^{7/8} = \sqrt[8]{7^7}$$

$$\sqrt[9]{4} \div \sqrt[6]{32} = \sqrt[9]{2^2} \div \sqrt[6]{2^5} = 2^{2/9 - 5/6} = 2^{4/18 - 15/18} = 2^{-11/18} = \frac{1}{\sqrt[18]{2^{11}}}$$

When the bases are **not like**, the rational number exponents must not be subtracted. If the indices are the same, the radicands are simply divided. For example,

$$\sqrt[5]{21} \div \sqrt[5]{7} = \sqrt[5]{3}$$

Raising a Power to a Power

To raise a power to a power, multiply the exponents.

$$(x^{4/7})^5 = x^{4/7 \cdot 5} = x^{20/7}$$
$$(x^{2/3})^{1/4} = x^{2/3 \cdot 1/4} = x^{1/6}$$

To simplify the power expression $(\sqrt[6]{7^5})^8$, remember that exponents are multiplied when raising a power to a power.

Begin by writing the radical as a rational number power.

$$(\sqrt[6]{7^5})^8 = (7^{5/6})^8$$

Now multiply the exponents.
$\tfrac{5}{6} \cdot 8 = \tfrac{20}{3}$

$$7^{5/6 \cdot 8} = 7^{20/3}$$

Change the fraction to a mixed number. $\tfrac{20}{3} = 6\tfrac{2}{3}$

$$7^{20/3} = 7^{6 + 2/3}$$

Change the rational number exponent to a radical.

$$7^{6 + 2/3} = 7^6 \sqrt[3]{7^2}$$

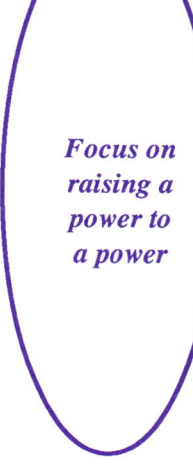

Focus on raising a power to a power

With all exponents the raising of a power to a power requires multiplying the exponents.

$$(x^{5/8})^{-3} = x^{5/8 \cdot -3} = x^{-15/8} = \frac{1}{x^{15/8}}$$

The cube (third root) of $\sqrt[7]{x^4}$ is found by raising the radical expression to the $\frac{1}{3}$ power.

$$\sqrt[3]{\sqrt[7]{x^4}} = (\sqrt[7]{x^4})^{1/3} = x^{4/7 \cdot 1/3} = x^{4/21} = \sqrt[21]{x^4}$$

Similarly, the square root of $\sqrt[5]{x^6}$ is found by raising the radical expression to the $\frac{1}{2}$ power.

$$\sqrt{\sqrt[5]{x^6}} = (\sqrt[5]{x^6})^{1/2} = x^{6/5 \cdot 1/2} = x^{3/5} = \sqrt[5]{x^3}$$

Unit 6 Exercise

Part A: Answers for all Part A problems are at the back of the book.

1. With counting number exponents, the multiplication of x^5 and x^7 is accomplished by finding the total number of x's.
$x^5 \cdot x^7 = (xxxxx)(xxxxxxx) = x^{12}$. There is no reason to memorize a rule for this process because the meaning of the exponents leads directly to the correct result. Use the meaning of the counting number exponents to simplify $x^{-7} \cdot x^5$.

2. If you remember the process for simplifying $x^{-7} \cdot x^5$, then the multiplication of $x^{1/2} \cdot x^{1/4}$ will require finding the sum of the exponents. $x^{1/2} \cdot x^{1/4} = x^{2/4} \cdot x^{1/4} = $ _____.

3. To simplify $x^{4/9} \cdot x^{10/9}$, recall what you would do with counting number exponents and then apply the same process. $x^{4/9} \cdot x^{10/9} = $ _____

4. With counting number exponents, the division of x^7 by x^4 is accomplished by finding the number of x's that are not eliminated by the cancelling process.

$$\frac{x^7}{x^4} = \frac{xxxxxxx}{xxxx} = \frac{\cancel{xxxx}xxx}{\cancel{xxxx}} = \frac{xxx}{1} = x^3$$

There is no reason to memorize a rule for this process because the meaning of the exponents leads directly to the correct result. Use the meaning of the counting number exponents to simplify the division of x^{-5} by x^3.

5. If you remember the process for simplifying the division of x^{-5} by x^3, then the division of $x^{3/10}$ by $x^{1/10}$ will require finding the difference of the exponents.

$$\frac{x^{3/10}}{x^{1/10}} = x^{3/10 - 1/10} = $$ _____

6. To simplify the division of $x^{1/2}$ by $x^{1/3}$, recall what you would do with counting number exponents and then apply the same process.

    $\dfrac{x^{1/2}}{x^{1/3}} = $ _____

7. With counting number exponents, the simplification of $(x^5)^4$ is accomplished by finding the total number of x's.
    $(x^5)^4 = (x^5)(x^5)(x^5)(x^5) = x^{20}$. There is no reason to memorize a rule for this process because the meaning of the exponents leads directly to the correct result. Use the meaning of the counting number exponents to simplify $(3^{-6})^2$.

8. If you remember the process for simplifying $(3^{-6})^2$, then the simplification of $(7^{1/5})^2$ will require finding the product of the exponents.
    $(7^{1/5})^2 = 7^{1/5 \cdot 2} = $ _____ .

9. To simplify $(x^{2/3})^4$, recall what you would do with counting number exponents and then apply the same process. $(x^{2/3})^4 = $ _____ .

10. To simplify $\sqrt[7]{x^4} \cdot \sqrt[5]{x^2}$ replace the radical numerals with power expressions involving rational number exponents.
    Simplify $\sqrt[7]{x^4} \cdot \sqrt[5]{x^2}$.

11. To simplify $\dfrac{\sqrt[4]{x^3}}{\sqrt{x}}$ replace the radical numerals with power expressions involving rational number exponents. Simplify $\dfrac{\sqrt[4]{x^3}}{\sqrt{x}}$.

12. To simplify $(\sqrt[5]{3^2})^3$ replace the radical numeral with a power expression involving a rational number exponent. Simplify $(\sqrt[5]{3^2})^3$.

Part B: Answers for odd-numbered problems of Part B are at the back of the book.

Simplify.

1. $x^{2/3} \cdot x^{1/4}$

2. $x^{4/7} \cdot x^{3/2}$

3. $\sqrt{x} \cdot \sqrt{x}$

4. $\sqrt{x^5} \cdot \sqrt[4]{x^3}$

5. $\sqrt[3]{x^2} \cdot \sqrt{x^3}$

6. $\sqrt[3]{x} \cdot \sqrt[4]{x}$

7. $(x^{5/4})^4$

8. $(x^{7/4})^2$

9. $(3^{2/3})^3$

10. $(y^{5/2})^3$

11. $(11^{1/2})^2$

12. $(x^{2/5})^{10}$

13. $\dfrac{x^2}{x^{1/2}}$

14. $\dfrac{x^{2/3}}{x^{1/4}}$

15. $\dfrac{x^{5/4}}{x}$

16. $\dfrac{x^{7/12}}{x^{3/4}}$

17. $\dfrac{\sqrt{x^7}}{\sqrt[6]{x^2}}$

18. $\dfrac{\sqrt[6]{x^7}}{\sqrt[8]{x^3}}$

19. $\dfrac{\sqrt[4]{x^{11}}}{\sqrt[4]{x^3}}$

20. $\left(\sqrt[4]{x^5}\right)^8$

21. $\left(\sqrt[8]{5^2}\right)^{12}$

22. $\left(\sqrt[3]{2^7}\right)^2$

23. $\left(\sqrt[4]{y^9}\right)^4$

24. $\left(\sqrt[6]{z^3}\right)^2$

25. $\left(\sqrt{x^5}\right)^2$

26. $\left(\sqrt[4]{3^2}\right)^6$

27. $\sqrt[5]{x^3} \cdot \sqrt[7]{x^4}$

28. $x^{5/8} \cdot x^{7/2}$

29. $\sqrt[3]{x} \cdot \sqrt{x}$

30. $\dfrac{\sqrt{x}}{\sqrt[4]{x}}$

31. $\dfrac{\sqrt[4]{x^3}}{\sqrt[5]{x^2}}$

32. $\left(\sqrt[3]{a^4}\right)^5$

Part C: No answers are given for these problems. However, each is accompanied by an ordered pair «C,U» showing the chapter and unit in which it was taught.

1. Graph the truth set for $2x(x + 5) > 0$ «1,4»

2. Solve $3x + 3\sqrt{2} = -4x + \sqrt{2}$ «2,5»

3. Simplify $\dfrac{3 + 5i}{1 - i}$ «3,3»

4. Use interval notation to describe the truth set for $|3x - 1| \geq 6$.   «4,4»

5. What is the range of $y = 3 - \sqrt{x - 2}$?   «5,3»

6. Find the slope of the line through the points $(-3,4)$ and $(-2,7)$.   «6,3»

7. Graph $y = x^2 + x - 6$ «7,2»

8. Evaluate $(-5)^{-4}$ «8,1»

9. Simplify $(x^7 y)^3$ «8,2»

10. Write $\sqrt[4]{x^{10} y^3}$ using positive rational exponents.   «8,5»

Unit 7: APPLICATIONS: RATE, TIME, AND PRODUCTION

Rates of Production

If a man can paint a house in 3 days, in one day he can paint $\frac{1}{3}$ of the house. In two days he can paint $\frac{2}{3}$ of the house. In three days he can paint $\frac{3}{3}$ of the house.

Similarly, if a laser beam can complete a cut through a steel plate in 5 minutes, the fraction of the cut that it can make in 1 minute is $\frac{1}{5}$; in 3 minutes it is $\frac{3}{5}$; and in 5 minutes it is $\frac{5}{5}$ or 1.

If a computer can run a program in 30 minutes, the fraction of the program it can run in 1 minute is $\frac{1}{30}$; in 15 minutes it is $\frac{15}{30}$; and in 30 minutes it is $\frac{30}{30}$ or 1.

Focus on finding a rate of production

The three previous examples show that when one (1) is placed over the length of time required to complete a job, the resulting rational number shows the fraction of the job completed in one unit of time. This fraction is called the **rate of production**.

Using variables, if x represents the hours required for the completion of a job by a person or machine, then $\frac{1}{x}$ represents the rate per hour or the fraction of the job completed in 1 hour; $\frac{2}{x}$ represents the fraction of the job completed in 2 hours, and $\frac{x}{x}$ or 1 is the fraction of the job completed in x hours.

People Working Together

If two people work on a project starting and completing it together, they will have each completed a fraction of the project, and the sum of the two fractions will add up to 1 (one) because one (1) job will have been completed.

For example, suppose that a master electrician can wire an apartment in 4 hours while his apprentice takes 12 hours to wire the same job. If the two men work together, the job can be completed in 3 hours because the master electrician will do $\frac{3}{4}$ and the apprentice $\frac{3}{12}$ of the job. The sum of the fractions is 1. $\frac{3}{4} + \frac{3}{12} = 1$.

Exponents 367

Focus on a relation between work, rate of production, and time

The solution of the problems of this unit depends on two facts.

1. The amount of work (w) completed on a project is the rate of production (r) multiplied by the time (t).
$$w = rt$$

2. If two people or machines work on a project until it is complete, each completing a fraction of the job, the sum of the fractions is one(1).

Solving A Work Problem

Problem: Bill can paint a house in 10 hours. When Sam works with Bill, they can paint the same size house in 6 hours. How long would it take Sam to paint the house by himself?

Solution

1. The problem should be carefully read and analyzed. The problem will be solved by writing open expressions for the fractions of the house painted by Bill and Sam in the 6 hours. The sum of the two fractions must be 1. This fact will provide the equation for the problem.

2. The quantities in the problem are listed.
    Bill's time to paint the house
    Sam's time to paint the house
    the fraction of the house Bill paints in 6 hours
    the fraction of the house Sam paints in 6 hours

3. Numbers or open expressions are assigned to the quantities listed in step 2.
    10 hours is Bill's time to paint the house
    t is Sam's time to paint the house
    $\frac{6}{10}$ is the fraction of the house Bill paints in 6 hours
    $\frac{6}{t}$ is the fraction of the house Sam paints in 6 hours

4. An equation is written expressing the number relations in the problem.

    $\left(\begin{array}{c} \text{The sum of the fractions of the} \\ \text{house painted by each man} \end{array} \right)$ is 1.

    $\frac{6}{10} + \frac{6}{t}$ = 1

5. The equation written in step 4 is solved and should be checked against the conditions of the problem. The solution of the equation is 15. It should be checked that Sam can paint the house in 15 hours.

Focus on another work problem

The following example is similar to the previous example except that the time for one of the machines to complete the job is not given. Instead the relation between the times of the two machines is given.

Problem: Two computers, working together, can process a set of cards in 8 minutes. How long does it take each computer to process the cards by itself if one machine runs twice as fast as the other?

Solution

1. If one machine can process the cards twice as fast as the other, the slower machine takes twice as much time as the faster machine to process the cards. Multiplying the fraction of the job that each machine can process in 1 minute by 8 will give an expression that will show how much of the job each machine can process in 8 minutes. The sum of the fractional expressions will be one (1) as one job is completed.

2. the time for the slower machine
   the time for the faster machine
   the fraction of the cards processed by the slow machine
   and the fraction of the cards processed by the fast machine

3. 2x is the slower machine's time

   x is the faster machine's time

   $\frac{8}{2x}$ is the fraction of the cards processed by the slower machine

   $\frac{8}{x}$ is the fraction of the cards processed by the faster machine

4. $\left(\begin{array}{c}\text{The sum of the fractions of the}\\\text{cards processed by each machine}\end{array}\right)$ is 1.

   $\frac{8}{2x} + \frac{8}{x} \qquad\qquad = \qquad 1$

5. The solution of $\frac{8}{2x} + \frac{8}{x} = 1$ is 12. The evaluation of 2x is 24. It should be checked that 12 minutes and 24 minutes will meet the conditions of the problem.

The Time Required When Several People or Machines Work on a Project

The previous problems were concerned with the question "how much time would a person or machine need to do a job by itself?" The problems illustrated below will ask "how long will it take to do a job when several people or machines work together on a project?" The solution of the problems will depend on the fact that work produced (w) equals rate of production (r) multiplied by the time (t).

$$w = rt$$

For example, if a machine can produce $\frac{1}{7}$ of a job in 1 hour, then

$2 \cdot \frac{1}{7} = \frac{2}{7}$ of the job is completed in 2 hours

$3 \cdot \frac{1}{7} = \frac{3}{7}$ of the job is completed in 3 hours

$7 \cdot \frac{1}{7} = \frac{7}{7}$ or 1 job is completed in 7 hours

and if x represents the hours, $x \cdot \frac{1}{7} = \frac{x}{7}$ of the job is completed in x hours.

Problem: Mr. Oat has a tractor that can plow a field in 7 hours and another that can plow the field in 6 hours. How long would it take to plow the field if both were used?

Solution

1. Since both tractors are plowing the field until it is completed, the operating times for the tractors will be equal. Each will have plowed a fraction of the field and the sum of the fractions will be 1.

2. the time that each tractor plows the field
   the fraction of the field plowed by the first tractor
   the fraction of the field plowed by the second tractor

3. x is the time that each tractor plows the field
   $\frac{x}{7}$ of the field is plowed by the first tractor in x hours
   $\frac{x}{6}$ of the field is plowed by the second tractor in x hours

4. The sum of the fractions   is   one (1).
   $\frac{x}{7} + \frac{x}{6}$ =   1

5. $3\frac{3}{13}$ is the solution of the equation above. It should be checked that $3\frac{3}{13}$ hours is the time that meets the conditions of the problem.

Focus on a combined work problem

Problem: Mr. Harty can complete an air conditioning job in 3 hours while his apprentice, Mr. Ampier, needs 6 hours to complete the same job. Once, when they were working on a similar job, Mr. Harty was called away before the job was complete and Mr. Ampier had to work 3 more hours by himself to complete the job. How long did they work together?

Solution

1. Mr. Harty completed a fraction of the job before he left. Mr. Ampier worked 3 hours more than Mr. Harty and completed a fraction of the job. The problem will be solved by using a variable to represent Mr. Ampier's time on the job. An equation will be written showing that the sum of the fractions of the job that each man completes is one (1).

2. Mr. Harty's time
   Mr. Ampier's time
   the fraction of the job completed by Mr. Harty
   the fraction of the job completed by Mr. Ampier

3. x is Mr. Harty's time

   $x + 3$ is Mr. Ampier's time

   $\frac{x}{3}$ is the fraction of the job completed by Mr. Harty

   $\frac{x + 3}{6}$ is the fraction of the job completed by Mr. Ampier

4. The sum of the fractions is one(1).
   $$\frac{x}{3} + \frac{x + 3}{6} = 1$$

5. The solution of the equation in step 4 is 1. Mr. Harty and Mr. Ampier spent 1 hour working together on the job. Mr. Ampier spent a total of 4 hours on the job.

Unit 7 Exercise

Part A: Answers for all Part A problems are at the back of the book.

1. If Dr. Karski can do a heart transplant in 5 hours, write the rational number that shows the fraction of the operation that he can do in 1 hour.

2. Write the rational number that represents the fraction of a roofing job that Mr. Tard can complete in 5 hours if he can complete the whole job in 17 hours.

3. Let y represent the number of hours needed by a computer to run a data processing program. Write open expressions to show the fraction of the program completed in 1 hour; in 2 hours; and in y hours.

4. If x represents the minutes that it takes for a red blood corpuscle to complete its tour of the circulatory system, write open expressions for the fraction of the tour that it completes in 1 minute; in 2 minutes; and in x minutes.

5. Working together, two technicians can complete a job in 2 hours. One of the men can do the job in 5 hours. How much time would be needed for the other technician to do the job by himself?

6. Karem can paint a bus in 144 minutes by himself, but when he works with Roosevelt on a similar bus, they can finish the job in 63 minutes. How long would Roosevelt need to paint the bus by himself?

7. When a furnace is used with a larger furnace that operates 3 times faster than the first furnace, they can process a batch of iron ore in 3 days. How many days would each furnace need to process a batch of ore by itself?

8. Future Electronics Corporation has two welding machines; one is twice as fast as the other. How long would each machine take to do a job if they take 4 hours to do the job when they are working together?

9. Two tilesetters are working in a kitchen. One could tile the kitchen in 10 hours while the other could do the job in 15 hours. If they both worked on the kitchen, how long would they spend on the job?

10. How long would it take two pumps working together to empty a tank if one of them could pump it out in 24 hours and the other larger pump could empty it in 8 hours?

11. Two computers were processing a set of data cards. Machine A could process all the cards in 9 hours. Machine B could process all the cards in 12 hours. Machine B broke down while running the cards and Machine A ran 1 more hour to finish. How long did the machines work together on the job?

12. Dirk, an apprentice electrician, could complete a job in 8 hours that his boss, Mr. Woltz, could do in 4 hours. Mr. Woltz started a similar job by himself. After a time, Dirk arrived, and they worked 2 more hours to finish the job. How long did Mr. Woltz work on the job by himself?

Part B: Answers for odd-numbered problems of Part B are at the back of the book.

1. Mary and Juanita are typing a book. Mary can type a chapter in 4 hours. If they both work on the chapter, they can type it in 3 hours. How long does Juanita need to type a chapter?

2. A roofer who can complete a roof in 6 hours can finish a similar roof in 4 hours when he works with a partner. How long would it take the partner to finish the roof by himself?

3. Ingrid can finish a production quota at a plant in 10 hours while Pearl needs 14 hours for the same job. If they work together, how long would they need to finish the work quota?

4. Three pumps are used to fill an oil tank. One can fill it in 15 hours, the second can fill it in 6 hours, and the last one needs 10 hours to fill the tank. When they are all used, how much time will they need to fill the tank?

5. For a power company, line crew A can complete a job in 5 days that would take line crew B 10 days to complete. If they worked together on the job and line crew B spent 3 times as long on the job as line crew A, how long would line crew A work?

6. The McGregor Rock Company has two rock crushers. One operates 3 times as fast as the other. Working together they can fill a truck in 3 minutes. How long would it take for the faster machine to fill the truck by itself?

7. An old-model drilling machine can complete a series of holes in 6 hours, but when used with a new model, the two machines can complete the same series in 1 hour. How long would it take the new machine to complete the holes by itself?

8. Three irrigation pumps are used together to irrigate a field. One of them can irrigate it in 36 hours, the second in 18 hours, and the third in 12 hours. How much time do they need to irrigate the field when all three pumps are used?

9. Bill and Sam are operating paving machines on a section of highway. Working together they paved the section in 6 hours. How long would it take Bill's machine to pave the section by itself if Sam's machine would take 5 hours more that Bill's machine to do the same job?

10. If Marcus can program a computer in y hours, what open expression represents the fraction of the program that he can do in 1 hour?

11. Write an open expression that shows the fraction of a job that Willie can complete in x hours if he can do $\frac{1}{5}$ of it in 1 hour.

12. In a factory, Inez can wire a receiver in 8 hours while Esmeralda can wire one in 6 hours. How long would it take them to wire a similar receiver if they worked on it together?

13. When a pump that can drain a swimming pool in 24 hours is used with a smaller pump, the two pumps working together can drain the pool in 15 hours. How long would it take the smaller pump to drain the pool by itself?

14. Cecil and Charles are painting a fence. Cecil could paint it in 10 hours, and Charles could do it in 5 hours by himself. After a few hours Charles had to quit and Cecil worked another hour to finish the job. How long did Charles work on the fence?

Part C: No answers are given for these problems. However, each is accompanied by an ordered pair «C,U» showing the chapter and unit in which it was taught.

1. Find the truth set for $x^2 + 4x = 12$. «2,6»

2. Use interval notation to write the truth set for $x^2 - 3x - 10 \leq 0$. «2,6»

3. Multiply and simplify $(7 + 3i)(3 - i)$. «3,3»

4. Find the truth set for $|5x + 2| = 3$. «4,3»

5. Write the linear equation $y = \frac{-2}{3}x + 2$ in standard form. «6,4»

6. Write the linear equation $5x - 2y = 10$ in y-intercept form. «6,4»

7. Graph $3x - 2y \geq -6$ and $x + y = 0$ «6,5»

8. Find the length of one leg of a right triangle if 12 inches is the length of the other leg and the hypotenuse is 13. «7,3»

9. Find the center and radius for the circle $x^2 + y^2 - 6x = 0$. «7,4»

10. Graph $y + 2 = (x - 5)^2$ «7,6»

Chapter 8 Test

«8,U» shows the unit in which this problem was studied in this chapter.

1. What is the base of the exponent 4 in $(-7x^4)^5$? «8,1»

2. Evaluate -3^4 «8,1»

3. Evaluate $(-3)^4$ «8,1»

4. True or false? $2^3 \cdot 4^2 = 8^5$ «8,2»

5. Simplify $5a^2b^4c^3 \cdot -3a^2c \cdot 2a^5$ «8,2»

6. Simplify $(x^5y^2)^3$ «8,2»

7. Simplify $\dfrac{-8x^3yz^5}{6x^4yz^3}$ «8,2»

8. Simplify $\left(\dfrac{-12x^{-2}y^5z}{18x^{-4}y^6z^3}\right)^{-3}$ «8,3»

9. Evaluate $\left(5 + \dfrac{2}{3}\right)^{-1}$ «8,3»

10. Evaluate $\left(3 - \dfrac{1}{4}\right)^0$ «8,3»

11. Is $(-5)^{2/6}$ equal to $(-5)^{1/3}$? «8,3»

Simplify and write the result with only positive exponents for problems 12-14.

12. $(x^6)^{-2}$ «8,4»

13. $\dfrac{18x^{-6}}{-4x^{-3}}$ «8,4»

14. $5x^{-3}y^4z^{-6} \cdot -3x^2y^{-5}z^6$ «8,4»

15. Write $\sqrt[5]{7^3}$ using rational exponents. «8,5»

16. Write $11^{4/3}$ using a radical sign. «8,5»

17. Simplify $\sqrt[4]{x^3} \cdot \sqrt[4]{x^2}$ «8,6»

18. Simplify $\dfrac{\sqrt[5]{x^3}}{\sqrt{x}}$ «8,6»

19. Simplify $(\sqrt[4]{2^3})^8$ «8,6»

20. Typist A would take 15 hours to type a report, and typist B could type the same report in 12 hours. If they worked together, how long would it take to complete the report? «8,7»

9
Exponential and Logarithm Functions

Unit 1: EXPONENTIAL FUNCTIONS

Some Types of Functions

Since Chapter 5 of this book, many of the topics studied have dealt specifically with functions — sets of ordered pairs in which each first component is used exactly one time as a first component.

Linear functions are those such as:
$$f(x) = -5x + 8$$

Polynomial functions are those such as:
$$f(x) = 5x^3 - 3x^2 + 4x - 7$$

Rational functions are those such as:
$$f(x) = \frac{4x - 7}{3x + 9}$$

In this unit, the variables appear in the position of exponents and the functions are exponential functions.

Variables in the Position of Exponents

In the expression 5^x, the variable is in the position of an exponent. To evaluate such an expression, the variable is replaced by a number.

If $x = 3$, then 5^x becomes

$$5^3 = 5 \cdot 5 \cdot 5 = 125$$

Focus on the evaluation of exponent (power) expressions

The expression 3^4 has an evaluation of $3 \cdot 3 \cdot 3 \cdot 3$ or 81.

The evaluation of $(\frac{1}{2})^3$ is $\frac{1}{2} \cdot \frac{1}{2} \cdot \frac{1}{2} = \frac{1}{8}$

The evaluation of 43^0 is 1 because any non-zero base with 0 as its exponent is 1.

The evaluation of $(\frac{2}{3})^{-2}$ is $\frac{3}{2} \cdot \frac{3}{2}$ or $\frac{9}{4}$

The Exponential Function f(x) = 3ˣ

$f(x) = 3^x$ is an exponential function with base 3. Replacements for x are used as exponents to create ordered pairs $(x, 3^x)$. For example, if x is replaced by 2 then

$f(x) = 3^x$ becomes $f(2) = 3^2$ and $f(2) = 9$.

Therefore, the ordered pair (2,9) is an element of the function.

As another example, if x is replaced by $\frac{1}{2}$ then

$f(x) = 3^x$ becomes $f(\frac{1}{2}) = 3^{1/2}$ and $f(\frac{1}{2}) = \sqrt{3}$.

Therefore, the ordered pair $(\frac{1}{2}, \sqrt{3})$ is an element of the function.

Exponential and Logarithm Functions

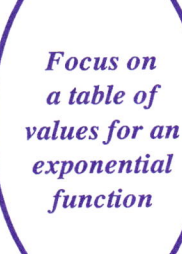

Focus on a table of values for an exponential function

Some of the ordered pairs of $f(x) = 3^x$ are shown in the table at the right.

If $x = 3$, $f(3) = 3^3 = 27$

If $x = -2$, $f(3) = 3^{-2} = \frac{1}{9}$

Notice that the values of $f(x)$ decrease as the values of x decrease.

x	f(x)
3	27
2	9
1	3
0	1
-1	$\frac{1}{3}$
-2	$\frac{1}{9}$
-3	$\frac{1}{27}$

Plotting the Points of the Table

The table of ordered pairs $(x, 3^x)$ found in the focus above was used to plot points on the x,y plane shown at the right.

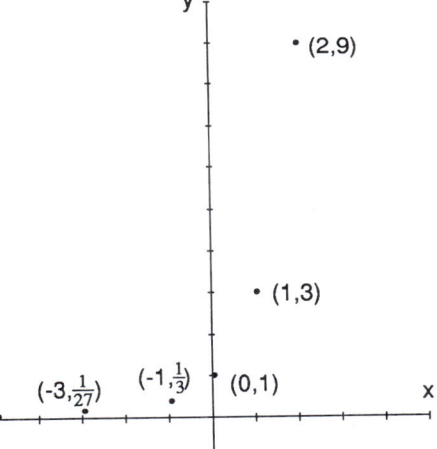

Two interesting conjectures should be made about the graph.
1. What would a smooth curve joining those points look like?
2. If non-integer replacements for x were used would the points lie along such a smooth curve?

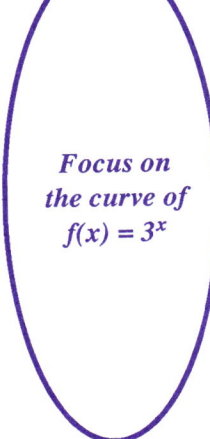

Focus on the curve of $f(x) = 3^x$

The curve at the right is the graph of $f(x) = 3^x$. Notice the following:
1. The curve is asymptotic to the negative x-axis.
2. The curve contains the three points $(1,3)$, $(0,1)$, and $\left(-1, \frac{1}{3}\right)$.
3. The curve contains points such as $\left(\frac{1}{2}, \sqrt{3}\right)$ and $\left(\frac{-2}{3}, \sqrt[3]{\frac{1}{9}}\right)$.
4. The curve suggests that $3^{.53107}$ is approximately equal to $3^{1/2}$ because $.53107$ is approximately $\frac{1}{2}$.

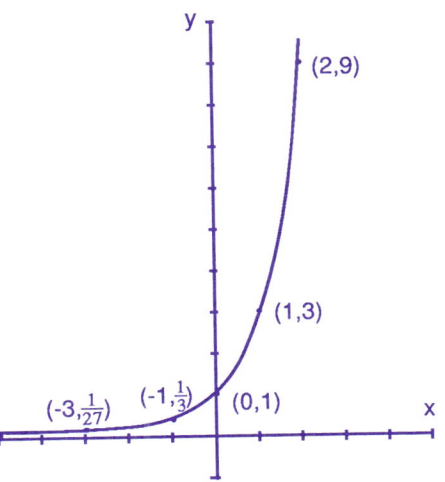

378   Chapter 9

Exponential Functions with Base Greater Than One

Functions of the form $f(x) = a^x$, $a > 1$ are exponential functions with base greater than 1. Each of these functions will have a graph similar in shape and position to the one shown on the preceding page for $f(x) = 3^x$.

Each graph of an exponential function with base greater than one will have:
1. A curve asymptotic to the negative y-axis.
2. A curve that contains the three points $(1,a)$, $(0,1)$, and $\left(-1,\frac{1}{a}\right)$.
3. A curve that suggests that $a^{1.73416}$ is approximately equal to $3^{7/4}$ because 1.73416 is approximately $\frac{7}{4}$.

Focus on the graph of functions $f(x) = a^x$ $a > 1$

The graph at the right shows the curve of $f(x) = 5^x$. Notice that since the base, 5, is greater than 1, the curve:

1. Is asymptotic to the negative y-axis.
2. Contains $(1,5)$, $(0,1)$, and $\left(-1,\frac{1}{5}\right)$.
3. The curve suggests approximations. $5^{.65143}$ is approximately equal to $5^{2/3}$ because .65143 is approximately equal to $\frac{2}{3}$.

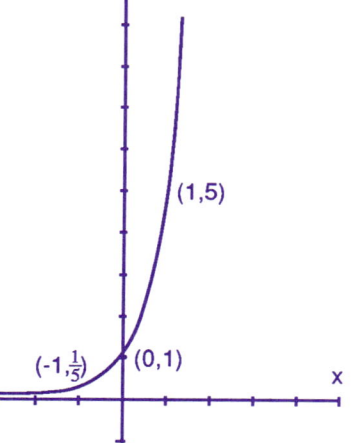

Exponential Functions with Base Less Than One

Earlier in this unit, relations of the form $f(x) = a^x$, $a > 1$ were stated to be exponential functions. Similarly, but with a different type of curve, relations of the form $f(x) = a^x$, $0 < a < 1$, are also exponential functions.

Some of the ordered pairs of $f(x) = \left(\frac{1}{3}\right)^x$ are shown in the table at the right.

If $x = 3$, $f(3) = \left(\frac{1}{3}\right)^3 = \frac{1}{27}$
If $x = -2$, $f(-2) = \left(\frac{1}{3}\right)^{-2} = 9$

Notice that the values of x decrease as the values of f(x) increase.

x	y
3	$\frac{1}{27}$
2	$\frac{1}{9}$
1	$\frac{1}{3}$
0	1
-1	3
-2	9

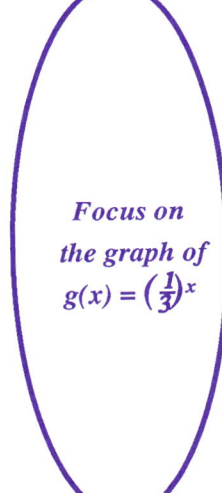

Focus on the graph of $g(x) = \left(\frac{1}{3}\right)^x$

Each graph of an exponential function with base greater than zero and less than one will have:

1. A curve asymptotic to the positive x-axis.
2. A curve that contains the three points (1,a), (0,1), and $\left(-1, \frac{1}{a}\right)$.
3. A curve that suggests that $a^{1.26489}$ is approximately equal to $a^{5/4}$ because 1.26489 is approximately $\frac{5}{4}$.

The curve at the right is the graph of $g(x) = \left(\frac{1}{3}\right)^x$.

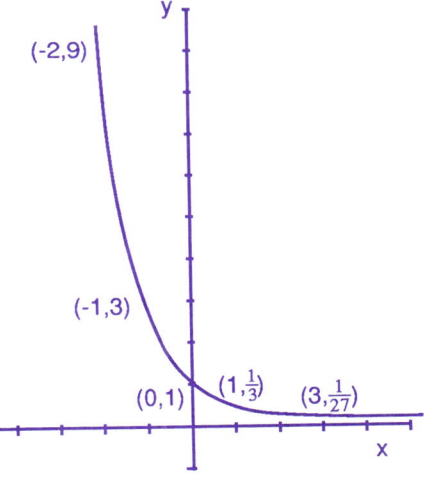

Relations with Bases Less Than Zero

Relations of the form $f(x) = a^x$, $a < 0$ are not defined for many rational number exponents. For example, $(-4)^{1/2}$ is not defined because a negative number has no real number square root. For this reason exponential functions are limited to positive bases.

Chapter 9

Unit 1 Exercise

Part A: Answers for all Part A problems are at the back of the book.

1. The evaluation of 3^2 is 3 • 3 or _____.
2. Evaluate 3^4
3. Evaluate 2^4
4. Evaluate $(\frac{1}{3})^2$
5. Any nonzero base with exponent zero gives 1. Evaluate 5^0
6. Evaluate $(\frac{3}{7})^0$
7. Any base with exponent 1 gives that base. Evaluate $(\frac{2}{3})^1$
8. Evaluate 8^1
9. Any nonzero base with exponent -1 gives the reciprocal of the base. Evaluate $(\frac{8}{5})^{-1}$
10. Evaluate 4^{-1}
11. Evaluate 2^x when x is replaced by 0.
12. Evaluate 2^x when x is replaced by $\frac{1}{2}$.
13. Evaluate 2^x when x = -1.
14. Evaluate 2^x when x = 5.
15. Evaluate 2^x when x = -5.
16. The set of points $(x, 2^x)$ can be designated as $f(x) = 2^x$. Is the relation f a function?
17. Complete the table at the right for $g(x) = 2^x$.

x	2^x
3	8
2	4
1	—
0	—
-1	—
-2	—
-3	—

18. Use the table of problem 17. Plot the seven points of relation g. Then draw a smooth curve from left to right through the seven points.

Part B: Answers for odd-numbered problems of Part B are at the back of the book.

1. Is the relation $g(x) = 3^x$ a function?
2. Is $k(x) = (\frac{3}{7})^x$ a function?
3. Is $h(x) = 5^x$ a function?
4. Is $y = 2^x$ an exponential function?
5. Is $y = x^2 - 5$ an exponential function?
6. Is $y = (\frac{7}{6})^x$ an exponential function?
7. Is $y = x^3$ an exponential function?
8. Graph $y = (\frac{9}{5})^x$

Exponential and Logarithm Functions 381

9. Graph $f(x) = \left(\frac{3}{8}\right)^x$

10. Graph $y = \left(\frac{1}{5}\right)^x$

11. Graph $f(x) = 1^x$

12. Graph the exponential function $y = 2^x + 3$ by translating each point of $y = 2^x$ three units upwards.

13. Graph the exponential function $y = 2^{x-3}$ by translating each point of $y = 2^x$ three units to the right.

14. Graph the exponential function $y = \left(\frac{7}{5}\right)^x - 2$.

15. Graph the exponential function $y = \left(\frac{3}{5}\right)^{x+2}$.

Part C: No answers are given for these problems. However, each is accompanied by an ordered pair «C,U» showing the chapter and unit in which it was taught.

1. Solve $x^6 + 64 = 0$ using the set of rational numbers. «1,5»

2. How much alcohol should be added to 80 gallons of 30% mixture to raise the percent of alcohol to 44%? «2,7»

3. Solve $5x^2 - 3x + 2 = 0$ «3,2»

4. Is $\frac{1}{2}$ in the neighborhood of $|x + 2| < \frac{3}{4}$? «4,4»

5. Is $\{(3,5),(-2,7),(4,6),(3,-5)\}$ a function? «5,3»

6. Graph $y = 3x - 1$ «6,5»

7. Graph $\frac{x^2}{16} + \frac{y^2}{9} = 1$ «7,7»

8. Evaluate $(4 + \frac{1}{3})^{-1}$ «8,3»

9. Is $(3 + \frac{1}{4})^0 = (4 + \frac{1}{3})^0$? «8,3»

10. Simplify $(\sqrt[5]{x^3})^{10}$ «8,6»

Unit 2: LOGARITHM FUNCTIONS

Solving Exponential Equations

The equation $5^x = 625$ has the variable in the position of an exponent. To find the truth set ask the question:
   What power of 5 is 625?

Since the answer to the question is 4, the root or solution of the equation is 4.
   $5^4 = 625$ is a true statement.

Focus on solving exponential equations

The exponential equation $2^x = \frac{1}{8}$ asks: What power of 2 is $\frac{1}{8}$?
The answer to the question is -3 and $\{-3\}$ is the truth set of the equation.
Similarly, the exponential equation $7^x = \sqrt{7}$ asks the question:
   What power of 7 is $\sqrt{7}$?
The answer to the question is $\frac{1}{2}$ and $\{\frac{1}{2}\}$ is the truth set of the equation.

Introducing Logarithm Notation

The equation $5^x = 625$ asks the same question as $\log_5 625 = x$. In both cases the question is: What power of 5 is 625?

Since $5^4 = 625$ is a true statement then $\log_5 625 = 4$ is also a true statement.

The notation "log" means "**logarithm**" and the small number to the lower right of "log" indicates its **base**. Notice that in both $5^x = 625$ and $\log_5 625 = x$ the base is 5.

Focus on the relationship between exponents and logarithms

$6^x = 36$ asks the same question as $\log_6 36 = x$. Both equations have 2 as the solution. $6^2 = 36$ and $\log_6 36 = 2$.

If $9^x = 3$ then its corresponding logarithm equation is $\log_9 3 = x$. Both equations have $\frac{1}{2}$ as their solutions. $9^{1/2} = 3$ and $\log_9 3 = \frac{1}{2}$.

Logarithm Functions

$g(x) = 7^x$ is an exponential function with base 7. It consists of an infinite number of ordered pairs $(x, 7^x)$. Three examples of ordered pairs in the function are $(2, 49)$, $(1, 7)$, and $(-1, \frac{1}{7})$.

Similarly, a logarithm function, $f(x) = \log_7 x$, can be defined. It will consist of an infinite number of ordered pairs $(x, \log_7 x)$. Three of the ordered pairs in this logarithm function are $(49, 2)$, $(7, 1)$, and $(\frac{1}{7}, -1)$.

Notice the relationship between ordered pairs of the logarithm function base 7 and ordered pairs of the exponential function base 7.

Remember the following relationship which is useful in actually computing with logarithm functions.
$$y = \log_a x \text{ if and only if } x = a^y$$

Focus on the relationship between $y = \log_a x$ and $x = a^y$

The value of $y = \log_5 125$ can be determined by solving the equation $125 = 5^y$. Since the solution is $y = 3$, then $\log_5 125 = 3$.

To determine the value of $y = \log_2 \frac{1}{32}$ solve the equation $\frac{1}{32} = 2^y$. Since $2^{-5} = \frac{1}{32}$ then $2^{-5} = 2^y$ and $y = -5$. Hence, $\log_2 \frac{1}{32} = -5$.

To determine the value of $y = \log_{10} 10{,}000$ solve the equation $10{,}000 = 10^y$.

Since $10^4 = 10{,}000$ then $\log_{10} 10{,}000 = 4$.

To determine the value of $\log_{1/3} 81$, use the fact that $y = \log_{1/3} 81$ is equivalent to $(\frac{1}{3})^y = 81$. Then the equation can be written with the same base on each side of the equality.

Hence, $\log_{1/3} 81 = -4$.

$y = \log_{1/3} 81$
$(\frac{1}{3})^y = 81$
$3^{-y} = 3^4$
$-y = 4$
$y = -4$

A Table of Values for y = log₁₀ x

The function $f(x) = \log_{10} x$ with base 10 is the **common** logarithm function and, before the hand held calculator became available, was often used to calculate answers for complex arithmetic problems.

At the right is shown a table of values for the common logarithm function, $f(x) = \log_{10} x$.

Each entry in the x column is used in the logarithm function to find its matching y ($\log_{10} x$) value.

Integer exponents with a base of 10 are relatively easy to use because the exponent indicates the number of decimal places in the numeral.

x	$\log_{10} x$
100,000	5
10,000	4
1,000	3
100	2
10	1
1	0
.1	-1
.01	-2
.001	-3
.0001	-4
.00001	-5

Focus on approximating values for the common logarithm $y = \log_{10} x$

The common logarithm of 946 ($\log_{10} 946$) is between 2 and 3 because $10^2 = 100 < 946$ and $10^3 = 1000 > 946$.

Since 946 is much closer to 1000 than it is to 100, a closer approximation for $\log_{10} 946$ would be: slightly less than 3.

Similarly, an approximation for the common logarithm of 0.00759 depends upon the following inequality:
$$10^{-3} = .001 < 0.00759 < .01 = 10^{-2}$$
The inequality above indicates:
$$-3 < \log_{10} 0.00759 < -2$$

Since 0.00759 is closer to 0.01 than it is to 0.001, a closer approximation for $\log_{10} 0.00759$ would be: slightly less than -2.

Exponential and Logarithm Functions

Unit 2 Exercise

Part A: Answers for all Part A problems are at the back of the book.

1. Find $\log_2 4$ by solving $4 = 2^x$.
2. Find $\log_2\left(\frac{1}{8}\right)$ by solving $\frac{1}{8} = 2^x$.
3. Find $\log_2 1$ by solving $1 = 2^x$.
4. Find $\log_3 3$ by solving $3 = 3^x$.
5. Find $\log_3\left(\frac{1}{3}\right)$ by solving $\frac{1}{3} = 3^x$.
6. The base of $f(x) = 3^x$ is 3 and the base of $\log_3 x$ is also _____.
7. The base of $\log_{1/2} x$ is _____.
8. The base of $\log_4 x$ is _____.
9. The $\log_3 x$ is related to 3^x. To find $\log_3 9$ solve $9 = 3^x$. $\log_3 9 =$ _____.
10. The $\log_{2/5} x$ is related to $\left(\frac{2}{5}\right)^x$. To find $\log_{2/5}\left(\frac{5}{2}\right)$ solve $\frac{5}{2} = \left(\frac{2}{5}\right)^x$. $\log_{2/5}\left(\frac{5}{2}\right) =$ _____.
11. Find $\log_2 8$ by solving its related equation.
12. Find $\log_2 32$ by solving its related equation.
13. Find $\log_2 2$ by solving its related equation.
14. $\sqrt{4} = 4^{1/2} = 2$. Find $\log_4 2$.
15. Complete the ordered pairs $(1,\_\_\_), (9,\_\_\_), \left(\frac{1}{9},\_\_\_\right)$ for the function $y = \log_3 x$.
16. Complete $(1,\_\_\_), (16,\_\_\_), \left(\frac{1}{8},\_\_\_\right)$ for $y = \log_2 x$.
17. Complete $(7,\_\_\_), (1,\_\_\_), \left(\frac{1}{7},\_\_\_\right)$ for $y = \log_7 x$.
18. Complete $(36,\_\_\_), (6,\_\_\_), (1,\_\_\_)$ for $y = \log_6 x$.
19. Complete $(36,\_\_\_), (6,\_\_\_), (1,\_\_\_)$ for $y = \log_{1/6} x$.
20. Complete $\left(\frac{1}{81},\_\_\_\right), \left(\frac{1}{9},\_\_\_\right), (1,\_\_\_)$ for $y = \log_{1/9} x$.

Part B: Answers for odd-numbered problems of Part B are at the back of the book.

1. $\log_4 64 =$ _____
2. $\log_3 243 =$ _____
3. $\log_4\left(\frac{1}{4}\right) =$ _____
4. $\log_{10} 10{,}000 =$ _____
5. $\log_3 27 =$ _____
6. $\log_2\left(\frac{1}{2}\right) =$ _____
7. $\log_2 64 =$ _____
8. $\log_3 81 =$ _____
9. $\log_3 1 =$ _____
10. $\log_3\left(\frac{1}{3}\right) =$ _____

11. $\log_4 16 =$ _____
12. $\log_4 4 =$ _____
13. $\log_5 25 =$ _____
14. $\log_{1/2} 8 =$ _____
15. $\log_{10} 100{,}000 =$ _____
16. $\log_{2/7} 1 =$ _____
17. $\log_6 6 =$ _____
18. $\log_9 \left(\frac{1}{9}\right) =$ _____
19. $\log_{1/5} 5 =$ _____
20. $\log_{3/5} \left(\frac{5}{3}\right) =$ _____
21. $\log_{1/5} 125 =$ _____

22. $\log_{\sqrt{3}} 9 =$ _____
23. $\log_{1/3} 81 =$ _____
24. $\log_{1/10} .0001 =$ _____
25. $\log_{10} 10^x =$ _____
26. $\log_{1/3} \left(\frac{1}{3}\right) =$ _____
27. For any real number $k > 0$, $k \neq 1$,
    $\log_k k =$ _____.
28. For any real number $k > 0$, $k \neq 1$,
    $\log_k 1 =$ _____.
29. Find the approximate value of $\log_{10} 2{,}345$.
30. Find the approximate value of $\log_{10} 0.0002123$.

Part C: No answers are given for these problems. However, each is accompanied by an ordered pair «C,U» showing the chapter and unit in which it was taught.

1. Simplify $-(5 + 3x) < 2x + 25$ «1,4»
2. Solve $\frac{x}{x+3} > x - 3$ «2,7»
3. Solve $(2 + i)(x + yi) + (3 - 5i) = (1 + i)$ «3,4»
4. Graph the truth set for $|x + 5| < 10$. «4,3»
5. Is $\{3,6,9,\ldots\}$ a finite set? «5,1»
6. Is $x^2 + y^2 = 81$ a function? «5,4»
7. Find the standard form of the linear equation with solutions $(5,-2)$ and $(-2,3)$. «6,4»
8. Graph $(x + 4)^2 + (y - 3)^2 = 1$ «7,4»
9. Simplify $\frac{-6x^5 y^{-2} z}{-9x^{-1} y^{-5} z^2}$ «8,3»
10. Sketch the graph of $f(x) = 2^x$. «9,1»

Exponential and Logarithm Functions   387

Unit 3: PLACE VALUE AND SCIENTIFIC NOTATION

Multiplying by Powers of 10

The decimal point in 178.3 is moved two places to the right when the number is multiplied by 100. $1.783 \cdot 100 = 178.3$

When multiplying 13.07 by 10, the decimal point will move one place to the right. $13.07 \cdot 10 = 130.7$

When multiplying a number by 10, 100, 1000, etc., the decimal should be moved to the right the same number of places as there are zeros in 10, 100, 1000, etc.
$$14.3 \cdot 1000 = 14{,}300.$$

Focus on multiplying by powers of 10

When multiplying a number by .1, .01, .001, etc., the decimal should be moved to the left the same number of places as there are digits to the right of the decimal point in .1, .01, .001, etc.
$$537. \cdot .1 = 53.7$$
$$537. \cdot .01 = 5.37$$
$$537. \cdot .001 = .537$$
$$537. \cdot .0001 = .0537$$

Sums of Powers of Ten

In the numeral 27 the 2 is in the tens place, because 27 is $2 \cdot 10 + 7$.

In the numeral 4631 the 4 is in the thousands place because 4631 is $4 \cdot 10^3 + 631$.

In the numeral 7.894 the 7 is in the ones place, because 7.894 is $7 \cdot 10^0 + .894$.

$$5284 = 5000 + 200 + 80 + 4$$
$$= 5 \cdot 1000 + 2 \cdot 100 + 8 \cdot 10 + 4 \cdot 1$$
$$= 5 \cdot 10^3 + 2 \cdot 10^2 + 8 \cdot 10^1 + 4 \cdot 10^0$$

$$396 = 300 + 90 + 6$$
$$= 3 \cdot 100 + 9 \cdot 10 + 6 \cdot 1$$
$$= 3 \cdot 10^2 + 9 \cdot 10^1 + 6 \cdot 10^0$$

Focus on writing numbers as sums of powers of 10

In the numeral 514.63 each digit represents a number that is to be multiplied by 10 with an appropriately chosen exponent. The digit 5 in 514.63 represents $5 \cdot 10^2$. The digit 6 represents $6 \cdot 10^{-1}$. The digit 3 represents $3 \cdot 10^{-2}$.

Similarly,
$$.278 = .2 + .07 + .008$$
$$= 2 \cdot \frac{1}{10} + 7 \cdot \frac{1}{100} + 8 \cdot \frac{1}{1000}$$
$$= 2 \cdot 10^{-1} + 7 \cdot 10^{-2} + 8 \cdot 10^{-3}$$

$$4136.2 = 4{,}000 + 100 + 30 + 6 + .2$$
$$= 4 \cdot 1000 + 1 \cdot 100 + 3 \cdot 10 + 6 \cdot 1 + 2 \cdot \frac{1}{10}$$
$$= 4 \cdot 10^3 + 1 \cdot 10^2 + 3 \cdot 10^1 + 6 \cdot 10^0 + 2 \cdot 10^{-1}$$

Writing Numbers in Scientific Notation

Any decimal numeral can be written as a number between 1 and 10 that is multiplied by 10 with a properly chosen exponent. The multiplication form of the numeral is its **scientific notation** form.

To write a numeral in **scientific notation**, it must be written as a number between 1 and 10 multiplied by 10 with an integer exponent. The scientific notation for 583 is $5.83 \cdot 10^2$. The scientific notation for 612 is $6.12 \cdot 10^2$. The scientific notation for 41.6 is $4.16 \cdot 10^1$. The scientific notation for .041 is $4.1 \cdot 10^{-2}$.

$$5{,}000 = 5 \cdot 10^3 \qquad 1 \leq 5 < 10$$
$$.004 = 4 \cdot 10^{-3} \qquad 1 \leq 4 < 10$$
$$30 = 3 \cdot 10^1 \qquad 1 \leq 3 < 10$$
$$5 = 5 \cdot 10^0 \qquad 1 \leq 5 < 10$$
$$7{,}000{,}000 = 7 \cdot 10^6 \qquad 1 \leq 7 < 10$$
$$40{,}000 = 4 \cdot 10^4 \qquad 1 \leq 4 < 10$$

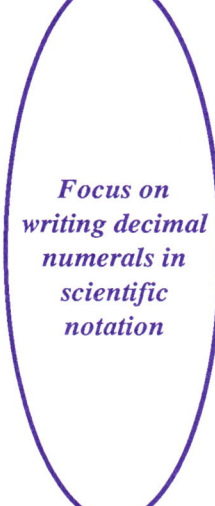

Focus on writing decimal numerals in scientific notation

Decimal numerals with more than one nonzero digit may also be written as a number between 1 and 10 multiplied by 10 with a properly chosen exponent. A number written as a number between 1 and 10 multiplied by a power of 10 is written in its scientific notation form.

Each of the following is an example of writing a number in its scientific notation form.

$$528 = 5.28 \cdot 100 = 5.28 \cdot 10^2$$
$$4{,}810{,}000 = 4.81 \cdot 10^6$$
$$5235 = 5.235 \cdot 10^3$$
$$.0541 = 5.41 \cdot 10^{-2}$$
$$5.63 = 5.63 \cdot 10^0$$
$$.0000487 = 4.87 \cdot 10^{-5}$$
$$.812 = 8.12 \cdot 10^{-1}$$

Unit 3 Exercise

Part A: Answers for all Part A problems are at the back of the book.

1. Numbers such as 1,000 and 100,000 can be written as counting number powers of 10. $1{,}000 = 10^3$ and $100{,}000 =$ _____.

2. 1,000,000 has six zeros. Write 1,000,000 with 10 as a base and the proper exponent.

3. Numbers such as 0.001 and 0.00001 can be written as negative integer powers of 10. $0.001 = 10^{-3}$ and $0.00001 =$ _____.

4. 0.000001 has six digits to the right of the decimal point and can be written as 10^{-6}. Write 0.0001 using 10 as the base and the proper exponent.

5. A product of a number and 1,000 can be written by moving the decimal point three places to the left because $1{,}000 = 10^3$. $4.8 \cdot 1{,}000 =$ _____

6. A product of a number and .000001 can be written by moving the decimal point six places to the right because $.000001 = 10^{-6}$. $1{,}284{,}636 \cdot .000001 =$ _____

7. In the numeral 41,376, 4 represents $4 \cdot 10^4$ and 1 represents $1 \cdot$ _____ .

8. In the numeral 6.1432, 6 represents $6 \cdot 10^0$ and 4 represents $4 \cdot$ _____ .

9. In the numeral 516,324.9, 6 represents $6 \cdot 10^3$ and 5 represents $5 \cdot$ _____ .

10. In the numeral .00417, 4 represents $4 \cdot 10^{-3}$ and 7 represents $7 \cdot$ _____ .

390   Chapter 9

11. 837 may be written as $8.37 \cdot 10^2$ because the 8 in 837 represents $8 \cdot 10^2$. Write 48,900 as a number between 1 and 10 multiplied by 10 with a properly chosen exponent.
48,900 = _____ .

12. .0235 may be written as $2.35 \cdot 10^{-2}$ because the 2 in .0235 represents $2 \cdot 10^{-2}$. Write .00617 as a number between 1 and 10 multiplied by 10 with a properly chosen exponent.

13. The scientific notation numeral for 87.3 is a number between 1 and 10 multiplied by a power of 10. The scientific notation numeral for 87.3 is _____ .

14. The scientific notation numeral for .0918 is a number between 1 and 10 multiplied by a power of 10. The scientific notation numeral for .0918 is _____ .

Part B: Answers for odd-numbered problems of Part B are at the back of the book.

Write scientific notation numerals for each of the following.

1. .4135 = _____
2. 716 = _____
3. 63,500 = _____
4. 61,000,000 = _____
5. .0008 = _____
6. 4.31 = _____
7. .00514 = _____
8. 6371.4 = _____
9. 800,000 = _____
10. .0528 = _____
11. .00362 = _____
12. 2.89 = _____
13. 63,500 = _____
14. .00041 = _____
15. .00008 = _____
16. .0047 = _____
17. 25,000,000 = _____
18. 408,000,000 = _____
19. 6,350 = _____
20. .405 = _____
21. .0013 = _____
22. 6.83 = _____
23. 550,000 = _____
24. .0006 = _____
25. 57 = _____
26. .49 = _____

27. 4,040 = _____

28. 61.5 = _____

29. 520 = _____

30. 61,900 = _____

31. .0417 = _____

32. .000512 = _____

33. 6.3 = _____

34. 4,700,000 = _____

35. .57 = _____

36. 63,800,000 = _____

Part C: No answers are given for these problems. However, each is accompanied by an ordered pair «C,U» showing the chapter and unit in which it was taught.

1. Using the set of rational numbers, solve $x^6 - 1$. «1,5»

2. Solve $\frac{1}{6}x - \frac{2}{3} = \frac{1}{4} + x$ «2,7»

3. The second side of a triangle is 3 times longer than the first side, and the third side is 5 inches longer than the second side. What is the length of the third side of the triangle if the perimeter is 54 inches? «3,5»

4. Use interval notation to describe the truth set of $|3x - 2| \geq 2$. «4,4»

5. Is $\frac{1}{2}$ in the neighborhood of $|x - 1| < \frac{3}{8}$? «4,4»

6. Graph $y = \frac{-2}{3}x + 2$ «6,1»

7. Graph the relation $(x - 3)^2 + (y + 2)^2 = 9$. «7,2»

8. How many turning points would you expect to find in the graph of the polynomial function, $y = x^3 - 3x^2 + x - 12$? «7,2»

9. Simplify $(-\sqrt[3]{2^4})^{-3}$ «8,6»

10. Evaluate $\log_3 81$ «9,2»

Unit 4: COMMON LOGARITHMS

The Logarithm Table

The **common logarithm** function is $f(x) = \log_{10} x$ and uses 10 as its base. The common logarithm of a number is the exponent of 10 such that 10 with that exponent would produce the given number.

The common logarithm of 100 is 2, because $10^2 = 100$. The common logarithm of .001 is -3, because $10^{-3} = .001$. Common logarithms for numbers such as 5 or 87 are not integers.

Tables have been made that show the common logarithms of real numbers. These tables are shown on pages 450 and 451 of this text.

This unit is an explanation of the use of these tables. Since all further work in this chapter is with common logarithms, the word "common" will be dropped and they will be referred to as simply "logarithms."

Focus on tables of common logarithms

Turn to the tables shown on pages 450 and 451. This is a table of logarithms. The capital letter N appears at the top of the first column on the left side of the tables.

The N column contains all the integers between 10 and 100. Find 40 in the N column. Directly to the right of 40 in the N column is the entry .60206. This is the exponent of 10 that would give 4. $10^{.60206} = 4$

Each two-digit number between 10 and 100 can be found in the N column. Find 65 in the N column. The numbers that appear to the right of 65 are called the **entries** in the 65 row.

There are 11 columns in the logarithm tables. The first column is headed by N, and the other ten are headed by digits 0, 1, 2, 3, 4, 5, 6, 7, 8, 9. Find the 5 column. The first entry below 5 in the 5 column is .02119. Find 30 in the N column. Find the column headed by 5. The entry of the table in the 30 row and 5 column is .48430.

Finding Entries in the Logarithm Table

The entries in the tables are logarithms. To find the logarithm of 4.36, use the 43 row and 6 column. The first two digits of 4.36 are 43 and the third digit of 4.36 is 6. The entry in the 43 row and 6 column is .63949.

$$\log_{10} 4.36 = .63949 \text{ and } 4.36 = 10^{.63949}$$

To find the logarithm of 5.04, use the 50 row and 4 column, because the first two digits of 5.04 are 50 and the third digit of 5.04 is 4. The entry in the 50 row, column 4 is .70243.

$$\log_{10} 5.04 = .70243 \text{ and } 5.04 = 10^{.70243}$$

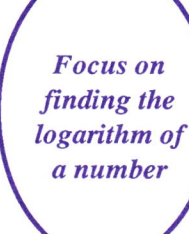

Focus on finding the logarithm of a number

To find the logarithm of 6, first affix two zeros to the right of the decimal point so that the three-digit numeral 6.00 is obtained. The logarithm of 6 or 6.00 is found by using the 60 row, 0 column. The entry in the 60 row, 0 column is .77815.

$$\log_{10} 6.00 = .77815 \text{ and } 6.00 = 10^{.77815}$$

The logarithm of 4.7 is found by using the 47 row, 0 column, because 4.7 = 4.70. The entry in the 47 row, 0 column is .67210.

$$\log_{10} 4.70 = .67210 \text{ and } 4.70 = 10^{.67210}$$

Logarithms for Numbers Between 1 and 10

The logarithm of 100 is 2, because $10^2 = 100$.
The logarithm of 1000 is 3, because $10^3 = 1000$.
The logarithm of .0001 is -4, because $10^{-4} = .0001$.

The logarithm of a number is the exponent of 10 that would produce the given number.

Every number greater than 1 and less than 10 will have an exponent between 0 and 1, because $10^0 = 1$ and $10^1 = 10$. Since 5.45 is a number between 1 and 10, its logarithm is a number between 0 and 1.

Focus on logarithms of numbers between 1 and 10

Each entry in the logarithm table is a five-digit decimal numeral for a number between 0 and 1. This is because the logarithm for any number between 1 and 10 is a number between 0 and 1.

The logarithm of 1.57 is .19590, which means $10^{.19590} = 1.57$.
The logarithm of 9.06 is .95713, which means that $10^{.95713} = 9.06$.

Mantissas and Characteristics

$x^5 \cdot x^2$ may be simplified to x^7, because the exponents are added in the simplification. This basic operation with exponents is involved in finding the **characteristic** and **mantissa** of a number's logarithm.

To find the logarithm of 512, the number must first be written in scientific notation. $512 = 5.12 \cdot 10^2$

The logarithm of 10^2 is 2, because that is the exponent of 10. The logarithm of 5.12 is found in the 51 row, 2 column of the logarithm table.

$$512 = 5.12 \cdot 10^2$$
$$= 10^{.70927} \cdot 10^2$$
$$= 10^{(.70927 + 2)}$$

The logarithm of 512 is (.70927 + 2). The five-place decimal .70927 is its **mantissa** and the integer 2 is its **characteristic**.

Focus on the mantissa and characteristic of a logarithm

To find the logarithm of .00008,
1. Write .00008 in scientific notation $.00008 = 8.00 \cdot 10^{-5}$
2. Find the logarithm of 8.00 $.00008 = 10^{.90309} \cdot 10^{-5}$
3. Write the logarithm as the sum of the exponents $\log_{10} .00008 = .90309 - 5$
4. The mantissa is .90309 and the characteristic is -5.

The log (logarithm) of .000417 is found in the following steps.
$$.000417 = 4.17 \cdot 10^{-4} = 10^{.62014} \cdot 10^{-4}$$
The log of .000417 is (.62014 − 4).

The logarithm of 6,010 is found in the following steps.
$$6,010 = 6.01 \cdot 10^3 = 10^{.77887} \cdot 10^3$$
The logarithm of 6,010 is (.77887 + 3).

The logarithm of .00417 is (.62014 − 3). It consists of two parts: a five digit decimal that is the mantissa and an integer that is the characteristic.

A Logarithm's Antilog

When the common logarithm of a number is given, the procedure for finding logarithms can be reversed and the result will be the number. This procedure for finding a number when given its logarithm is called finding **antilogs**.

For example, if $(.84634 + 3)$ is the logarithm of a number then the following procedure would be used for finding its number (antilog).
1. $(.84634 + 3)$ consists of two parts:
$$?????? = 10^{.84634} \cdot 10^3$$
2. Find the entry .84634 in the table on page 451 and use the row and column to write a 3-digit number that is between 1 and 10. (70 row, column 2)
$$?????? = 7.02 \cdot 10^3$$
3. Multiply
$$7{,}020 = 7.02 \cdot 10^3$$
Therefore, the antilog of $(.84634 + 3)$ is 7,020.

Focus on finding antilogs

To find the antilog of $(.31387 + 6)$, first find the mantissa in the log tables. The mantissa of $(.31387 + 6)$ is in the 20 row, 6 column.

$$10^{.31387} \cdot 10^6 = 2.06 \cdot 10^6 = 2{,}060{,}000$$

$(.31387 + 6)$ is the logarithm of 2,060,000, and 2,060,000 is the antilog of $.31387 + 6$.

The antilog of $(.47712 - 4)$ is found by first locating the mantissa in the tables. The mantissa of $(.47712 - 4)$ is in the 30 row, 0 column.

$$10^{.47712} \cdot 10^{-4} = 3.00 \cdot 10^{-4} = .0003$$

The antilog of $(.47712 - 4)$ is .0003.

Approximating Antilogs

The mantissa of (.37056 + 3) does not appear in the table. The two mantissas closest to .37056 appear in the 23 row, columns 4 and 5. .37056 is closer to .37107.

$.37056 + 3 \doteq .37107 + 3$ ($\doteq$ means approximately equal to)
Antilog of $.37056 + 3 \doteq 2{,}350$

When a mantissa does not appear in the table, choose the row and the column that most closely approximates its value.

.72510 does not appear in the table, but its value is closest to the mantissa in the 53 row, 1 column.

Focus on approximating antilogs

The mantissa of (.78204 + 5) does not appear in the logarithm table, but it is closest to the mantissa in the 60 row, 5 column. To find the antilog of (.78204 + 5),

$$10^{.78204} \cdot 10^5 \doteq 6.05 \cdot 10^5 = 605{,}000$$

The mantissa of (.90125 − 2) is not in the table, but it is closest to the mantissa in the 79 row, 7 column. To find the antilog of (.90125 − 2),

$$10^{.90125} \cdot 10^{-2} \doteq 7.97 \cdot 10^{-2} = .0797$$

Unit 4 Exercise

Part A: Answers for all Part A problems are at the back of the book.

1. Use the logarithm table to find the entry in:
    a. the 70 row and 4 column.
    b. the 52 row and 9 column.
    c. the 13 row and 3 column.
    d. the 83 row and 7 column.

2. To find the logarithm of 61 use the 61 row, 0 column. What entry is in the 61 row, 0 column?

3. To find the logarithm of 5.17, use the _____ row, _____ column.

4. To find the logarithm of 8.06, use the _____ row, _____ column.

5. To find the logarithm of 4.15, use the _____ row, _____ column.

6. Find the logarithm of 4.75 to complete:
    $10^? = 4.75$

7. Find the logarithm of 6.18 to complete:
    $10^? = 6.18$

8. To find the antilog of (.29505 + 2), the first step is to find the mantissa in the table that is closest to .29505. This mantissa is in the _____ row, _____ column.

9. To find the antilog of (.77503 – 6), first locate the mantissa closest to .77503. This mantissa is in the _____ row, _____ column.

10. Find the antilog of (.84255 + 1) by first finding the mantissa closest to .84255.

11. Find the antilog of (.38195 + -2) by first finding the mantissa closest to .38195.

Part B: Answers for odd-numbered problems of Part B are at the back of the book.

In problems 1-20 find logs (logarithms) for each number.

1. 7
2. 6.5
3. 1.07
4. 9.61
5. 5.4
6. 417,000
7. .00132
8. 2730
9. .96
10. 5.27
11. .107
12. 37.8
13. 601
14. 75.7
15. 8880
16. 460,000,000
17. .53
18. .0003
19. 4.3
20. 350

In problems 21-40 find antilogs for each number.

21. (.78104 – 3)
22. (.82543 + 5)
23. (.27184 + 1)
24. (.89873 + 0)
25. (.57989 – 1)
26. (.95856 + 4)
27. (.17590 + 8)
28. (.56703 + 2)
29. (.74819 + 4)
30. (.11394 + 0)
31. (.82944 + 4)
32. (.79069 + 1)
33. (.93543 – 3)
34. (.36740 + 2)
35. (.53925 + 4)
36. (.92430 – 1)
37. (.57634 + 1)
38. (.10950 – 6)
39. (.64835 + 2)
40. (.85187 + 3)

Part C: No answers are given for these problems. However, each is accompanied by an ordered pair «C,U» showing the chapter and unit in which it was taught.

1. Graph the truth set for $(x – 3)(3x – 1) < 0$. «1,6»

2. Use interval notation to write the truth set for $x^2 – 3x – 10 \leq 0$. «2,6»

3. Simplify $(5 – 3i)^2$ «3,3»

4. Draw a Venn diagram for $A \cup \overline{B}$. «5,5»

5. Write $-3x + 2y = 8$ in y-intercept form. «6,4»

6. Graph $y + 2 = (x – 4)^2$ «7,6»

7. Simplify $\sqrt[3]{x^2} \cdot \sqrt[4]{x^3}$ and write the answer in radical form. «8,6»

8. Simplify $(\sqrt[9]{x^4})^5$ «8,6»

9. Sketch the graph of $f(x) = \frac{1}{3}x$. «9,1»

10. Write 0.00342 using scientific notation. «9,3»

Unit 5: COMPUTING USING LOGARITHMS

Calculators Versus Computing

In this age, problems such as:

$$43.5 \times .000784$$
$$91.8 \div 87,500$$
$$\sqrt[5]{(13.2)^3}$$

are often done with the assistance of a calculator and nothing in this unit is intended to denigrate the value of the technology that is now available.

However, the individual who relies on a calculator as if it were a magic box that produces accurate answers can easily be victimized by the technology. The individual who understands the processes that are being handled by the technology can enslave the technology rather than become its victim.

This unit is intended to provide an understanding of the computations necessary to find answers for complex arithmetic problems. This knowledge will greatly improve your ability to correctly estimate answers. It will also give you a chance to figure out answers when (not if) the batteries fail.

In Multiplying, Logarithms are Added

Logarithms are exponents. Since exponents are added when multiplying like bases, the common logarithms of two factors can be added to find the logarithm of the product.

For example, the product of 426 • 3820 can be found by adding the logarithms of 426 and 3820 and then finding the antilog of the sum. To find 426 • 3820,

1. Find the sum of the logarithms
$$\begin{aligned}\log_{10}(426 \cdot 3820) &= \log_{10} 426 + \log_{10} 3820 \\ &= (.62941 + 2) + (.58206 + 3) \\ &= (1.21147 + 5) \\ &= (0.21147 + 6)\end{aligned}$$

2. Find the antilog of the sum of the logarithms
Antilog of $(0.21147 + 6) = 1.63 \cdot 10^6 = 1{,}630{,}000$

3. Therefore, 426 • 3820 = 1,630,000. The answer is correct to its first three digits.

Because exponents are added when multiplying like bases, the product of two positive real numbers x and y can be calculated using logarithms as:
$\log_{10} xy = \log_{10} x + \log_{10} y$

Focus on multiplying using logarithms

To multiply two numbers by using logarithms:
1. Find the logarithm of each number.
2. Add the logarithms.
3. Find the antilog of the sum.

To multiply 406,000 by .0592 using logarithms,
1. Find the logarithm of each number.
$\log_{10} 406{,}000 = .60853 + 5$ and $\log_{10} .0592 = .77232 - 2$
2. Add the logarithms.
$$\begin{aligned}\log_{10} 406{,}000 + \log_{10} .0592 &= (.60853 + 5) + (.77232 - 2) \\ &= (.60853 + .77232) + (5 - 2) \\ &= (1.38085 + 3) \\ &= (0.38085 + 4)\end{aligned}$$
3. Find the antilog of the sum.
The antilog of .38085 + 4 is 24,000.
Therefore, 406,000 • .0592 = 24,000

Logarithms are Subtracted When Dividing

Logarithms are exponents. Since exponents are subtracted when dividing like bases, the common logarithms can be subtracted to find the logarithm of the quotient.

For example, to divide 81.7 by .0062 subtract the logarithm of .0062 from the logarithm of 81.7 and then find the antilog of the difference. To find $81.7 \div .0062$,

1. Find the difference of the logarithms
$$\begin{aligned}\log_{10}(81.7 \div .0062) &= \log_{10} 81.7 - \log_{10} .0062 \\ &= (.91222 + 1) - (.79239 - 3) \\ &= .91222 + 1 - .79239 + 3 \\ &= (.91222 - .79239) + 1 + 3 \\ &= (.11983 + 4)\end{aligned}$$

2. Find the antilog of the difference of the logarithms
Antilog of $(.11983 + 4) = 1.32 \cdot 10^4 = 13{,}200$

3. Therefore, $81.7 \div .0062 = 13{,}200$. The answer is correct to three digit accuracy.

Because exponents are subtracted when dividing like bases, the quotient of two positive real numbers x and y can be calculated using logarithms as:
$$\log_{10} \frac{x}{y} = \log_{10} x - \log_{10} y$$

Focus on dividing using logarithms

To divide two numbers by using logarithms:
1. Find the logarithm of each number.
2. Subtract the logarithms. (Dividend minus divisor)
3. Find the antilog of the difference.

To divide 3,740 by 124 using logarithms,
1. Find the logarithm of each number.
$\log_{10} 3{,}740 = .57287 + 3$ and $\log_{10} 124 = .09342 + 2$
2. Subtract the logarithms.
$$\begin{aligned}\log_{10} 3{,}740 - \log_{10} 124 &= (.57287 + 3) - (.09342 + 2) \\ &= (.57287 - .09342) + 3 - 2 \\ &= (.47945 + 1)\end{aligned}$$
3. Find the antilog of the difference.
The antilog of $.47945 + 1$ is 30.2
Therefore, $3{,}740 \div 124 = 30.2$

The mantissa of a logarithm is always positive. This means that whenever the subtraction of two mantissas would result in a negative number, then 1 must be borrowed from the characteristic. Note the borrowing in the second step of the following example.

To divide 27,000 by 4.7 using logarithms,
1. Find the logarithm of each number.
   $\log_{10} 27{,}000 = .43136 + 4$ and $\log_{10} 4.7 = .67210 + 0$
2. Subtract the logarithms.
   $$\begin{aligned}\log_{10} 27{,}000 - \log_{10} 4.7 &= (.43136 + 4) - (.67210 + 0)\\ &= (.43136 - .67210) + 4 - 0\\ &= ([1.43136 - 1] - .67210) + 4\\ &= (1.43136 - .67210) + 4 - 1\\ &= (.75926 + 3)\end{aligned}$$
3. Find the antilog of the difference.
   The antilog of .75926 + 3 is 5,740
   Therefore, 27,000 ÷ 4.7 = 5,740

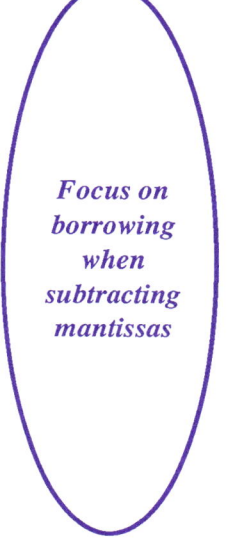

Focus on borrowing when subtracting mantissas

Raising A Number to a Power

Because exponents are multiplied when raising a power to a power, the calculation of x^n using logarithms is:

$$\log_{10} x^n = n \cdot \log_{10} x$$

To compute $(76.1)^8$ using logarithms, find the logarithm of 76.1, multiply it by 8, and finally take the antilog of the product.

$$\begin{aligned}\log_{10} (76.1)^8 &= 8 \cdot \log_{10} 76.1\\ &= 8(.88138 + 1)\\ &= 7.05104 + 8\\ &= .05104 + 15\end{aligned}$$
$(76.1)^8 \doteq 1{,}120{,}000{,}000{,}000{,}000$

To evaluate $(.094)^5$
1. Find the logarithm of .094
2. Multiply it by 5.
3. Find the antilog.
   $$\begin{aligned}\log_{10} (.094)^5 &= 5 \cdot \log_{10} .094\\ &= 5(.97313 - 2)\\ &= 4.86565 - 10\\ &= .86565 - 6\end{aligned}$$

The antilog of (.86565 − 6) is .00000734

Focus on raising a number to a power

Finding Roots Using Logarithms

$\sqrt[7]{51}$ can be written with a fractional exponent as $(51)^{1/7}$. Since multiplication by 1/7 is the same as division by 7, the computation of $\sqrt[7]{51}$ depends upon the division of the logarithm of 51 by 7.

To compute $\sqrt[7]{51}$,

1. Find the logarithm of 51. $\log_{10} 51 = .70757 + 1$

2. **Decrease** the characteristic so it is evenly divisible by 7 and add that amount to the mantissa. $\log_{10} 51 = 1.70757 + 0$

3. Divide the logarithm by 7. If necessary round off the mantissa at five decimal places.
$$\frac{\log_{10} 51}{7} = \frac{1.70757 + 0}{7}$$
$$\frac{\log_{10} 51}{7} = .24394 + 0$$

4. Find the antilog of .24394 + 0 $\sqrt[7]{51} = 1.75$

$$\log_{10} \sqrt[n]{x} = \frac{1}{n} \cdot \log_{10} x$$

Focus on computing roots of numbers

To compute $\sqrt[5]{.0081}$,

1. Find the logarithm of .0081 $\log_{10} .0081 = .90849 - 3$

2. Decrease the characteristic so it is evenly divisible by 5 and add that amount to the mantissa. $\log_{10} .0081 = 2.90849 - 5$

3. Divide the logarithm by 5. If necessary round off the mantissa at five decimal places.
$$\frac{\log_{10} .0081}{5} = \frac{2.90849 - 5}{5}$$
$$\frac{\log_{10} .0081}{5} = .58170 - 1$$

4. Find the antilog of .58170 − 1 $\sqrt[5]{.0081} = .382$

Writing Logarithm Expressions for Complex Problems

To compute $\dfrac{4.9 \cdot (6.45)^7}{.0067 \sqrt[7]{829}}$ there are five separate problems to complete. The numerator and denominator are computed separately and the division is the final step.

The numerator is computed as: $\log_{10} 4.9 + 7 \cdot \log_{10} 6.45$

The denominator is computed as: $\log_{10} .0067 + \dfrac{1}{7} \cdot \log_{10} 829$

The complete computation can be described by the following logarithm expression:
$(\log_{10} 4.9 + 7 \cdot \log_{10} 6.45) - (\log_{10} .0067 + \dfrac{1}{7} \cdot \log_{10} 829)$

Focus on writing a logarithm expression for a complex problem

To compute $\dfrac{(67.2)^4 \cdot (.973)^5}{85,000 \sqrt{18.6}}$ find logarithm answers for the numerator and denominator. Then subtract the logarithms to find the antilog of the final result.

The numerator is computed as: $4 \cdot \log_{10} 67.2 + 5 \cdot \log_{10} .973$

The denominator is computed as: $\log_{10} 85,000 + \dfrac{1}{2} \cdot \log_{10} 18.6$

The complete computation can be described by the following logarithm expression:
$(4 \cdot \log_{10} 67.2 + 5 \cdot \log_{10} .973) - (\log_{10} 85,000 + \dfrac{1}{2} \cdot \log_{10} 18.6)$

Unit 5 Exercise

Part A: Answers for all Part A problems are at the back of the book.

1. To multiply 62,000 and 49 using logarithms, write the expression
$$\log_{10} 62{,}000 + \log_{10} 49$$
Write the expression needed to calculate $370 \cdot 23$ using logarithms.

2. Multiply 428 by 123 using logarithms.

3. To divide 54.6 by 4.3 using logarithms, write the expression
$$\log_{10} \tfrac{54.6}{4.3} = \log_{10} 54.6 - \log_{10} 4.3$$
Write the expression needed to calculate $.0362 \div 21.6$ using logarithms.

4. The mantissa of a logarithm must be positive. Divide 275 by .0083 using logs. (It will be necessary to borrow 1 from the characteristic of the log of the dividend.)

5. To raise 41.2 to the 3rd power using logarithms, write the expression
$$\log_{10}(41.2)^3 = 3 \cdot \log_{10} 41.2$$
Write the expression needed to calculate $(5.3)^4$ using logarithms.

6. Use logarithms to calculate $(20.5)^4$.

7. To find the 5th root of 837 using logarithms, write the expression
$$\log_{10} \sqrt[5]{837} = \tfrac{1}{5} \cdot \log_{10} 837$$
Write the expression needed to calculate $\sqrt[7]{91.4}$ using logarithms.

8. To divide a logarithm by 6, decrease the characteristic so it is evenly divisible by 6 and add that amount to the mantissa. Calculate $\sqrt[6]{4350}$ using logs. (It will be necessary to adjust the characteristic and mantissa for division by 6.)

9. If $x \cdot y = z$, which of the following is true?
   a. $\log_{10} z = \log_{10} x \cdot \log_{10} y$
   b. $\log_{10} z = \log_{10} x + \log_{10} y$
   c. $\log_{10} z = \log_{10}(x + y)$

10. If $x \div y = z$, which of the following is true?
    a. $\log_{10} z = \log_{10} x - \log_{10} y$
    b. $\log_{10} z = \log_{10} x \div \log_{10} y$
    c. $\log_{10} z = \log_{10} y + \log_{10} x$

11. If $x^y = z$, which of the following is true?
    a. $\log_{10} z = \log_{10} x + \log_{10} y$
    b. $\log_{10} z = y \cdot \log_{10} x$
    c. $\log_{10} z = x \cdot \log_{10} y$

12. If $\sqrt[k]{x} = z$, which of the following is true?
    a. $\log_{10} z = k \cdot \log_{10} x$
    b. $\log_{10} z = \tfrac{1}{k} \cdot \log_{10} x$
    c. $\log_{10} z = \log_{10} x + \log_{10} k$

Exponential and Logarithm Functions

Part B: Answers for odd-numbered problems of Part B are at the back of the book.

For problems 1-20, calculate using logarithms.

1. $47{,}000{,}000 \div .0085$
2. $.0415 \cdot .0064$
3. $(31.6)^5$
4. $\sqrt{.00000412}$
5. $6.19 \cdot .00056$
6. $5{,}900{,}000 \div .00034$
7. $\sqrt[3]{1250}$
8. $(.65)^{10}$
9. $72{,}300 \cdot .00041$
10. $49{,}000 \div 63$
11. $(1.05)^{100}$
12. $\sqrt[4]{256}$
13. $505 \div .00017$
14. $.00031 \cdot .00127$
15. $\sqrt[5]{.0047}$
16. $(23)^6$
17. $63{,}000 \cdot .0055$
18. $.0075 \div .00089$
19. $(.0055)^4$
20. $\sqrt[7]{.00000036}$

For each of the following, write a logarithm expression that could be used to calculate the answer.

21. $\dfrac{(4.17)^6}{.546 \cdot 93.7}$

22. $\dfrac{40{,}700 \cdot 890}{\sqrt[3]{780{,}000}}$

23. $\dfrac{\sqrt[5]{73}}{230 \cdot (9.27)^6}$

24. $\dfrac{260 \cdot .0032}{(1.7)^9}$

25. $\dfrac{\sqrt[3]{.047}}{4{,}820 \cdot .0316}$

26. $\dfrac{(72.3)^5 \sqrt[3]{78}}{8.420 \cdot 369}$

27. $\dfrac{(0.135)^5}{602 \div .00017}$

28. $\dfrac{(.012)^3}{2.46 \cdot \sqrt[5]{12.7}}$

Part C: No answers are given for these problems. However, each is accompanied by an ordered pair «C,U» showing the chapter and unit in which it was taught.

1. Factor $x^2 - 3x(a-b) + 2(a-b)^2$ «1,4»

2. Simplify $-6\sqrt[3]{108}$ «2,3»

3. Simplify $\dfrac{3-7i}{3+7i}$ «3,3»

4. Use interval notation to describe the truth set of $|3x - 1| = 8$. «4,4»

5. Write the linear function $y = \dfrac{-2}{5}x - 3$ in standard form. «6,4»

6. Find the center and radius for the circle $x^2 - 6x + y^2 + 8x = -9$. «7,4»

7. John can do a job in 6 hours, and Bob can do the same job in 8 hours. Working together on the job, how long will it take them? «8,7»

8. Evaluate $\log_{1/2} 8 =$ _____ «9,2»

9. Write 32,406 using scientific notation. «9,3»

10. Evaluate $4.12 \cdot 10^{-2}$ «9,3»

406   Chapter 9

Unit 6: APPLICATIONS: APPROXIMATING SOLUTIONS AND FINDING EQUIVALENT UNITS

Finding Approximate Solutions by Using Logarithms

In this unit problems that might be encountered in physics or chemistry will be solved by using logarithms. These problems were difficult to solve before the advent of the calculator and were commonly solved using logarithms. The intent here is to understand the arithmetic involved and to provide clues for figuring out such problems if your batteries fail.

Problem: What is the radius of a sphere that has a volume of 278 cubic feet?

Solution

The relation between the volume(V) and radius(r) of a sphere is given by the formula $V = \frac{4}{3}\pi r^3$. When V is replaced with 278 and $\frac{22}{7}$ is used as an approximation for pi, the equation above becomes $278 = \frac{4}{3} \cdot \frac{22}{7} \cdot r^3$.

The equation is solved using the steps below.

$$278 = \frac{4}{3} \cdot \frac{22}{7} \cdot r^3$$
$$278 = \frac{88}{21} \cdot r^3$$
$$\frac{21}{88} \cdot 278 = \frac{21}{88} \cdot \frac{88}{21} \cdot r^3$$
$$\frac{21 \cdot 278}{88} = r^3$$
$$r = \sqrt[3]{\frac{21 \cdot 278}{88}}$$

The value of r can be found using the following logarithm expression.

$$\log_{10} r = \tfrac{1}{3}(\log_{10} 21 + \log_{10} 278 - \log_{10} 88)$$

The radius of the sphere, rounded to the nearest foot, is 4 feet.

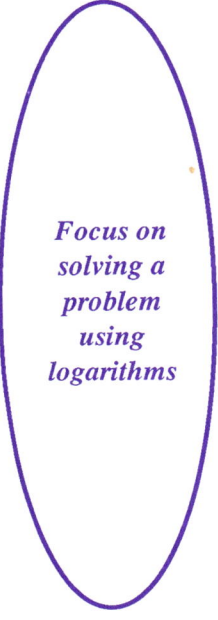

Focus on solving a problem using logarithms

Problem: When an object is dropped, the relationship between the distance (D) that it falls and the elapsed time (t) in seconds after its release is shown by the equation $D = 16t^2$. Find the time for an object to fall 555 feet from the top of the Washington Monument.

Solution
D is replaced by 555 in the equation $D = 16t^2$ and the equation solved for t as shown in the steps below.

$$555 = 16t^2$$
$$\frac{555}{16} = t^2$$
$$t = \sqrt{\frac{555}{16}}$$

The time (t) is found by computing the logarithm expression:
$$\log_{10} t = \tfrac{1}{2}(\log_{10} 555 - \log_{10} 16)$$
To the nearest tenth, t = 5.9 seconds.

Using Logarithms When Changing Units

Many times in science courses, it is necessary to change a measurement from one unit to another. The conversion can be made by multiplying the measurement by expressions equivalent to one (1) that will produce the desired units.

Each of the following ratios is equal to 1. Notice that the labels for the numerators and denominators are different and, in fact, the ratios are equal to 1 because of the labels.

$$\frac{5280 \text{ feet}}{1 \text{ mile}} \qquad \frac{3 \text{ feet}}{1 \text{ yard}} \qquad \frac{1 \text{ minute}}{60 \text{ seconds}} \qquad \frac{1 \text{ cubic foot}}{1728 \text{ cubic in.}}$$

The example on the following page shows how fractions equal to one can be multiplied to convert miles per hour into feet per second.

To convert 78 miles per hour into feet per second, it is necessary to complete the following proportion.

$$\frac{78 \text{ miles}}{1 \text{ hour}} = \frac{?? \text{ feet}}{?? \text{ seconds}}$$

The labels of the numerators of the proportion are "miles" and "feet." The labels of the denominators of the proportion are "hour" and "seconds." Begin by writing two ratios equal to 1 that have those labels.

$$\frac{1 \text{ mile}}{5280 \text{ feet}} \qquad \frac{1 \text{ hour}}{60 \cdot 60 \text{ seconds}}$$

Since each of these new ratios is equal to 1, they may be multiplied by the left side of the proportion.

$$\frac{78 \text{ miles}}{1 \text{ hour}} \cdot \frac{1 \text{ mile}}{5280 \text{ feet}} \cdot \frac{1 \text{ hour}}{60 \cdot 60 \text{ seconds}} = \frac{?? \text{ feet}}{?? \text{ seconds}}$$

To make the conversion the labels in the numerators and denominators must be identical except that the numerator must have "feet" and the denominator must have "seconds." This can be accomplished by inverting one of the ratios equal to 1.

$$\frac{78 \text{ miles}}{1 \text{ hour}} \cdot \frac{5280 \text{ feet}}{1 \text{ mile}} \cdot \frac{1 \text{ hour}}{60 \cdot 60 \text{ seconds}} = \frac{?? \text{ feet}}{?? \text{ seconds}}$$

The multiplication of ratios now gives:

$$\frac{78 \cdot 5280 \text{ feet } (1 \text{ mile} \cdot 1 \text{ hour})}{60 \cdot 60 \text{ seconds } (1 \text{ mile} \cdot 1 \text{ hour})} = \frac{?? \text{ feet}}{?? \text{ seconds}}$$

and this ratio may be simplified using the logarithm expression
$$(\log_{10} 78 + \log_{10} 5280) - (\log_{10} 60 + \log_{10} 60)$$

Rounded off to the nearest foot, the evaluation of the expression above is 114 feet per second. 78 miles per hour was converted to feet per second when it was multiplied by expressions equivalent to 1.

Focus on solving a problem using logarithms

Problem: Find the weight of one gallon of water by using the following conversion facts. 1 cubic foot contains 1728 cubic inches. 1 cubic foot of water weighs 62.4 pounds. 1 gallon contains 231 cubic inches.

Solution
To solve the problem the following equality needs to be completed:
$$1 \text{ gallon of water} = ?? \text{ pounds}$$

Each of the conversion facts is used to write a ratio equal to 1.

$$\frac{1 \text{ gallon of water}}{1} \cdot \frac{1 \text{ cubic foot}}{1728 \text{ cu in}} \cdot \frac{1 \text{ cubic foot}}{62.4 \text{ lbs}} \cdot \frac{1 \text{ gallon of water}}{231 \text{ cu in}} = \frac{?? \text{ pounds}}{1}$$

Two fractions need to be inverted.

$$\frac{1 \text{ gallon of water}}{1} \cdot \frac{1 \text{ cubic foot}}{1728 \text{ cu in}} \cdot \frac{62.4 \text{ pounds}}{1 \text{ cubic foot}} \cdot \frac{231 \text{ cu in.}}{1 \text{ gallon of water}} = \frac{?? \text{ pounds}}{1}$$

Now the labels match the desired result.

$$\frac{1 \cdot 1 \cdot 62.4 \cdot 231 \text{ pounds (gallon of water} \cdot \text{cubic foot} \cdot \text{cubic in.)}}{1 \cdot 1728 \cdot 1 \cdot 1 \text{ (gallon of water} \cdot \text{cubic foot} \cdot \text{cubic in.)}} = ?? \text{ pounds}$$

The computation may be completed using the logarithm expression below.
$$(\log_{10} 62.4 + \log_{10} 231) - \log_{10} 1728$$

A gallon of water weighs approximately 8.33 pounds.

Unit 6 Exercise

Part A: Answers for all Part A problems are at the back of the book.

1. The equation $V = s^3$ shows the relationship between the volume (V) of a cube and the length of one of its edges (s). Use logarithms to find the edge of a cube that has a volume of 47.3 cubic centimeters.

2. Use logarithms and the equation $D = 16t^2$ to find the seconds (t) required for an object to fall a distance (D) of 37,500 feet.

3. Use logarithms to find the radius of a sphere that has a volume of 5379 cubic feet. Use $\frac{22}{7}$ as an approximation for π.

4. Write two fractions equal to 1 for the fact that 1 cubic foot contains 1728 cubic inches.

5. Write two fractions equal to 1 for the fact that 1 gallon contains 231 cubic inches.

6. Write two fractions equal to 1 for the fact that 1 minute equals 60 seconds.

7. Use logarithms and the conversion facts of problems 4-6 to find how many cubic feet per second are equivalent to a flow of 563 gallons per minute.

8. Write fractions equal to 1 for each of the following.
   a. 1 cubic inch of steel = 0.283 pounds
   b. 1 kilometer is 1000 meters.
   c. 1 minute has 60 seconds.
   d. 1 cubic foot = 1728 cubic inches
   e. 1 mile is 5280 feet.
   f. There are 60 minutes in an hour.

9. Use some of the conversion facts from problem 8 to find the weight of one cubic foot of steel.

10. Use some of the conversion facts from problem 8 along with the fact that 1 foot is approximately .305 meters. Find how many miles per hour are equivalent to 48 kilometers per hour.

11. The speed of sound is 1089 feet per second at sea level. Use some of the conversion facts from problem 8 to find the equivalent speed in miles per hour.

Part B: Answers for odd-numbered problems of Part B are at the back of the book.

1. Use the formula $V = \frac{4}{3}\pi r^3$ to find the radius of a sphere when the volume is 68.2 cubic centimeters. Use $\frac{22}{7}$ as an approximation for pi.

2. Use the formula $V = \frac{4}{3}\pi r^3$ to find the radius of a sphere when the volume of the sphere is 45,200 cubic meters. Use $\frac{22}{7}$ as an approximation for pi.

3. Use the formula $D = 16t^2$ to find the time necessary for an object to fall 3260 feet.

4. Use the formula $V = s^3$ to find the length of the edge of a cube that has a volume of 137 cubic feet.

5. Use the formula $V = s^3$ to find the length of the edge of a cube that has a volume of 3.6 cubic inches.

Exponential and Logarithm Functions   411

6. Find the volume of a cube when the edge of the cube is 1.24 inches long.

7. Find the approximate solution to the nearest tenth for the equation $37.8 = 1.8x^4$.

8. Use the formula $V = s^3$ to find the length of the edge of a cube that has a volume of 25.8 cubic centimeters.

9. Use the formula $V = \frac{4}{3}\pi r^3$ to find the radius of a sphere that has a volume of 0.459 cubic feet. Use $\frac{22}{7}$ as an approximation for π.

For the remaining problems of this exercise, use the equivalences following #15.

10. Find how many feet per second are equivalent to 66 miles per hour.

11. Find the weight of one gallon of gasoline.

12. Find the number of yards per second that is equivalent to 45 miles per hour.

13. Find the weight in pounds of one gallon of milk.

14. Find the number of Btu's per pound that is equivalent to 25 kilocalories per gram.

15. The mean density of the earth is 5.52 grams per cubic centimeter. How many pounds per cubic foot are equivalent to 5.52 grams per cubic centimeter?

1 gallon = 231 cubic inches
1 pound = 16 ounces
1 minute = 60 seconds
1 cubic foot = 1728 cubic inches
1 cubic foot of gasoline weighs 41.2 pounds
1 cubic inch = 6.45 cubic centimeters
1 mile = 5280 feet
1 Btu = 0.252 kilocalories
1 yard = 3 feet
1 ounce = 28.4 grams
1 cubic foot of milk weighs 64.2 pounds
1 pound = 454 grams
1 hour = 60 minutes

Part C: No answers are given for these problems. However, each is accompanied by an ordered pair «C,U» showing the chapter and unit in which it was taught.

1. Is $\frac{-3}{0}$ a rational number?   «1,3»

2. Solve $x^2 - 8x = 0$ using the set of rational numbers.   «1,5»

3. Simplify $-7\sqrt[5]{64}$ «2,6»

4. $3i^5 \cdot -5i^3 = $ _____   «3,1»

5. The sum of the digits of a two digit number is 15. The number with the digits reversed is 9 less than the number. Find the original number.   «3,5»

6. Graph $y = -3x + 6$ «6,1»

7. Find the slope and y-intercept of $5x - 2y = 8$.   «6,2»

8. Graph the circle $(x - 2)^2 + (y + 1)^2 = 4$. Sketch its image circle after the translation that moves the center to (0,0) and write the equation of the new circle.   «7,1»

9. Graph $\frac{x^2}{4} - \frac{y^2}{9} = 1$ «7,7»

10. Find the volume of a sphere whose radius is 2.7 inches. (Use 3.14 as an approximation for pi.)   «9,6»

Chapter 9 Test

«9,U» shows the unit in which this problem was studied in this chapter.

1. Sketch the graph of $f(x) = \left(\frac{1}{2}\right)^x$ «9,1»

2. Sketch the graph of $f(x) = 3^x$ «9,1»

3. Is $y = x^2$ an exponential function? «9,1»

4. Evaluate $\log_2 32 =$ _____. «9,2»

5. Evaluate $\log_7 \left(\frac{1}{7}\right) =$ _____. «9,2»

6. Evaluate $\log_{\frac{1}{3}} 9 =$ _____. «9,2»

Write problems 7-9 using scientific notation.

7. .0317 «9,3»

8. 3,420,000 «9,3»

9. 5.9 «9,3»

Evaluate problems 10-14 using logarithms.

10. $3{,}620 \cdot .0132$ «9,5»

11. $384 \div .00217$ «9,5»

12. $(19.4)^5$ «9,5»

13. $\sqrt[3]{46.5}$ «9,5»

14. $\dfrac{41.3\sqrt{112}}{(1.8)^2}$ «9,5»

15. Use the formula $V = \frac{4}{3}\pi r^3$ to find the radius of a sphere when the volume is 47.8 cubic centimeters. Use $\frac{22}{7}$ as an approximation for pi. «9,6»

16. Use the formula $V = e^3$ to find the edge of a cube that has a volume of 2190 cubic inches. «9,6»

17. Use the formula $D = 16t^2$ to find the time it takes an object to fall 893 feet. «9,6»

18. Solve $3.7x^3 = 104$ «9,6»

Use the following equivalences for problems 19-20.

100 cm = 1 meter
1 hour = 60 minutes
1 minute = 60 seconds
1 mile = 5280 feet
1 mile = 1609 meters

19. Find the number of feet per second that is equivalent to 55 miles per hour. «9,6»

20. Convert 54 centimeters per second to miles per minute. «9,6»

10
Systems of Equations

Unit 1: SYSTEMS OF TWO LINEAR EQUATIONS

Linear Equations with Two Variables

The equation $3x - 5y = 11$ is a linear equation with two variables, x and y. Some ordered pairs (x,y) are solutions of the equation and others are not. (7,2) is a solution because when x is replaced by 7 and y by 2, the equation $3x - 5y = 11$ becomes the true statement. $3 \cdot 7 - 5 \cdot 2 = 11$. (5,-1) is not a solution because when x is replaced by 5 and y by -1, the equation becomes the false statement $3 \cdot 5 - 5 \cdot -1 = 11$.

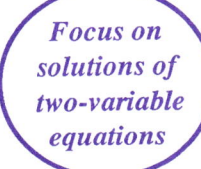
Focus on solutions of two-variable equations

(3,5) is a solution for the two-variable equation $2x + y = 11$ because the equation becomes the true statement $2 \cdot 3 + 1 \cdot 5 = 11$.
Note: (3,5) is an ordered 2-tuple.

(4,1) is **not** a solution for $2x + y = 11$ because the equation becomes the false statement $2 \cdot 4 + 1 = 11$.

> **Definition** **A System of Equations**
>
> A set of equations is a **system of equations** **if and only if** it consists of at least two equations.

Focus on the meaning of a system of equations

Two examples of systems of equations are shown at the right.

$$3x + y = 7$$
$$x - y = 5$$

The top example is a system of two linear equations and the second system has one linear and one quadratic equation.

$$4x - y = -8$$
$$y = x^2 - 6$$

> **Definition** **Common Solution of a System**
>
> An ordered pair (x,y) is a **common solution for a system of equations** **if and only if** it is a solution for each equation in the system.

Focus on the meaning of a common solution for a system of equations

$(3,-4)$ is a common solution for the system of two equations shown at the right because both

$$2 \cdot 3 + (-4) = 2 \text{ and } 3 - (-4) = 7$$

are true numerical statements.

$$2x + y = 2$$
$$x - y = 7$$

$(-2,7)$ is a common solution for the system of equations shown at the right because both of the following are true numerical statements.

$$4 \cdot -2 + 3 \cdot 7 = 13 \text{ is true.}$$
$$7 = (-2)^2 + 3 \text{ is true.}$$

$$4x + 3y = 13$$
$$y = x^2 + 3$$

Graphing to Find a Common Solution

The common solution for the system
$$2x + y = 7$$
$$x - y = 5$$
can be found by graphing. Each equation is graphed separately. The line marked $2x + y = 7$ represents all the ordered pair solutions of its equation. Similarly, the line of $x - y = 5$ represents all the solutions of its equation. The only point common to the two lines is the intersection point $(4,-1)$.

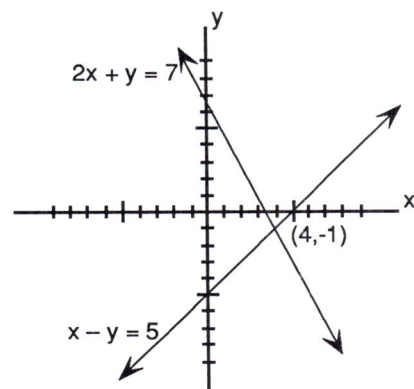

Although the graphing method for finding common solutions clearly shows the idea, graphing is not the easiest method for solving such problems. The easier method is to generate a new system of equations for which the common solution is obvious.

Generating New Equations for a System of Equations

To find the common solution for $3x - y = 7$ and $2x + y = 3$, the left sides of the equations are added and the right sides of the equations are added. This process allows us to include the new equation, $x = 2$, in the system and assures that the common solution of the original system is also a solution of the new system.

$$3x - y = 7$$
$$2x + y = 3$$
$$\overline{5x + 0 = 10}$$
$$x = 2$$

This means that we now have a system of three equations.
$$3x - y = 7, \quad 2x + y = 3, \quad \text{and} \quad x = 2$$

Because of the equation $x = 2$ we know that the common solution is of the form $(2,y)$. Consequently, by replacing x by 2 in either original equation, the common solution $(2,-1)$ is found.

$$3x - y = 7$$
$$3(2) - y = 7$$
$$6 - y = 7$$
$$-y = 1$$
$$y = -1$$

Focus on finding common solutions by the addition method

One technique for solving a system of two linear equations is the **addition method**. This method involves adding the left and right sides of the equations to generate a new equation which has only one variable. In other words, one of the variables is eliminated by the addition. The example below illustrates the addition method.

To find a common solution for the pair of equations shown at the right, it is necessary to eliminate either the x's or the y's. To do this, both equations must be multiplied by numbers so that the coefficients of one of the variables are opposites. The coefficients of the y's in the equations are 2 and -3. Since 6 is the smallest number that can be evenly divided by 2 and -3, both equations should be multiplied by numbers so that the coefficients for y will be 6 and -6.

$5x + 2y = 11$
$2x - 3y = 12$

1. Multiply the first equation by 3. $3(5x + 2y) = 3 \cdot 11$ $15x + 6y = 33$
2. Multiply the second equation by 2. $2(2x - 3y) = 2 \cdot 12$ $4x - 6y = 24$
3. The equations are added to eliminate the y's and determine that x = 3. $19x = 57$
   $x = 3$
4. The system now consists of 3 equations: $5x + 2y = 11$, $2x - 3y = 12$, and $x = 3$
5. Since x = 3, a replacement is made in one of the original equations to find y. The common solution is (3,-2).   If x = 3, then $5x + 2y = 11$ becomes $15 + 2y = 11$ or $2y = -4$ or $y = -2$.

The Substitution Method

A second technique for solving a system of equations is the **substitution method**. This technique requires that one of the variables be replaced by an open expression.

For example, to find the common solution for the pair of equations shown at the right, substitution is the easiest method.

$2x - 3y = 5$
$x = 4y$

Since x equals 4y, x can be replaced by (4y) in the first equation to generate the new equation y = 1.

$x = 4y$

$2x - 3y = 5$
$2(4y) - 3y = 5$
$8y - 3y = 5$
$5y = 5$
$y = 1$

Since y = 1 and x = 4y, then x = 4 • 1 or x = 4. The common solution for the system of equations is (4,1).

Systems of Equations

Focus on finding common solutions by the substitution method

To find the common solution for $y = 2x - 1$ and $3x - y = 8$ the fact that the first equation is solved for y makes it easy to use the substitution method.

Replace y in the second equation by $(2x - 1)$. Notice that parentheses are used in making the substitution.

$$y = 2x - 1$$
$$3x - y = 8$$
$$3x - (2x - 1) = 8$$
$$3x - 2x + 1 = 8$$
$$x + 1 = 8$$
$$x = 7$$

The system now has three equations: $y = 2x - 1$, $3x - y = 8$, and $x = 7$
Substituting 7 for x in $y = 2x - 1$ and solving for y gives the result $y = 13$.

Therefore, the common solution for the system of equations is (7,13).

Systems Without Unique Common Solutions

It is possible for a system of equations to have no common solution.

The system consisting of $y = 2x - 3$ and $6x - 3y = -12$ is represented graphically by parallel lines. Attempts to find a simple, new equation for the system, result in false numerical statements.

If $y = 2x - 3$ then $6x - 3y = -12$ becomes
$$6x - 3(2x - 3) = -12$$
$$6x - 6x + 9 = -12$$
$$9 = -12$$

The false numerical statement $9 = -12$ indicates there is no common solution.

Focus on a system with an infinite number of common solutions

The system of equations at the right consists of two equivalent equations (all of the solutions are common) and therefore it has an infinite number of common solutions.

$$x - 3y = 12 \qquad y = \frac{x}{3} - 4$$
$$x - 3\left(\frac{x}{3} - 4\right) = 12$$
$$x - x + 12 = 12$$
$$12 = 12$$

An attempt to solve such a system leads to a true numerical statement. The true numerical statement indicates that the two original equations are equivalent. Every solution of one is also a solution of the other.

Unit 1 Exercise

Part A: Answers for all Part A problems are at the back of the book.

1. Multiply both sides of the first equation by the LCM of the denominators. Then find the common solution.
   $\frac{1}{4}x - \frac{1}{3}y = 3$
   $2x + y = 13$

2. Multiply both sides of the first equation by the LCM of the denominators. Then find the common solution.
   $\frac{1}{2}x + \frac{2}{3}y = \frac{1}{6}$
   $2x + 5y = 10$

3. In using substitution to solve the system $4x - 3 = 7y$ and $x = y - 5$, what open expression can be used to replace x in the first equation?

4. In using substitution to solve the system $x - 2y = 6$ and $5x = y$, what open expression can be used to replace y in the first equation?

5. In solving the system $7x - 2y = 13$ and $5x - 9 = 4y$, which method, addition or substitution, would you choose?

6. In solving the system $5x - 9 = y$ and $6x + 7y = 16$, which method, addition or substitution, would you choose?

Part B: Answers for odd-numbered problems of Part B are at the back of the book.

Find common solutions.

1. $5x - 2y = 0$
   $2x + 3y = 19$

2. $y = 5x$
   $x + 2y = 22$

3. $5x - 4y = -2$
   $2x + 3y = 13$

4. $x + 3y = -6$
   $x - 2y = 14$

5. $4x - 3y = 5$
   $-3x + 2y = -2$

6. $2x - 7y = 15$
   $3x - 2y = 14$

7. $x + 6y = 0$
   $y = -2x$

8. $3x - 2y = 8$
   $x = y + 3$

9. $2x - 8y = 2$
   $x - 5y = 0$

10. $4x + 3y = 15$
    $2x - 5y = -25$

11. $x = 3y$
    $2x + y = 14$

12. $2x + y = 6$
    $3x - \frac{1}{4}y = 2$

13. $7x - 3y = 6$
    $3x - 2y = -1$

14. $3x - 2y = -5$
    $y = 4x$

15. $x = y + 2$
    $2x - y = 3$

16. $3x - 5y = -3$
    $x + y = 7$

17. $x - \frac{1}{3}y = 1$
    $2x + y = 7$

18. $3x - 2y = -8$
    $x = 2y$

19. $2x + y = 1$
    $x + y = 3$

20. $3x - y = 11$
    $x + y = 5$

21. $y = 2x - 7$
    $3x - 4y = 18$

22. $2x + 3y = -13$
    $x - y = 1$

23. $x + 3y = 19$
    $x + 2y = 13$

24. $y = 4x$
    $5x - 2y = 9$

25. $3x - 2y = 9$
    $2x + 3y = -7$

Part C: No answers are given for these problems. However, each is accompanied by an ordered pair «C,U» showing the chapter and unit in which it was taught.

1. Multiply $(\sqrt{3} - \sqrt{7})$ by its conjugate. «2,2»

2. Solve $7x^2 + 34x - 5 = 0$ «3,2»

3. Use interval notation to describe the truth set for $|4x - 3| > 12$. «4,4»

4. Graph $y = -3$ «6,2»

5. Graph $y = x^2 + 5x - 6$ «7,2»

6. Find the length of the hypotenuse of a right triangle when both legs are 6 inches. «7,3»

7. Simplify $(2 - \frac{3}{5})^{-2}$ «8,3»

8. Evaluate $(3\sqrt{2} - 7)^0$ «8,3»

9. Write 658 using scientific notation. «9,3»

10. Use logarithms to evaluate $\frac{7.4 \sqrt{165}}{(2.3)^3}$ «9,5»

Unit 2: SYSTEMS WITH QUADRATIC EQUATIONS

Quadratic Equations with Two Variables

The equation $y = (x - 3)^2 - 5$ is a quadratic equation with two variables, x and y. From Chapter 8 the equation should be recognized as describing a conic section — in this case a parabola.

(5,-1) is a solution for $y = (x - 3)^2 - 5$ because the equation becomes the true statement $-1 = (5 - 3)^2 - 5$ when x is replaced by 5 and y is replaced by -1. (2,1) is **not** a solution for $y = (x - 3)^2 - 5$ because the equation becomes a false statement $1 = (2 - 3)^2 - 5$ when x is replaced by 2 and y is replaced by 1.

Focus on possible intersections for the graphs of a linear and a quadratic equation

The graph at the right shows the parabola of $y = (x - 3)^2 - 5$ and the lines of three linear equations: $x + 2y = 14$, $2x - y = 12$, and $x + y = -6$.

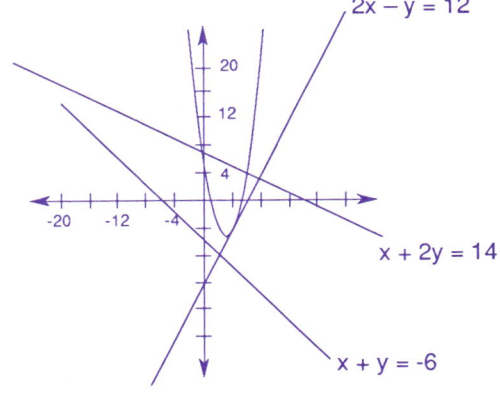

Notice that the parabola intersects at two points with $x + 2y = 14$ and therefore has two common solutions as marked on the graph.

The parabola intersects at only one point with $2x - y = 12$ and therefore has one common solution, (4,-4).

The parabola intersects at no point with $x + y = -6$ and therefore has no common solution, (x,y) where x and y are real numbers.

Whenever a common solution for a linear equation and a quadratic equation (each with two variables) is sought there will be three possible outcomes: two common solutions, one common solution, or no common solution.

Intersecting A Line and A Circle

The common solution for the system containing $2x - y = 5$ and $x^2 + y^2 = 25$ can be found by graphing. The first equation is linear and has a straight line as its graph. The second equation is a circle with center $(0,0)$ and radius 5.

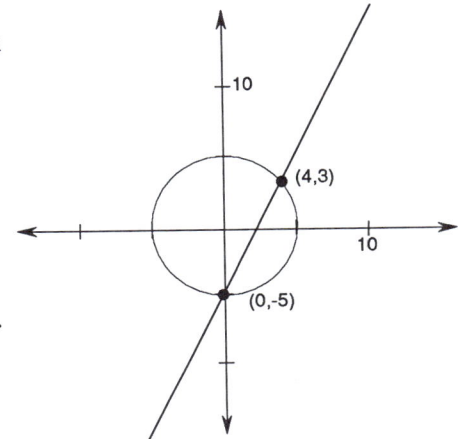

Each equation is graphed separately. A carefully drawn graph will show two points of intersection which are the common solutions, $(0,-5)$ and $(4,3)$.

Focus on solving a system algebraically

The system of equations at the right contains one linear equation and one quadratic equation. The common solutions for the system can be found algebraically in the following steps.

$$2x - y = 5$$
$$x^2 + y^2 = 25$$

1. Solve the linear equation for one of its variables. In this case, solving for y avoids any fractions and makes the computations simpler.

   $$2x - y = 5$$
   $$\underline{-2x \qquad\quad -2x}$$
   $$-y = -2x + 5$$
   $$y = 2x - 5$$

2. Substitute $(2x - 5)$ for y in the quadratic equation.

   $$x^2 + (2x - 5)^2 = 25$$
   $$x^2 + 4x^2 - 20x + 25 = 25$$
   $$5x^2 - 20x + 25 = 25$$
   $$5x^2 - 20x = 0$$
   $$5x(x - 4) = 0$$
   $$x = 0 \text{ or } x = 4$$

3. Solve the quadratic equation which now has only one variable, x.

4. Complete the ordered pairs $(0,\underline{})$ and $(4,\underline{})$ for the common solutions of the system. Either original equation can be used for this purpose, but the linear equation will be easier.

   $$2x - y = 5$$
   $$2 \cdot 0 - y = 5 \text{ or } 2 \cdot 4 - y = 5$$
   $$-y = 5 \text{ or } -y = -3$$
   $$y = -5 \text{ or } y = 3$$

5. The common solutions are $(0,-5)$ and $(4,3)$.

Intersecting A Line and An Ellipse

The common solutions for $x + y = 1$ and $5x^2 + 7y^2 = 35$ are the intersections of the straight line and ellipse shown at the right.

$x + y = 1$ is a linear equation and has a straight line as its graph. $5x^2 + 7y^2 = 35$ is the equation of an ellipse with center (0,0), major axis $2\sqrt{7}$, and minor axis $2\sqrt{5}$.

Each equation is graphed separately. A carefully drawn graph indicates only approximations for the common solutions because the elements of the ordered pairs are not integers. In fact they are irrational numbers.

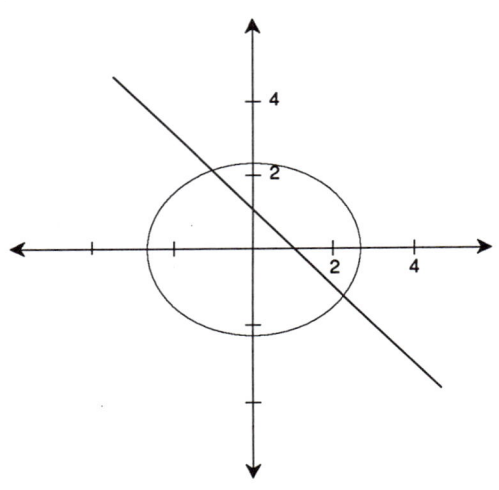

Focus on solving a system algebraically

The system of equations shown at the right contains one linear equation and one quadratic equation. The common solutions for the system can be found algebraically in the following steps.

$x + y = 1$
$5x^2 + 7y^2 = 35$

1. Solve the linear equation for one of its variables. In this case, it makes little difference whether it is solved for x or y.

$$\begin{aligned} x + y &= 1 \\ -y &= -y \\ \hline x &= -y + 1 \end{aligned}$$

2. Substitute $(-y + 1)$ for x in the second equation.

$5(-y + 1)^2 + 7y^2 = 35$
$5(y^2 - 2y + 1) + 7y^2 = 35$
$5y^2 - 10y + 5 + 7y^2 = 35$
$12y^2 - 10y - 30 = 0$
$6y^2 - 5y - 15 = 0$

3. Solve the quadratic equation using the quadratic formula.

$$y = \frac{5 \pm \sqrt{385}}{12}$$

4. Since $y = \frac{5 \pm \sqrt{385}}{12}$ and $x = -y + 1$, then $x = -\left(\frac{5 \pm \sqrt{385}}{12}\right) + 1$ and the common solutions are $\left(\frac{7 - \sqrt{385}}{12}, \frac{5 + \sqrt{385}}{12}\right)$ and $\left(\frac{7 + \sqrt{385}}{12}, \frac{5 - \sqrt{385}}{12}\right)$.

Intersecting A Line and A Hyperbola

The common solutions for $2x+y = 3$ and $4x^2 - 9y^2 = 36$ can be found by graphing as shown at the right. The first equation is linear and has a straight line as its graph. The second equation is a hyperbola with center $(0,0)$, vertices at $(3,0)$ and $(-3,0)$, and branches opening to the sides.

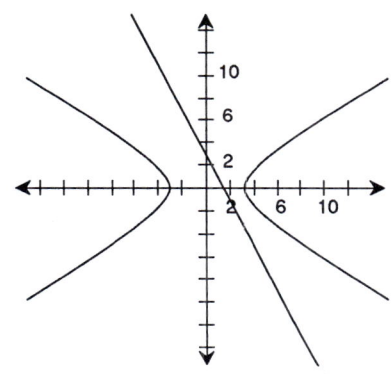

Each equation is graphed separately. A carefully drawn graph shows there are no intersections and this means there is no common solution for the two equations.

Focus on solving a system algebraically

The system of equations at the right can be solved algebraically in the following steps.

$2x + y = 3$
$4x^2 - 9y^2 = 36$

1. Solve the linear equation for one of its variables. In this case, it is easier to solve for y and avoid a result that contains fractions.

$2x + y = 3$

$y = -2x + 3$

2. Substitute $(-2x + 3)$ for y in the second equation.

$4x^2 - 9(-2x + 3)^2 = 36$
$4x^2 - 9(4x^2 - 12x + 9) = 36$
$4x^2 - 36x^2 + 108x - 81 = 36$
$-32x^2 + 108x - 117 = 0$
$32x^2 - 108x + 117 = 0$

3. Solve the quadratic equation using the quadratic formula.

$$y = \frac{108 \pm \sqrt{(-108)^2 - 4 \cdot 32 \cdot 117}}{2 \cdot 32}$$

4. The solutions will not be real numbers because $b^2 - 4ac = (-108)^2 - 4 \cdot 32 \cdot 117$ is a negative number. This means there are no ordered pairs of real numbers (x,y) that are common solutions for the system.

Unit 2 Exercise

Part A: Answers for all Part A problems are at the back of the book.

1. If two straight lines are drawn on the x,y plane how many times will they intersect? Make a drawing to explain the answer.

2. If a circle and a straight line are drawn on the x,y plane how many times will they intersect? Make a drawing to explain the answer.

3. If a parabola and a straight line are drawn on the x,y plane how many times will they intersect? Make a drawing to explain the answer.

4. If an ellipse and a straight line are drawn on the x,y plane how many times will they intersect? Make a drawing to explain the answer.

5. If a hyperbola and a straight line are drawn on the x,y plane how many times will they intersect? Make a drawing to explain the answer.

6. Find values for x for the equations $y = 4x + 5$ and $y = x^2$ by substituting $(4x + 5)$ for y in the second equation and solving the quadratic equation.

7. Complete the ordered pairs (-1,___) and (5,___) as solutions for $y = 4x + 5$.

8. Determine whether (-1,1) and (5,25) are solutions of $y = x^2$.

9. Solve $2x - y = 1$ for y to find the open expression that can be substituted for y.

10. Substitute $(2x - 1)$ for y in the equation $y = (x - 2)^2 + 4$ and find a value for x that will be used in the common solution for the system.

11. Complete the ordered pair (3,___) as a solution for $2x - y = 1$.

12. Determine whether (3,5) is a solution of $y = (x - 2)^2 + 4$.

Part B: Answers for odd-numbered problems of Part B are at the back of the book.

Find common solutions for each system.

1. $y = (x + 5)^2 - 2$ and $x - y = -5$

2. $x^2 + y^2 = 49$ and $2x - y = 8$

3. $x^2 - 4(y + 1)^2 = 36$ and $3x - y = 2$

4. $2x - y = 1$ and $3x^2 + (y - 2)^2 = 3$

5. $(x - 3)^2 + (y + 4)^2 = 25$ and $x + y = 0$

6. $3x - y = 4$ and $y = x^2 + 4x + 7$

7. $x^2 + 4x + 2y^2 = -2$ and $x + 2y = 3$

8. $3x - y = 2$ and $4x^2 - 9y^2 = 36$

Systems of Equations   425

9. $x + 2y = 6$ and $y = (x - 7)^2 + 1$

10. $(x + 7)^2 + (y - 3)^2 = 16$ and $x - y = -6$

11. $3x^2 - y^2 = 3$ and $x + y = 4$

12. $x - y = 1$ and $3x^2 + y^2 = 3$

13. $3x - y = 4$ and $(x - 1)^2 + y^2 = 9$

14. $2(x - 3)^2 - 5y^2 = 10$ and $x + 2y = 1$

15. $4(x - 5)^2 + (y - 3)^2 = 4$ and $x + y = 6$

16. $y = (x - 3)^2 - 7$ and $3x - y = 4$

Part C: No answers are given for these problems. However, each is accompanied by an ordered pair «C,U» showing the chapter and unit in which it was taught.

1. Solve $3x^2 - 9x = 0$ using the set of rational numbers. «1,5»

2. Simplify $2\sqrt{72} - \sqrt{8} - 5\sqrt{18} + 3\sqrt{2}$ «2,3»

3. Simplify $\dfrac{\sqrt{7} - 3\sqrt{5}}{2\sqrt{7} + \sqrt{5}}$ «3,3»

4. Solve $|3x - 2| = 0$ «4,2»

5. Find the rule of correspondence that would match the elements of $\{1,2,3,\ldots\}$ with the set $\{5,7,9,\ldots\}$. «5,1»

6. Write the linear equation with the solutions (-2,3) and (5,2) in standard and y-intercept forms. «6,4»

7. Graph $x \leq -2$ or $y > 2x - 1$ «6,5»

8. Find the center and radius for the circle $x^2 + y^2 - 2x = 3$. «7,4»

9. Simplify $(\sqrt[3]{5^2})^6$ «8,6»

10. $\text{Log}_3 \underline{\quad\quad} = 4$ «9,2»

Unit 3: SYSTEMS WITH THREE EQUATIONS

Three Dimensional Graphs

The x,y plane is two dimensional. It has width and length, but no depth. The x,y plane is similar to the extended surface of a table. Each point on the surface has an ordered pair (x,y), but no point above or below the surface can be described completely by such an ordered pair.

To describe points not on the surface of the x,y plane another dimension is needed. Graphically, this may be accomplished by use of a z-axis that is perpendicular to the x,y plane through the point (0,0). A point on the z-axis that is five units above (0,0) is described by the ordered triple (0,0,5). A point on the z-axis that is three units below (0,0) is described by the ordered triple (0,0,-3).

Similarly, a point on a line perpendicular to the x,y plane through the point (7,-2) will be described by an ordered triple of the form (7,-2,z) where the real number replacing z indicates the directed distance above or below the x,y plane.

Focus on ordered triples for a three-dimensional space

The set of all ordered triples (x,y,z) where the variables are replaced by real numbers represents a three-dimensional space. The human experience is with this type of geometric environment. It is a three-dimensional space in which objects have length, width, and depth.

Each physical object in our environment has three dimensions. Some idealized concepts such as points, lines, and planes have less dimensions.
        A single point has no dimensions.
        A line has one dimension, length.
        A plane has two dimensions, length and width.

Planes in Three-Space

Each plane in 3-space is a two-dimensional figure described by an equation of the form $ax + by + cz = d$ where not all three of the letters a,b,c are zero.

For example, the equation $3x + y - 2z = 9$ is a plane in 3-space. Every ordered triple (x,y,z) that is a solution of $3x + y - 2z = 9$ lies on its plane. Every ordered triple that is not a solution of $3x + y - 2z = 9$ does not lie on the plane.

The equation $0x + 0y + z = 0$ is a plane in 3-space. It is the plane containing all ordered triples of the form $(x,y,0)$ and is, in fact, the plane previously referred to as the x,y plane. The plane whose equation is $z = 0$ is the x,y plane.

The equation $x + 0y + 0z = 5$ is another plane in 3-space. It is the plane perpendicular to the x,y plane which contains the line $x = 5$ on the x,y plane. All ordered triples of the form $(5,y,z)$ are solutions of the equation and points on the plane of $x = 5$.

Focus on planes in 3-space

The x,y plane in 3-space has $0x + 0y + z = 0$ or $z = 0$ as its equation.

Every plane parallel to the x,y plane will have $0x + 0y + z = d$ or $z = d$ as its equation (d is the directed distance between the plane of $z = d$ and the x,y plane). If the x,y plane ($z = 0$) is considered to be a horizontal plane, then every other horizontal plane has an equation of the form $z = d$. The plane of $z = 3$ is parallel to the x,y plane and the distance between the two planes is 3 units.

Similarly, the equations $x = 0$ and $y = 0$ are, respectively, the y,z and x,z planes. If the x,y plane ($z = 0$) is considered to be a horizontal plane, then each equation of the form $x = d$ has a plane parallel to the y,z plane and each equation of the form $y = d$ has a plane parallel to the x,z plane. The plane of $y = 5$ is parallel to the x,z plane and the distance between the two planes is 5 units.

All equations of the form $ax + by + 0z = d$ where not both a,b are zero have planes perpendicular to the x,y plane. The planes of $9x - 7y + 0z = 4$ and $x + 3y + 0z = -2$ are perpendicular to the x,y plane.

428   Chapter 10

Intersections of Two Planes

The intersection of two planes creates three possibilities:

1. The two planes may be parallel and, if so, their intersection is empty. For example, the ceiling and floor of a well-constructed room should be in parallel planes (they will have no points in common).

2. The two planes may be identical and then the intersection is the same plane.

3. If the intersection of two distinct planes is not empty then the intersection is always a straight line. For example, the floor and a wall of a well-constructed room meet in a straight line and have no other common points.

Focus on the intersection of two planes

Two planes are parallel when they have equations $ax + by + cz = d$ and $ax + by + cz = e$ where $d \neq e$. For example, each of the following pairs of equations have parallel planes.
   $5x - 3y + 7z = 12$ and $5x - 3y + 7z = -5$
   $x + 4y - z = 4$ and $x + 4y - z = 9$

Two equations have the same plane if the equations are of the form $ax + by + cz = d$ and $kax + kby + kcz = kd$ where $k \neq 0$. For example, each of the following pairs of equations have identical planes.
   $5x - 3y + 7z = 12$ and $10x - 6y + 14z = 24$ where $k = 2$
   $x + 4y - z = 4$ and $-3x - 12y + 3z = -12$ where $k = -3$

Two planes which are neither parallel nor identical will always have a straight line as their intersection. For example, each of the following pairs of equations have an intersection that is a straight line.
   $6x - y + 4z = 8$ and $2x + 6y - 5z = 3$
   $x + 4y - z = 4$ and $5x + y + 8z = -7$
   $2x + y - 3z = 7$ and $z = 0$

The last example is of particular interest. First, the intersection will be a line on the x,y plane ($z = 0$). Second, the line must have $2x + y = 7$ as its equation on the x,y plane.

Intersections of Three Planes

Three distinct planes may intersect in four different ways:

1. If any two of the planes are parallel, their intersection is empty and consequently the intersection of the three planes is empty.

2. One pair of planes intersects in a straight line which is parallel to the third plane. If so, the intersection of the three planes is empty.

3. One pair of planes may intersect in a straight line and the intersection of a different pair of planes may be the same line. If so, the intersection is the line and there will be an infinite set of common solutions.

4. One pair of planes intersects in a straight line which intersects the third plane in a single point. If so, the intersection of the three planes is that single point.

Focus on the intersection of three planes

The planes of $5x - 3y + 7z = 12$ and $5x - 3y + 7z = -5$ are parallel. Consequently, the intersection of those two planes with any third plane will be empty.

The planes of $6x - y + 7z = 8$ and $6x - y - 3z = 6$ are not parallel because the coefficients of z are different. However, the intersections of those two planes with the x,y plane (z = 0) are the parallel lines $6x - y = 8$ and $6x - y = 6$. Consequently, the intersection of $6x - y + 7z = 8$, $6x - y - 3z = 6$, and the x,y plane is the empty set even though each pair of planes intersects in a straight line.

No pair of planes in the system $x + 4y - z = 4$, $x + 4y + 5z = 4$, and $x + 4y - 4z = 4$ are parallel because the coefficients of z are different. However, the intersections of the three planes with the x,y plane are the line $x + 4y = 4$. Consequently, the intersection of the three planes is a line. The three planes have an infinite set of common solutions, (x,y,0) where $x + 4y = 4$.

The planes of $3x - 2y - z = 7$, $x + 2y - z = -3$, and $x + 4y + z = 3$ intersect in a single point (3,-1,4). The intersection of the first two planes is a line which intersects the third plane in exactly one point, (3,-1,4).

Solving Systems with Three Equations

To solve a system of three equations with three variables, the basic idea is to generate a new system of equations with only two variables. When a common solution for the new system is found, it can be used to solve the original three-equation system.

To solve a system of three equations,
$$2x - y + z = -7 \quad (a)$$
$$x + y + z = -2 \quad (b)$$
$$x - y - 2z = 3 \quad (c)$$

1. Use any pair of equations and eliminate one of the variables. In step 1, equations (a) and (b) were added to eliminate the y's.

   1. $2x - y + z = -7 \quad (a)$
   $x + y + z = -2 \quad (b)$
   $\overline{3x + 2z = -9} \quad (d)$

2. Use any different pair of equations and eliminate the same variable. In step 2 equations (b) and (c) were used to eliminate the y's.

   2. $x + y + z = -2 \quad (b)$
   $\underline{x - y - 2z = 3} \quad (c)$
   $2x - z = 1 \quad (e)$

3. Equations (d) and (e) now form a 2 equation-2 variable system. This system is solved to determine values for x and z.

   3. $3x + 2z = -9 \quad (d)$
   $2x - z = 1 \quad (e)$

   $3x + 2z = -9$
   $\underline{4x - 2z = 2}$
   $7x = -7$
   $x = -1$

   If x = -1 then 2x – z = 1 becomes
   -2 – z = 1 or z = -3.

4. The value for y is found by substituting x = -1 and z = -3 in one of the original equations.

   If x = -1 and z = -3 then 2x – y + z = -7 becomes -2 – y – 3 = -7 or y = 2.

5. Confirm (-1,2,-3) as the common solution of the original system.

   2x – y + z = -7 becomes -2 – 2 – 3 = -7 or -7 = -7
   x + y + z = -2 becomes -1 + 2 – 3 = -2 or -2 = -2
   x – y – 2z = 3 becomes -1 – 2 + 6 = 3 or 3 = 3

Focus on solving a system with 3 equations and 3 variables

To solve the system at the right,

$$3x - 5y - 4z = 9 \quad \text{(a)}$$
$$2x + 4y - 5z = 59 \quad \text{(b)}$$
$$x - 3y + 2z = -27 \quad \text{(c)}$$

1. Use any pair of equations and eliminate one of the variables. Equations (a) and -3 times (c) were added to eliminate the x's.

   $$\begin{array}{rl} 3x - 5y - 4z = 9 & \text{(a)} \\ -3x + 9y - 6z = 81 & -3 \cdot \text{(c)} \\ \hline 4y - 10z = 90 & \text{(d)} \\ 2y - 5z = 45 & \text{(d)} \end{array}$$

2. Use any different pair of equations and eliminate the same variable.

   $$\begin{array}{rl} 2x + 4y - 5z = 59 & \text{(b)} \\ -2x + 6y - 4z = 54 & -2 \cdot \text{(c)} \\ \hline 10y - 9z = 113 & \text{(e)} \end{array}$$

3. Equations (d) and (e) now form a 2 equation-2 variable system which is solved for z and y.

   $$\begin{array}{rl} -10y + 25z = -225 & -5 \cdot \text{(d)} \\ 10y - 9z = 113 & \text{(e)} \\ \hline 16z = -112 \\ z = -7 \end{array}$$

   If $z = -7$ then $2y - 5z = 45$
   becomes $2y + 35 = 45$ or $y = 5$.

4. Substitute $z = -7$ and $y = 5$ in one of the original equations.

   If $z = -7$ and $y = 5$
   then $x - 3y + 2z = -27$
   becomes $x - 15 - 14 = -27$
   or $x - 29 = -27$ or $x = 2$.

5. $(2, 5, -7)$ is the common solution of the original system of 3 equations-3 variables. This should be confirmed using the three original equations.

Systems Without Unique Solutions

A three equation system may have a unique solution, no solution, or an infinite number of solutions.

The system at the right has an infinite number of solutions because equations (a) and (c) are equivalent ($2 \cdot a = c$).

$2x + 4y - 3z = 1$ (a)
$5x - 3y + 4z = 6$ (b)
$4x + 8y - 6z = 2$ (c)

An attempt to solve this system will result in a numerical true statement such as $5 = 5$. When such statements are generated the system has an infinite number of solutions.

Focus on a system with no solutions

The system at the right has no common solutions. Equations (a) and (b) have parallel planes (identical coefficients).

$9x - 4y + 7z = 5$ (a)
$9x - 4y + 7z = 8$ (b)
$7x + 6y - 2z = 4$ (c)

An attempt to solve the system will result in a numerically false statement such as $11 = -8$. When such a result is found, it indicates that the system has no common solution.

Solving Systems of N Equations

If the equation $5x - y + 2z = 7$ represents a 3 dimensional object then it follows that the equation $4x + 7y - 2z + 6w - 3u = 8$ must represent a 5 dimensional object. That may seem impossible for physical objects where our senses seem limited to 3 dimensions, but other concepts such as weather, personality, or politics may involve far more than 3 dimensions. Some of the mathematics that could be applied to many-dimensioned problems is only an extension of the processes used with 3 dimensional problems.

The system at the right consists of 5 equations with 5 variables. To solve such a system, first generate a new system with 4 variables and 4 equations. Then generate another system with 3 variables and 3 equations. Solve the system and, by substitution, find values for the remaining variables.

$4x + 7y - 2z + 6w - 3u = 8$
$8x + 3y + z - 3w - 2u = 0$
$6x - 5y + 4z - 8w - u = 3$
$-x + 3y + 2z - 5w + 9u = 7$
$3x - 3y - 4z + 7w - 8u = 5$

The basic method of solving a system of n equations with n variables is to generate a new system with n – 1 equations and n – 1 variables.

To solve a system with n equations and n variables,
1. Use any pair of equations to eliminate one of the variables.
2. Use any different pair of equations and eliminate the same variable.
3. Repeat step 2 until a system with n – 1 equations and n – 1 variables is generated.
4. Solve the n – 1 equation, n – 1 variable system. (Perhaps this requires repeating steps 1-3 until the system consists of 3 or less equations with 3 or less variables.

For example a system of 10 equations, 10 variables is used to generate a system with 9 equations and 9 variables. The system is then reduced in size until it can be solved as a 2 equation-2 variable system.

Unit 3 Exercise

Part A: Answers for all Part A problems are at the back of the book.

1. Both (3,1,-3) and (1,3,0) are solutions of $2x + 5y - 2z = 17$. Is (-3,3,-4) a solution of $2x + 5y - 2z = 17$?
2. Is (-1,-2,1) a solution of $-x - 3y + z = 8$?
3. For the system shown below, use the first and second equations to eliminate x.
   $2x - 3y + z = -1$ (1)
   $x - 4y - 2z = -8$ (2)
   $-x + 2y + z = 5$ (3)
4. For the system shown below, use the second and third equations to eliminate x.
   $2x - 3y + z = -1$ (1)
   $x - 4y - 2z = -8$ (2)
   $-x + 2y + z = 5$ (3)
5. Find the common solution/s or state that none exist for the two equations obtained in problems 3 and 4.
6. Use the values for y and z obtained in problem 5 with any one of the three original equations in problem 3 to find the common solutions for the 3 equation-3 variable system.
7. Check the solution for all 3 equations in problem 3.

Part B: Answers for odd-numbered problems of Part B are at the back of the book.

Find a unique common solution for each system of equations. If a unique common solution does not exist, state why.

1. $x + 4y - z = -15$
   $3x - 3y + z = 18$
   $4x - y - z = 16$

2. $x + y - z = -4$
   $x - y - z = -2$
   $2x - y - 3z = -7$

3. $2x - y + 3z = 5$
   $x + y + 7z = -2$
   $6x - 3y + 9z = 15$

4. $4x + y - 2z = -5$
   $2x - y - 2z = -4$
   $6x + 3y + z = 3$

5. $x + y - z = 9$
   $2x + y - 3z = 19$
   $3x - 5y + z = -1$

6. $3x + 2y + z = 17$
   $x - y + 4z = 12$
   $5x + y - z = 6$

7. $x - y + 3z = 1$
   $3x - y + z = 2$
   $6x - 2y + 2z = 1$

8. $2x + 7y + z = 8$
   $x - 9y + 3z = 9$
   $x - y - 2z = -1$

9. $x + y - x = 10$
   $x - y - z = 6$
   $y + z = -3$

10. $x - 2y + 3z = 1$
    $3x - 6y + 9z = 3$
    $x + y - z = 6$

11. $y + z = -2$
    $x + z = -1$
    $x + y = 3$

12. $5x - 2y + 7z = 32$
    $6x + 2y - 3z = -3$
    $4x - 2y - 5z = -27$

13. $x - 2y + 3z = 1$
    $3x - 6y + 9z = 3$
    $x + y - z = 6$

14. $3x - z = 3$
    $2x + 3y + 7z = -1$
    $4x - 2y - 3z = 6$

15. $3x - y + 2z = -1$
    $x - 2y + z = 1$
    $-6x + 2y - 4z = 12$

Part C: No answers are given for these problems. However, each is accompanied by an ordered pair «C,U» showing the chapter and unit in which it was taught.

1. Graph the truth set for $(3x - 2)(x + 7) \leq 0$. «1,6»

2. Solve $2x - \frac{1}{5} = 4 - \frac{1}{2}x$ «2,7»

3. The second side of a triangle is $\frac{1}{2}$ the length of the first side, and the third side is 5 inches more than the first side. What is the length of the first side of the triangle that has a perimeter of 60 inches? «3,5»

4. Graph the truth set for $|2x - 5| > 9$. «4,4»

5. Find the slope of the line through the points $(-4,-1)$ and $(1,-3)$. «6,3»

6. Find the length of one side of a right triangle if the other side is 12 inches and the hypotenuse is 15 inches. «7,3»

7. Write the equation of a circle with a radius of 100 and its center is at the origin. «7,4»

8. Simplify $\sqrt[3]{x^2} \cdot \sqrt[3]{x^3}$ «8,6»

9. Is $y = 2x^2$ an exponential function? «9,1»

10. Evaluate $\log_3$ _____ $= 4$. «9,2»

Unit 4: MATRICES AND DETERMINANTS

A Matrix of Real Numbers

A matrix is a rectangular array of real numbers. Frequently, the array is enclosed with square brackets. At the right is shown a system of three equations and three variables. Below that system is a rectangular array of real numbers (a matrix) which contains the coefficients of the system of equations.

$$5x - y + 3z = 4$$
$$2x + 7y = 3$$
$$3x - 2y + z = 15$$

$$\begin{bmatrix} 5 & -1 & 3 \\ 2 & 7 & 0 \\ 3 & -2 & 1 \end{bmatrix}$$

Notice that the second equation has no z term and its entry in the matrix is a zero.

Focus on the meaning of a coefficient matrix

At the right are shown two systems of equations with their associated coefficient matrices (plural of matrix).

x coefficients are in one column, y's in another, and z's in a third.

$$3x - 7y + z = 24$$
$$9x + 8y - 2z = 14$$
$$4x - y = 11$$

$$\begin{bmatrix} 3 & -7 & 1 \\ 9 & 8 & -2 \\ 4 & -1 & 0 \end{bmatrix}$$

$$2x + 3y = 7$$
$$x - y = 2$$

$$\begin{bmatrix} 2 & 3 \\ 1 & -1 \end{bmatrix}$$

Square Matrices

A system of n linear equations with n variables will always have an associated coefficient matrix which has n rows (one for each equation) and n columns (one for each variable).

For example, the system shown at the right contains 3 equations and 3 variables. Its associated coefficient matrix has 3 rows and 3 columns.

Any matrix which has an equal number of rows as columns is a **square** matrix.

$$5x - y + 3z = 4$$
$$2x + 7y = 3$$
$$3x - 2y + z = 15$$

$$\begin{bmatrix} 5 & -1 & 3 \\ 2 & 7 & 0 \\ 3 & -2 & 1 \end{bmatrix}$$

Focus on the meaning of a square matrix

Two matrices are shown at the right. Each is an example of a **square** matrix.

$$\begin{bmatrix} 7 & 5 & 0 & 4 \\ 6 & -4 & 7 & -8 \\ 9 & -2 & -3 & 0 \\ 8 & -3 & 5 & 1 \end{bmatrix}$$

The top matrix is a 4 x 4 matrix and the bottom matrix is a 2 x 2 matrix. A matrix which has n rows and m columns is called an n x m matrix.

$$\begin{bmatrix} -1 & 4 \\ -6 & -7 \end{bmatrix}$$

Definition   **Determinant of a 2 x 2 Matrix**

Every 2 x 2 matrix of the form shown at the right has a real number called its **determinant** assigned to it.

$$\begin{bmatrix} a & b \\ c & d \end{bmatrix}$$

The real number D **if and only if** $D = ad - bc$
is the **determinant** of a 2 x 2 matrix

Square brackets ([and]) are used to enclose a matrix, but vertical lines, like absolute value symbols, are used to indicate a matrix's determinant.

$$\begin{vmatrix} a & b \\ c & d \end{vmatrix} = ad - bc$$

The determinant of a square matrix A is denoted as | A |.

Focus on the determinant of a 2 x 2 matrix

The determinant of a 2 x 2 matrix is a real number. It is found by multiplying the two entries on the diagonal from upper left to lower right and then subtracting the product of the entries on the diagonal from upper right to lower left.

$$\begin{vmatrix} a & b \\ c & d \end{vmatrix} = ad - bc$$

Two examples of the application of the formula are shown at the right.

$$\begin{vmatrix} 7 & -3 \\ 4 & -1 \end{vmatrix} = 7 \cdot -1 - -3 \cdot 4 = 5$$

Notice that the determinant of each matrix is a real number. One determinant is 5; the other -24.

$$\begin{vmatrix} -2 & -4 \\ -5 & 2 \end{vmatrix} = -2 \cdot 2 - -4 \cdot -5 = -24$$

The Diagonals of a Square Matrix

The computation of the determinant of a square matrix larger than 2 x 2 is dependent upon using "diagonals" of the matrix. The word "diagonals" is in quotes because its meaning with regard to matrices is slightly changed.

$$\begin{bmatrix} a & b & c & d \\ e & f & g & h \\ i & j & k & m \\ n & p & q & r \end{bmatrix}$$

There are four upper left to lower right "diagonals" for the 4 x 4 matrix shown above. The entries a-f-k-r lie along the most obvious "diagonal." A second "diagonal" contains the entries b-g-m-n. A third "diagonal" contains the entries c-h-p-i. The fourth upper left to lower right "diagonal" has the entries d-q-j-e.

Focus on upper right to lower left diagonals

A 4 x 4 matrix has 4 upper left to lower right "diagonals" and it also has 4 upper right to lower left "diagonals."

$$\begin{bmatrix} a & b & c & d \\ e & f & g & h \\ i & j & k & m \\ n & p & q & r \end{bmatrix}$$

The most obvious upper right to lower left "diagonal" contains the entries d-g-j-n. A second such "diagonal" contains c-f-i-r. A third "diagonal" contains b-e-q-m. And the fourth upper right to lower left "diagonal" contains a-p-k-h.

The pattern shown here for "diagonals" of a 4 x 4 matrix can be followed for any n x n matrix where n > 2. In all such cases there will be n upper left to lower right "diagonals" and n upper right to lower left "diagonals."

> **Definition**   **The Determinant of an N x N Matrix, N > 2**
>
> Given an n x n matrix with n > 2 the number D is its **determinant** **if and only if** D is the sum of the products on the upper left to lower right diagonals minus the sum of the products on the upper right to lower left diagonals.
>
> To find the determinant for the 3 x 3 matrix shown below,
> 1. Find the sum of the products along the upper left to lower right diagonals.
> $aei + bfg + chd$
> 2. Find the sum of the products along the upper right to lower left diagonals.
> $ceg + bdi + ahf$
> 3. Write the difference of the sums found in steps 1 and 2.
> $(aei + bfg + chd) - (ceg + bdi + ahf)$
>
> $$\begin{bmatrix} a & b & c \\ d & e & f \\ g & h & i \end{bmatrix}$$

Finding the Determinant of a 3 x 3 Matrix

To find the determinant of the 3 x 3 matrix shown at the right, it is necessary to find the products of the entries along the 6 "diagonals."

$$\begin{bmatrix} 3 & -2 & -1 \\ 0 & -2 & 1 \\ 4 & 0 & 5 \end{bmatrix}$$

$D = (3 \cdot -2 \cdot 5 + -2 \cdot 1 \cdot 4 + -1 \cdot 0 \cdot 0) - (-1 \cdot -2 \cdot 4 + -2 \cdot 0 \cdot 5 + 3 \cdot 0 \cdot 1)$

$D = (-30 + -8 + 0) - (8 + 0 + 0)$

$D = (-38) - (8)$

$D = -46$

In the example below the determinant of a 4 x 4 matrix is found. Again, the sum of the products along "diagonals" from upper right to lower left will be subtracted from the sum of the products along "diagonals" from upper left to lower right.

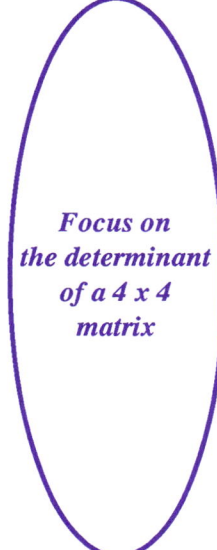

Focus on the determinant of a 4 x 4 matrix

$$\begin{vmatrix} -1 & 4 & -2 & 0 \\ 0 & -2 & 3 & -2 \\ 1 & 6 & -2 & 4 \\ 5 & 0 & 0 & -2 \end{vmatrix} = \begin{array}{l} (-1 \cdot -2 \cdot -2 \cdot -2 + 4 \cdot 3 \cdot 4 \cdot 5 + -2 \cdot -2 \cdot 0 \cdot 1 + 0 \cdot 0 \cdot 6 \cdot 0) \\ - (0 \cdot 3 \cdot 6 \cdot 5 + -2 \cdot -2 \cdot 1 \cdot -2 + 4 \cdot 0 \cdot 0 \cdot 4 + -1 \cdot 0 \cdot -2 \cdot -2) \end{array}$$

$$= (8 + 240 + 0 + 0) - (0 + -8 + 0 + 0)$$
$$= (248) - (-8)$$
$$= 256$$

Unit 4 Exercise

Part A: Answers for all Part A problems are at the back of the book.

1. Write the coefficient matrix for the system $7x - 9y = 8$ and $4x + 2y = 7$.

2. Write the coefficient matrix for the system $3x - 4y + z = 0$, $2x + 6y + 3z = 4$, and $-5x + 8y - 2z = 4$.

3. Write the 3 x 3 coefficient matrix for the system $5x - z = 3$, $x + 4z = 9$, and $3y - 5z = 6$.

4. To find the coefficient matrix for $6x = y - 2$ and $x + 5 = 3y$ the equations first must be written in the form $ax + by = c$. Find the coefficient matrix for the system.

5. The determinant $\begin{vmatrix} a & b \\ c & d \end{vmatrix}$ is $ad - bc$.

    Find the determinant $\begin{vmatrix} s & m \\ j & k \end{vmatrix}$

6. The determinant $\begin{vmatrix} 5 & 1 \\ 7 & 3 \end{vmatrix}$ is $5 \cdot 3 - 1 \cdot 7$ or $15 - 7$ or 8.

    Find the determinant $\begin{vmatrix} 6 & 5 \\ -3 & -2 \end{vmatrix}$

7. The determinant $\begin{vmatrix} a & b & c \\ d & e & f \\ g & h & i \end{vmatrix}$ is

$(aei + bfg + chd) - (ceg + bdi + ahf)$.
Find the determinant below.

$\begin{vmatrix} k & m & q \\ j & r & s \\ t & u & n \end{vmatrix}$

8. The determinant $\begin{vmatrix} 4 & -1 & 1 \\ 3 & 1 & 2 \\ 0 & -2 & 3 \end{vmatrix}$ is

$(12 + 0 + -6) - (0 + -9 + -16)$ or 31.
Find the determinant below.

$\begin{vmatrix} 2 & 5 & -1 \\ -1 & 0 & 4 \\ 1 & 2 & -2 \end{vmatrix}$

9. Is the determinant $\begin{vmatrix} a & b \\ c & d \end{vmatrix}$ equal to the determinant below?

$\begin{vmatrix} -a & -b \\ -c & -d \end{vmatrix}$

10. Is the determinant $\begin{vmatrix} a & b & c \\ d & e & f \\ g & h & i \end{vmatrix}$ equal to the determinant below?

$\begin{vmatrix} -a & -b & -c \\ -d & -e & -f \\ -g & -h & -i \end{vmatrix}$

Part B: Answers for odd-numbered problems of Part B are at the back of the book.

For problems 1-10, find the determinant.

1. $\begin{vmatrix} 2 & 5 \\ 1 & 3 \end{vmatrix}$

2. $\begin{vmatrix} 3 & 11 \\ 4 & -5 \end{vmatrix}$

3. $\begin{vmatrix} 3 & -2 \\ 1 & 2 \end{vmatrix}$

4. $\begin{vmatrix} 3 & -9 \\ 2 & -7 \end{vmatrix}$

5. $\begin{vmatrix} -3 & 3 \\ -4 & -3 \end{vmatrix}$

6. $\begin{vmatrix} -2 & 0 & 1 \\ 3 & -2 & 1 \\ -2 & -1 & -1 \end{vmatrix}$

7. $\begin{vmatrix} 4 & 2 & 7 \\ -2 & 1 & -1 \\ 0 & 5 & 0 \end{vmatrix}$

8. $\begin{vmatrix} 0 & -2 & -1 \\ 4 & -1 & 3 \\ 1 & 4 & -2 \end{vmatrix}$

9. $\begin{vmatrix} 1 & 0 & -3 \\ 2 & -1 & -2 \\ 4 & -2 & -1 \end{vmatrix}$

10. $\begin{vmatrix} 5 & -1 & 4 \\ -2 & 1 & 2 \\ 1 & 0 & 3 \end{vmatrix}$

For problems 11-20, write coefficient matrices and then find determinants.

11. $7x - 3y = 6$
    $3x - 2y = -1$

12. $8x + y = 6$
    $3x - \frac{1}{4}y = 2$

13. $x = y + 2$
    $2x - y = 3$

14. $3x - 5y = -3$
    $x + y = 7$

15. $x - \frac{1}{3}y = 1$
    $2x + y = 7$

16. $x + y - z = -4$
    $x - y - z = -2$
    $2x - y - 3z = -7$

17. $2x - y + 3z = 5$
    $x + y + 7z = -2$
    $6x - 3y + 9z = 15$

18. $4x + y - 2z = -5$
    $2x - y - 2z = -4$
    $6x + 3y + z = 3$

19. $x + y - z = 0$
    $y + 2x - 3z = 0$
    $4x - y - 2z = -3$

20. $3x + 2y + z = 0$
    $x - y + 4z = 4$
    $5x + 3z = 7$

Part C: No answers are given for these problems. However, each is accompanied by an ordered pair «C,U» showing the chapter and unit in which it was taught.

1. Solve the equation $5x^2 - 8x = -4$. «1,4»

2. Use the table of square roots to find the decimal approximation for $\frac{8 - \sqrt{5}}{3}$. «2,4»

3. The sum of the digits of a two digit number is 9. The number with the digits reversed is 27 less than the number. Find the number. «3,5»

4. Write an open sentence that indicates the distance between the number represented by x and -7 is 12. «4,4»

5. True or false? $(-1, 5) \in \{x, y \mid y = x^2 - 4x + 5\}$ «5,2»

6. Find the standard form of the linear equation with solutions (3,-2) and (-1,-4). «6,4»

7. Graph the polynomial function $y = x^2 + 4x$. «7,2»

8. It takes Lou 12 hours to type a report and Mary can type the report in 8 hours. If they both worked on the report, how long would it take them to complete it. «8,7»

9. Using logarithms, evaluate $\frac{17.8(1.4)^2}{\sqrt{29}}$ «9,5»

10. Find the common solutions.
    $x + y + z = 4$
    $x - y + 3z = 10$
    $2x + y - z = 3$ «10,3»

Unit 5: CRAMER'S RULE

Solving a System of Equations

To find a value for x in the system of equations at the right, the following steps could be used.

$ax + by = e$
$cx + dy = f$

1. Multiply the first equation by d. $adx + bdy = de$

2. Multiply the second equation by -b. $-bcx + -bdy = -bf$

3. Add the new equations eliminating y. $adx - bcx = de - bf$

4. Factor the left side of the equation. $(ad - bc)x = de - bf$

5. Divide by $(ad - bc)$. $x = \frac{de - bf}{ad - bc}$

Notice that the denominator of $\frac{de-bf}{ad-bc}$ is the determinant of the coefficient matrix. Notice also that if the coefficients of x were replaced by the constants e and f, the numerator of the fraction would be the determinant of the coefficient matrix.

Focus on solving the system for y

To find a value for y in the system of equations at the right, the following steps could be used.

$ax + by = e$
$cx + dy = f$

1. Multiply the first equation by -c. $-acx - bcy = -ce$

2. Multiply the second equation by a. $acx + ady = af$

3. Add the new equations eliminating x. $ady - bcy = af - ce$

4. Factor the left side of the equation. $(ad - bc)y = af - ce$

5. Divide by $(ad - bc)$. $y = \frac{af - ce}{ad - bc}$

Notice that the denominator of $\frac{af-ce}{ad-bc}$ is the determinant of the coefficient matrix. Notice also that if the coefficients of y were replaced by the constants e and f, the numerator of the fraction would be the determinant of the coefficient matrix.

Matrices and Systems of Equations

The two preceding examples can be generalized as follows:

For any system of two linear equations as shown at the right, three matrices can be formed.

$$ax + by = e$$
$$cx + dy = f$$

M_c is the coefficient matrix.

$$M_c = \begin{bmatrix} a & b \\ c & d \end{bmatrix}$$

M_x is the new coefficient matrix when the coefficients of x are replaced by e and f.

$$M_x = \begin{bmatrix} e & b \\ f & d \end{bmatrix}$$

M_y is the new coefficient matrix when the coefficients of y are replaced by e and f.

$$M_y = \begin{bmatrix} a & e \\ c & f \end{bmatrix}$$

The values of x and y in the common solution of the system can be found as quotients of the determinants of the matrices.

$$x = \frac{|M_x|}{|M_c|} \text{ and } y = \frac{|M_y|}{|M_c|}$$

Focus on solving a system of equations using determinants

For the system of equations shown at the right,

$$5x + 6y = 8$$
$$2x - 3y = 9$$

1) M_c is $\begin{bmatrix} 5 & 6 \\ 2 & -3 \end{bmatrix}$ 2) M_x is $\begin{bmatrix} 8 & 6 \\ 9 & -3 \end{bmatrix}$ 3) M_y is $\begin{bmatrix} 5 & 8 \\ 2 & 9 \end{bmatrix}$

In each case, the coefficients and constants of the equations are used as matrix entries. M_c is the basic matrix and uses the coefficients of x and y in the same order as they appear in the equations. M_x and M_y are edited versions of M_c with the constants 8 and 9 replacing the coefficients of x and y, respectively.

The components of the common solution are found using determinants.

$$x = \frac{|M_x|}{|M_c|} = \frac{-78}{-27} = \frac{78}{27} = \frac{26}{9} \text{ and } y = \frac{|M_y|}{|M_c|} = \frac{29}{-27} = \frac{-29}{27}$$

$(\frac{26}{9}, \frac{-29}{27})$ is the common solution of the system $5x + 6y = 8$ and $2x - 3y = 9$.

Solving a 3 x 3 System Using Determinants

The system shown at the right can be solved using determinants by first writing four matrices.

$ax + by + cz = m$
$dx + ey + fz = n$
$gx + hy + jz = p$

M_c is the coefficient matrix.

$$M_c = \begin{bmatrix} a & b & c \\ d & e & f \\ g & h & j \end{bmatrix}$$

M_x is the new coefficient matrix when the coefficients of x are replaced by m,n,p.

$$M_x = \begin{bmatrix} m & b & c \\ n & e & f \\ p & h & j \end{bmatrix}$$

M_y is the new coefficient matrix when the coefficients of y are replaced by m,n,p.

$$M_y = \begin{bmatrix} a & m & c \\ d & n & f \\ g & p & j \end{bmatrix}$$

M_z is the new coefficient matrix when the coefficients of z are replaced by m,n,p.

$$M_z = \begin{bmatrix} a & b & m \\ d & e & n \\ g & h & p \end{bmatrix}$$

Focus on using determinants with a 3 x 3 system

A system of 3 equations-3 variables has 4 associated matrices: one to represent the coefficients of the variables, and one for each of the variables.

For the system of equations shown at the right, the four matrices are listed below.

$5x - y - 2z = -8$
$4x - 2y + 6z = 1$
$3x - y + 4z = -1$

$$M_c = \begin{bmatrix} 5 & -1 & -2 \\ 4 & -2 & 6 \\ 3 & -1 & 4 \end{bmatrix} \quad M_x = \begin{bmatrix} -8 & -1 & -2 \\ 1 & -2 & 6 \\ -1 & -1 & 4 \end{bmatrix} \quad M_y = \begin{bmatrix} 5 & -8 & -2 \\ 4 & 1 & 6 \\ 3 & -1 & 4 \end{bmatrix} \quad M_z = \begin{bmatrix} 5 & -1 & -8 \\ 4 & -2 & 1 \\ 3 & -1 & -1 \end{bmatrix}$$

The x, y, z solutions are found as quotients of determinants.

$$x = \frac{|M_x|}{|M_c|} = \frac{32}{-16} = -2, \; y = \frac{|M_y|}{|M_c|} = \frac{48}{-16} = -3, \text{ and } z = \frac{|M_z|}{|M_c|} = \frac{-8}{-16} = \frac{1}{2}$$

The common solution for the system is $(-2, -3, \frac{1}{2})$.

Cramer's Rule

The use of determinants to solve systems of n equations with n variables is called Cramer's Rule. Although Cramer's Rule was in use long before the computer, the method is an excellent example of the type of problem where computers have greatly expanded the power of mathematics in solving problems. Solving a 10 x 10 system of equations with pencil and paper is an awesome task in terms of time and accuracy. Once programmed, the computer task is straight-forward calculations.

To use Cramer's Rule on the system at the right,

$$3x - 4y = 13$$
$$-x + 6y = -2$$

1. The determinant of M_x is divided by the determinant of M_c to solve for x.

$$x = \frac{|M_x|}{|M_c|} = \frac{\begin{vmatrix} 13 & -4 \\ -2 & 6 \end{vmatrix}}{\begin{vmatrix} 3 & -4 \\ -1 & 6 \end{vmatrix}} = \frac{70}{14} = 5$$

2. The determinant of M_y is divided by the determinant of M_c to solve for y.

$$y = \frac{|M_y|}{|M_c|} = \frac{\begin{vmatrix} 3 & 13 \\ -1 & -2 \end{vmatrix}}{\begin{vmatrix} 3 & -4 \\ -1 & 6 \end{vmatrix}} = \frac{7}{14} = \frac{1}{2}$$

The common solution of the system above is $(5, \frac{1}{2})$.

Focus on solving a 3 x 3 system with Cramer's Rule

Cramer's Rule may be applied to a system of n equations, n variables. For each variable, k, the determinant of M_k is divided by the determinant of M_c.

$$x + 3y - z = 16$$
$$2x - y + z = 9$$
$$x - 2y + z = 0$$

$$x = \frac{|M_x|}{|M_c|} = \frac{\begin{vmatrix} 16 & 3 & -1 \\ 9 & -1 & 1 \\ 0 & -2 & 1 \end{vmatrix}}{\begin{vmatrix} 1 & 3 & -1 \\ 2 & -1 & 1 \\ 1 & -2 & 1 \end{vmatrix}} = \frac{7}{1} = 7$$

$$y = \frac{|M_y|}{|M_c|} = \frac{\begin{vmatrix} 1 & 16 & -1 \\ 2 & 9 & 1 \\ 1 & 0 & 1 \end{vmatrix}}{\begin{vmatrix} 1 & 3 & -1 \\ 2 & -1 & 1 \\ 1 & -2 & 1 \end{vmatrix}} = \frac{2}{1} = 2$$

$$z = \frac{|M_z|}{|M_c|} = \frac{\begin{vmatrix} 1 & 3 & 16 \\ 2 & -1 & 9 \\ 1 & -2 & 0 \end{vmatrix}}{\begin{vmatrix} 1 & 3 & -1 \\ 2 & -1 & 1 \\ 1 & -2 & 1 \end{vmatrix}} = \frac{-3}{1} = -3$$

The common solution of the system is (7,2,-3).

Systems of Equations

Unit 5 Exercise

Part A: Answers for all Part A problems are at the back of the book.

1. Find the matrix M_c for: $4x - 7y = 1$ and $2x - 5y = 0$.

2. Write M_c for $2x + 9y = 8$ and $x = 3y + 4$ by first writing the second equation in the form $ax + by = c$.

3. Write M_c and M_x for: $6x - 5y = 7$ and $3x + 8y = 4$.

4. Write M_c and M_x for $5x - 7 = y$ and $4y = 3 - x$ by first writing both equations in the form $ax + by = c$.

5. Find the x component of the common solution for the system $x + 2y = 10$ and $2x - y = 0$ by dividing $|M_x|$ by $|M_c|$.

6. Find the y component of the common solution for the system $x + 2y = 10$ and $2x - y = 0$ by dividing $|M_y|$ by $|M_c|$.

7. Find the x component of the common solution for the system $x + 3 = y$ and $2x + 3y = 4$ by dividing $|M_x|$ by $|M_c|$.

8. Find the y component of the common solution for the system $x + 3 = y$ and $2x + 3y = 4$ by dividing $|M_y|$ by $|M_c|$.

9. Write the matrices M_x, M_y, M_z, and M_c for the system $x - 2y - z = 11$, $3x + y + z = 3$, and $2x - 3y + z = 25$.

10. Write the matrices M_x, M_y, M_z, and M_c for the system $4x + y = 11$, $3x + z = 7$, and $-3y + z = 4$ by first writing each equation in the form $ax + by + cz = d$.

11. Find the x component of the common solution for the system $2x - 3y + z = 25$, $x - 2y - z = 11$, and $3x + y + z = 3$ by dividing $|M_x|$ by $|M_c|$.

12. Find the y component of the common solution for the system $2x - 3y + z = 25$, $x - 2y - z = 11$, and $3x + y + z = 3$ by dividing $|M_y|$ by $|M_c|$.

13. Find the z component of the common solution for the system $2x - 3y + z = 25$, $x - 2y - z = 11$, and $3x + y + z = 3$ by dividing $|M_z|$ by $|M_c|$.

Chapter 10

Part B: Answers for odd-numbered problems of Part B are at the back of the book.

Find common solutions for the following systems using Cramer's Rule.

1. $x - y = 8$ and $2x + 3y = 1$

2. $2x - y + 3z = -8$, $x + 5y + z = 5$, and $-x + 2z = 8$

3. $6x - 2y = 0$ and $5x - y = 1$

4. $6x - 5y = 7$ and $2x + 9y = 4$

5. $5x - 3y = 7$ and $2x + y = 4$

6. $8x - 5y = 0$ and $3x - y = 1$

7. $x = 2y + 5$ and $2x - y = 1$

8. $3x + 5y = 4$ and $2x + 10 = 3y$

9. $2x + 5y = -29$ and $3x - y = -1$

10. $x - 2y = -2$ and $3x + y = 8$

11. $x + 2y - z = 2$, $x - y + z = 0$, and $2x - 3y - z = 3$

12. $2x + 2y - z = -9$, $x - y + z = 4$, and $3x - 2y + 2z = 7$

13. $2x + 3y - z = 16$, $y + 2z = -8$, and $x - y = 1$

14. $-2x - 3y = 6$, $2x - 4y - z = 3$, and $3x + y = -2$

15. $x + y - z = -2$, $x - y + z = 4$, and $2x - y - z = 1$

Part C: No answers are given for these problems. However, each is accompanied by an ordered pair «C,U» showing the chapter and unit in which it was taught.

1. Find the truth set for $3x^2 + 7x = 6$. «2,6»

2. If the purchase of 3 cars and 2 trucks cost $40,800, and the purchase of 7 cars and 1 truck cost $66,600, what is the price of each car and each truck? «3,5»

3. Graph the truth set of $|2x - 5| = -7$. «4,3»

4. Find the slope for $2x - 7y = -14$. «6,3»

5. Graph $2x + y > 4$ and $x - 2y < 7$. «6,5»

6. Find the distance between the points (-5,0) and (2,-4). «7,3»

7. Simplify $(\sqrt[6]{3^4})^3$ «8,6»

8. Evaluate $\log_3 243 = $ _____ «9,2»

9. Find the common solution for:
$2x - 3y = 6$
$y = x - 4$ «10,1»

10. Find the common solution for:
$y = x^2 - 3$
$x + 3y = 1$ «10,2»

Chapter 10 Test

«10,U» shows the unit in which the problem was studied in this chapter.

1. Which ordered pairs are solutions of $4x - y = 2$?
   a. (0,2) b. (1,2) c. (0,2)
   d. $\left(\frac{-1}{2}, 0\right)$ e. (1,6)   «10,1»

2. Which ordered triples are solutions of $2x - y + z = 5$?
   a. (1,1,4) b. $\left(\frac{-1}{2}, 1, -4\right)$ c. $\left(\frac{1}{2}, 5, -1\right)$
   d. (3,0,-1). e. (-1,-2,5)   «10,3»

Find common solutions for problems 3-10 or state that none exists.

3. $3x - y = 4$
   $2x + y = 1$ «10,1»

4. $4x - y = -5$
   $x + 2y = 10$ «10,1»

5. $y = 2x$
   $3x - y = 2$ «10,1»

6. $x = y - 1$
   $3x + y = 5$ «10,1»

7. $x + y = 2$
   $y = x^2$ «10,2»

8. $(x - 1)^2 + (y + 2)^2 = 25$
   $x + 3y = 10$ «10,2»

9. $x + y + z = 3$
   $x - y + 3z = 7$
   $2x + y - z = 0$ «10,3»

10. $y + z = 5$
    $x - y + z = 0$
    $x - z = -4$ «10,3»

Find the common solution for each system in problems 11-14 using Cramer's Rule.

11. $2x - 3y = 17$
    $3x + 4y = -17$ «10,5»

12. $x - y = 5$
    $3x - y = 1$ «10,5»

13. $x + y + 3z = -8$
    $2x + y - z = 5$
    $x - y + z = -2$ «10,5»

14. $2x + 3y = -8$
    $x + z = -4$
    $x - y - z = 4$ «10,5»

Logarithm Table

N	0	1	2	3	4	5	6	7	8	9
10	.00000	.00432	.00860	.01284	.01703	.02119	.02531	.02938	.03342	.03743
11	.04139	.04532	.04922	.05308	.05690	.06070	.06446	.06819	.07188	.07555
12	.07918	.08279	.08636	.08991	.09342	.09691	.10037	.10380	.10721	.11059
13	.11394	.11727	.12057	.12385	.12710	.13033	.13354	.13672	.13988	.14301
14	.14613	.14922	.15229	.15534	.15836	.16137	.16435	.16732	.17026	.17319
15	.17609	.17898	.18184	.18469	.18752	.19033	.19312	.19590	.19866	.20140
16	.20412	.20683	.20952	.21219	.21484	.21748	.22011	.22272	.22531	.22789
17	.23045	.23300	.23553	.23805	.24055	.24304	.24551	.24797	.25042	.25285
18	.25527	.25768	.26007	.26245	.26482	.26717	.26951	.27184	.27416	.27646
19	.27875	.28103	.28330	.28556	.28780	.29003	.29226	.29447	.29667	.29885
20	.30103	.30320	.30535	.30750	.30963	.31175	.31387	.31597	.31806	.32015
21	.32222	.32428	.32634	.32838	.33041	.33244	.33445	.33646	.33846	.34044
22	.34242	.34439	.34635	.34830	.35025	.35218	.35411	.35603	.35793	.35984
23	.36173	.36361	.36549	.36736	.36922	.37107	.37291	.37475	.37658	.37840
24	.38021	.38202	.38382	.38561	.38739	.38917	.39094	.39270	.39445	.39620
25	.39794	.39967	.40140	.40312	.40483	.40654	.40824	.40993	.41162	.41330
26	.41497	.41664	.41830	.41996	.42160	.42325	.42488	.42651	.42813	.42975
27	.43136	.43297	.43457	.43616	.43775	.43933	.44091	.44248	.44404	.44560
28	.44716	.44871	.45025	.45179	.45332	.45484	.45637	.45788	.45939	.46090
29	.46240	.46389	.46538	.46687	.46835	.46982	.47129	.47276	.47422	.47567
30	.47712	.47857	.48001	.48144	.48287	.48430	.48572	.48714	.48855	.48996
31	.49136	.49276	.49415	.49554	.49693	.49831	.49969	.50106	.50243	.50379
32	.50515	.50651	.50786	.50920	.51055	.51188	.51322	.51455	.51587	.51720
33	.51851	.51983	.52114	.52244	.52375	.52504	.52634	.52763	.52892	.53020
34	.53148	.53275	.53403	.53529	.53656	.53782	.53908	.54033	.54158	.54283
35	.54407	.54531	.54654	.54777	.54900	.55023	.55145	.55267	.55388	.55509
36	.55630	.55751	.55871	.55991	.56110	.56229	.56348	.56467	.56585	.56703
37	.56820	.56937	.57054	.57171	.57287	.57403	.57519	.57634	.57749	.57864
38	.57978	.58092	.58206	.58320	.58433	.58546	.58659	.58771	.58883	.58995
39	.59106	.59218	.59329	.59439	.59550	.59660	.59770	.59879	.59988	.60097
40	.60206	.60314	.60423	.60531	.60638	.60746	.60853	.60959	.61066	.61172
41	.61278	.61384	.61490	.61595	.61700	.61805	.61909	.62014	.62118	.62221
42	.62325	.62428	.62531	.62634	.62737	.62839	.62941	.63043	.63144	.63246
43	.63347	.63448	.63548	.63649	.63749	.63849	.63949	.64048	.64147	.64246
44	.64345	.64444	.64542	.64640	.64738	.64836	.64933	.65031	.65128	.65225
45	.65321	.65418	.65514	.65610	.65706	.65801	.65896	.65992	.66087	.66181
46	.66276	.66370	.66464	.66558	.66652	.66745	.66839	.66932	.67025	.67117
47	.67210	.67302	.67394	.67486	.67578	.67669	.67761	.67852	.67943	.68034
48	.68124	.68215	.68305	.68395	.68485	.68574	.68664	.68753	.68842	.68931
49	.69020	.69108	.69197	.69285	.69373	.69461	.69548	.69636	.69723	.69810
50	.69897	.69984	.70070	.70157	.70243	.70329	.70415	.70501	.70586	.70672

N	0	1	2	3	4	5	6	7	8	9
50	.69897	.69984	.70070	.70157	.70243	.70329	.70415	.70501	.70586	.70672
51	.70757	.70842	.70927	.71012	.71096	.71181	.71265	.71349	.71433	.71517
52	.71600	.71684	.71767	.71850	.71933	.72016	.72099	.72181	.72263	.72346
53	.72428	.72509	.72591	.72673	.72754	.72835	.72916	.72997	.73078	.73159
54	.73239	.73320	.73400	.73480	.73560	.73640	.73719	.73799	.73878	.73957
55	.74036	.74115	.74194	.74273	.74351	.74429	.74507	.74586	.74663	.74741
56	.74819	.74896	.74974	.75051	.75128	.75205	.75282	.75358	.75435	.75511
57	.75587	.75664	.75740	.75815	.75891	.75967	.76042	.76118	.76193	.76268
58	.76343	.76418	.76492	.76567	.76641	.76716	.76790	.76864	.76938	.77012
59	.77085	.77159	.77232	.77305	.77379	.77452	.77525	.77597	.77670	.77743
60	.77815	.77887	.77960	.78032	.78104	.78176	.78247	.78319	.78390	.78462
61	.78533	.78604	.78675	.78746	.78817	.78888	.78958	.79029	.79099	.79169
62	.79239	.79309	.79379	.79449	.79518	.79588	.79657	.79727	.79796	.79865
63	.79934	.80003	.80072	.80140	.80209	.80277	.80346	.80414	.80482	.80550
64	.80618	.80686	.80754	.80821	.80889	.80956	.81023	.81090	.81158	.81224
65	.81291	.81358	.81425	.81491	.81558	.81624	.81690	.81757	.81823	.81889
66	.81954	.82020	.82086	.82151	.82217	.82282	.82347	.82413	.82478	.82543
67	.82607	.82672	.82737	.82802	.82866	.82930	.82995	.83059	.83123	.83187
68	.83251	.83315	.83378	.83442	.83506	.83569	.83632	.83696	.83759	.83822
69	.83885	.83948	.84011	.84073	.84136	.84198	.84261	.84323	.84386	.84448
70	.84510	.84572	.84634	.84696	.84757	.84819	.84880	.84942	.85003	.85065
71	.85126	.85187	.85248	.85309	.85370	.85431	.85491	.85552	.85612	.85673
72	.85733	.85794	.85854	.85914	.85974	.86034	.86094	.86153	.86213	.86273
73	.86332	.86392	.86451	.86510	.86570	.86629	.86688	.86747	.86806	.86864
74	.86923	.86982	.87040	.87099	.87157	.87216	.87274	.87332	.87390	.87448
75	.87506	.87564	.87622	.87679	.87737	.87795	.87852	.87910	.87967	.88024
76	.88081	.88138	.88195	.88252	.88309	.88366	.88423	.88480	.88536	.88593
77	.88649	.88705	.88762	.88818	.88874	.88930	.88986	.89042	.89098	.89154
78	.89209	.89265	.89321	.89376	.89432	.89487	.89542	.89597	.89653	.89708
79	.89763	.89818	.89873	.89927	.89982	.90037	.90091	.90146	.90200	.90255
80	.90309	.90363	.90417	.90472	.90526	.90580	.90634	.90687	.90741	.90795
81	.90849	.90902	.90956	.91009	.91062	.91116	.91169	.91222	.91275	.91328
82	.91381	.91434	.91487	.91540	.91593	.91645	.91698	.91751	.91803	.91855
83	.91908	.91960	.92012	.92065	.92117	.92169	.92221	.92273	.92324	.92376
84	.92428	.92480	.92531	.92583	.92634	.92686	.92737	.92788	.92840	.92891
85	.92942	.92993	.93044	.93095	.93146	.93197	.93247	.93298	.93349	.93399
86	.93450	.93500	.93551	.93601	.93651	.93702	.93752	.93802	.93852	.93902
87	.93952	.94002	.94052	.94101	.94151	.94201	.94250	.94300	.94349	.94399
88	.94448	.94498	.94547	.94596	.94645	.94694	.94743	.94792	.94841	.94890
89	.94939	.94988	.95036	.95085	.95134	.95182	.95231	.95279	.95328	.95376
90	.95424	.95472	.95521	.95569	.95617	.95665	.95713	.95761	.95809	.95856
91	.95904	.95952	.95999	.96047	.96095	.96142	.96190	.96237	.96284	.96332
92	.96379	.96426	.96473	.96520	.96567	.96614	.96661	.96708	.96755	.96802
93	.96848	.96895	.96942	.96988	.97035	.97081	.97128	.97174	.97220	.97267
94	.97313	.97359	.97405	.97451	.97497	.97543	.97589	.97635	.97681	.97727
95	.97772	.97818	.97864	.97909	.97955	.98000	.98046	.98091	.98137	.98182
96	.98227	.98272	.98318	.98363	.98408	.98453	.98498	.98543	.98588	.98632
97	.98677	.98722	.98767	.98811	.98856	.98900	.98945	.98989	.99034	.99078
98	.99123	.99167	.99211	.99255	.99300	.99344	.99388	.99432	.99476	.99520
99	.99564	.99607	.99651	.99695	.99739	.99782	.99826	.99870	.99913	.99957
100	.00000	.00043	.00087	.00130	.00173	.00217	.00260	.00303	.00346	.00389

Square Root Table

Number	Square Root	Number	Square Root	Number	Square Root
1	1.000	36	6.000	71	8.426
2	1.414	37	6.083	72	8.485
3	1.732	38	6.164	73	8.544
4	2.000	39	6.245	74	8.602
5	2.236	40	6.325	75	8.660
6	2.449	41	6.403	76	8.718
7	2.646	42	6.481	77	8.775
8	2.828	43	6.557	78	8.832
9	3.000	44	6.633	79	8.888
10	3.162	45	6.708	80	8.944
11	3.317	46	6.782	81	9.000
12	3.464	47	6.856	82	9.055
13	3.606	48	6.928	83	9.110
14	3.742	49	7.000	84	9.165
15	3.873	50	7.071	85	9.220
16	4.000	51	7.141	86	9.274
17	4.123	52	7.211	87	9.327
18	4.243	53	7.280	88	9.381
19	4.359	54	7.348	89	9.434
20	4.472	55	7.416	90	9.487
21	4.583	56	7.483	91	9.539
22	4.690	57	7.550	92	9.592
23	4.796	58	7.616	93	9.644
24	4.899	59	7.681	94	9.695
25	5.000	60	7.746	95	9.747
26	5.099	61	7.810	96	9.798
27	5.196	62	7.874	97	9.849
28	5.292	63	7.937	98	9.899
29	5.385	64	8.000	99	9.950
30	3.477	65	8.062	100	10.000
31	5.568	66	8.124		
32	5.657	67	8.185		
33	5.745	68	8.246		
34	5.831	69	8.307		
35	5.916	70	8.367		

Answers

Chapter 0 Pre-Test

1. true
2. $12, \sqrt{3}, -15, \frac{4}{9}, \sqrt{\frac{3}{10}}$
3. $\{-5,14\}$
4. $2, (x-5), (3x+2)$
5. $3xy^2 - 3x^2y + 1$
6. $4x^2 - 1$
7. $6x^2 - 5x - 21$
8. $15x^3 - 7x^2 - 7x - 1$
9. $\frac{9}{2}x - \frac{3}{10}$
10. $\frac{-1}{2}$
11. $\{\frac{28}{9}\}$
12. $\{-6\}$
13. $\{-4\}$
14. $\{-8\}$
15. $\{\frac{-8}{45}\}$
16. $\{\frac{1}{8}\}$
17. 9, 15, 18
18. no
19. $x > \frac{9}{2}$
20. $x > 2$
21. $-3(5x + 1)$
22. $(3x - 2)(4x - 3)$
23. $(c + d)(r + s)$
24. $x(x^2 - 3x + 1)$
25. $(b + c)(a - x)$
26. $(x - 4)(x - 10)$
27. $(x + 7)(x - 6)$
28. $(x + 7)(x - 7)$
29. $(2x - 9)(2x + 1)$
30. $(x + 2)(3x + 4)$

Chapter 0, Unit 1
Part A

1. $\{2,15,37\}$
2. $7 \notin \{1,3,5,6\}$
3. no
4. yes
5. $\{1,2,3,\ldots\}$
6. no
7. counting number
8. no
9. yes
10. yes
11. 17, 31
12. 17, 31, -3, -9
13. $17, 31, -3, -9, \frac{4}{5}$
14. $17, 31, -3, -9, \frac{4}{5}, \sqrt{3}, \pi$
15. $\{x \mid x$ is a whole number between 2 and 7$\}$
16. $\{x \mid x$ is an even number between 5 and 15$\}$

Chapter 0, Unit 1
Part B

1. true
3. true
5. true
7. no
9. $\{11,12,13,\ldots\}$
11. $\{9,17\}$
13. $\{1,2,3,4,5,6,7\}$
15. 0, -4, 19, -13
17. $\pi, 4, \frac{-2}{3}, \sqrt{11}, -17, \frac{0}{5}$
19. 27, 7

Chapter 0, Unit 2
Part A

1. $8x - 16y$
2. $a + b$
3. $-15xy^2z$
4. $28a^4b^3c^2$
5. $7y + 35$
6. $24x + 1$
7. $-4x + 11$
8. $14x - 7$
9. $5x^4 + 3x^3 + 6x - 8$
10. $-6x^4 + 9x^3 - 5x^2 - 6x - 3$
11. $3x^2 - 4x - 15$
12. $4x^2 - 3x - 10$
13. $63x^2 - 76x + 21$
14. $\frac{15}{4}x - \frac{3}{5}$
15. $7x + \frac{17}{42}$
16. $\frac{2}{5}x - \frac{1}{30}$
17. $\frac{14}{3}x + \frac{2}{5}$
18. $\frac{-1}{3}$
19. $-6x + \frac{29}{24}$
20. $\frac{5}{8}x - \frac{13}{10}$
21. $x^3 - 10x^2 + 24x - 9$
22. $x^3 + 5x^2 - 3x + 18$
23. $x^3 - 125$
24. $x^3 + 1$

Chapter 0, Unit 2
Part B

1. $-3x + 2y$
3. $3xy^2 - 5x$
5. $-5xy$
7. $-20x^4y^5z^2$

9. $30 + 20x$
11. $5x + 10$
13. $x^2 - 5x - 24$
15. $25x^2 - 10x + 1$
17. $4x^2 - 20x + 25$
19. $30x^2 + 11x - 30$
21. $-x^3 - 7x^2 - 2$
23. $3x - \frac{11}{24}$
25. $\frac{25}{3}x + \frac{5}{6}$
27. $\frac{-5}{3}x + \frac{9}{20}$
29. $x^3 - 7x^2 + 13x - 15$
31. $6x^3 - 11x^2 + 5x - 3$
33. $2x^3 - 15x^2 + 14x + 24$
35. $8x^3 + 27$

Chapter 0, Unit 3
Part A

1. $\{9\}$
2. $\{\ \}$
3. $\{-5\}$
4. $\{\ \}$
5. $\{\frac{8}{3}\}$
6. $\{\frac{-45}{8}\}$
7. -5
8. $+3$
9. $+9x$
10. $\frac{-1}{12}$
11. 60
12. $56x$
13. $(2,7)$
14. $\left(\frac{3}{11}, \frac{1}{11}\right)$

Chapter 0, Unit 3
Part B

1. $\{8\}$
3. $\{9\}$
5. $\{7\}$
7. $\{\ \}$
9. $\{\ \}$
11. $\{7\}$
13. $\{1\}$
15. $\{\ \}$
17. $\{\ \}$
19. $\{-7\}$
21. $\{\frac{-3}{4}\}$
23. $\{\frac{52}{105}\}$
25. $\{\frac{-7}{2}\}$
27. $\{\frac{105}{17}\}$
29. $\{\frac{29}{18}\}$
31. $(5,-1)$
33. $(4,-3)$
35. $(2,-5)$

Chapter 0, Unit 4
Part A

1. yes
2. yes
3. yes
4. 6 and 9
5. 10 and 8
6. yes
7. no
8. yes
9. no
10. $x \geq -5$
11. $x > 10$
12. $x < -9$
13. $x \leq 13$
14. $5x > x + 8$
15. $4x > 8$
16. $x > 2$
17. $x < 4$
18. $x \leq 5$
19. $x \leq \frac{-5}{4}$
20. $x < 3$

Chapter 0, Unit 4
Part B

1. $x \geq 2$
3. $x < -3$
5. $x > \frac{-16}{3}$
7. $x \leq 2$
9. $x > 3$
11. $x \geq 2$
13. $x < -1$
15. $x > 5$
17. $x < -4$
19. $x < \frac{-4}{3}$

Chapter 0, Unit 5
Part A

1. $7(x - 3)$
2. $3xy(x - 4y)$
3. $x(x^2 - 2x - 1)$
4. $-2(5x + 3)$
5. $(x + 5)$
6. $(x + 2)(3x + 7)$
7. $(x - 6)(5x^2 - x + 4)$
8. $(t + s)(r + q)$
9. $(x + 6)(x - 3)$
10. $(3x - 2)(2x + 1)$

Chapter 0, Unit 5
Part B

1. $2xy^2z(4x^2 - 9z)$
3. $xy(x + 1)$
5. $2(4x^3y - 5z)$
7. $-3xy^2(3y - 5)$
9. $2(x^2 - 5x + 7)$
11. $(4x - 1)(x^2 - 6)$
13. $(2x - 1)(4x + 1)$
15. $(x + 4)(x^2 + x + 4)$
17. $(x + 3)(2x^2 + 6x - 5)$
19. $(u + r)(m + t)$
21. $(a + b)(x^2 + y^2)$
23. $-14abc^3(2ac - 3b)$
25. $5xz^3(3x - 5z)$
27. $(x - 3)(8x + 5)$
29. $(y + z)(x - a)$
31. $(x + 7)(x + 5)$
33. $(x + 7)(x - 2)$
35. prime
37. $(x - 9)(x - 6)$
39. prime
41. $(2x + 7)(3x + 1)$
43. $(x + 2)(3x - 5)$
45. prime
47. $(x - 3)(3x - 4)$
49. prime

Chapter 0 Test

1. false
2. no
3. $\{1,2,4,5,7,11\}$
4. $5x, -3$
5. $6x - y$
6. $35x - 40$
7. $x^2 - 4x - 45$
8. $12x^3 - 11x^2 + 8x - 4$
9. $\frac{20}{3}x - \frac{1}{5}$
10. $\frac{3}{5}$
11. $\{18\}$
12. $\{7\}$
13. $\{-2\}$
14. $\{\frac{-12}{5}\}$
15. $\{\frac{12}{11}\}$
16. $\{\frac{-11}{4}\}$
17. yes
18. yes
19. $x \leq -2$
20. $x \leq -4$
21. $xy(x + 1)$
22. $2(x^2 - 4x - 6)$
23. $(x + 5)(x^2 + 5x + 1)$
24. prime
25. $-14abc(2ac^4 - 3)$
26. $(x - 8)(x + 4)$
27. $(x + 11)(x + 1)$
28. $(2x - 3)(3x + 4)$
29. prime
30. $(x + 5)(3x + 2)$

Chapter 1, Unit 1
Part A

1. $\{1,2,3,\ldots\}$
2. no
3. counting
4. Commutative
5. yes
6. $17 \cdot 9x$
7. yes
8. Associative
9. is not
10. $(9 + 4)x = 13x$
11. $7y + 35$
12. $24 + 16x$
13. $\{8\}$
14. $\{\ \}$
15. $\{9\}$
16. $\{1,2,3,\ldots\}$
17. $\{20,21,22,\ldots\}$
18. $\{1,2,3,\ldots,35\}$

Chapter 1, Unit 1
Part B

1. true
3. true
5. counting
7. is not
9. is not
11. a
13. $5z + 30$
15. $(5 + 8)xy^2 = 13xy^2$
17. $\{19\}$
19. $\{3\}$
21. $\{\ \}$
23. $\{\ \}$
25. $\{1,2,3,\ldots\}$
27. $\{\ \}$
29. $\{1,2,3,\ldots,20\}$

Chapter 1, Unit 2
Part A

1. $\{\ldots,-3,-2,-1,0,1,2,3,\ldots\}$
2. integer
3. Closure
4. no
5. yes
6. $-6z + 5y$
7. $(x + 8)(4x - 3y)$
8. $6(4x + 7y)$
9. $6y + (-4x - 4z)$
10. $(5 \cdot -4)y$
11. $[3(x + 4)](x - 6)$
12. zero
13. $4a - 5b$

14. -1y
15. zero
16. {16}
17. { }
18. {-2}
19. {-3,-2,-1,0, . . .}
20. {-5,-4,-3, . . .}

Chapter 1, Unit 2
Part B

1. no
3. 3 – 5x
5. -4 + (x + 15)
7. yes
9. {-7}
11. {-6}
13. { }
15. {9}
17. {. . . -3,-2,-1,0,1,2,3, . . .}
19. {7}
21. {-3,-4,-5, . . .}
23. {-2,-1,0, . . .}
25. {-9,-10,-11, . . .}

Chapter 1, Unit 3
Part A

1. no
2. yes
3. $\frac{-1}{0}$
4. $\frac{-3}{4}$
5. $\frac{4}{3}$
6. $\frac{-9}{5}$
7. $\frac{1}{9}$
8. $\frac{-1}{7}$
9. $\frac{-1}{x}$

10. yes
11. 30x – 27 = 20
12. 40 + 21x = 22
13. 5x > 19
14. $x < \frac{-4}{7}$
15. 9x < -5
16. -4x > 11
17. $x > \frac{5}{6}$
18. x < -2

Chapter 1, Unit 3
Part B

1. Distributive Law of Multiplication over Addition
3. Inverse Law of Addition
5. Inverse Law of Multiplication
7. Closure Law of Multiplication
9. Closure Law of Addition
11. Inverse Law of Multiplication
13. Distributive Law of Multiplication over Addition
15. Inverse Law of Addition
17. Multiplication Law of Negative One
19. $\{\frac{43}{3}\}$
21. { }
23. $\{\frac{5}{6}\}$
25. $\{\frac{-35}{12}\}$
27. x > -2
29. $x \leq \frac{-14}{3}$

Chapter 1, Unit 4
Part A

1. (3x + 4)(2x + 1)
2. (x – 4)(x + 3)
3. prime

4. $5(x - 4)(x - 3)$
5. $(x - y + 8)(x - y - 1)$
6. $(x - 5 + 3y)(x - 5 + y)$
7. $(a + b - 5d)(a + b + d)$
8. $(7 - a - d)(1 - a - d)$
9. $(r - s)(r^2 + rs + s^2)$
10. $(a - 4)(a^2 + 4a + 16)$
11. $(x + 1)(x^2 - x + 1)$
12. $(2x + 5)(4x^2 - 10x + 25)$
13. $(5y - 2x)(25y^2 + 10yx + 4x^2)$
14. $(4a - 3b)(16a^2 + 12ab + 9b^2)$
15. $(x + z)(x^2 - xz + z^2)$

Chapter 1, Unit 4
Part B

1. $(x - 4y)(x - 3y)$
3. $(x + 3y)^2$
5. $(x - 5)(x^2 + 5x + 25)$
7. $(x + 6y)(x - y)$
9. $(x + 10y)(x + 5y)$
11. $(x + 2)(x^2 - 2x + 4)$
13. $(2x - 1)(4x^2 + 2x + 1)$
15. $(x + y + 4a - 4b)(x + y - a + b)$
17. $(x + 5y)(x - 5y)$
19. $(x + 11y)(x - 3y)$
21. $(x - 10)(x^2 + 10x + 100)$
23. $4(x - 5)(x + 2)$
25. $(x + 14)(x - 1)$
27. $(10 - 3b)(100 + 30b + 9b^2)$
29. $(y - x - a)^2$
31. $(10y - 1)(100y^2 + 10y + 1)$
33. $(1 - a + r)(1 - 4a + 4r)$
35. $(a + 2b)(a^2 - 2ab + 4b^2)$
37. $(x - 7 + 8y)(x - 7 + 2y)$
39. $(x + a - 10)(x + a - 1)$
41. prime
43. $(x + 8y)(x - 2y)$
45. $(x - y + 2z)(x - y + z)$
47. $(D^4 + E^2)(D^8 - D^4E^2 + E^4)$
49. $(xy + z)(x^2y^2 - xyz + z^2)$

Chapter 1, Unit 5
Part A

1. $\{-7, 7\}$
2. $\{0, 4\}$
3. $\{0, -2\}$
4. $\{0, \frac{5}{4}\}$
5. $\{0, \frac{3}{2}\}$
6. $x^2 - 8x + 10$ is prime
7. $\{\ \}$
8. $\{\frac{-7}{3}, \frac{7}{3}\}$
9. $\{-2\}$
10. $\{-4, 4, -1, 1\}$

Chapter 1, Unit 5
Part B

1. $\{0, 5\}$
3. $\{0, 9\}$
5. $\{0, \frac{7}{2}\}$
7. $\{-3, 2\}$
9. $\{\frac{-1}{2}, -3\}$
11. $\{5, -4\}$
13. $\{\ \}$
15. $\{\frac{1}{2}, -3\}$
17. $\{\frac{2}{3}, 4\}$
19. $\{2, -2, 1, -1\}$
21. $\{\frac{5}{4}\}$

Chapter 1, Unit 6
Part A

1. negative
2. $>$
3. $<$
4. $x - 3 > 0$ and $x + 7 > 0$ or
   $x - 3 < 0$ and $x + 7 < 0$

5. <
6. >
7. x − 4 > 0 and x − 9 < 0   or
   x − 4 < 0 and x − 9 > 0
8. a
9. b
10. the endpoints are open in the first one, and they are solid (closed) in the second one
11. the first is two half-lines, and the second one is a line segment

Chapter 1, Unit 6
Part B

1. ⟵──○────────○────⟶ x
       -6   0

3. ⟵──┼──●────────●──⟶ x
        0 3   7

5. ⟵──○──┼────────○──⟶ x
       -4 0   7

7. ⟵──●──┼────────●──⟶ x
       -3 0   10

9. ⟵──●──●──┼────⟶ x
       -7 -3 0

11. ⟵──○────────○──⟶ x
         1   6

13. ⟵──●────┼──┼──⟶ x
        -5   0 1

15. ⟵──○────┼○──⟶ x
        -5   0 $\frac{2}{3}$

17. ⟵──●──●────⟶ x
        0 $\frac{4}{9}$

19. ⟵──○──┼────────○──⟶ x
       $\frac{-8}{3}$ 0   8

Chapter 1, Unit 7
Part A

1. 3x
2. 45(y + 8)
3. .05n + .10(2n) = .25n
4. 3t + 4(t + 2) = 7t + 8

5. 300 adults, 400 children
6. 67 @ 12¢, 77 @ 15¢
7. 71 lbs @ 80¢ and 29 lbs @ 60¢
8. 6 @ $1,800 and 12 @ $3,000
9. 50 gallons @ 80¢ and 150 gallons @ 60¢
10. 55 lbs @ 40¢ and 45 lbs @ 80¢
11. 250 gallons @ $6.00 and 750 gallons @ $4.00
12. 400 lbs @ 70¢ and 300 lbs @ 42¢

Chapter 1, Unit 7
Part B

1. 240 adults and 560 children
3. 360 qt @ $4 and 540 qt @ $5
5. 42 lbs @ 80¢
7. 25 quarters and 50 dimes
9. 400 grams
11. .08x + .05(3x − 1) = .23x − .05
13. 15 red and 17 firebrick

Chapter 1 Test

1. no
2. same evaluation for any replacement(s) of the variable(s)
3. Commutative Law of Addition
4. 3xy(3x − 5)
5. yes
6. no
7. (3x − 4y)(x + y)
8. (x − 5y)(x − 2y)
9. (4a + 5)(16a^2 − 20a + 25)
10. 4(x^2 − 2x − 15) = 4(x + 3)(x − 5)
11. prime
12. (a − 5x − 5y)(a + 2x + 2y)
13. (a + 4 − 4b)(a + 4 + 4b)
14. $x \geq \frac{7}{5}$
15. $x > \frac{-1}{5}$
16. $\{\frac{22}{15}\}$

17. $\{\frac{6}{5}\}$
18. $\{0,7\}$
19. $\{\frac{-4}{3},1\}$
20. $\{\ \}$
21. $\{2,-2,3,-3\}$
22. $\{2,-2\}$
23. ←——o———o——→ at -3 and 0
24. ←——●———●——→ at $\frac{-5}{2}$ and 5
25. 30 gallons @ $3.00; 70 gallons @ $4.00

Chapter 2, Unit 1
Part A

1. B
2. C
3. B
4. number line with point at π between 2 and 4
5. rational
6. yes
7. yes
8. yes
9. yes
10. yes
11. no
12. yes
13. real
14. real number line
15. yes
16. no
17. true
18. $(a + b) + c$
19. a
20. $(17 \cdot 2) \cdot \sqrt{5}$

Chapter 2, Unit 1
Part B

1. yes
3. yes
5. yes
7. yes
9. real number line
11. true
13. false
15. real
17. true
19. $\pi + 3$
21. $b \cdot a$
23. $a \cdot (b \cdot c)$
25. x
27. 0
29. w
31. $-\sqrt{68}$
33. -t
35. π
37. 0
39. $4\sqrt{3} + a\sqrt{3}$

Chapter 2, Unit 2
Part A

1. rational
2. irrational
3. $2\sqrt{6}$
4. $6\sqrt{2}$
5. $\frac{5}{4}$
6. $-\sqrt{15}$
7. $\frac{\sqrt{6}}{3}$
8. $\frac{2\sqrt{7} + 5\sqrt{21}}{7}$
9. $8\sqrt{3}$
10. no
11. $8\sqrt{2}$
12. $-24 + 9\sqrt{2}$
13. $10 + 5\sqrt{7} - 2\sqrt{3} - \sqrt{21}$
14. $-8 + 9\sqrt{5}$
15. $(3\sqrt{5} + 2\sqrt{6})$
16. 21
17. $\frac{28 + 17\sqrt{6}}{50}$
18. $\frac{11 + 8\sqrt{2}}{7}$

Chapter 2, Unit 2
Part B

1. $4\sqrt{3}$
3. $5\sqrt{3}$
5. $5\sqrt{5}$
7. $\frac{3}{10}$
9. 51
11. $6\sqrt{5}$
13. $-12\sqrt{6}$
15. $\frac{7\sqrt{2}}{2}$
17. $\frac{5\sqrt{2}}{6}$
19. $\frac{-\sqrt{2}}{2}$
21. $\frac{2\sqrt{6} - 3\sqrt{30}}{6}$
23. $\frac{\sqrt{21}}{3}$
25. $\frac{\sqrt{11}}{11}$
27. $-4\sqrt{2}$
29. $2 + 4\sqrt{3}$
31. $-2\sqrt{5} + \sqrt{2}$
33. $\sqrt{2} + \sqrt{5}$
35. $12 - \sqrt{5}$
37. $9 + 2\sqrt{30}$
39. $158 + 27\sqrt{6}$
41. $31 + \sqrt{15}$
43. 164
45. $7 - \sqrt{15}$
47. $-6 + \sqrt{35}$
49. $-10 + 7\sqrt{2}$

Chapter 2, Unit 3
Part A

1. 125, 216, 343
2. rational
3. irrational
4. irrational
5. rational
6. $2\sqrt[3]{2}$
7. $5\sqrt[3]{2}$
8. $-3\sqrt[3]{3}$
9. $-2\sqrt[3]{6}$
10. 256 and 625
11. $2\sqrt[4]{3}$
12. $-15\sqrt[4]{2}$
13. $\frac{4\sqrt[3]{15}}{9}$
14. $\frac{-3}{2}$
15. $-2 + 5\sqrt[3]{2}$

Chapter 2, Unit 3
Part B

1. 1, 8, 27, 64, 125
3. $5\sqrt[3]{5}$
5. -48
7. $2\sqrt[4]{2}$
9. $\sqrt[4]{4}$
11. $\frac{10\sqrt[4]{27}}{3}$
13. $\sqrt[4]{60} - \sqrt[4]{3} + 20$

Chapter 2, Unit 4
Part A

1. a. .6 b. .85
2. a. $.\overline{142857}$ b. $.\overline{18}$
3. a. $\frac{173}{1000}$ b. $\frac{39}{200}$
   c. $\frac{643}{900}$ d. $\frac{213}{990}$
4. a. 7.550 b. 5.745
   c. 8.944 d. 9.798
5. a. $10\sqrt{2} = 14.14$
   b. $3\sqrt{30} = 16.431$
   c. $4\sqrt{5} = 8.944$
   d. $4\sqrt{10} = 12.648$
6. 3.3165
7. 4.3595
8. 8.646
9. 2.354
10. .786

Chapter 2, Unit 4
Part B

1. .9375
3. $.\overline{857142}$
5. $.\overline{6}$
7. $.\overline{5}$
9. $.8\overline{3}$
11. $\frac{104}{125}$
13. $\frac{2168}{495}$
15. $\frac{4175}{9999}$
17. 9.165
19. 8.544
21. 21.213
23. 12.123
25. 10.634
27. $\frac{1.764}{3} = .588$
29. $\frac{-3.243}{4} = -.811$

Chapter 2, Unit 5
Part A

1. yes
2. yes
3. no
4. yes
5. $x - 7 = 2$
6. yes
7. $x - 14$
8. 9
9. yes
10. no
11. yes
12. yes
13. yes
14. yes
15. no

Chapter 2, Unit 5
Part B

1. 12
3. $\frac{-\sqrt{42} - 8\sqrt{7} - 3\sqrt{6} - 24}{58}$
5. $\frac{2\sqrt{21}}{3}$
7. $\frac{-21}{32}$
9. $\frac{2 - 5\sqrt{13}}{20}$
11. $\frac{15 + 3\sqrt{3}}{2}$
13. $\frac{44}{3}$
15. $\frac{6\sqrt{3} - \sqrt{42}}{3}$
17. $\frac{-5\sqrt{7} + 5\sqrt{21}}{2}$
19. $\frac{5 + \sqrt{17}}{2}$
21. $x > -3\sqrt{3}$
23. $x \leq \frac{-5\sqrt{5} - 60}{139}$
25. $x \leq \frac{4 + \sqrt{26}}{5}$
27. $x \leq \sqrt{3} - 4$

29. $x > \frac{\sqrt{7}+\sqrt{5}}{5}$

Chapter 2, Unit 6
Part A

1. {5,3}
2. {0,-8}
3. {2}
4. {10,4}
5. $\{4+\sqrt{11}, 4-\sqrt{11}\}$
6. {1,-9}
7. { }
8. (-∞,7)
9. [-3,8)
10. {17}
11. no
12. $\sqrt{(x+12)(x-3)} = x+4$
13. {52}
14. (-∞,-1) ∪ (4,∞)
15. $[-1, \frac{3}{2}]$

Chapter 2, Unit 6
Part B

1. {4,-3}
3. $\{1+\sqrt{11}, 1-\sqrt{11}\}$
5. {5,-11}
7. $[\frac{-1}{5}, 3]$
9. { }
11. all real numbers
13. $\{\frac{7+\sqrt{29}}{2}, \frac{7-\sqrt{29}}{2}\}$
15. $\{\frac{5+\sqrt{13}}{6}, \frac{5-\sqrt{13}}{6}\}$
17. (-1,5)
19. all real numbers
21. {1,2}
23. {-7,7}
25. {86}

27. {8}
29. {6,2}
31. $\{\frac{7+\sqrt{13}}{2}, \frac{7-\sqrt{13}}{2}\}$

Chapter 2, Unit 7
Part A

1. $25x - 12 = 16$
2. 30
3. $20x - 24 = 25$
4. $\frac{-1}{5}$
5. $(\frac{-16}{15}, 0)$
6. $(-\infty, \frac{-35}{4}] \cup (0, \infty)$
7. 1, -1
8. $\frac{4}{3}$
9. $\frac{-26}{5}$
10. (-∞,-3) ∪ (0,3)
11. (6,∞) ∪ (-4,-3)
12. (5,7) ∪ (-∞,-4]

Chapter 2, Unit 7
Part B

1. {1}
3. $\{\frac{-7}{4}\}$
5. $\{\frac{6}{5}\}$
7. $(-\infty, \frac{-3}{2}]$
9. $(-\infty, \frac{44}{15})$
11. $\{\frac{-1}{9}\}$
13. $\{\frac{-67}{11}\}$
15. $\{\frac{13}{9}\}$
17. (0,∞) ∪ (-∞,-4]

19. (0,∞) ∪ $(-\infty, \frac{-3}{10}]$
21. (0,∞) ∪ $(-\infty, \frac{-30}{11}]$
23. $(\frac{12}{7}, \infty)$
25. $(0, \frac{22}{5}]$
27. {6}
29. {5,-2}
31. {3,-3}
33. {-5,3}
35. {-3,-1}
37. $(\frac{-13}{2}, -6)$
39. [3,∞) ∪ (1,2]

Chapter 2, Unit 8
Part A

1. 2.8
2. 46
3. .10x
4. .02(y + 10)
5. .05(x + 3)
6. 500 lbs
7. 20 lbs
8. 666.67 lbs
9. 80 gallons
10. $66\frac{2}{3}$ lbs
11. $2\frac{2}{3}$ gallons
12. 2 gallons

Chapter 2, Unit 8
Part B

1. 4 grams
3. 30 gallons
5. 900 pounds

7. 100 pounds
9. $6\frac{2}{3}$ gallons
11. 10 grams
13. 740 gallons
15. 4 gallons

Chapter 2 Test

1. real number line
2. $r^2 - 12 = 0$
3. $-\sqrt{30}$
4. -13
5. $\frac{14 - 9\sqrt{3}}{47}$
6. $-10\sqrt[3]{3}$
7. $-20 + 2\sqrt[3]{2}$
8. $-10\sqrt[4]{5}$
9. 17.110
10. .292
11. $-.035$
12. 8.944
13. 29.044
14. $\{-5,3\}$
15. $\{3\sqrt{2},-3\sqrt{2}\}$
16. $\left\{\frac{3+\sqrt{14}}{5}, \frac{3-\sqrt{14}}{5}\right\}$
17. $\{-4 + 2\sqrt{5}, -4 - 2\sqrt{5}\}$
18. $\{-2\}$
19. $\frac{11}{16}$
20. $\frac{-86}{3}$
21. $(-\infty,-5) \cup (1,\infty)$
22. $[-3,3]$
23. $(4,\infty)$
24. $33\frac{1}{3}$ gallons
25. 10 gallons

Chapter 3, Unit 1
Part A

1. no
2. no
3. -1
4. i
5. $5i$
6. real
7. $7i$
8. negative
9. positive
10. $5i$
11. $2i$
12. $8i$
13. $3i\sqrt{3}$
14. -20
15. $-i$
16. $-i$
17. $6i - 10$ or $-10 + 6i$
18. $-2 - 3i$
19. $-9 + 20i$

Chapter 3, Unit 1
Part B

1. $4i$
3. $2i\sqrt{10}$
5. $-6i\sqrt{6}$
7. $i\sqrt{85}$
9. -18
11. 36
13. 48
15. -40
17. $-\sqrt{130}$
19. $-\sqrt{165}$
21. -81
23. -4
25. -7
27. i
29. i
31. -8
33. $-24i$
35. -15
37. $-6 + 14i$
39. $3 + 12i$
41. $15 + 4i$
43. $10 - 11i$
45. $10 + 4i$
47. 25
49. 50

Chapter 3, Unit 2
Part A

1. $\{4i,-4i\}$
2. $\{8i,-8i\}$
3. $\{5i,-5i\}$; $(5i)^2 = 25i^2 = 25 \cdot -1 = -25$
4. $\{i\sqrt{19},-i\sqrt{19}\}$; $(i\sqrt{19})^2 = -1 \cdot 19 = -19$
5. $\{1 + i, 1 - i\}$
6. $2i$
7. $2i - 2 - 2i + 2 = 0$
8. $\left\{\frac{3+\sqrt{21}}{2}, \frac{3-\sqrt{21}}{2}\right\}$
9. $\frac{-9 - 3\sqrt{21}}{2}$
10. $\frac{15 + 3\sqrt{21}}{2}$
11. $\frac{15 + 3\sqrt{21}}{2} - 3\left(\frac{3+\sqrt{21}}{2}\right) - 3 = 0$
12. $\left\{\frac{3+i\sqrt{7}}{2}, \frac{3-i\sqrt{7}}{2}\right\}$
13. $\frac{-9 - 3i\sqrt{7}}{2}$
14. $\frac{1 + 3i\sqrt{7}}{2}$
15. $\frac{1 + 3i\sqrt{7}}{2} + \frac{-9 - 3i\sqrt{7}}{2} + 4 = 0$

Chapter 3, Unit 2
Part B

1. $\{10i, -10i\}$
3. $\{i\sqrt{7}, -i\sqrt{7}\}$
5. $\{2i\sqrt{2}, -2i\sqrt{2}\}$
7. $\{i, -i\}$
9. $\{3i\sqrt{5}, -3i\sqrt{5}\}$
   Check: $(3i\sqrt{5})^2 = 9i^2 \cdot 5 = 45i^2 = -45$
11. $\{-1 + i\sqrt{6}, -1 - i\sqrt{6}\}$
    Check:
    $5 - 2i\sqrt{6} - 2 + 2i\sqrt{6} + 7 = 0$
13. $\left\{\frac{3 + i\sqrt{15}}{2}, \frac{3 - i\sqrt{15}}{2}\right\}$
    Check:
    $\frac{-6 + 6i\sqrt{15}}{4} + \frac{-9 - 3i\sqrt{15}}{2} + 6 = 0$
15. $\left\{\frac{5 + i\sqrt{7}}{2}, \frac{5 - i\sqrt{7}}{2}\right\}$
17. $\{1 + i\sqrt{2}, 1 - i\sqrt{2}\}$
19. $\left\{\frac{1 + i\sqrt{3}}{2}, \frac{1 - i\sqrt{3}}{2}\right\}$
21. $\left\{\frac{1 + i\sqrt{39}}{4}, \frac{1 - i\sqrt{39}}{4}\right\}$

Chapter 3, Unit 3
Part A

1. yes
2. yes
3. yes
4. imaginary
5. imaginary
6. $-3i$
7. yes
8. $0 - 5i$
9. $5 + 9i$
10. $6 + 17i$
11. $(8 + 5i)$
12. $(4 + 6i)$
13. $(-4 + 10i)$
14. $(-7 - 3i)$
15. $\frac{-11 - 41i}{34}$
16. T
17. F
18. T
19. F
20. T

Chapter 3, Unit 3
Part B

1. yes
3. $-7 + 0i$
5. $\frac{4}{17} + 0i$
7. $3 + 2i$
9. $0 + 0i = 0$
11. $-2 + 6i$
13. $0 + 2i = 2i$
15. $8 + 0i = 8$
17. $0 + 0i = 0$
19. $15 - 4i$
21. $-2 + 26i$
23. $5 - 14i$
25. $-1 - 17i$
27. 5
29. 13
31. $\frac{13 - 9i}{5}$
33. $\frac{34 - 2i}{20} = \frac{17 - i}{10}$
35. $\frac{-27 + 36i}{45} = \frac{-3 + 4i}{5}$
37. $\frac{11 + 2i}{5}$
39. $\frac{11 + 23i}{13}$

Chapter 3, Unit 4
Part A

1. 4
2. $n = q$
3. 5
4. $x = -3, y = -4$
5. $x = 0, y = \frac{5}{7}$
6. $x = \sqrt{5}, y = 2$
7. $7 + 9i$
8. $3 - 5i$
9. $x = 3, y = 3$
10. $x = -3, y = -10$
11. $\frac{1}{-2 - 5i}$
12. $\frac{1}{3 - 2i}$
13. $x = \frac{-1}{13}, y = \frac{8}{13}$
14. $x = \frac{19}{26}, y = \frac{17}{26}$
15. $x = \frac{12}{5}, y = \frac{9}{5}$

Chapter 3, Unit 4
Part B

1. $x = 11, y = -1$
3. $x = 5, y = 16$
5. $x = 3, y = -2$
7. $x = -4, y = -3$
9. $x = \frac{9}{34}, y = \frac{19}{34}$
11. $x = \frac{-26}{25}, y = \frac{-7}{25}$
13. $x = 9, y = -5$
15. $x = \frac{-4}{5}, y = \frac{-2}{5}$
17. $x = \frac{67}{53}, y = \frac{49}{53}$
19. $x = \frac{1}{10}, y = \frac{-3}{10}$

Chapter 3, Unit 5
Part A

1. $6 per book, $2 per magazine
2. 6 inches, 6 inches, and 9 inches
3. $367 and $271
4. 57
5. plane 600 mph; wind 60 mph
6. 1 liter of 90% and 9 liters of 50%
7. scull, 12 mph; current 4 mph
8. 600 – 15% alc.; 400 – 80% alc.

Chapter 3, Unit 5
Part B

1. l = 20 inches; w = 17 inches
3. 42
5. 600 gal. of 1% additives, 1400 gal. of 3% additives
7. $5 per bat, $6 per glove
9. $3,200 and $3,600
11. 82
13. bonds, $3000; savings, $1000

Chapter 3 Test

1. negative
2. $4i\sqrt{2}$
3. $-10i$
4. $23 - 2i$
5. $2i\sqrt{5}, -2i\sqrt{5}$
6. $\frac{-1 + i\sqrt{7}}{2}, \frac{-1 - i\sqrt{7}}{2}$
7. $1 + 2i, 1 - 2i$
8. $\frac{1 + i\sqrt{23}}{6}, \frac{1 - i\sqrt{23}}{6}$
9. $(11 - i)$
10. $(21 - i)$
11. 29
12. $\frac{-2 + 16i}{13}$
13. $(3 - 5i)$
14. $\frac{1}{2 + 3i}$
15. $x = 11, y = -4$
16. $x = 0, y = -2$
17. $x = 8, y = -1$
18. 74
19. 12 inches
20. $20 per stuffed toy, $110 per train set

Chapter 4, Unit 1
Part A

1. 6
2. 5
3. $|-8|$
4. zero
5. 43
6. 6, -6
7. distance cannot be negative
8. yes
9. no
10. ←●————————●→ x
    -6   0   6
11. ←——●——+——●——→ x
       -3 0 3

Chapter 4, Unit 1
Part B

1. 4
3. 28
5. 7
7. 18
9. 27
11. {13,-13}
13. { }
15. {12,-12}
17. { }
19. {10,-10}
21. ←—○——+——○—→ x
      -10   0   10
23. ←——●——+——●——→ x
       -6   0   6
25. all real numbers
    ←————+————→ x
         0
27. ←——○——+——○——→ x
       -3 0 3
29. ←——●————+————●——→ x
       -11   0   11
31. ←——○——+——○——→ x
       -2 0 2
33. zero
    ←————●————→ x
         0
35. empty set
    ←————+————→ x
         0

Chapter 4, Unit 2
Part A

1. $|13 - 2|$ and $|2 - 13|$
2. $|-5 - 7|$ and $|7 + 5|$
3. -3
4. 10
5. -2, 8
6. -11, 3
7. no
8. when x is negative (x < 0)
9. negative (x < 0)
10. negative (x < 0)
11. nonnegative (x ≥ 0)
12. $\{3, \frac{-1}{3}\}$
13. $\{\frac{-9}{5}, -1\}$
14. $-(x - 8) = 2$ or $-x + 8 = 2$

15. +(7x + 2) = 8 or
    -(7x + 2) = 8
16. $\{\frac{7}{2}, -2\}$
17. False, 7 + 2 ≠ 5
18. False, 7 ≠ -7

Chapter 4, Unit 2
Part B

1. 9
3. 23
5. 2
7. 2
9. 13
11. 9
13. 4
15. 17
17. {13,3}
19. {-10,-8}
21. {0,-16}
23. {3,7}
25. {-8,-4}
27. { }
29. {-10,-2}
31. $\{4, \frac{-2}{3}\}$
33. $\{-2, \frac{-5}{2}\}$
35. { }
37. $\{2, \frac{1}{3}\}$
39. {1,-6}

Chapter 4, Unit 3
Part A

1. 5 < x < 13
2. x < -13 or x > 3
3. (-∞,5) ∪ (13,∞)
4. [-4, 6]
5. a
6. d
7. b
8. e
9. c
10. 8 ≤ x ≤ 13 and [8,13]
11. -8 < x ≤ 14 and (-8,14]
12. -6 < x < 2 and (-6,2)
13. -8 < x < -5 and (-8,-5)
14. -3 < x < 11 and (-3,11)
15. 0 ≤ x ≤ 9 and [0,9]
16. -13 ≤ x ≤ -8 and [-13,-8]

Chapter 4, Unit 3
Part B

1. (number line with closed dots at -9 and -1)
3. (number line with closed dots at -3 and 7)
5. empty set
7. whole number line
9. (number line with closed dots at 1 and 7/3)
11. x < -6 or x > 4
13. empty set
15. x < 2 or x > 14
17. $\frac{-13}{3} \leq x \leq 1$
19. x < -1 or x > $\frac{-1}{3}$
21. (-∞,-7) ∪ (-3,∞)
23. [1,13]
25. (-∞,∞)
27. $(\frac{-9}{2}, 1)$
29. empty set

Chapter 4, Unit 4
Part A

1. | x – 2 | = 1
2. | x – 5 | = 3
3. | x + 1 | = 3
4. -2, 6
5. | x – 3 | = 7
6. 3, 5
7. -6, 4
8. | x – 2 | = 4
9. (number line with open circles at 3 and 5) →x
10. 2.9, 3, 3.19
11. $\frac{1}{10}$, 1
12. | x – 5 | < .03

Chapter 4, Unit 4
Part B

1. | x – 3 | = 8
3. -1, 3
5. -1, $\frac{3}{2}$
7. {-1,7}
9. 2; 3
11. yes
13. yes
15. -1; 3
17. .5; 4
19. 13 and 23
21. -4 and 2
23. 3 ft and 7 ft

Chapter 4 Test

1. yes
2. yes
3. (number line with closed dots at -4 and 4) →x

4. [number line: open circles at -3 and 3]
5. false
6. 0
7. 2,8
8. -3,6
9. { }
10. [number line: closed circles at -6 and -2]
11. [number line: open circles at 0/2 and 8]
12. [number line: closed circles at -5 and -1]
13. no solution [number line at 0]
14. [number line: open at 0/2, closed at 5]
15. $(-\infty,-2) \cup (3,\infty)$
16. $(-\infty,\infty)$
17. $\left(\frac{-22}{5}, \frac{8}{5}\right)$
18. $\left[-2, \frac{8}{3}\right]$
19. $|x+3|=4$
20. yes

Chapter 5, Unit 1
Part A

1. {9,51,96,103}
2. no
3. true
4. true
5. no
6. no
7. false
8. 3
9. yes
10. 40
11. 49
12. 12
13. 18
14. no
15. $x \Rightarrow 3x$
16. $x \Rightarrow 3x+1$
17. yes
18. no
19. infinite
20. infinite

Chapter 5, Unit 1
Part B

1. false
3. false
5. false
7. false
9. false
11. false
13. false
15. false
17. false
19. false
21. $x \Rightarrow x+57$
23. $x \Rightarrow \frac{1}{x}$
25. $x \Rightarrow x+7$
27. $x \Rightarrow -2x$
29. $x \Rightarrow \frac{2x+1}{5}$

Chapter 5, Unit 2
Part A

1. a. false
   b. false
2. a. true
   b. true
3. {-4,-3,-2,...,2,3,4}
4. {-5,-4,-3,$\sqrt{14}$,-$\sqrt{14}$}
5. no
6. (50,10)
7. {(0,2),(0,-2),(2,0),(-2,0), (1,-1),(-1,1),(1,1),(-1,-1)}
8. {7,15,19,20,22}
9. {15}
10. {-9,3}
11. {-4,-7}
12. $\left\{\frac{-7+\sqrt{61}}{2}, \frac{-7-\sqrt{61}}{2}\right\}$
13. {8,11}
14. { }
15. {-3}
16. yes
17. yes
18. true
19. true

Chapter 5, Unit 2
Part B

1. true
3. false
5. true
7. false
9. false
11. true
13. true
15. false
17. true
19. false
21. false
23. true
25. false
27. true
29. false
31. false
33. true
35. true
37. (-4,-30)
39. (-3,9)
41. (6,3)
43. (-9,4)

Chapter 5, Unit 3
Part A

1. yes
2. no
3. a, d
4. c
5. c
6. {5,4,7,-5}
7. {2,3,4,5,6}
8. {x | x ∈ Reals}
9. {y | y ∈ Reals}
10. 3
11. 0
12. $x < 6$
13. $y < 0$
14. 3
15. -2
16. x has no restrictions
17. $y < -6$

Chapter 5, Unit 3
Part B

1. relation; D = {2,4}; R = {3}
3. not a relation
5. relation; D = {2}; R = {3,5}
7. relation; D = {1,2,8}; R = {3,0}
9. not a relation
11. not a relation
13. relation; x unrestricted, $y \geq 0$
15. relation; $x \neq -10$; $y \neq 2$
17. relation; $x \geq 4$; $y \geq 0$
19. relation; $x \geq 9$; $y \geq 0$
21. relation; $x \neq -2$; $y \neq 0$
23. relation; $x \leq 5$; $y \geq 0$
25. relation; x unrestricted, $y \geq 0$
27. relation; x unrestricted, $y \geq 1$
29. relation; $x \geq 0$; $y \geq -11$
31. relation; x unrestricted, $y \geq -4$
33. relation; $x \neq \frac{4}{3}$; $y \neq 6$
35. relation; x unrestricted, $y \geq 4$
37. relation; x unrestricted, $y \leq 6$
39. relation; $-3 \leq x \leq 3$; $y \geq -3$
41. relation; $x \neq -1$ and $x \neq 1$; $y \neq 0$
43. relation; $x \neq 0$; $y \neq -7$
45. relation; x unrestricted, $y \geq 8$
47. relation; $-5 \leq x \leq 5$; $-5 \leq y \leq 5$
49. relation; $-3 \leq x \leq 11$; $-7 \leq y \leq 7$

Chapter 5, Unit 4
Part A

1. 5
2. yes, 4
3. no
4. once (one time)
5. yes
6. yes
7. no
8. one
9. yes
10. no
11. one
12. yes
13. yes, (9,3)
14. no
15. no
16. yes
17. yes (6,8)
18. no
19. is not
20. b
21. -8
22. 9

Chapter 5, Unit 4
Part B

1. 4
3. no
5. yes; {2,3,5,8}
7. yes; all reals
9. yes; all reals
11. yes; $x \neq \frac{3}{2}$
13. yes; all reals
15. no
17. yes; $x \geq 0$
19. no
21. -13
23. 4
25. $\frac{5}{13}$
27. $2\sqrt{2}$
29. -2 is not in the domain
31. $\frac{2}{3}$
33. $\frac{23}{3}$
35. 64
37. 11
39. -20
41. $\frac{-2}{7}$
43. 51
45. $\sqrt{13}$
47. $\frac{-7}{5}$
49. $\frac{13}{20}$

Chapter 5, Unit 5
Part A

1.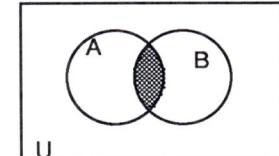

468 Answers

2.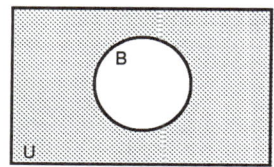

3.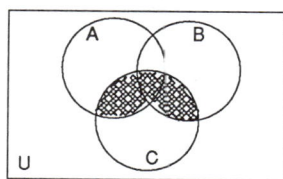

4.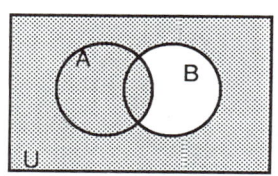

Chapter 5, Unit 5
Part B

1.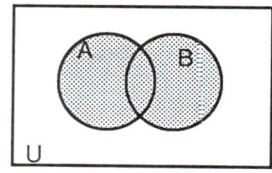

3.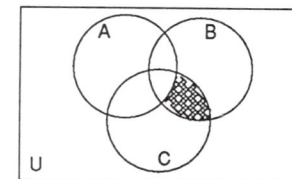

5.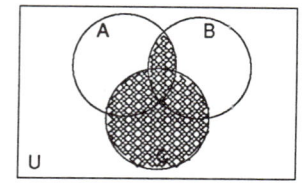

7.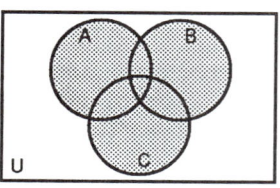

9.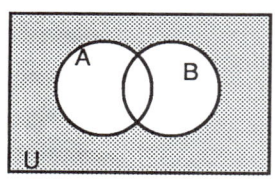

11.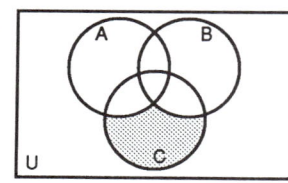

13.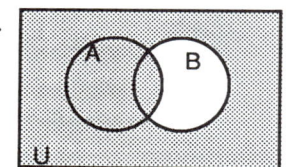

15.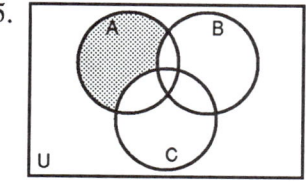

17.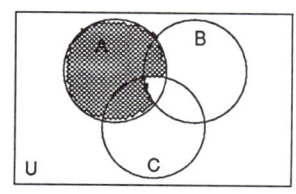

19. a. 1   b. 23   c. 1

Chapter 5 Test

1. yes
2. no
3. $x \Rightarrow 3x$
4. $x \Rightarrow x + 4$
5. $x \Rightarrow \frac{-1}{x^2}$
6. true
7. false
8. (-3,<u>19</u>)
9. (5,<u>0</u>)
10. (-2,<u>3</u>)
11. {3,-1,5,1}
12. {-3,5,-2}
13. $x \neq \frac{7}{2}$
14. $y \geq 4$
15. $y \geq -4$
16. yes
17. no
18. -1
19.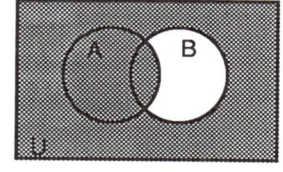

20.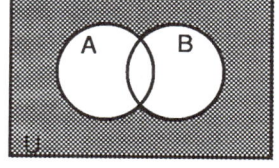

Chapter 6, Unit 1
Part A

1. yes
2. yes
3. no
4. yes
5. no
6. (2,<u>7</u>)
7. (1,<u>2</u>)
8. (-3,<u>2</u>), (0,<u>-4</u>), (3,<u>-10</u>)
9. Every point on the line represents a solution for $y = -2x - 4$.
10. (-4,<u>1</u>), (0,<u>3</u>), (4,<u>5</u>)
11. Every point on the line represents a solution for $y = \frac{1}{2}x + 3$.
12.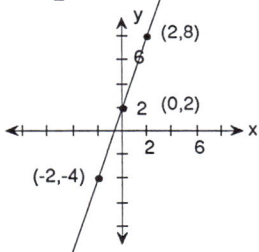
13.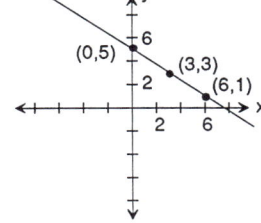
14. yes
15.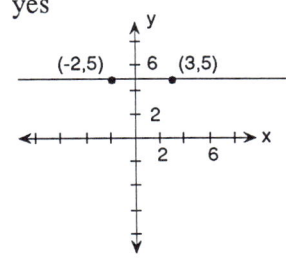

Chapter 6, Unit 1
Part B

1. yes
3. yes
5. no
7. yes
9. yes
11.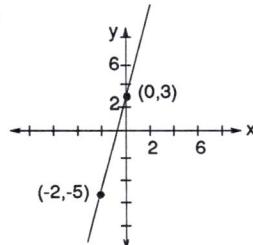
13.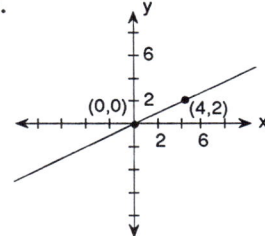
15.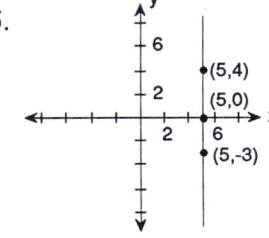
17.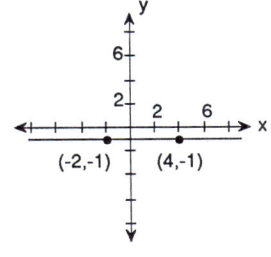
19.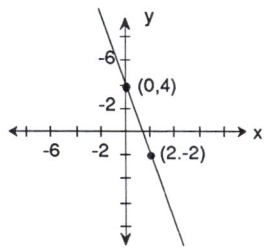

Chapter 6, Unit 2
Part A

1. yes
2. yes
3. no
4. yes
5. positive
6. negative
7. negative
8. positive
9. (0,<u>-7</u>)
10. (0,<u>-1</u>)
11. $y = 2x - 5$; (0,-5)
12. $y = 2x + 6$; (0,6)

Chapter 6, Unit 2
Part B

1. linear function
   (0,3)
   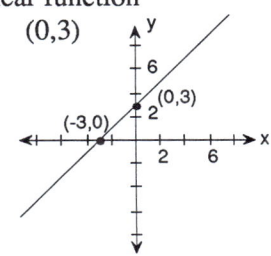
   slope is positive

3. linear function
   (0,−5)
   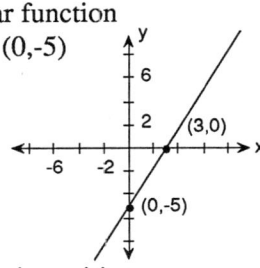
   slope is positive

5. linear function
   $\left(0, \tfrac{5}{3}\right)$
   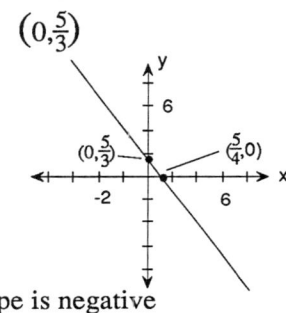
   slope is negative

7. linear function
   (0,−4)
   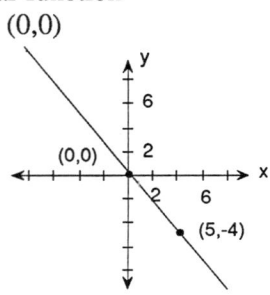
   slope is zero

9. not a linear function

11. linear function
    (0,0)
    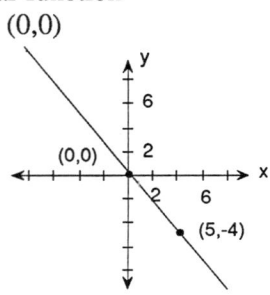
    slope is negative

13. linear function
    (0,−7)
    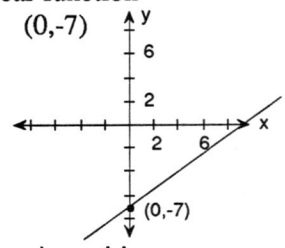
    slope is positive

15. linear function
    (0,5)
    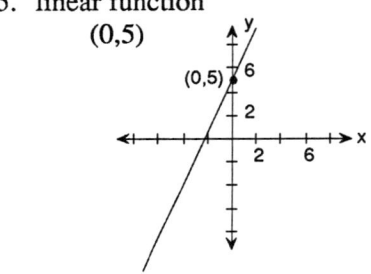
    slope is positive

17. linear function
    (0,−6)
    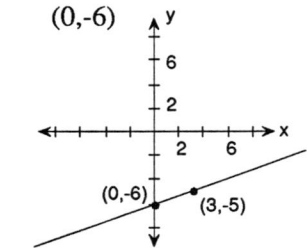
    slope is positive

19. linear function
    (0,−10)
    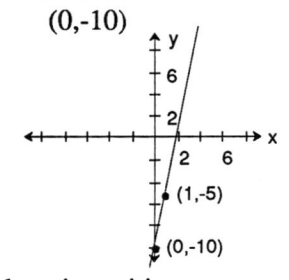
    slope is positive

21. linear function
    (0,5)
    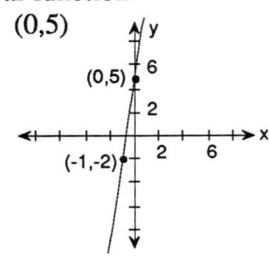
    slope is positive

23. linear function
    (0,1)
    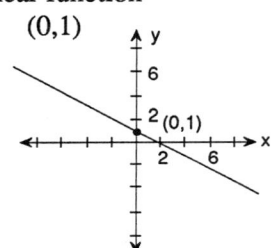
    slope is negative

25. linear function
    (0,2)
    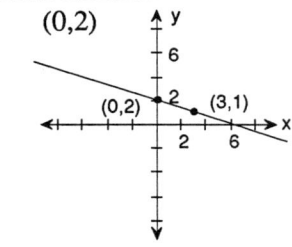
    slope is negative

27. linear function
    (0,0)
    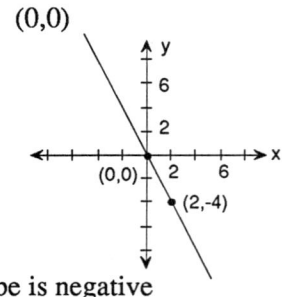
    slope is negative

Chapter 6, Unit 3
Part A

1. increases by 6
2. decreases by 2
3. increases by 2
4. increases by 5
5. $\frac{2}{4}$ or $\frac{1}{2}$
6. $\frac{3}{1}$ or 3
7. $\frac{1}{1}$ or 1
8. $\frac{4}{-9}$ or $\frac{-4}{9}$
9. $\frac{14}{-1}$ or -14
10. y
11. x – axis
12. y
13. undefined
14. undefined

Chapter 6, Unit 3
Part B

1. 1
3. 1
5. undefined
7. $\frac{-7}{3}$
9. $\frac{-1}{2}$
11. -1
13. $\frac{-6}{5}$
15. $\frac{-5}{7}$
17. $\frac{-17}{4}$
19. undefined
21. 0
23. 6
25. undefined
27. 0
29. undefined

Chapter 6, Unit 4
Part A

1. $y = 3x - 5$
2. $y = -2x - 4$; m = -2, (0,-4)
3. $3x - 5y = -10$
4. $3x - 5y = -27$
5. $9x - 2y = 33$
6. $y = \frac{3}{5}x - 4$
7. $3x + 4y = 11$
8. $3x - 7y = -34$
9. $m = \frac{-5}{3}$; (0,-1)
10. $m = 2\frac{2}{3}$; (0,-2)
11. $m = \frac{2}{3}$; $(0, \frac{-8}{3})$
12. $m = -2$; (0,4)
13. $m = 0$; (0,-13)
14. no

Chapter 6, Unit 4
Part B

1. $x - y = -10$
3. $4x - 7y = 21$
5. $4x - 5y = 59$
7. $3x - 4y = 4$
9. $7x - 3y = -5$
11. $y = -2x + 7$
13. $y = \frac{1}{2}x + \frac{5}{7}$
15. $y = \frac{4}{3}x - \frac{55}{12}$
17. $15x + 5y = 8$
19. $10x - 15y = 87$
21. $y = x - 5$
23. $y = 2x$
25. $y = -x + 1$
27. $2x + y = 1$
29. $2x + y = -9$

Chapter 6, Unit 5
Part A

1. Many ordered pairs have the same first component.
2. The solutions of $3x + y \geq 5$ include the line of $3x + y = 5$.
3. $x + y < 3$
4. The line itself ($3x + 5y = 4$), the points to the upper right of the line ($3x + 5y > 4$), and the points to the lower left of the line ($3x + 5y < 4$).
5. To show the boundary of the points determined by $y < \frac{-2}{5}x + 3$.

6.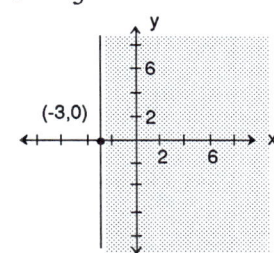

7.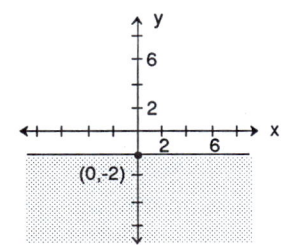

8.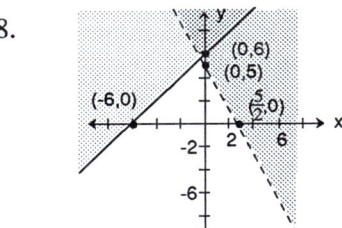

9. intersection
10.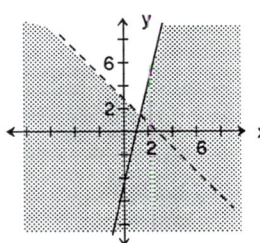
11. union
12. all
13. all

Chapter 6, Unit 5
Part B

1.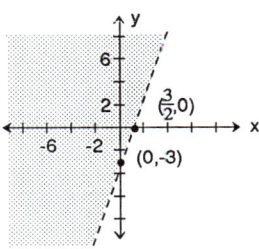

3.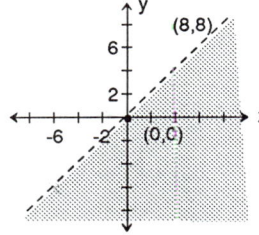

5.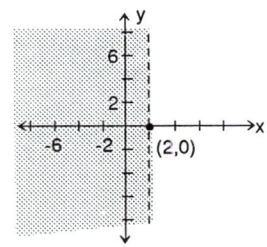

7.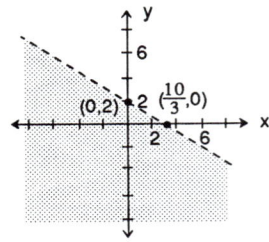

9.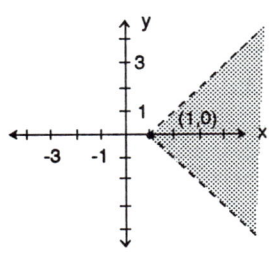

11.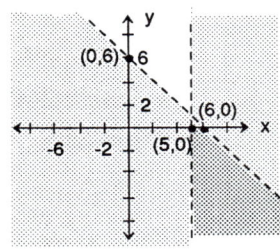

13.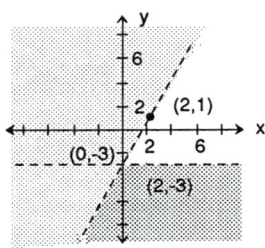

15.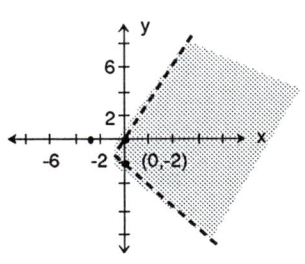

Chapter 6 Test

1. -6
2. 1
3. yes
4.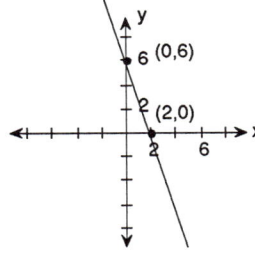

5.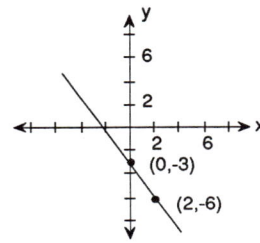

6.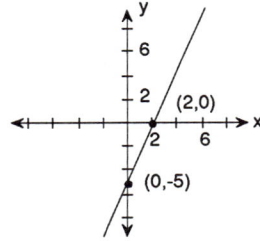

7. slope is $\frac{5}{2}$
8. y-intercept is (0,-5)
9.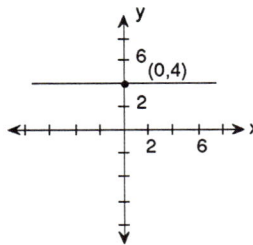

10. slope 0; y-intercept (0,4)
11. y; undefined

12. $\frac{-2}{3}$
13. 1
14. undefined
15. $3x + y = 15$
16. $5x + 2y = -11$
17. $y = \frac{-3}{2}x + 5$
18. $2x - 3y = 15$
19. $y = \frac{3}{4}x - 3$
20. $4x + y = 26$
21. $3x - 7y = 26$
22.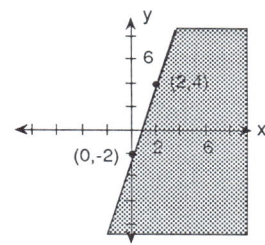
23.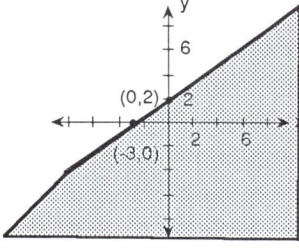
24.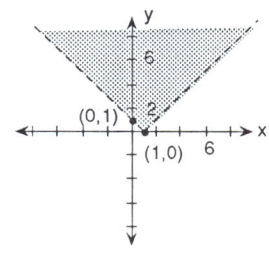

25.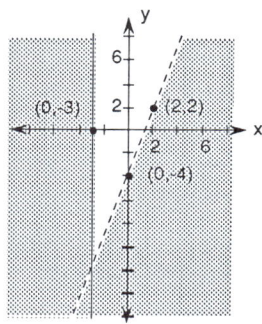

Chapter 7, Unit 1
Part A

1. $(-1, 3)$
2.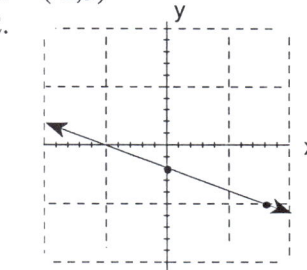
3. $x = x' + 3;\ y = y' - 9$
4. $2x - 5y = 0$
5. yes
6. $(x + 3)^2 + (y - 5)^2 = 9$
7. yes
8. $y - 2 = x^2$
9. $x = x' - 2;\ y = y' + 8$
10. $(0, -5)$
11.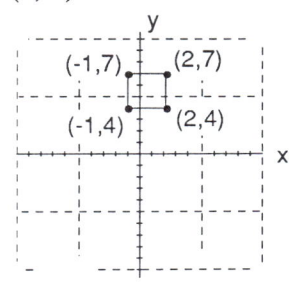

Focus on Intermediate Algebra   473

12.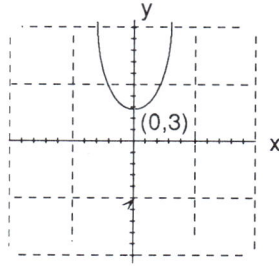

13.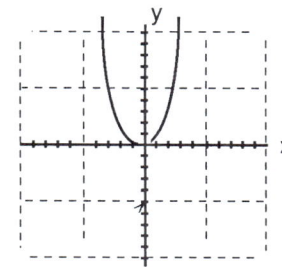

14.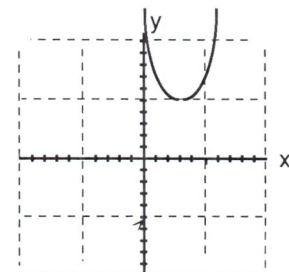

15. yes

Chapter 7, Unit 1
Part B

1. $x = x' - 2;\ y = y' + 5$
3. $x = x' + 3;\ y = y' - 4$
5. $x = x' - 2;\ y = y' - 6$
7. $x = x' + 2;\ y = y' - 7$
9. $x = x' - 2;\ y = y' - 1$

474 Answers

11.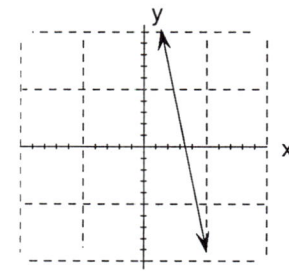

13.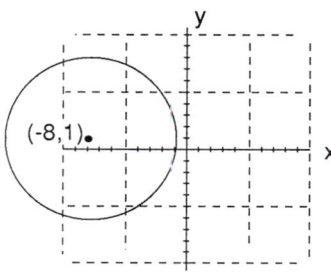

15.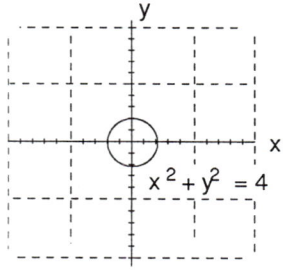

17. $6(x-2)^2 + 3(y+7)^2 = 18$
19. $y + 6 = (x-4)^2$
21. $y = x^2$
23. $(7,8)$
25.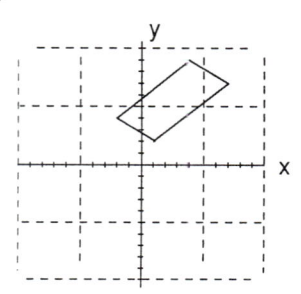

Chapter 7, Unit 2
Part A

1. no
2. U-shaped curve
3.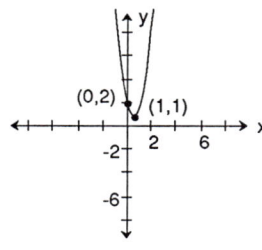
4. 2
5.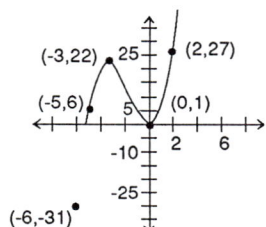
6.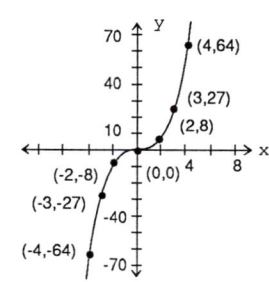
7. 3
8.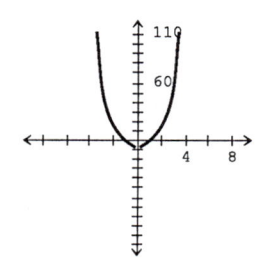

9.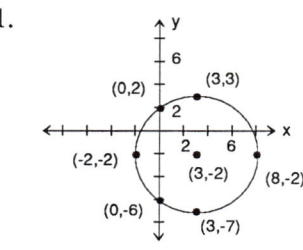

10. $-2 \leq x \leq 8$
11.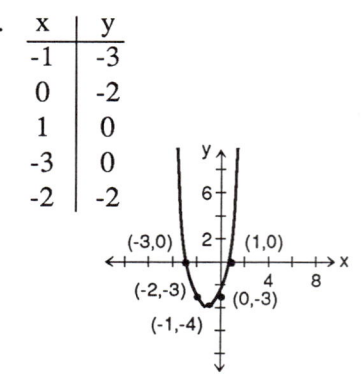

Chapter 7, Unit 2
Part B

1.
x	y
0	-7
3	-16
7	0
-1	0

3.
x	y
-1	-3
0	-2
1	0
-3	0
-2	-2

5.
x	y
0	0
-1	1
2	4
-2	4

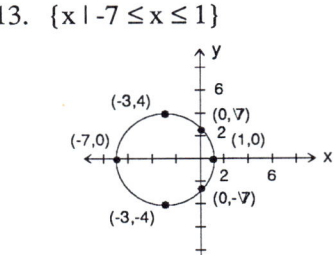

7.
x	y
3	-4
0	5
1	0
5	0

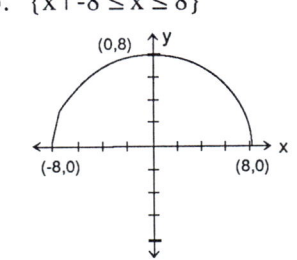

9.
x	y
0	0
3	0
-2	0

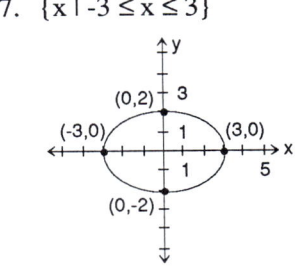

11. $\{x \mid -6 \leq x \leq 6\}$

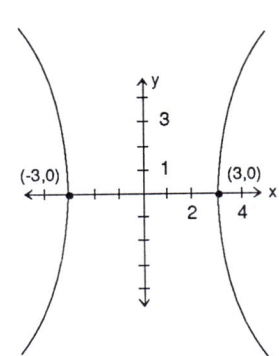

13. $\{x \mid -7 \leq x \leq 1\}$

15. $\{x \mid -8 \leq x \leq 8\}$

17. $\{x \mid -3 \leq x \leq 3\}$

19. $\{x \mid 3 \leq |x|\}$

Chapter 7, Unit 3
Part A

1. 13
2. $\sqrt{41}$
3. (2,7) or (5,-1)
4. 4 and 9
5. $\sqrt{97}$
6. $\sqrt{73}$
7. $\sqrt{97}$
8. no
9. $(x-8)^2 + (y+2)^2 = 25$
10. $\sqrt{(x-4)^2 + (y-3)^2} = \sqrt{(9-5)^2 + (-2-7)^2}$

Chapter 7, Unit 3
Part B

1. 13
3. $2\sqrt{5}$
5. $3\sqrt{2}$
7. 10
9. $\sqrt{53}$
11. $7x + 3y = -14$
13. $\left(\frac{-51}{28}, 0\right)$
15. $(x-5)^2 + (y+2)^2 = 16$
17. (-2,0)
19. (10,3)

Chapter 7, Unit 4
Part A

1.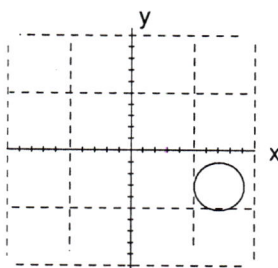

2. $\sqrt{(x-0)^2 + (y-0)^2} = 2$
3. $x^2 + y^2 = 4$
4.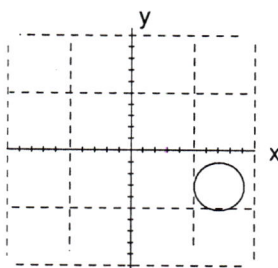

5. yes
6. $(x-7)^2 + (y+3)^2 = 4$
7. $x^2 + y^2 = 49$
8. $(x+4)^2 + (y+2)^2 = 25$
9. $(-2,5); 8$
10. $(-7,3); \sqrt{19}$
11. -9 is less than 0
12. $(x + \frac{1}{2})$ is not squared
13. $(y - \sqrt{7})^3$ is not squared
14. minus sign
15. $(-3, \frac{7}{2}), 5$
16. 16
17. $\frac{81}{4}$

Chapter 7, Unit 4
Part B

1. yes
3.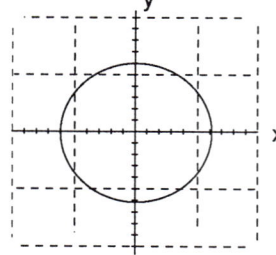

5.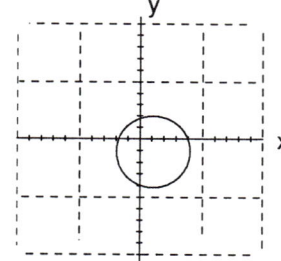

7.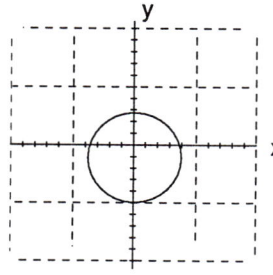

9.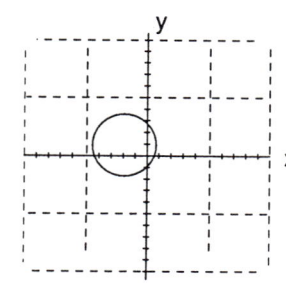

11. $x^2 + y^2 = 25$
13. $(x+1)^2 + (y-3)^2 = 1$
15. $(x-7)^2 + (y+2)^2 = 8$
17. $(x+6)^2 + (y - \frac{5}{8})^2 = 9$
19. $(x-2)^2 + (y - \frac{1}{2})^2 = 5$
21. $(x+5)^2 + y^2 = 49;$
    $(-5,0); r = 7$
23. $(x + \frac{3}{2})^2 + (y-6)^2 = 39;$
    $(\frac{-3}{2}, 6); r = \sqrt{39}$
25. $(x - \frac{7}{2})^2 + (y+3)^2 = 19;$
    $(\frac{7}{2}, -3), \sqrt{19}$

Chapter 7, Unit 5
Part A

1.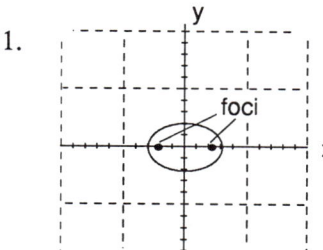

2.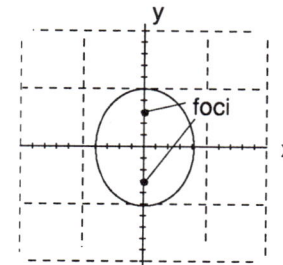

3.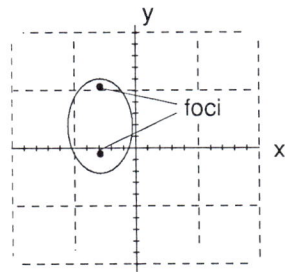

4.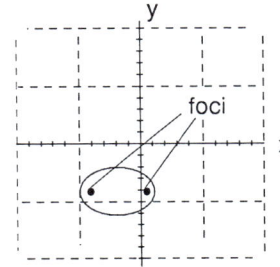

5. Yes
6. $\frac{x^2}{5^2} + \frac{y^2}{4^2} = 1$
7. $\frac{x^2}{4^2} + \frac{y^2}{5^2} = 1$
8. $\frac{(x-4)^2}{49} + \frac{(y+1)^2}{21} = 1$
9. major axis is horizontal in one and vertical in the other.
10. $\frac{(x-4)^2}{100} + \frac{(y+1)^2}{16} = 1$
11. circle; horizontal axis is the same length as vertical axis
12. minus sign
13. exponent of 3 on $(y-5)$
14. $\frac{(x+3)^2}{4} + \frac{(y-\frac{7}{2})^2}{3} = 1$

    $(-3, \frac{7}{2})$; horizontal axis = 4, and vertical axis = $2\sqrt{3}$
15. 25
16. 150
17. $(5,3)$; $2\sqrt{2}$ and $\sqrt{6}$
18. yes

Chapter 7, Unit 5
Part B

1. no
3.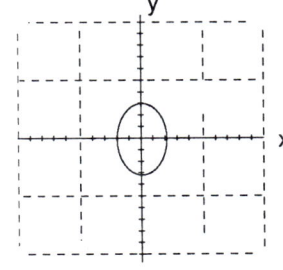

5.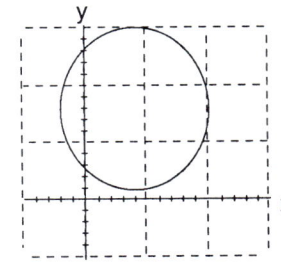

7.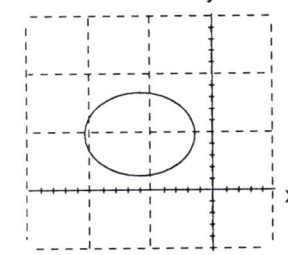

9. $\frac{x^2}{25} + \frac{y^2}{9} = 1$
11. $\frac{x^2}{36} + \frac{y^2}{20} = 1$
13. $\frac{x^2}{3^2} + \frac{y^2}{7^2} = 1$
15. $\frac{(x-5)^2}{16} + \frac{(y+6)^2}{100} = 1$
17. $\frac{(x+2)^2}{49} + \frac{(y-7)^2}{4} = 1$

Focus on Intermediate Algebra 477

19. $\frac{(x+3)^2}{6} + \frac{(y-0)^2}{8} = 1$
21. $\frac{(x+1)^2}{4} + \frac{y^2}{5} = 1$
23. $\frac{(x-3)^2}{36} + \frac{(y+2)^2}{9} = 1$
25. $\left(x - \frac{7}{2}\right)^2 + \frac{(y+3)^2}{2} = 1$

Chapter 7, Unit 6
Part A

1.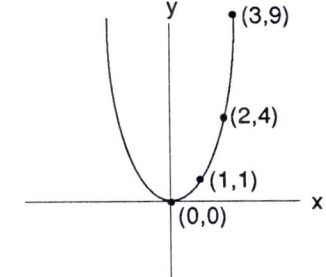

2.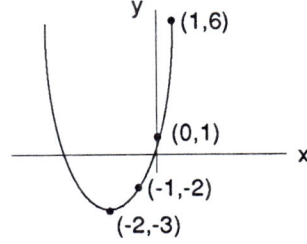

3.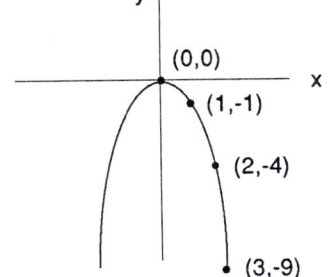

478   Answers

4.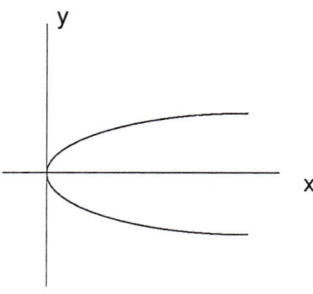

5. yes
6. $y + \frac{3}{2} = \frac{1}{6}(x - 3)^2$
7. $\frac{1}{4}(x - 4)^2 = y - 2$
8. $y = x^2$
9. $(0, \frac{1}{4})$, $y = \frac{-1}{4}$
10. $y = x^2$
11. $y + 1 = (x - 4)^2$
12. downwards
13. because $(y + 3)$ is squared
14. $y + 13 = (x + 4)^2$
15.

16. Open to the left side of the x,y plane.
17. (0,0), (-7,5), (3,3)
18. 45° counter-clockwise

Chapter 7, Unit 6
Part B

1.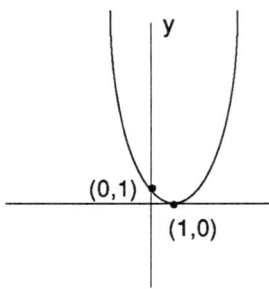

3.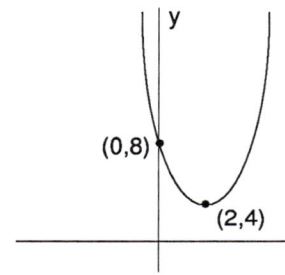

5.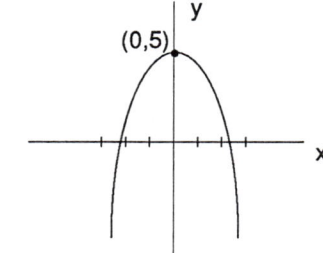

7.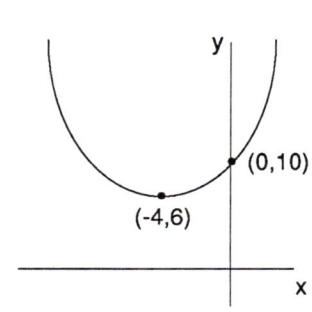

9. $y - 1 = \frac{-1}{4}(x + 4)^2$
11. $x + 1 = \frac{1}{4}(y - 4)^2$
13. $y = 3x^2$
15. $x = \frac{5}{9}(y - 3)^2$
17. The curve of $y = x^2 + 3$ is 3 units above the curve of $y = x^2$.
19. $y = 2x^2$ has the same vertex as $y = x^2$ but the curve is narrower.
21. $y + \frac{9}{2} = \frac{5}{6}(x + 2)^2$
23. $y - \frac{5}{3} = \frac{-8}{3}(x - 1)^2$
25. $y + 9 = 2(x + \frac{5}{2})^2$

Chapter 7, Unit 7
Part A

1.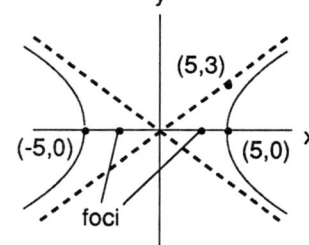

2.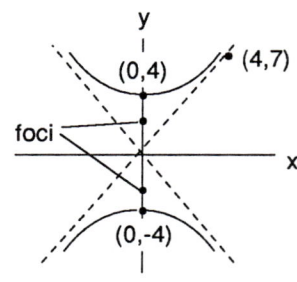

3.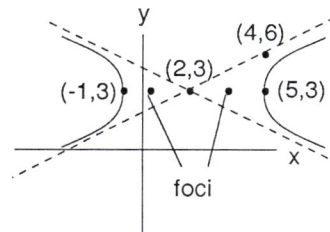

4.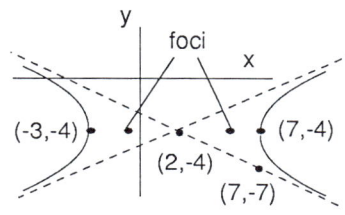

5. yes
6. $\frac{x^2}{4} - \frac{y^2}{5} = 1$
7. (6,0); (-6,0)
8. (0,12); (0,-12)
9. The first equation has branches opening to the sides. The second's branches open up and down.
10. yes
11. no, it is an ellipse
12. no, (y − 5) is not squared
13. $\frac{(x - \frac{5}{2})^2}{5} - \frac{(y-5)^2}{2} = 1$
14. 9
15. 36
16. $\frac{(x-3)^2}{2} - (y-3)^2 = 1$
17. yes

Chapter 7, Unit 7
Part B

1. no
3.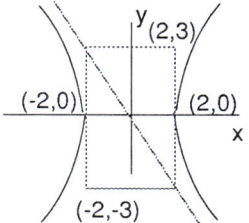

5.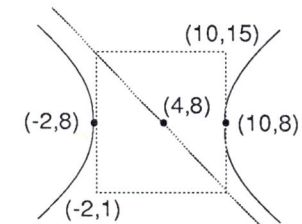

7.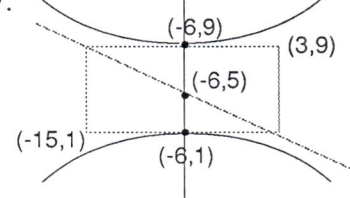

9.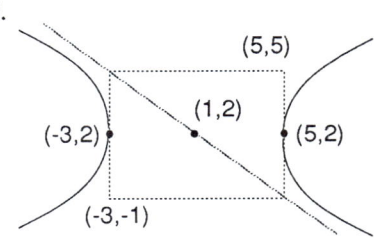

11. $x^2 - \frac{y^2}{15} = 1$
13. $\frac{(x-2)^2}{4} - \frac{(y-3)^2}{21} = 1$
15. $\frac{x^2}{9^2} - \frac{y^2}{7^2} = 1$
17. $\frac{x^2}{6^2} - \frac{y^2}{4^2} = 1$
19. $\frac{(x+4)^2}{16} - \frac{(y-5)^2}{25} = 1$
21. $\frac{(x+2)^2}{4} - \frac{y^2}{5} = 1$
23. $\frac{(x-3)^2}{4} - (y-2)^2 = 1$
25. $\frac{(y+3)^2}{5} - \frac{(x-4)^2}{2} = 1$

Chapter 7 Test

1. $2x - 3y = 8$
2.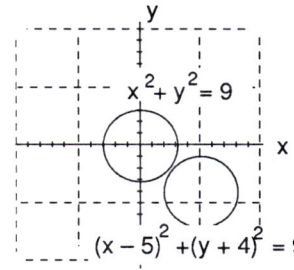

3. 3
4.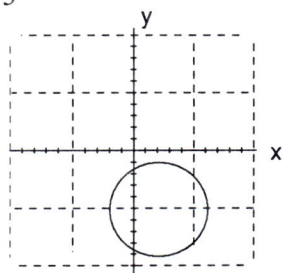

5.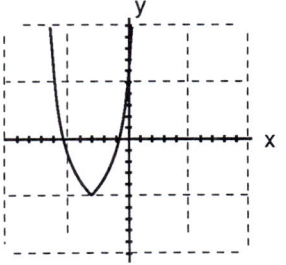

480 Answers

6.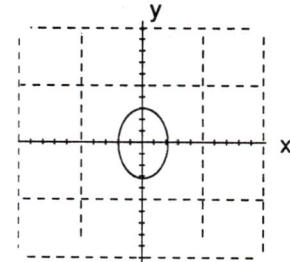

7.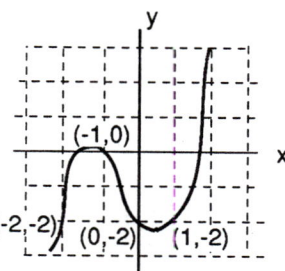

8. 13 feet
9. 8 inches
10. $6\sqrt{5}$
11. 5
12. $2x - y = 3$
13. yes
14. $x^2 + y^2 = 9$
15. $(x - 4)^2 + (y - 1)^2 = 16$
16. $(-3,0)$; 3
17. $(2,-1)$; 5
18.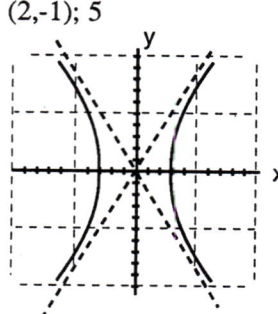

19.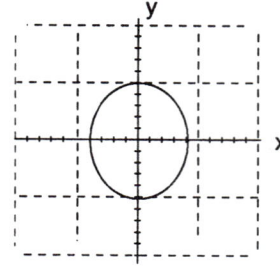

20.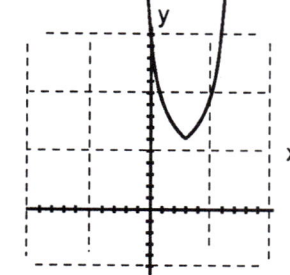

21.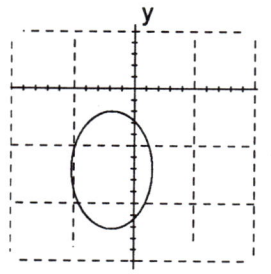

22.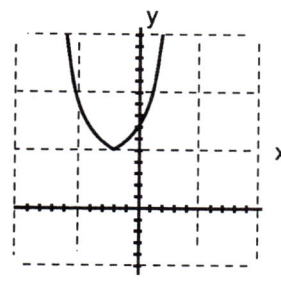

23.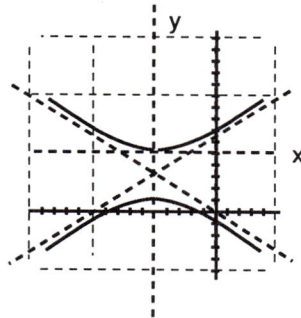

24.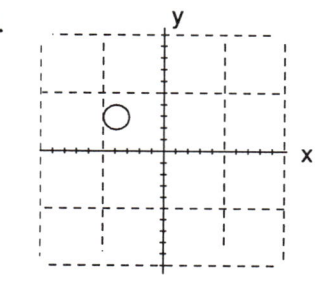

Chapter 8, Unit 1
Part A

1. 7 and 5
2. 8
3. 3
4. 1
5. 5
6. no
7. no
8. $-3x^5$
9. x
10. 1
11. $81x^4$
12. -32

Chapter 8, Unit 1
Part B

1. 5, -2
3. 5
5. 4 • x • x
7. 1
9. false
11. -27
13. -64
15. 81
17. 16
19. -125
21. 9
23. 256
25. -16
27. -24
29. 250
31. 81
33. 1
35. -1

Chapter 8, Unit 2
Part A

1. no
2. yes
3. no
4. no
5. no
6. 3
7. 9
8. 5
9. 6
10. 243
11. 72
12. no
13. x^{12}
14. $5x^3$
15. 3, denominator
16. $\frac{1}{x^7}$
17. 8, numerator
18. x^5

Chapter 8, Unit 2
Part B

1. false
3. false
5. true
7. true
9. 6^9
11. x^{13}
13. 2^9
15. x^7
17. $-3x^7$
19. $6x^7$
21. $12x^4y^2z^6$
23. $-6x^2yz^5$
25. $-6x^7y^7z^7$
27. $30x^6y^7$
29. $-15x^3y^5$
31. $-25x^2y^2z^2$
33. -40
35. 3^{12}
37. x^6
39. x^{15}
41. x^{20}
43. x^{30}
45. x^{40}
47. x^{40}
49. x^{20}
51. x^{30}
53. $16x^4y^2$
55. $16x^8y^{12}$
57. $-x^{15}y^5z^{20}$
59. x^{35}
61. x^7
63. $-27x^6$
65. x^6
67. $\frac{1}{xy^4}$
69. $3xy^4z$
71. $-3z^4$
73. $\frac{-4y^2z^7}{3}$
75. $\frac{24yz}{x}$
77. $\frac{9}{8x^4y^3z^6}$
79. $\frac{1}{3x^3y}$

Chapter 8, Unit 3
Part A

1. 1
2. 1
3. undefined
4. $\frac{1}{8}$
5. $\frac{1}{8^3}$
6. 9^7
7. 1
8. y^{-8}
9. reciprocal
10. yes
11. no

Chapter 8, Unit 3
Part B

1. $\frac{1}{7}$
3. $\frac{125}{64}$
5. $\frac{1}{81}$
7. 1
9. $\frac{7}{59}$

482   Answers

11. $\dfrac{1}{x^6}$
13. $\dfrac{1}{y^7}$
15. x^8
17. y^{42}
19. y^3
21. $\dfrac{1}{y^8}$
23. $\dfrac{1}{x^{17}}$
25. $\dfrac{1}{z^5}$
27. z^{15}
29. $\dfrac{1}{x^{71}}$

Chapter 8, Unit 4
Part A

1. x^8
2. x^{-8}
3. x^{-2}
4. x^4
5. x^{-5}
6. x^2
7. x^{21}
8. x^{-15}
9. x^{24}
10. $16x^{20}y^{28}$
11. $\dfrac{-x^{25}}{32y^{15}}$
12. $\dfrac{y^{18}}{x^{24}}$
13. $\dfrac{-125y^6 z^{15}}{64x^9}$

Chapter 8, Unit 4
Part B

1. x^3
3. x^{20}
5. x^3
7. $\dfrac{1}{x^8}$
9. 1
11. $\dfrac{-3}{2x^4}$
13. $\dfrac{x^6}{9}$
15. $\dfrac{-7}{x^3 y^2}$
17. $\dfrac{1}{x^{40}y^{25}}$
19. $-12x^2 y^3$
21. $\dfrac{y^{20}}{256x^{12}}$
23. $\dfrac{-64x^6 y^9}{27}$
25. $\dfrac{36x^9}{y^{14}}$
27. $\dfrac{-64}{x^9 y^{18}}$
29. $\dfrac{-y^{20}}{32x^5}$

Chapter 8, Unit 5
Part A

1. $\sqrt{34}$
2. 37
3. 28
4. $\sqrt[9]{12}$
5. $\sqrt[8]{46}$
6. $\sqrt[8]{5^3}$
7. $\sqrt[5]{9^2}$
8. $2;\ 3^4 = 81,\ (-3)^4 = 81$
9. $1;\ (-2)^3 = -8$
10. $1;\ 3^5 = 243$
11. none
12. $4^{6/5}$
13. $\sqrt[5]{(-5)^3}$

Chapter 8, Unit 5
Part B

1. $\sqrt[4]{7^3}$
3. $\sqrt{35}$
5. $\sqrt[5]{(-7)^8}$
7. $\sqrt[5]{12}$
9. $\sqrt[7]{9^4}$
11. "cube root of 6 to the 4th power"
13. "eighth root of 5 cubed"
15. "seventh root of 6 cubed"
17. "tenth root of 8 to the 4th power"
19. "cube root of 8 to the 5th power"
21. $4^{9/3}$
23. $x^{4/3}$
25. $3^{4/5}$
27. $5^{2/3}$
29. $5^{3/8}$
31. $\sqrt[12]{11^5}$
33. $\sqrt{14}$
35. $\sqrt[3]{12^5}$
37. $\sqrt[8]{5}$
39. $\sqrt{5}$

Chapter 8, Unit 6
Part A

1. x^{-2} or $\frac{1}{x^2}$
2. $x^{3/4}$
3. $x^{14/9}$
4. x^{-8} or $\frac{1}{x^8}$
5. $x^{2/10}$
6. $x^{1/6}$
7. 3^{-12} or $\frac{1}{3^{12}}$
8. $7^{2/5}$
9. $x^{8/3}$
10. $x^{34/35}$
11. $x^{1/4}$
12. $3^{6/5}$

Chapter 8, Unit 6
Part B

1. $x^{11/12} = \sqrt[12]{x^{11}}$
3. x
5. $x^{13/6} = x^2\sqrt[6]{x}$
7. x^5
9. 3^2 or 9
11. 11
13. $x^{3/2} = x\sqrt{x}$
15. $x^{1/4} = \sqrt[4]{x}$
17. $x^{19/6} = x^3\sqrt[6]{x}$
19. x^2
21. 5^3
23. y^9
25. x^5
27. $x^{41/35} = x^3\sqrt[35]{x^6}$
29. $x^{5/6} = \sqrt[6]{x^5}$
31. $x^{7/20} = \sqrt[20]{x^7}$

Chapter 8, Unit 7
Part A

1. $\frac{1}{5}$
2. $\frac{5}{17}$
3. $\frac{1}{y}\frac{2}{y}\frac{y}{y}$
4. $\frac{1}{x}\frac{2}{x}\frac{x}{x}$
5. $3\frac{1}{3}$ hours
6. 112 minutes
7. 4 days, 12 days
8. 6 hours, 12 hours
9. 6 hours
10. 6 hours
11. $4\frac{4}{7}$ hours
12. 1 hour

Chapter 8, Unit 7
Part B

1. 12 hours
3. $5\frac{5}{6}$ hours
5. 2 days
7. $1\frac{1}{5}$ hours
9. 10 hours
11. $\frac{5}{x}$
13. 40 hours

Chapter 8 Test

1. x
2. -81
3. 81
4. false
5. $-30a^9b^4c^4$
6. $x^{15}y^6$
7. $\frac{-4z^2}{3x}$
8. $\frac{-27y^3z^6}{8x^6}$
9. $\frac{3}{17}$
10. 1
11. no
12. $\frac{1}{x^{12}}$
13. $\frac{-9}{2x^3}$
14. $\frac{-15}{xy}$
15. $7^{3/5}$
16. $\sqrt[3]{11^4}$
17. $x^2\sqrt[20]{x^3}$
18. $\sqrt[10]{x}$
19. 2^6 or 64
20. $6\frac{2}{3}$ hours

Chapter 9, Unit 1
Part A

1. 9
2. 81
3. 16
4. $\frac{1}{9}$
5. 1
6. 1
7. $\frac{2}{3}$
8. 8
9. $\frac{5}{8}$
10. $\frac{1}{4}$
11. 1
12. $\sqrt{2}$
13. $\frac{1}{2}$
14. 32
15. $\frac{1}{32}$
16. yes
17.
x	y
1	2
0	1
-1	$\frac{1}{2}$
-2	$\frac{1}{4}$
-3	$\frac{1}{8}$

18.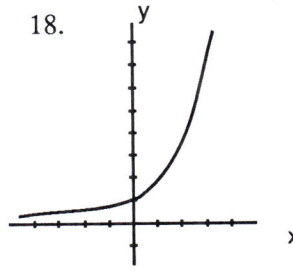

Chapter 9, Unit 1
Part B

1. yes
3. yes
5. no
7. no
9.

484 Answers

11.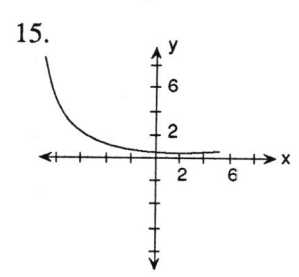

13. (graph)

15. (graph)

14. $\frac{1}{2}$
15. $(1,\underline{0}), (9,\underline{2}), (\frac{1}{9},\underline{-2})$
16. $(1,\underline{0}), (16,\underline{4}), (\frac{1}{8},\underline{-3})$
17. $(7,\underline{1}), (1,\underline{0}), (\frac{1}{7},\underline{-1})$
18. $(36,\underline{2}), (6,\underline{1}), (1,\underline{0})$
19. $(36,\underline{-2}), (6,\underline{-1}), (1,\underline{0})$
20. $(\frac{1}{81},\underline{2}), (\frac{1}{9},\underline{1}), (1,\underline{0})$

Chapter 9, Unit 2
Part A

1. 2
2. -3
3. 0
4. 1
5. -1
6. 3
7. $\frac{1}{2}$
8. 4
9. 2
10. -1
11. 3
12. 5
13. 1

Chapter 9, Unit 2
Part B

1. 3
3. -1
5. 3
7. 6
9. 0
11. 2
13. 2
15. 5
17. 1
19. -1
21. -3
23. -4
25. x
27. 1
29. slightly greater than 3

Chapter 9, Unit 3
Part A

1. 10^5
2. 10^6
3. 10^{-5}
4. 10^{-4}
5. 4,800
6. 1.284636
7. 10^3
8. 10^{-2}
9. 10^5
10. 10^{-5}
11. $4.89 \cdot 10^4$
12. $6.17 \cdot 10^{-3}$
13. $8.73 \cdot 10^1$
14. $9.18 \cdot 10^{-2}$

Chapter 9, Unit 3
Part B

1. $4.135 \cdot 10^{-1}$
3. $6.35 \cdot 10^4$
5. $8.0 \cdot 10^{-4}$
7. $5.14 \cdot 10^{-3}$
9. $8 \cdot 10^5$
11. $3.62 \cdot 10^{-3}$
13. $6.35 \cdot 10^4$
15. $8.0 \cdot 10^{-5}$
17. $2.5 \cdot 10^7$
19. $6.35 \cdot 10^3$
21. $1.3 \cdot 10^{-3}$
23. $5.5 \cdot 10^5$
25. $5.7 \cdot 10^1$
27. $4.04 \cdot 10^3$
29. $5.2 \cdot 10^2$
31. $4.17 \cdot 10^{-2}$
33. $6.3 \cdot 10^0$
35. $5.7 \cdot 10^{-1}$

Chapter 9, Unit 4
Part A

1. a. .84757
   b. .72346
   c. .12385
   d. .92273
2. .78533
3. 51, 7
4. 80, 6
5. 41, 5
6. .67669
7. .79099
8. 19, 7
9. 59, 6
10. 69.6
11. .0241

Chapter 9, Unit 4
Part B

1. .84510 + 0
3. .02938 + 0
5. .73239 + 0
7. .12057 − 3
9. .98227 − 1
11. .02938 − 1
13. .77887 + 2
15. .94841 + 3
17. .72428 − 1
19. .63347 + 0
21. .00604
23. 18.7
25. .380
27. 150,000,000
29. 56,000
31. 67,500
33. .00862
35. 34,600
37. 37.7
39. 445

Chapter 9, Unit 5
Part A

1. $\log_{10} 370 + \log_{10} 23$
2. 52,600

3. $\log_{10} .0362 - \log_{10} 21.6$
4. 33,100
5. $4 \cdot \log 5.3$
6. 177,000
7. $\frac{1}{7} \cdot \log 91.4$
8. 4.04
9. b
10. a
11. b
12. b

Chapter 9, Unit 5
Part B

1. 5,530,000,000
3. 31,500,000
5. .00347
7. 10.8
9. 29.6
11. 132
13. 2,970,000
15. .342
17. 346
19. .000000000915
21. $(6\log_{10} 4.17) - (\log_{10} .546 + \log_{10} 93.7)$
23. $\left(\frac{1}{5}\log_{10} 73\right) - (\log_{10} 230 + 6\log_{10} 9.27)$
25. $\left(\frac{1}{3}\log_{10} .047\right) - (\log_{10} 4820 + \log_{10} .0316)$
27. $(5\log_{10} .135) - (\log_{10} 602 - \log_{10} .00017)$

Chapter 9, Unit 6
Part A

1. 3.62 cm
2. 48.4 sec
3. 10.9 ft

4. $\frac{1 \text{ cu ft}}{1728 \text{ cu in}}$ and $\frac{1728 \text{ cu in}}{1 \text{ cu ft}}$
5. $\frac{1 \text{ gal}}{231 \text{ cu in}}$ and $\frac{231 \text{ cu in}}{1 \text{ gal}}$
6. $\frac{1 \text{ min}}{60 \text{ sec}}$ and $\frac{60 \text{ sec}}{1 \text{ min}}$
7. 1.25 cu ft/sec
8. a. $\frac{1 \text{ cu in of steel}}{.283 \text{ lb}}$
   b. $\frac{1 \text{ km}}{1 \text{ m}}$
   c. $\frac{1 \text{ min}}{60 \text{ sec}}$
   d. $\frac{1 \text{ cu ft}}{1728 \text{ cu in}}$
   e. $\frac{1 \text{ mile}}{5280 \text{ ft}}$
   f. $\frac{60 \text{ min}}{1 \text{ hr}}$
9. 489 lb
10. 29.8 mph
11. 743 mph

Chapter 9, Unit 6
Part B

1. 2.54 cm
3. 14.3 sec
5. 1.53 in
7. 2.14
9. .478 ft
11. 5.51 lbs
13. 8.58 lbs
15. 136 lbs

Chapter 9 Test

1.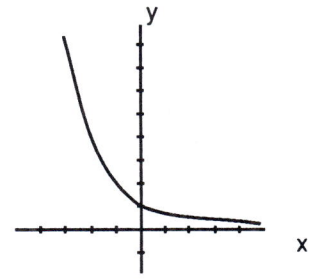
2.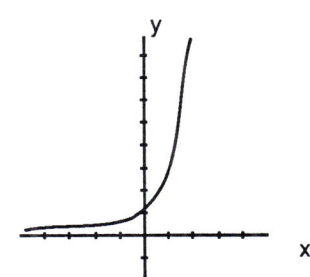
3. no
4. 5
5. -1
6. -2
7. $3.17 \cdot 10^{-2}$
8. $3.42 \cdot 10^{6}$
9. $5.9 \cdot 10^{0}$
10. 47.8
11. 177,000
12. 2,750,000
13. 3.60
14. 135
15. 2.25 cm
16. 13.0 in
17. 7.47 ft
18. 3.04
19. 80.6 ft
20. .0201 miles per minute

Chapter 10, Unit 1
Part A

1. (8,-3)
2. (-5,4)
3. (y − 5)
4. (5x)
5. Addition
6. Substitution

Chapter 10, Unit 1
Part B

1. (2,5)
3. (2,3)
5. (-4,-7)
7. (0,0)
9. (5,1)
11. (6,2)
13. (3,5)
15. (1,-1)
17. (2,3)
19. (-2,5)
21. (2,-3)
23. (1,6)
25. (1,-3)

Chapter 10, Unit 2
Part A

1. 0 or 1
2. 0, 1, or 2
3. 0, 1, or 2
4. 0, 1, or 2
5. 0, 1, or 2
6. x = -1 or x = 5
7. (-1,1), (5,25)
8. yes; both are solutions
9. y = 2x -1
10. x = 3

11. (3,5)
12. yes

Chapter 10, Unit 2
Part B

1. (-3,2), (-6,-1)
3. no solution
5. (0,0), (7,-7)
7. no solution
9. no solution
11. $\left(\frac{-4 \pm 3\sqrt{6}}{2}, \frac{12 \mp 3\sqrt{6}}{2}\right)$
13. $\left(\frac{13 \pm \sqrt{89}}{10}, \frac{-1 \pm 3\sqrt{89}}{10}\right)$
15. $\left(\frac{21}{5}, \frac{9}{5}\right)$, (5,1)

Chapter 10, Unit 3
Part A

1. yes
2. yes
3. 5y + 5z = 15 or y + z = 3
4. -2y − z = -3
5. y = 0, z = 3
6. (-2, 0, 3)
7. -4 − 0 + 3 = -1 is true
   -2 − 0 − 6 = -8 is true
   2 + 0 + 3 = 5 is true

Chapter 10, Unit 3
Part B

1. (2,-5,-3)
3. Equations 1 and 3 have the same plane. Infinite common solutions.
5. (4,2,-3)

7. Equations 2 and 3 have parallel planes. No common solutions.
9. (3,2,-5)
11. (2,1,-3)
13. Equations 1 and 2 have the same plane. Infinite common solutions.
15. Equations 1 and 3 have parallel planes. No common solutions.

Chapter 10, Unit 4
Part A

1. $\begin{bmatrix} 7 & -9 \\ 4 & 2 \end{bmatrix}$

2. $\begin{bmatrix} 3 & -4 & 1 \\ 2 & 6 & 3 \\ -5 & 8 & -2 \end{bmatrix}$

3. $\begin{bmatrix} 5 & 0 & -1 \\ 1 & 0 & 4 \\ 0 & 3 & -5 \end{bmatrix}$

4. $\begin{bmatrix} 6 & -1 \\ 1 & -3 \end{bmatrix}$

5. sk − mj
6. -12 − (-15) = 3
7. (krn + mst + quj) − (qrt + mjn + kus)
8. -4
9. yes
10. no (opposites)

Chapter 10, Unit 4
Part B

1. 1
3. 8
5. 21
7. -50
9. -3

11. $\begin{bmatrix} 7 & -3 \\ 3 & -2 \end{bmatrix}$, -5

13. $\begin{bmatrix} 1 & -1 \\ 2 & -1 \end{bmatrix}$, 1

15. $\begin{bmatrix} 1 & -\frac{1}{3} \\ 2 & 1 \end{bmatrix}$, $\frac{5}{3}$

17. $\begin{bmatrix} 2 & -1 & 3 \\ 1 & 1 & 7 \\ 6 & -3 & 9 \end{bmatrix}$, 0

19. $\begin{bmatrix} 1 & 1 & -1 \\ 2 & 1 & -3 \\ 4 & -1 & -2 \end{bmatrix}$, -7

Chapter 10, Unit 5
Part A

1. $\begin{bmatrix} 4 & -7 \\ 2 & -5 \end{bmatrix} = M_c$

2. $\begin{bmatrix} 2 & 9 \\ 1 & -3 \end{bmatrix} = M_c$

3. $\begin{bmatrix} 6 & -5 \\ 3 & 8 \end{bmatrix} = M_c$

$\begin{bmatrix} 7 & -5 \\ 4 & 8 \end{bmatrix} = M_x$

4. $\begin{bmatrix} 5 & -1 \\ 1 & 4 \end{bmatrix} = M_c$

$\begin{bmatrix} 7 & -1 \\ 3 & 4 \end{bmatrix} = M_x$

5. $x = \frac{-10}{-5} = 2$
6. $y = \frac{-20}{-5} = 4$
7. $x = \frac{-5}{5} = -1$
8. $y = \frac{10}{5} = 2$

9. $M_x = \begin{bmatrix} 11 & -2 & -1 \\ 3 & 1 & 1 \\ 25 & -3 & 1 \end{bmatrix}$

$M_y = \begin{bmatrix} 1 & 11 & -1 \\ 3 & 3 & 1 \\ 2 & 25 & 1 \end{bmatrix}$

$M_z = \begin{bmatrix} 1 & -2 & 11 \\ 3 & 1 & 3 \\ 2 & -3 & 25 \end{bmatrix}$

$M_c = \begin{bmatrix} 1 & -2 & -1 \\ 3 & 1 & 1 \\ 2 & -3 & 1 \end{bmatrix}$

10. $M_x = \begin{bmatrix} 11 & 1 & 0 \\ 7 & 0 & 1 \\ 4 & -3 & 1 \end{bmatrix}$

$M_y = \begin{bmatrix} 4 & 11 & 0 \\ 3 & 7 & 1 \\ 0 & 4 & 1 \end{bmatrix}$

$M_z = \begin{bmatrix} 4 & 1 & 11 \\ 3 & 0 & 7 \\ 0 & -3 & 4 \end{bmatrix}$

$M_c = \begin{bmatrix} 4 & 1 & 0 \\ 3 & 0 & 1 \\ 0 & -3 & 1 \end{bmatrix}$

11. $x = 2$
12. $y = -6$
13. $z = 3$

Chapter 10, Unit 5
Part B

1. (5,-3)
3. $\left(\frac{1}{2}, \frac{3}{2}\right)$
5. $\left(\frac{19}{11}, \frac{6}{11}\right)$
7. (-1,-3)
9. (-2,-5)
11. (1, 0, -1)
13. $\left(\frac{31}{11}, \frac{20}{11}, \frac{54}{11}\right)$
15. (1,-1,2)

Chapter 10 Test

1. b
2. a, d, e
3. (1,-1)
4. (0,5)
5. (2,4)
6. (1,2)
7. (-2,4),(1,1)
8. (4,2),(1,3)
9. (1,0,2)
10. (-1,2,3)
11. (1,-5)
12. (-2,-7)
13. (1,0,-3)
14. (-1,-2,-3)

Appendix – Derivation of the Quadratic Formula

To solve $ax^2 + bx + c = 0$, $\qquad ax^2 + bx + c = 0$

1. Add -c to both sides of the equation.

$$ax^2 + bx = -c$$

2. Multiply both sides of the equation by $\frac{1}{a}$ to make the coefficient of x^2 equal to 1.

$$\frac{1}{a}(ax^2 + bx) = \frac{1}{a} \cdot -c$$

$$x^2 + \frac{b}{a}x = \frac{-c}{a}$$

3. Add a number to both sides of the equation to form a perfect square trinomial. Factor.

$$x^2 + \frac{b}{a}x + \frac{b^2}{4a^2} = \frac{-c}{a} + \frac{b^2}{4a^2}$$

$$(x + \frac{b}{2a})^2 = \frac{b^2 - 4ac}{4a^2}$$

4. Write two linear equations.

$$x + \frac{b}{2a} = \sqrt{\frac{b^2 - 4ac}{4a^2}} \quad \text{or} \quad x + \frac{b}{2a} = -\sqrt{\frac{b^2 - 4ac}{4a^2}}$$

5. Solve the linear equations.

$$x = \frac{-b}{2a} + \frac{\sqrt{b^2 - 4ac}}{2a} \quad \text{or} \quad x = \frac{-b}{2a} - \frac{\sqrt{b^2 - 4ac}}{2a}$$

The quadratic formula is often written as $x = \frac{-b \pm \sqrt{b^2 - 4ac}}{2a}$ where a, b, and c stand for the coefficients and constants of $ax^2 + bx + c = 0$ and the symbol "±" means that one root is found using the plus symbol and the other root is found using the minus symbol.

The quadratic formula can be used to solve any quadratic equation in the form $ax^2 + bx + c = 0$. To solve $5x^2 - 9x + 2 = 0$, a = 5, b = -9, and c = 2.

To use the quadratic formula, the numerical values of a, b, and c for $ax^2 + bx + c = 0$, must be used.

Index

A

Absolute Value	170
Addends	5
Addition Expression	27
Addition Laws for Real Numbers	86
Addition of Complex Numbers	155
Addition of Imaginary Numbers	146
Addition of Real Numbers	4
Additive Inverse	44
Additive Property of Equality	15, 111
Additive Property of Inequality	112
And	23-24, 186, 206
Antilogs	395
Associative Law of Addition	36, 42, 49, 84
Associative Law of Multiplication	36, 43, 49, 84
Asymptote	320-325
Axis of a Double Cone	294

B

Base of an Exponent	333
Base of a Logarithm	382
Binomial	11, 28
Braces	2

C

Center of a Circle	295, 297
Center of an Ellipse	302, 303
Center of a Hyperbola	320-321
Characteristic of a Logarithm	394
Circles	294-299
Closure Law	36, 42-43, 49, 84
Coefficient	8, 90, 333
Common Factor	27
Common Logarithms	392-397
Common Solution	18, 414
Commutative Law of Addition	36, 42, 49, 84
Commutative Law of Multiplication	36, 43, 49, 84
Complementation	224
Completeness Property	85
Completing the Square	118
Complex Numbers	154-156
Congruent	267
Conic Sections	294-325
Conjugates	93, 155
Counting Numbers	4, 36
Counting Number Exponents	331-342
Cramer's rule	445-446
Cube Roots	97
Cubic Equation	65

D

Decimals	102-107
Determinant of a Matrix	436-439
Difference	5
Difference of Two Cubes	58
Directrix	311-312
Distance	172, 178
Distance on x,y Plane	286-291
Distance Formula	288
Distributive Law of Multiplication over Addition	36, 43, 49, 84
Dividend	5
Division	50
Division of Radical Expressions	361
Division of Power Expressions	339, 349
Division of Real Numbers	5
Division with Logarithms	400
Divisor	5
Domain	212

E

Element	2, 207
Ellipse	302-307
Empty Set, { }	14
Equal Complex Numbers	159

Equal Ordered Pairs	200
Equal Sets	199
Equivalent Equations	15, 110
Equivalent Inequalities	24-25, 112
Equivalent Open Expressions	8, 36
Exponent Expression	331
Exponential Functions	375-381
Exponents	8, 331-374
Extraneous Root	130

F

Factors	5, 331
Factoring Polynomials	27-31, 55-59
Factored Form	27
Finite Set	202
First Degree Equation	231
Foci (Focus Points)	302
Focus Point	302, 303, 311, 320-321
Function	217-219

G

Graphing an Absolute Value Equation	174
Graphing an Absolute Value Inequality	174
Graphing a 2-Variable Equation	232-234
Graphing Circles	296-298
Graphing Conics	296-325
Graphing Ellipses	304-307
Graphing Exponential Functions	377-379
Graphing Hyperbolas	322-325
Graphing Linear Functions	229-254
Graphing Parabolas	313-316
Graphing Polynomial Functions	277-281
Graphing Relations	281-283
Graphing 2-Variable Inequalities	257-261
Grouping	29

H

Half-line	174, 185
Hyperbola	320-325
Hypotenuse of a Right Triangle	286

I

i, imaginary number	144
Identity Element for Addition	42
Identity Law of Addition	42, 49, 84
Imaginary Numbers	143-146
Index	354
Inequality Symbols	21
Infinite Set	202
Infinity	122, 189
Integers	4, 42
Integer Exponents	343-353
Intercept Points	234
Intersection of Sets	3, 223
Interval Notation	122, 189
Inverse Law of Addition	42, 49, 84
Inverse Law for Multiplication	49, 84
Irrational Numbers	82

L

LCM	17
Laws of Counting Numbers	36
Laws of Integers	42-43
Laws of Rational Numbers	49
Laws of Real Numbers	84-85
Least Common Multiple (LCM)	17
Leg of a Right Triangle	242, 286
Like Terms	8, 38
Line segment	185
Linear Equation	14-18, 110-112
Linear Equation with 2 Variables	229
Linear Function	219, 237-240
Linear Inequality	21-25, 112-114
Linear Inequality with 2 Variables	257-261
Logarithm Functions	382-386

M

Mantissa of a Logarithm	394
Matrices	435
Matrix	435
Member	2
Minuend	5
Multiplication Expression	27
Multiplication Laws of Real Numbers	86-87
Multiplication of Complex Numbers	155
Multiplication of Imaginary Numbers	144
Multiplication of Radical Expressions	360
Multiplication of Real Numbers	5
Multiplication of Power Expressions	336, 348
Multiplication Property of Equality	111
Multiplication Property of Inequality	113
Multiplication with Logarithms	399
Multiplicative Inverse	50
Multiplication Law of One	36, 43, 49, 84
Multiplication Law of Negative One	43, 49, 84
Multiplication Law of Zero	43, 49, 84

N

Negative Inequality	68
Neighborhood	194
Non-Terminating, Non-Repeating Decimals	105
Number Line	81
Number Ray	21

O

Open Expression	35, 42, 48, 84
Open Segment	174, 185
Open Sentence	3
Opposites	15, 42, 155
Or	23, 187
Order of Operations	6
Ordered Pair (x,y)	18, 200, 207
Ordered Triple (x,y,z)	426
Ordered Pair Solutions (Equations)	231
Origin (0,0)	265

P

Parentheses	8
Parabola	311-317
Perfect Cubes	97
Perfect Fourth Powers	98
Perfect Squares	90
Polynomial over the Integers	10
Polynomial Function	219
Positive Inequality	67
Power Expression	336
Prime Polynomial	28
Principle Root	355
Product	5
Pythagorean Theorem	286

Q

Quadratic Equation	62-65, 116-123, 149-151
Quadratic Formula	118
Quadratic Inequality	67-69, 123
Quartic Equation	64
Quotient	5

R

Radical Expression	90-100
Radical Signs	90
Radicands	90, 354
Radius	295, 297
Range	213
Rate of Production	366
Ratio	242
Rational Function	219
Rational Numbers	4, 48, 82
Rational Number Exponents	354-363
Rationalizing Denominators	91, 92, 94, 99, 156
Ray	184, 186
Real Number Line	81
Real Numbers	4, 81-141
Reciprocal	17, 50
Relation	211-214
Removing Parentheses	8

Repeating Decimals	103
Right Triangle	242
Root	14
Roster Method	200, 205
Rotations	272
Rule of Correspondence	202

S

Scientific Notation	388
Second Degree Equation	231
Set	2
Set Selector Notation (Method)	3, 205-208
Simplifying	8-11, 90-100, 145-146, 351
Slant of a Double Cone	294
Slope of a Linear Function	238, 242-246
Slope-Intercept Form	250
Solution of an equation	14
Soluton of a 2-variable equation	231
Solving Absolute Value Equations	177-181
Solving Absolute Value Inequalities	184-189
Solving Exponential Equations	382
Solving Pairs of Linear Equations	18
Solving Linear Equations	14-18, 110-114
Solving Linear Inequalities	21-25, 110-114
Solving Quadratic Equations	116-121, 149-151
Solving Quadratic Inequatlities	123
Solving Systems of Equations	413-434
Square Matrix	435
Square Roots	90
Standard Form for Linear Equation	250
Substitution	18, 56, 416
Subtraction	43
Subtraction of Real Numbers	4
Subtrahend	5
Sum	5
Sum of Two Cubes	59
Systems of Equations	413-434

T

Terminating Decimals	102
Terms	5
Three Dot Set Notation	2, 205
Total	5
Translations	268
Trinomial	11, 29
Truth set	14, 39
Two-Point Circle	194

U

Undefined Slope	246
Union of Sets	3, 223
Unit Circle	296
Universal Set	223

V

Variable	35, 42, 48, 84
Vector	266
Venn Diagram	223-225
Vertex of a Double Cone	294
Vertex of an Ellipse	302
Vertex of a Parabola	311, 314
Vertices (Vertex Points)	302, 304, 314, 320

W

Whole Number	3
Word Problem Procedure	72

X

x,y plane	232
x-axis	234

Y

y-axis	234
y-intercept	238

Z

z-axis	426
Zero as an Exponent	344